Sustainable Bioenergy Production - An Integrated Approach

Hans Ruppert • Martin Kappas • Jens Ibendorf
Editors

Sustainable Bioenergy Production - An Integrated Approach

Springer

Editors
Hans Ruppert
Department of Sedimentology
and Environmental Geology
University of Göttingen
Göttingen, Germany

Martin Kappas
Department of Cartography, GIS and
Remote Sensing
Institute of Geography
University of Göttingen
Göttingen, Germany

Jens Ibendorf
Interdisciplinary Centre for Sustainable
Development
University of Göttingen
Göttingen, Germany

ISBN 978-94-007-6641-9 ISBN 978-94-007-6642-6 (eBook)
DOI 10.1007/978-94-007-6642-6
Springer Dordrecht Heidelberg New York London

Library of Congress Control Number: 2013941356

Printed on acid-free paper

Springer is part of Springer Science+Business Media (www.springer.com)

Preface

This volume, entitled *Sustainable Bioenergy Production: An Integrated Approach*, focuses primarily on the advantages and implications of sustainable bioenergy production in terms of ensuring a more sustainable world despite its growing energy demands.

This book addresses a new concept that focuses on the interactions between different uses of agricultural land (e.g., agriculture for food, forage or energy and nature conservation) and their ecological, economic and societal impacts. This research concept provides new insights into the competition for resources and the synergies between different land uses. It seeks to improve people's understanding of bioenergy's potentials and the future of land use management and biomass management. To date, the transition towards renewable energy has been misunderstood as only an economic demand, rather than as a means to gain various societal and ecological advantages. Today, biomass is produced to generate energy and renewable raw materials, while simultaneously benefitting soil resources, water resources and biodiversity. The transition to a 'greener' economy is an important precondition to achieve the sustainable development of societies.

Chapter 1, *Sustainable bioenergy production: An integrated perspective,* by Ruppert, Kappas and Ibendorf provides an overview of the controversial issue of sustainable bioenergy production and sets the background for the subsequent chapters.

Chapter 2, *Bioenergy villages in Germany: Applying the Göttingen Approach of Sustainability Science to promote sustainable bioenergy projects,* by Schmuck, Eigner-Thiel, Karpenstein-Machan, Sauer, Ruppert and Roland provides a retrospective overview of the early development of sustainable bioenergy projects in Germany. Bioenergy villages such as Jühnde serve as best practice examples. This approach points the way to the development of a future sustainable energy supply. By means of integrated research, a holistic perspective is provided of the issues concerning sustainable bioenergy production.

The future use of biomass as an energy type needs to be assessed comprehensively and will require careful management of natural land resources such as soil and water. Unsustainable biomass use would undermine bioenergy's climate-related advantages. Precise estimations of the planet's bioenergy potentials are needed on a global, regional and local scale.

Chapter 3 by Kappas provides a short review of global bioenergy potentials and their contribution to the world's future energy demand. While there are many estimations of the future biomass potential, it is clear that bioenergy will play an important part in our future energy supply if we compare the average global bioenergy production potential in 2050 with the highest predictions of global primary energy consumption in 2050. This chapter provides a framework for further estimations of the bioenergy potentials in a region (Germany as a whole) and in a specific site (the local perspective).

State-of-the-art knowledge of the biomass potentials in Germany can be provided by using the process-based vegetation model as described by Tum, Günther and Kappas in Chap. 4. The chapter also presents an approach to estimate sustainable straw energy potentials by means of a modelled Net Primary Productivity (NPP) product, which has been validated by empirical data on the managed area and mean yields of the main crops in Germany. The Biosphere Energy Transfer Hydrology Model (BETHY/DLR) is the theoretical framework to estimate the NPP of the agricultural areas in Germany. The regional estimations of the bioenergy potentials throughout Germany provide a basis to assess the bioenergy potentials on a local scale (site-specific biomass potentials).

Chapter 5, *Modelling site-specific biomass potentials* by Bauböck, offers a new tool to assess local-scale biomass potentials. For the assessment of site-specific and larger area biomass potentials in Lower Saxony, a carbon-based crop model – BioSTAR – was developed at the University of Göttingen. The first validations of the model by means of measured agricultural harvest data from different farms in Lower Saxony are providing convincing results. Chapter 5 concludes the topic of bioenergy potentials and shows that the tools and estimations used to assess bioenergy potentials are already available and that they deliver robust results for future planning.

The next topic – the environmentally sound optimisation of bioenergy production – is addressed in Chaps. 6 and 7.

Chapter 6, *Integrative energy crop cultivation as a way to a more nature-orientated agriculture* by Karpenstein-Machan, focuses on the vision of integrative energy cultivation concepts, which contribute to a more diverse and sustainable rural landscape, keep nature in balance and conserve ecosystems. Integrated cultivation concepts are introduced that should harmonise the relationship between utilisation/production and landscape protection. This integrated concept shows convincing ways to prevent monocultures (e.g., of maize) through a diversified energy crop cultivation system.

Chapter 7 by Saathoff, von Haaren and Rode focuses on the *scale-relevant impacts of biogas crop production: A methodology to assess environmental impacts and farm management capacities*. Typical research questions are: To what extent can the ecological impacts of local biogas crop production be solved by integrated farm management at the farm level? Can potential obstacles to species-friendly and climate-friendly land management be reduced by providing optimal site-specific information about the potential advantages and disadvantages of implemented conservation measures? The outcomes of Chaps. 6 and 7 are a view of the environmentally sound optimisation of bioenergy production, which is important for economic and social linkages.

Environmental and social costs are part of the economic system and include external environmental costs. External diseconomies should be considered when pursuing sustainable consumption and production. Hence the next topic "The economic optimisation of bioenergy production." For example, Chap. 8 by Daub, Uhlemair, Ruwisch and Geldermann optimises *bioenergy villages' local heat supply networks* and provides important advice for a more decentralised energy supply.

According to the examination of the current and future bioenergy potentials (Chaps. 3, 4 and 5), the environmentally and ecologically sound optimisation of bioenergy production (Chaps. 6 and 7) and the economic optimisation of a heat distribution network relying on wood and crops (Chap. 8), the next contemporary issue is the socially acceptable optimisation of bioenergy production (Chaps. 9, 10, 11 and 12).

Chapter 9 by Granozewski, Reise, Spiller and Musshoff first considers the *growth of biogas production in German agriculture by providing an analysis of farmers' investment behaviour*. In German agriculture, renewable energy production from biogas has undergone a dynamic expansion over the past years, which is still continuing. However, with regard to biogas plants, farmers differ in their investment behaviour. A better understanding of farm-level decision-making structures is particularly important for policy-makers and local authorities to estimate biogas production's future investment potential. This chapter also analyses the conflicts in German agriculture regarding land rate leases and land use competition.

The question of willingness to invest in future bioenergy production is coupled with the *social acceptance of bioenergy use and the success factors of communal bioenergy projects* in Chapter 10 by Wüste and Schmuck. This chapter provides insights into the highly dynamic development of bioenergy production facilities in Germany, which are not all in line with sustainability criteria. A growing number of people in Germany's rural areas are directly or indirectly affected by increased bioenergy utilisation. In many cases, bioenergy plants are mainly built for economic considerations, without involving the local population and other stakeholders. Growing fears caused by the local population's lack of information often lead to conflicts, resistance and lower acceptance of bioenergy projects. Chapter 10 addresses potential avenues to sustainable bioenergy projects that local populations will support.

Chapter 10's results lead to the central challenge of *applying the sustainability science principles of the Göttingen approach to initiate renewable energy solutions in three German districts* as Schmuck, Karpenstein-Machan and Wüste describe in Chapter 11. This chapter summarises an interdisciplinary and transdisciplinary action research project that reports on the application of sustainability science principles to convert the energy supply in three German districts of Lower Saxony into renewables.

Finally, different bioenergy concepts regarding sustainable development are evaluated by Eigner-Thiel, Schmehl, Ibendorf and Geldermann in Chapter 12. This chapter focuses on a sustainability assessment of different concepts of biomass energy use in order to provide decision support that takes environmental, economic, social and technical perspectives into consideration. Bioenergy concepts in rural areas are of particular interest; possible technical and organisational concepts can, for example, be a biogas plant operated by electric service providers, or a single biogas plant owned by one farmer or a bioenergy village owned by a village cooperative. This chapter describes the development of suitable ecological, economic, social and technical criteria with which to assess the sustainability of different concepts and the adaptation of existing indicator systems to the special requirements of sustainable biomass use for energy. The results of this sustainability assessment illustrate different biomass concepts' advantages and disadvantages according to multi-criteria decision analysis methods. This decision support tool will facilitate mayors, district administrators, farmers and investors' decision process regarding the most sustainable concept for a certain area.

Two specific topics – the combustion of wood and straw and producing bioenergy on degraded soils – are addressed to complete the book's holistic perspective.

Chapter 13 by Seidel, Orasche, Ruppert and Schnelle-Kreis examines the organic and inorganic emissions during the burning of wood and straw in heat systems. The hazardous potential of the emitted pollutants' particulate matter is not at all well-known and is important for future acceptance of solid biomass sources such as wood and straw.

Contaminated soils should not be used for the production of food or forage crops. In Chapter 14, Sauer and Ruppert argue that energy plants should be grown in these polluted areas. Since the process of the phytoremediation of soils contaminated by heavy metal to acceptable low values requires several thousand years, it is more feasible to leave the toxic elements in the soil. The metal transfer from the different polluted soils to different energy plants was tested to find crops with low transfer factors. The advantages of using such crops are that the fermentation process in the biogas plant will not be impaired by heavy metals and that the residues of the biogas production can be recycled in the fields from which the plants were harvested without exceeding the maximum permissible values for heavy metal.

Energy alternatives based on locally available renewable resources, such as bioenergy, are crucial to the creation of a new energy mix. At the same time, increasing the energy efficiency of the whole economy and of all energy alternatives is an essential precondition to transition to a renewable energy system and a society oriented to sustainability.

To develop a modern, forward-looking energy supply from biomass, such as biomass for heat and power generation and liquid biofuels for transport, there should be a balance between the amount of biomass required for food production and for material purposes. Crop types, production methods and conversion technologies need to be matched with local conditions within the different landscapes to establish a national transformation plan, to reduce the increasing land use competition between food/fodder and energy crop production as well as the use of forests for energy.

Rethinking the linkages between bioenergy, climate change (limiting global temperature change to 2 °C), land use and water requires an integrated assessment of the energy, land and water nexus.

The advantages of sustainable bioenergy production use should always outweigh the effect of its possible environmental damage. The current book is an outcome of ongoing research in Lower Saxony, Germany, to provide an integrated approach to sustainable bioenergy development.

Göttingen, Christmas 2012

Martin Kappas
Hans Ruppert
Jens Ibendorf

Acknowledgments

We are thankful for the financial support for our bioenergy research group.

The FNR (Fachagentur Nachwachsende Rohstoffe, Agency for Renewable Resources, Gülzow, Germany) gratefully provided on behalf of the BMELV (Bundesministerium für Ernährung, Landwirtschaft und Verbraucherschutz, German Federal Ministry of Food, Agriculture and Consumer Protection, Berlin) the financial basis for the interdisciplinary lighthouse project "Bioenergy Village" from 2000 to 2008 (Chap. 2).

The Ministry of Science and Culture of Lower Saxony (Hannover, Germany) kindly supported since 2009 the interdisciplinary project "Sustainable use of bioenergy: bridging climate protection, nature conservation and society" (Chaps. 1, 3, 4, 5, 6, 7, 8, 9, 10, 11, 12, 13 and 14).

Inside our research project, we like to thank all team members and contributors to the current book for their fruitful discussions and cooperative work to finalize our book.

We are also gratefully obliged to the very constructive contributions of the members of the academic advisory committee and of the external reviewers.

Hans Ruppert
Martin Kappas
Jens Ibendorf

Contents

Abbreviations

AHP	Analytic hierarchy process
BImSchG	German Bundes-Immissionsschutzgesetz
BioSTAR	Biomass Simulation Tool for Agricultural Resources
BioSt-NachV	German Biomasse-Strom-Nachhaltigkeitsverordnung
CED	Cumulative energy demand
CHE	Condensing heat exchanger
CHP	Combined heat and power plant
CSD	Commission on Sustainable Development
DPSIR	Driving force, Pressure, State, Impact, Response analysis
DSS	Decision support system
EC	Elemental carbon
EEG	German Renewable Energy Sources Act
EJ	Exajoule = 10^{18} Joule (J) = 1,000 Petajoule
ESP	Electrostatic precipitator
FFH EU	Flora and Fauna Habitat directive
GEMIS	Globales Emissions Modell Integrierter Systeme
GHG	Greenhouse gas emissions
GIS	Geographical information system
GPP	Global primary production
ha	hectare
LAI	Leaf Area Index
LCA	Life-cycle assessment
ICP-OES	Inductively coupled plasma optical emission spectrometer
ICP-MS	Inductively coupled plasma mass spectrometer
IPCC	Intergovernmental Panel on Climate Change
IZNE	Interdisciplinary Centre for Sustainable Development (Interdisziplinäres Zentrum für Nachhaltige Entwicklung)
MADM	Multi-attribute decision-making

MAVT	Multi-attribute value theory
MAUT	Multi-attribute utility theory
MCDA	Multi-criteria decision analysis
MCDM	Multi-criteria decision-making
MCDSS	Multi-criteria decision support system
MIP	Mixed integer program
MJ	Megajoule $= 10^6$ Joule
MODM	Multi-objective decision-making
MWh	MegaWatt hours
NPP	Net primary productivity
NPV	Net present value
NUTS	Nomenclature des Unites Territoriales Statistiques
OC	Organic carbon
PAH	Polycyclic aromatic hydrocarbon
PAR	Photosynthetically Active Radiation
PJ	Petajoule $= 10^{15}$ Joule
PM	Particulate matter
RUE	Radiation use efficiency
SCOPE	Scientific Committee on Problems of the Environment
SETAC	Society of Environmental Toxicology and Chemistry
SOA	Secondary organic aerosol
TEQ	Toxic equivalent
TF	Transfer factor
TSP	Total suspended particulate matter (the total of all particles in the air)
TTC	Threshold of toxicological concern
UN	United Nations
UNCED	United Nations Conference on Environment and Development
WBGU	German Advisory Council Global Change (Wissenschaftlicher Beirat der Bundesregierung Globale Umweltveränderungen)
WCED	World Commission on Environment and Development
WP	Water productivity coefficient
WSOC	Water-soluble organic compounds

Part I
Setting the Scene

Chapter 1
Sustainable Bioenergy Production: An Integrated Perspective

Hans Ruppert, Martin Kappas, and Jens Ibendorf

Abstract Energy from crops, wood and biological residues should always be viewed together with other renewable energy sources, such as wind and hydro power, photovoltaic power, concentrated solar power, solar heating, etc. All of them and the introduction of efficiency measures allow to attenuate climate change as well as the future shortage and increasing price of conventional fossil and nuclear energy sources. Bioenergy has several advantages that make it strategically important: It can provide heat, electricity as well as liquid and gaseous fuels; it can be stored and used when needed; and it can balance fluctuations in the electricity grid.

Owing to widespread malnutrition and a rapidly growing global population with increasing calorie and livestock product requirements, estimates of the potential areas where energy plants could be grown vary widely. A reduction in livestock production on fertile soils, in harvest and postharvest losses and in the waste of food would open up large areas for more sustainable farming as well as for energy crop production. Bioenergy should not lead to monoculture; instead, it should increase biodiversity in agricultural areas and enrich the landscape, thereby improving the

H. Ruppert (✉)
Department of Sedimentology and Environmental Geology, University of Göttingen, Goldschmidtstrasse 1, D-37077 Göttingen, Germany

Interdisciplinary Centre for Sustainable Development (IZNE), University of Göttingen, Goldschmidtstrasse 1, D-37077 Göttingen, Germany
e-mail: hrupper@gwdg.de

M. Kappas
Department of Cartography, GIS and Remote Sensing, Institute of Geography, University of Göttingen, Goldschmidtstrasse 5, D-37077 Göttingen, Germany
e-mail: mkappas@uni-goettingen.de; mkappas@gwdg.de

J. Ibendorf
Interdisciplinary Centre for Sustainable Development (IZNE), University of Göttingen, Goldschmidtstrasse 1, D-37077 Göttingen, Germany
e-mail: jibendo@gwdg.de

H. Ruppert et al. (eds.), *Sustainable Bioenergy Production - An Integrated Approach*,
DOI 10.1007/978-94-007-6642-6_1, © Springer Science+Business Media Dordrecht 2013

population's acceptance thereof. Owing to the state-of-the-art technology, greenhouse gas savings – especially from biogas production for combined power and heat plants – can be quite a bit higher than 70 %. The nutrients that are recycled back into the fields with the digestate maintain soil fertility and save money.

In addition, this chapter discusses the application of Germany's Renewable Energy Source Act and the resulting National Biomass Action Plan. Moreover, it highlights the quota system and feed-in tariffs as promoters of the successful expansion of renewable energy forms.

Keywords Bioenergy potential • Efficiency • Greenhouse gas • Meat versus bioenergy • Biodiversity • Feed-in tariff

1.1 Arguments for Renewable Energy Production

Together with population growth, energy and water shortages, declining arable land and forests, soil degradation, climate change, ocean acidification, decreasing resources and food insecurity are the major challenges of the twenty-first century. To counter these unfavourable trends, concerted actions have to be taken to achieve climate stabilisation, to conserve biodiversity, soil, and forests, to ensure the availability and quality of water, economic and social development, human well-being, global security and a secure, low-carbon energy supply, all of which should be focussed on sustainable development on a local, national, and international level.

Today, fossil fuels (crude oil, natural gas, coal and uranium) account for approximately 85 % of the global primary energy production (DERA 2011). In the next few decades, renewable types of energy, including wind, water, solar power and bioenergy, have to become the major energy suppliers, because:

- The reserves and resources of fossil energy sources (especially natural oil and gas) are nearly depleted, even if shale gas and unconventional oil resources are included in these reserves. Depending on the underlying models and estimations of the global consumption of oil and gas, and on technical improvements and prices, oil production may have passed its global peak or will do so in the near future. This has important consequences for the supply and price of fossil energy sources.
- As implied by the name, renewables, which are based on solar energy, will be available as long as the reactor sun gives off energy; that is, for several billion years to come. It has taken just two centuries for man to seriously deplete the earth's fossil fuels formed over millions of years in the sediments of the outer earth crust.
- Under optimal circumstances, renewables' CO_2 emissions are far lower than those of their fossil fuel counterparts, such as coal, oil and gas. The increase of renewable energy sources counteract the climate change by reducing greenhouse gas emissions.
- Municipalities save on fossil fuel costs, create jobs, collect taxes and lease revenues by installing renewable energy systems (Mühlenhoff 2010).

Strong efficiency measures should be in place when transforming energy systems into renewables. In 2009, 80 % of the emitted greenhouse gases in Europe originated from the energy sector (Boßmann et al. 2012). However, this sector has a strong potential for decarbonisation as it offers a variety of technologies ranging from carbon-neutral electricity generation by means of highly efficient energy conversion processes to energy saving options (Boßmann et al. 2012). Europe's building sector has the highest final energy saving potential. The electricity and petroleum sector has the highest potential for financial benefits. By 2050, the overall final energy demand could be 57 % lower than the baseline projection, which translates into cost savings of about 500 billion EUR per year (based on the 2005 value) (Boßmann et al. 2012). This will increase supply security while decreasing Europe's external fuel bill and enhancing its competitiveness in the global economy. Therefore, the European Union's Energy Roadmap's target is to reduce greenhouse gas emissions by at least 80 % by 2050 (Boßmann et al. 2012). With regard to heating, house insulation and the functionality of stoves and furnaces should be improved. More effective electricity devices should be used to reduce the waste of electrical power. With regard to bioenergy, additional efficiency measures include the energetic use of organic wastes and plant residues as well as the energetic recycling of used materials, such as furniture, construction wood, fibres, etc., which may be reused or co-fired at the end of their useful live (see 'Cascade use' in Box 1.1; WBGU 2008). Generating electricity from biogas should always involve the use of 'waste' heat for household or industrial processes, such as drying. Thus, combined heat and power (CHP) plants can offer double the total usable energy (FNR 2009; FNR/GIZ 2010; see Box 1.1).

The energy transformation of fossil into renewable energy can also promote sustainable development and better incomes in developing countries if similar supporting transformations occur within the political, economic and social systems, and the local, national and international authorities support them (WBGU 2011).

Box 1.1 Bioenergy Glossary

Sources: modified from REN21 (2012), WBGU (2008, 2011), IEA (2011).

Biomass is any organic living or dead material of biological origin, excluding fossil fuels or peat. Biomass comes in solid or liquid forms. Examples are wood, energy crops derived from dedicated plantations, wastes, organic residues from industrial and municipal sources and manure. These materials can be converted into biofuels, biogas or biomethane.

Bioenergy/Biomass energy refers to the final or useful energy derived from biomass. *Traditional bioenergy* is produced by burning solid biomass, including agricultural residues, animal dung, forest products, gathered fuel wood and charcoal. These types of biomass are often burnt inefficiently in open fireplaces, stoves, or furnaces to provide heat energy for cooking, comfort, and small-scale agricultural and industrial processing, usually in rural areas of developing countries. The emission of air pollutants during burning often cause

(continued)

Box 1.1 (continued)

health hazards due to incomplete combustion. In 2008, it was estimated that 2.7 billion people were dependent on biomass for cooking (Chum et al. 2011). Therefore, about 85.6 % of global bioenergy is used in a traditional way (WBGU 2008). *Modern biomass energy* is derived from solid, liquid, and gaseous biomass fuels used, for example, for heat and power generation (space heating and electricity) as well as for transportation. Modern bioenergy involves burning biomass directly or converting it into more convenient fuels, for example, through the pyrolysis and gasification of solid biomass to produce liquid and gaseous fuels, the anaerobic digestion of suitable biomass materials to produce biogas, the transesterification of vegetable oils to produce biodiesel, and the fermentation of sugars to produce ethanol.

Biofuels comprise a wide range of liquid and gaseous fuels derived from biomass – including the liquid fuels bioethanol and biodiesel as well as biogas. Biofuels can be combusted in vehicle engines as transport fuels and in stationary engines for heat and electricity generation. These fuels can also be used for domestic heating and cooking. Today, *first-generation biofuels* are mostly used. These include sugar cane ethanol, starch-based ethanol, biodiesel, fatty acid methyl ester (FAME) and straight vegetable oil (SVO) as well as biogas/biomethane. Feedstocks typically used for the production of liquid biofuels include: sugar cane and sugar beet, starch-bearing grains, like corn and wheat, oil crops, like canola and palm, and, in some cases, animal fats. *Advanced or second-generation biofuels* comprise different emerging and novel conversion technologies that are currently in the research and development, pilot, demonstration or early commercial phases. Advanced biofuels include synthetic biofuels, such as Fischer-Tropsch diesel, biomethane and biohydrogen, which is produced by thermochemical processes, such as gasification and pyrolysis.

Biodiesel is a diesel-equivalent, processed biofuel used in diesel engines of cars, trucks, buses and other vehicles. It can also be used for stationary heat and power generation. Through the process of transesterification (a chemical process that removes the glycerine from the oil), biodiesel is produced from oilseed crops, such as soya bean (*Glycine max*), rapeseed (*Brassica napus*; cultivar canola), oil palm (*Elaeis guineensis*) and, from other oil sources, such as waste cooking oil and animal fats.

Bioethanol, which is mostly used as a gasoline substitute, is produced by fermenting any biomass high in carbohydrates with the aid of yeast or bacteria. Today, ethanol is made from starches and sugars (usually corn, sugar cane, or small cereals/grains), but second generation technologies will produce it from cellulose and hemicellulose. Small amounts of bioethanol can be used to substitute gasoline for use in ordinary spark ignition engines (stationary or in vehicles), or can be used in stronger blends (usually up to 85 % ethanol, or 100 % in Brazil) in slightly modified engines (flexible fuel vehicles).

(continued)

Box 1.1 (continued)

Biogas/Biomethane: Biogas is a mixture of methane and carbon dioxide produced through the anaerobic bacterial degradation of organic matter (fermentation), such as agricultural and food industry wastes, sewage sludge, biological remnants in municipal waste and – especially in Germany – purposely cultivated energy crops (after ensiling). The digestion (biological transformation) of organic material into biogas occurs in a fermentation plant. The methane is the fraction of biogas that can be utilised for energy recovery. Biomethane is nearly pure methane, which is separated from biogas by removing carbon dioxide, hydrogen sulphide, siloxanes, water and some other impurities. Biomethane can be injected into natural gas networks and used as a substitute for natural gas. Biogas and biomethane can be burnt to produce heat and power. Biomethane can also be used as a fuel for cars.

Greenhouse gases are those gaseous constituents of the atmosphere that, due to their selective absorption of thermal radiation, cause warming of the lower atmosphere. The primary anthropogenic greenhouse gases are carbon dioxide (CO_2), nitrous oxide (N_2O) and methane (CH_4). Other greenhouse gases are traffic-caused ozone (O_3) and industrial gases, such as hydrofluorocarbons (HFCs), perfluorocarbons (PFCs), sulphur hexafluoride (SF6) and the ozone-depleting chlorofluorocarbons (CFCs).

Carbon dioxide (CO_2) is a naturally occurring greenhouse gas. It is a product of burning fossil energy carriers and biomass. Deforestation, wetland transformation (peat, swamps, etc.) in agricultural areas, other land-use changes and industrial processes, such as cement production, all lead to additional CO_2 emissions.

Methane (CH_4) is a greenhouse gas emitted mainly from livestock and rice cultivation. It is the principal component of natural gas and biogas.

Nitrous oxide (N_2O) is a persistent greenhouse gas, which is mainly emitted by nitrogen fertilisers in agriculture, the livestock sector (primarily cows, chickens and pigs) and by the burning of fossil fuels.

Carbon dioxide equivalents ($CO_{2\ eq}$) are a measure of the degree to which a mixture of gases contributes to global warming. With the aid of a conversion factor, the global warming potential of non-CO_2 greenhouse gases is expressed as the quantity of CO_2 that would cause the equivalent warming effect. For example, in a 100-year time horizon, methane (CH_4) will trap a lot of heat, warming up the lower atmosphere 25 times and nitrous oxide (N_2O) 298 times more than the same amount of CO_2. The calculation of $CO_{2\ eq}$ makes it possible to include all greenhouse gases in one unit and allows for a comparison of their individual impacts on global warming.

Cascade use refers to a strategy seeking to use resources, or products made from such resources, for as long as possible within the economic cycle. The

(continued)

Box 1.1 (continued)

material passes through as many phases of use as possible. This approach boosts the overall value creation and improves environmental performance. In the case of biomass, cascading can mean that the biomass is first used as an industrial feedstock, after which its energy content is recovered at the end of the product cycle. For example, furniture or wood used in construction can be co-fired in a heat or power plant at the end of its service life.

Combined heat and power (CHP)/Cogeneration plants are facilities that not only generate electricity during the combustion of fuel, but their waste heat can also be used for space heating purposes (as in district heating systems) or for heat/cool-dependent production processes in industry.

Ecosystem services are the benefits people gain from ecosystems. These include supply services, such as food, water or energy, regulatory services, such as carbon sequestration, protection against flooding, or against the spread of disease, cultural or recreational services and support services, such as nutrient cycles, as well as seed and pollen dispersal, all of which maintain the Earth's life-support systems.

Energy is the ability to do work. It comes in different forms, including thermal, radiant, kinetic and electrical energy. *Primary energy* is the energy embodied in natural resources, such as coal, natural gas, biological materials and other renewable sources. *Final energy* is the energy that is available to the final consumer in a usable form (such as electricity from an electrical outlet), where it can provide services such as lighting, refrigeration, etc.

Energy crops are cultivated to extract energy from their biomass. This may involve using either a specific part of the crop (e.g., maize grain or vegetable oil extracted from seeds), or the entire above-ground biomass (e.g., field crops or grass used for biogas installations, woody species, such as poplar or willow for heat and power production). For more information, see Chap. 6.

Energy efficiency or energy conversion efficiency is the ratio between the useful energy output of an energy conversion machine and the expended energy input. It answers the question of how much input energy should be applied to produce a certain amount of electric power, mechanical work, heat, fuel, etc. The ratio is always smaller than one.

Joule/Kilojoule/Megajoule/Gigajoule/Terajoule/Petajoule/Exajoule: A joule (J) is a unit of work or energy and is equal to the energy expended to produce one watt of power for one second. A kilojoule (KJ) is a unit of energy equal to one thousand (10^3) joules; 1 megajoule (MJ) = 1 million (10^6) joules; 1 gigajoule (GJ) = 1 billion (10^9) joules; 1 terajoule = 1 trillion (10^{12}) joules; 1 petajoule = 1quadrillian (10^{15}) joules; 1 exajoule = 1 quintillion (10^{18}) joules. One barrel (159 l) of oil can store approximately 6 GJ of potential chemical energy.

(continued)

Box 1.1 (continued)

Land-use changes (LUC): The term land use refers to the human use of an area of land for a certain purpose, while land-use changes refer to changes in such human use. These include logging, afforestation, sealing, drainage, the conversion of cropland to grassland (and vice versa), or the conversion of cropland to fallow land. Land-use changes can take place directly, for instance, when forests are cleared and the land is used to cultivate energy crops. It is more difficult to identify land-use changes induced by indirect mechanisms. When food crops are replaced with energy plants, the agricultural production that previously took place on this cropland has to take place elsewhere. We refer to such situations as indirect land use change (ILUC)

Pellets are a solid biomass fuel produced by compressing pulverised dry biomass, such as waste wood and agricultural residues. Pellets are usually cylindrical in shape with a diameter of around 1 cm and a length of 3–5 cm. They are easy to handle, store and transport; they are used as fuel for heating and cooking applications as well as for electricity and combined heat and power generation.

Renewable energy includes solar, wind, hydro, oceanic, geothermal, biomass and other sources of energy derived from "sun energy", which is thus renewed indefinitely as a course of nature. Forms of useable energy include electricity, hydrogen, fuels, thermal energy and mechanical force. More broadly speaking, renewable energy is derived from non-fossil and non-nuclear sources in ways that can be replenished, are sustainable and have no harmful side effects. The ability of an energy source to be renewed also implies that its harvesting, conversion and use occur in a sustainable manner, i.e. avoiding negative impacts on the viability and rights of local communities and natural ecosystems.

Short-rotation plantations (SRPs) refer to the cultivation of fast-growing tree species (e.g., poplar and willow) on agricultural land to produce biomass. The concept derives from coppicing, a method traditionally used to produce firewood. The rotation period extends from the growth period until the trees are cut; its duration thus depends on the use of the wood. For pulpwood or for woodchip production, the trees are harvested after 3–5 years. The below-ground root mass remains in the soil, enabling the growth of coppice shoots the following year.

Transformation refers to the initiation and progression of an active transition or a change. The German Advisory Council on Global Change (WBGU 2011) uses the term great transformation as "the modification of both the national and the global economy within planetary guard rails in order to avoid irreversible damages on the Earth system and its ecosystems, and the impact of these damages on human kind". Such guard rails prevent the mean global temperature from increasing more than 2 °C above the pre-industrial level, protect soil and biodiversity, etc., in order to avoid risks and catastrophies and to preserve the Earth systems resources and services and secure humankind's natural life support system and sustainable development. While transformation is primarily

(continued)

Box 1.1 (continued)

focussed on the analysis of social, economic, cultural, and political changes, it must also consider sustainable technological and ecological/environmental improvements to ensure a sustainable future. In a broad sense, the WBGU (2008, 2011) regards transformation as a chance to achieve the sustainability goals of mitigating climate change by shifting from fossil fuel to renewable energy sources and to overcome energy poverty in developing countries.

Watt: A watt is a unit of power that measures the rate of energy conversion or transfer. Power is the rate at which energy is consumed or generated. For example, a light bulb with a power rating of 100 watts (W) that is switched on for one hour consumes 100 Watt-hours (Wh) of energy, 0.1 kilowatt-hour (kWh), or 360 kilojoules (kJ).

1.2 Bioenergy – Pros and Cons

Biomass, especially wood, was the first energy form applied by man. It has been used since just after the Ice Age (10,000 BC) and is still the most important heat supplier in developing countries. IEA (2011), however, estimates that 1.3 billion people have no access to electricity and 2.7 billion people are without clean and effective cooking facilities. Of these people, 84 % live in rural areas. In 2008, their traditional use of biomass, mainly wood, dung, etc. amounted to 8.5 % of the global final energy consumption (REN21 2012), or 10.2 % of the primary energy supply (Chum et al. 2011).

1.2.1 Bioenergy Pros

Bioenergy has three main advantages over other renewables:

- *Reservable*: Bioenergy is easy to store and can be used as required. It can therefore balance the fluctuation of wind and solar power (regulating energy).
- *Different usable forms*: Plant material can be used in a solid (e.g., wood), liquid (biodiesel and bioethanol) or gaseous state (biogas); the liquid and gaseous states are easily obtained through chemical transformation processes.
- *Versatility*: The different states can be used for heat and power production, or as fuel for mobility and other purposes. The other renewables produce mostly electricity.

Bioenergy production has additional advantages:

- *Promotes biodiversity*: Energy plant cropping may increase the biodiversity of arable land if energy plantation concepts are realised as double cropping during the year, or as the cultivation of plant mixtures instead of monocultures. Weeds can also be used if they do not lower yields in general. In addition, short-rotation cropping or agroforestry can be incorporated into energy crop farming (see Chap. 6).

- *Ensuring good yields*: These diversification concepts ensure energy plants' yields, decrease soil erosion and increase the attractiveness of the environment by providing more diversified landscapes.
- *Element recycling for fertilisation*: If remnants of the energetic use of crops, such as the residual digestate from biogas plants or wood ashes, are recycled to the areas from which the plants were taken, a nearly perfect recycling of the elements is possible (except for nitrogen). This fertilisation can be done when the growing plants need nutrients. It saves money and fertiliser resources (an important example is phosphorous, whose extraction maximum should be reached in 2030; Cordell et al. 2009; Gilbert 2009).
- *Monetary advantages*: Bioenergy offers local farmers new income opportunities, which could also reduce rural exodus and alleviate poverty, thereby decreasing the gap between the rich and the poor in developing countries (WBGU 2011). Bioenergy production can also decrease dependence on imported fossil fuels, thus improving countries' foreign exchange balances and energy security. Furthermore, it can expand access to modern energy services and bring infrastructure, such as roads, telecommunications, schools and health centres, to poor rural areas (GBEP 2011; WBGU 2011).
- *Job creation*: The introduction of bioenergy may create new jobs. Growing, harvesting and distributing bioenergy feedstock are specifically very labour-intensive. Additionally, biomass, biofuels and biogas production have created approximately 2.5 million technological jobs globally (REN21 2012).

To ensure bioenergy's sustainable production, the Global Bioenergy Partnership created 24 sustainability indicators with clear advice on how to handle them (GBEP 2011). Stakeholders and decision makers should be encouraged to use them as an analytical tool and facilitate decisions on and planning for sustainable bioenergy development. These indicators are based on interrelated environmental, social and economic pillars. The pillars comprise indicators, such as greenhouse gas and pollutant emissions, the productive capacity of the land and ecosystems, water availability, biological diversity, land-use changes, access to land, jobs, labour conditions, social development, human health and safety; efficiencies in bioenergy, production, conversion, distribution and end-use; as well as economic, technological, and logistic development and energy security.

1.2.2 Bioenergy Cons

Despite these benefits, the use of bioenergy has some limitations:

- *Land use conflicts and food-fuel competition*: The production of energy plants on farmland leads to a competition for arable land for the production of food and animal fodder.
- *Monoculture*: The production of only one high-yield plant, such as maize, in consecutive years leads to an area poor in biodiversity, decreases the landscape's

attractiveness, degrades soils through humus losses, increases the erosion risk and requires substantial fertilisation.

- *Acceptance*: In Germany, the increase in maize for energy use has decreased the acceptance of bioenergy production. Moreover, the comfort of people who live near a biogas plant might be affected due by increased traffic during the harvest season.
- *Greenhouse gas balance*: The greenhouse gas balance is not neutral, especially if the strong greenhouse gas methane escapes from fermentation plants during biogas production. Furthermore, the intensified application of nitrogen to increase energy crop yields produces the very strong climate gas nitrous oxide (N_2O).
- *Emissions of toxic compounds*: The ineffective burning of wood or charcoal in developing countries, but also in old fireplaces in industrialised countries, emits toxic compounds into the atmosphere.
- *Financial implications*: Besides breathing life into rural economies and the creation of new jobs, the competition for land increases the price of comestible goods if the production of food plants decreases due to increased energy croplands. Additionally, the rent for farmland may increase.

Some of the limitations described above, such as the environmental implications, should be compared with the conditions arising from the burning of oil, gas, or coal. In the following section, we examine how some of the emerging conflicts can be de-escalated.

1.2.3 Evaluating and Reducing Emerging Conflicts

1.2.3.1 Land Use Conflicts and the Food-Fuel Competition

To date, the amount of crops used for energy production is still small. In 2008, about 74 % of the world's agricultural production, which totalled about 10 billion tonnes, was allocated to animal food (fodder), while 18 % was used for food, only 3.7 % was used for energy and 4.3 % for biomaterial production (Raschka and Carus 2012). In that year, 260 million hectares (ha) of the 1,445 million ha of arable land were used for food, 1,030 million ha were used for animal feedstuff, 55 million ha were used for energy crops and 100 million ha were used for biomaterial production (Raschka and Carus 2012).

It is difficult to expand the area suitable for agricultural use, since agriculture has already cleared or converted 70 % of the grassland worldwide, 50 % of the savannah, 45 % of the temperate deciduous forest and 27 % of the tropical forest biomes (Foley et al. 2011). This conversion of land for agriculture has tremendous impacts; it reduces habitats and biodiversity, depletes the humus in soils, soil fertility, the freshwater available and the water quality. In turn, critical ecosystem services are being depleted. It is hard to find areas for the production of energy plants that do not compete with the production of food, fodder or plants for the material or industrial sectors. In addition, the prevailing natural resources used by man have already exceeded the Earth's carrying capacity (Rees 2006; WWF 2012). The pressure will increase further in the near future, as agricultural production needs to increase by 60 % over the next 40 years to meet the rising food demand (OECD/FAO 2012).

On a global and a national scale, the estimates of the bioenergy potentials that are available as part of a future sustainable energy supply are very contradictory. This is because the estimates depend on many parameters with unknown influences (e.g., Chum et al. 2011; Haberl et al. 2011). Therefore, the following questions need to be answered:

(a) How can the efficiency of crops, irrigation, nutrient supplies and cycling be increased in order to close the yield gap (see Chaps. 5 and 6)?
(b) To what extent does increased CO_2 fertilisation influence yields?
(c) To what extent will fertiliser (especially phosphorus) be available in future?
(d) How will soil degradation by means of salinisation, erosion, pollution, etc. develop?
(e) How accurate is the evaluation of degraded and marginal land's potential?
(f) Will improvements in management and technology, including the improved use of biomass (plant residue and cascade utilisation), lead to more efficient systems for agriculture or for consumers?
(g) What will the growth rate of the global population be?
(h) How will the competition between bioenergy and animal feed or bio-based material production develop?
(i) How will rising temperatures, changing rainfall patterns (amount and distribution) and the increased frequency of extreme events influence cultivation and plant yields?

Owing to these constraints, bioenergy could be a bridging technology for the transformation from fossil-based energy systems to future energy systems potentially based on wind and solar energy.

Chum et al. (2011) describe some of the difficulties of providing a global outlook for bioenergy potentials in 2050. They estimate that, in the median scenario, bioenergy will contribute 120–155 EJ/year to the global primary energy supply and can contribute up to 265–300 EJ/year in the highest deployment scenarios. This upper limit decreases to approximately 100 EJ/year if policy frameworks and enforcing mechanisms are not introduced, or if there is strong competition with biomaterials. These numbers should be compared to the total global primary energy supply, which was 492 EJ in 2008.

The German National Academy of Science (Leopoldina) estimates that, in Europe (EU25) and Germany, the potential for bioenergy is negligible, that it will only meet a small percentage of the country's primary energy needs and that it will mainly rely on waste (Haberl et al. 2012). The same authors furthermore assume that almost all the biomass that can be sustainably harvested worldwide will be required for human food, animal feed, construction materials, or as a basis for chemicals, leaving very little room for the use of biomass as an energy source, apart from wastes.

In contrast, by means of model calculations Zeddies et al. (2012) estimate that, in Germany, 2.4 million ha of land are available for 2020 for bioenergy production in addition to the 2.1million ha that are already being used for energy crops (out of 12 million ha of arable land in Germany). By 2050, this area could increase to 7.5 million ha even if 2.4 million ha are used for food export. Since they believe that it will be possible to obtain enough land to provide food security, the authors estimate that, by 2050, 200–300 million ha of land will be globally available for energy crop production.

Fritsche et al. (2012) estimate that advanced biofuels (see 'Biofuels' in Box 1.1.) could cover up to 70 % of the fuel demand for all modes of transportation in Germany by 2050, bearing in mind that fuel demand will be significantly reduced by then and that biofuel demand will be met without land use competition or additional imports. According to these authors, biofuels will be produced either from residual biomass, or from agricultural areas that have been made available for the production thereof. This biofuel production ought not to have a negative impact on biodiversity, nor lead to the conversion of pastureland or meadows into cropland. Additionally, this production ought not to reduce Germany's self-sufficiency in food supply.

The world's technical and sustainable biomass supply potentials as well as the expected demand for biomass (primary energy), which are based on global energy models and estimates of the world's total primary energy demand in 2050 (see Chap. 3), are shown in Fig. 1.1. The current world biomass use and the primary energy demand are shown for comparative purposes.

In addition to bioenergy, there are other competitors for food and fodder production areas: The replacement of all organic compounds in Germany's chemical industry will require yields from 58 % of the country's arable land, while the replacement of lubricants with bio-lubricants will require a further 11 % thereof (Bringezu et al. 2009).

On a global scale, the production of animal food (meat, milk and eggs) is by far the most restricting factor with regard to the availability of fertile arable land area for organic farming or energy crop production. Animal food production covers 1,030 million ha of farmland – about four times more than vegetable food production (Raschka and Carus 2012). Livestock production covers an additional 3,550 million ha, which are mostly used as pasture and grazing lands. Meat production is growing strongly. It has tripled between 1970 and 2009 from about 100 million to 300 million tonnes per year (OECD/FAO 2012).

A more specific description of the real land use conflicts does not cover food-energy competition, but meat-energy competition. While grazing land that is unsuitable for food production may ensure additional calories and proteins for people, highly productive areas used for animal feed are a net drain on the potential global supply of food (Foley et al. 2011) or bioenergy. Global meat consumption differs quite dramatically: In about 20 developing countries, the yearly meat

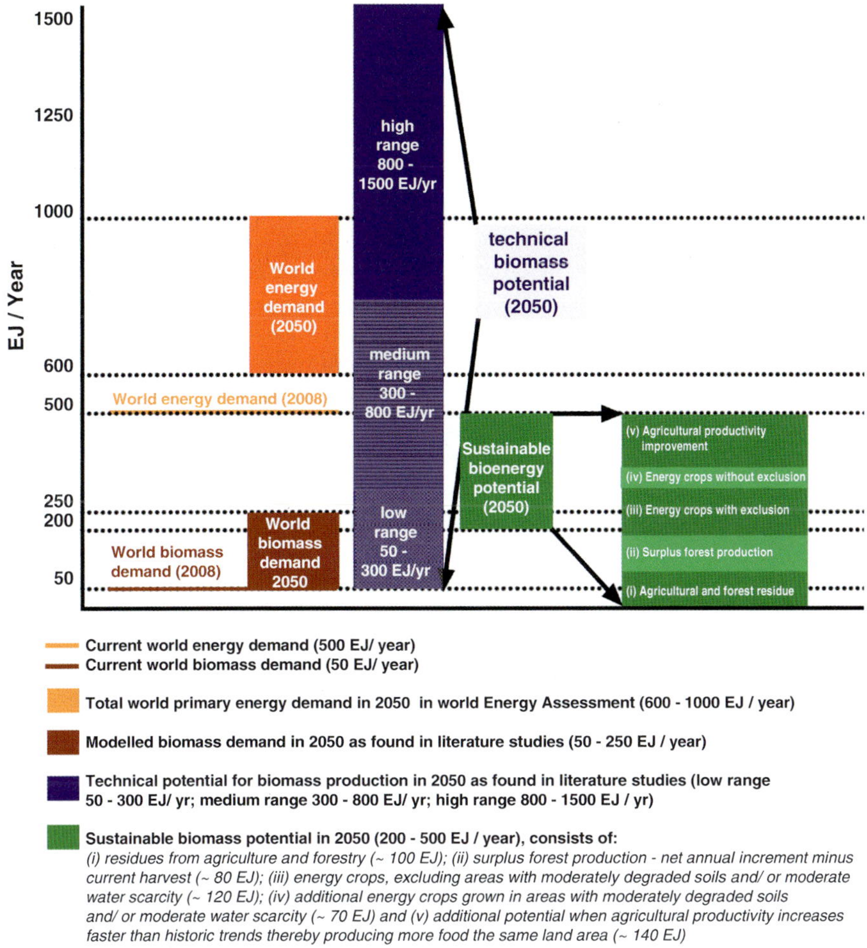

Fig. 1.1 Visualization of the technical and sustainable biomass supply potentials, expected demand for biomass (primary energy) based on global energy models, expected total world primary energy demand in 2050 (Bauen et al. 2009) and current world biomass use and primary energy demand (Dornburg et al. 2008; figure modified from Bauen et al. 2009)

consumption per capita is below 10 kg compared to an average of 80 kg in developed countries. Additionally, meat consumption in developing countries is increasing constantly (FAO 2012).

1.2.3.2 Land Use Change Through Meat Production

The expansion of land for livestock development has been the main driving force behind deforestation in, for example, Latin America and the Caribbean, while overgrazing is prevalent in other regions (FAO 2012). In Argentina and Brazil, for example, soya beans are cultivated to provide proteins for industrial livestock farming in European countries, such as Germany. To meet the requirement to generate 62 % of Germany's energy from vegetal food products, about 10 million tonnes of carbon are harvested from biomass every year. In contrast, annually, the feeding of animals to produce meat, milk and egg products requires about 53 million tonnes of carbon from plant biomass, which are either harvested or grazed, as well as about 9 million tonnes of imported carbon (about 5.4 million tonnes from soya beans and soya bean products, 2.3 million tonnes from rapeseed and 1.2 million tonnes from maize) (Haberl et al. 2012). Between 2008 and 2010, Germany used approximately 7 million additional ha of land outside Europe, thereby virtually increasing Germany's agricultural area from 17 to 24 million ha (Bringezu et al. 2009; Witzke et al. 2011). If the production of meat were to be reduced, the cultivation of other food products, such as protein-rich plants, would have to increase to ensure balanced human diets. However, this would require a much smaller area than that which is currently used for livestock production. The meat footprint of the average German is more than 1,000 m^2 (including 230 m^2 for soya production in overseas areas). This means that more than 8 million ha are used for meat production alone. Moreover, in Germany, 60 % of the grain and 70 % of the oilseed plants are used as fodder (Witzke et al. 2011).

1.2.3.3 Environmental Impact of Intensive Meat Production

Besides consuming much of the land and water resources, intensive animal husbandry or factory farming leads to many other problems:

1. Domestic animals emit large amounts of the greenhouse gases, methane and nitrous oxide.
2. Fields become overfertilised with manure.
3. Surface water becomes contaminated with nitrate and phosphate (causing eutrophication).
4. Nitrate and ammonia filter into the groundwater (posing a threat to human health when used as drinking water).
5. Pharmaceutical ingredients (hormones, antibiotics, etc.) are transferred from animals to the water.
6. Ammonia is released into the atmosphere where it is transformed into nitric acid, which in turn acidifies rainwater and soils.

In Germany, approximately 2 of the 11.7 tonnes of CO_2-equivalent emissions that are annually released per capita, are released during the production, processing, packaging, transport, marketing and consumption of agricultural products. An additional 0.5 tonnes of CO_2-equivalent emissions are annually released per capita due to land use changes. In the value-added nutrition chain, 204 million tonnes of CO_2 equivalents are annually emitted in Germany. About two-thirds of these greenhouse gas emissions can be attributed to animal products (meat, milk, eggs, etc.) and one-third to plant products (Noleppa 2012). In addition, importing soya material as a cheap source of protein for animal feed is an important aspect of factory farming in Europe. Approximately 85 % of the world's total soya production is used for animal feed. The international production of soya beans has increased by a factor of 2 over the last 30 years; in South America, it has increased by a factor of 4, with the strongest growth in Argentina and Brazil. These two countries are currently also the largest producers of soya beans (Reenberg and Fenger 2011).

Germany uses approximately 2.6 million ha of land outside the European Community to satisfy the need for soya products for livestock production (Witzke et al. 2011). In Brazil and Argentina, soya cultivation has expanded to land previously used for grazing or for natural habitat. This leads to direct land use changes by affecting the local savannahs and to indirect land use changes by exerting pressure on the tropical rainforest of Amazonia causing negative effect on the global carbon dioxide budget and on biological diversity (Barona et al. 2010; Reenberg and Fenger 2011). Additionally, the majority of soya bean crops in the USA and South America are genetically modified (GM) and become part of the meat and milk production, even in countries like Germany, where GM food is not accepted on the market. Another aspect is animal welfare: Nearly 75 % of the world's poultry production, more than 50 % of its pork production, and 60 % of all egg production occur in large-scale intensive industrial production systems (FAO 2009). Factory-farmed animals are usually confined to small pens, cages, sheds, or indoor stalls. They therefore mostly do not have access to pastures, fresh air and sunlight, and are unable to perform many of their natural behaviours (MacDonald 2012).

All of these arguments suggest that high meat consumption needs to be questioned. The reduction of meat production will:

- ease the pressure on the land areas used by man,
- open up space for a more ecological agriculture or for bioenergy plants,
- reduce emissions of greenhouse gases,
- reduce mostly inhumane industrial livestock farming,
- improve people's health and food security, and
- may lower the cost of basic food.

Greatly altered human behaviour is a prerequisite for change and is a crucial challenge. Creating awareness of the consequences of the disproportionate consumption of animal products is one of the prerequisites for a sustainable future.

Besides the meat-fuel competition, the following aspects also influence the availability of land areas:

1.2.3.4 Postharvest Food Losses and Food Wasting

Postharvest food losses occur during threshing, grading, packaging, transport, storage, processing, distribution and marketing, or due to biological or chemical contamination and deterioration. Between one-fifth and one-half of produced food is estimated to be lost early in the supply chain segments (the dominant form in developing countries), or wasted at the consumer end (the dominant retail and household levels in industrialised countries), globally amounting to about 1.3 billion tonnes per year (Parfitt et al. 2010; Grethe et al. 2011; Gustavsson et al. 2011). Every German throws away an average of more than 80 kg of comestible food every year (Noleppa and von Witzke 2012; Kranert et al. 2012; Noleppa 2012). This equates to 25 billion EUR, 2.4 of the 16.9 million ha of agricultural land and to 40 million tonnes of CO_2 equivalents per year.

1.2.3.5 Health Aspects

Between 1980 and 2008, the worldwide prevalence of obesity (overweightness) almost doubled (WHO 2012). By 2008, 10 % of men and 14 % of women in the world were obese. The reasons for this include a higher consumption of calories and increased consumption of animal products, salt, sugar and processed and fried foods. From a health perspective, significant reductions in meat consumption in the OECD would be preferable (McMichael et al. 2007; Witzke et al. 2011). In Germany, the adoption of more healthy, less animal-biased nutrition would release 1.8 Mio ha of productive land (Noleppa and von Witzke 2012) and would lower greenhouse gas emissions by 27 million tonnes per year (Noleppa 2012). A stronger awareness of the health aspects and optimal portion sizes of meat, egg and dairy in people's daily diet not only helps fight obesity, but also opens up areas for food and fodder production.

1.2.3.6 Land Deals for Bioenergy Production

Land deals (grabbing) with Africa, Latin America and Southeast Asia for the large-scale production of food, cash crops and biofuels have increased in recent years. Domestic and transnational companies, governments and individuals' land grabbing quickly escalated after the increase in food prices in 2007–2008 – especially in sub-Saharan Africa. In many cases, this has led to the expropriation or displacement of rural populations and to an increase in food prices in these countries. Countries' institutional infrastructure may be ill equipped to handle an upsurge in investor interest. Together with weak land protection rights, this may lead to uncompensated land loss due to the land users' exiting businesses or land being given away or sold at well below its true social and economic value. Over 46 million ha of large-scale farmland acquisitions or negotiations were announced between October 2008 and August 2009 (Deininger and Byerlee 2011). To compensate for the negative effects, we have, amongst others, formulated the following principles for responsible agro-investments:

- respect land and resource rights,
- ensure food security,
- ensure transparency,

- ensure that all the affected people are consulted and that they participate in all the important decisions,
- safeguard social sustainability by making investments that have a desirable social and distributional impact, and
- promote environmental sustainability by minimising and mitigating the risk and magnitude of the negative impacts.

If these principles are followed, there is a chance that countries with large tracts of currently uncultivated land suitable for cultivation, or with large gaps between potential and actual yields, may increase their outputs and welfare. Private investors may then provide their farmers with knowledge, technology, infrastructure, market access and relevant institutions.

To summarise, energy crops should only be cultivated in areas in which the food demand has been satisfied and crops can be produced without their having a harmful impact on the forests, wetlands, nature conservation areas, etc. Bioenergy production should not jeopardise food security, but rather strengthen it. Food security and environmental sustainability have to be integral parts of energy crop production. It should also be taken into account that, for millennia, bioenergy (especially in the form of heat), together with food and water, has been essential for human survival and that many people in developing countries still rely on it.

1.2.3.7 Monoculture and Acceptance

There is a strong movement against bioenergy in Germany. This is especially due to many people's unwillingness to accept the increase in maize crops (a high productivity energy plant) – a process to which they refer to as 'Vermaisung' (a verb formed from the German noun "Mais" (maize)). This is mainly because the plants grow up to 2.5 m, blocking people's views. However, crop rotation, inter-cropping, double cropping and introducing agroforestry or short-rotation cropping can enhance the landscape's appearance, thus increasing people's acceptance of energy cropping. Other concepts for increasing biodiversity through energy plants are described in Chaps. 6 and 7.

1.2.3.8 Greenhouse Gas Balance (GHG) and Other Environmental Impacts

Agriculture contributes 30–35 % of global greenhouse gas emissions. In particular, land use change – in the form of deforestation, grassland transformation, methane emissions from livestock and rice cultivation, nitrous oxide emissions from fertilised soils and unsustainable water withdrawals – contributes to this (Foley et al. 2011).

Theoretically, bioenergy is carbon dioxide-neutral. Crops and trees take up carbon dioxide (CO_2) from the atmosphere and water (H_2O) from the soil and transform it by turning the external energy source sunlight into organic compounds (simplified CH_2O) and oxygen (O_2). This process is called photosynthesis. The very simplified reaction is:

$$CO_2 + H_2O + \text{sunlight } (h\nu) \rightarrow CH_2O \text{ (biomass)} + O_2.$$

Inversely, biomass can be oxidised (e.g., the burning of wood) and thus produce carbon dioxide, water and energy. Another way to produce energy from biomass is to include the intermediate step of producing methane or liquid biofuel, which can be burnt in a second step. In the case of methane, biomass is fed into an anaerobic digester (biogas plant). During several stages, bacteria systematically transform the biomass into methane (CH_4) and carbon dioxide. This is done according to the following simplified reaction:

$$2\,CH_2O \rightarrow CH_4 + CO_2.$$

In a gas burner or engine, the methane can be oxidised by air oxygen to carbon dioxide and water, releasing energy:

$$CH_4 + 2\,O_2 \rightarrow CO_2 + 2\,H_2O.$$

In the end, all the carbon that plants take from the atmosphere is quantitatively released back into the atmosphere. This process is completely carbon-dioxide-neutral and produces no additional greenhouse gas.

Criticism of the greenhouse gas balance, especially with regard to biogas production, arises due to:

- methane leaks in biogas plants,
- the release of greenhouse gases (methane and nitrous oxide) during and after the application of nitrogen fertiliser and the liquid biogas digestate to soils and
- the direct and indirect land use change of forests, grasslands and wetlands.

Methane and nitrous oxide have a strong warming potential. According to Haberl et al. (2012) and of EMPA (2012), bioenergy generated from specially grown energy crops instead of waste or plant residues (domestic and industrial organic waste, manure, straw, wood remnants, etc.) releases more CO_2-equivalent emissions for the production of bioethanol and biodiesel than fossil fuel combustion does. The reasons for this are the application of nitrogen fertilisers, which produce nitrous oxide, as well as direct and indirect land use change, which releases CO_2, etc.

An exception is the production of biogas from crops and grass that have a slightly more favourable greenhouse gas balance than the production of biodiesel and bioethanol (EMPA 2012). A systematic review by Liebetrau et al. (2012) reveals that, under typical conditions in Germany, the generation of electricity from biogas emits 70 % less greenhouse gas than the production of conventional fossil electricity. If the technical installation is optimised and biological and crop residues are used, this percentage can increase to 90 %. The energetic use of biomass in stationary combined power and heat plants is economically and ecologically more advantageous than the production of liquid fuels (Bringezu et al. 2009). A mid-class car would be able to travel 67,000 km with the biogas fuel produced from one ha of land, while it would achieve 41,000 km with biodiesel (including its by-products) produced from the same amount of land, and only 36,000 km with bioethanol (valid for agricultural yields in Germany; FNR 2012). Biogas fuels therefore have a much better greenhouse gas balance.

On the other hand, land use changes in terms of forests or grasslands transformed into croplands for energy crop production have a very unfavourable greenhouse gas balance. Replacing forested areas with, for example, oil palm plantations will lead to substantial greenhouse gas emissions and considerable biodiversity losses. Conversely, if degraded areas are transformed into cropland for energy plants, such as oil palm, jatropha, sunflowers, etc., the plants and the soils could act as a carbon sink. However, the land use change argument can not be applied to countries in which biomass crops are produced on already existing farmland. These countries would not have to cut down forests or transform grassland and wetland into arable land, nor does the argument apply to countries where wood for energy is gathered from forest residues or offcuts from the wood processing industry.

The EMPA study (2012) lists additional negative environmental impacts, such as the eutrophication and acidification of soils and water, which increase their toxicity (e.g., contaminating them with nitrates and ammonium), the enrichment of toxic particulate matter in the air during biomass burning, the depletion of water resources in scarcity areas and the consumption of fertilisers. If, on the other side, greenhouse gas effects are considered in addition to these harms, the environmental sustainability of traditional biomass production is clearly better than conventional oil and gas utilisation.

Therefore, areas' individual greenhouse gas balance and their real environmental impact should be estimated. Bioenergy systems' impact assessments should be compared to those of replaced systems, which are usually based on fossil fuel combustion, but also to the impact systems of replaced crops cultivated for food or fodder. The digestion of manure in biogas plants is much more environmentally friendly than storing it in a container and applying it directly to fields. The digestion process significantly reduces manure's methane, nitrous oxide, ammonia and its unpleasant smell. Moreover, using the residues of a fermentation plant will decrease the need for additional fertilisation, because the nutrients are quantitatively recycled. This digestate can be applied to the farmland from which the crops were originally harvested. The negative environmental impact of the extraction, transport and processing of fossil oil and gas should also be considered when calculating environmental damage. The assessment should also include water consumption and pollution, methane emissions installations, transport, accidents, such as oil spills, oil tanker collisions, etc. Embodied energy (= the total amount of energy used for buildings, machines, fertiliser, pesticides, transport, etc. in order to produce something) should also be taken into account when assessing the impact of bioenergy. A total life cycle assessment, including the direct and indirect impacts, is necessary to compare the production and use of bioenergy and fossil energy.

1.2.3.9 Emissions of Toxic Compounds During the Combustion of Biomass (for More Information, See Chap. 13)

There is a lack of systematic comparative studies of toxic emissions from fossil fuel and bioenergy sources. It can be assumed that the burning of biogas and biomethane for heat and power generation emits similar, but negligible, amounts of harmful substances as natural gas. The situation regarding biofuels and fossil liquid fuels, which are burnt in engines or in household oil heaters, may be slightly different: Fossil oil contains a higher concentration of vanadium, molybdenum, nickel, cobalt, zinc, copper and sulphur (Jungbluth 2007) than liquid biofuels do. Liquid fossil fuel has a higher emission rate of these elements when burnt if they are not removed when the oil is processed. The emission of critical organic substances is assumed to be low in both biofuels and fossil oil if optimum burning conditions prevail. When burnt to generate heat and electricity, the emissions of the solid fuels lignite and coal can be compared with emissions from the burning of wood. As a former plant material, coal contains critical elements such as antimony, lead, cadmium, cobalt, copper, nickel mercury, sulphur, selenium, vanadium, zinc, thallium, uranium and tin (for data on coal, see Dones et al. 2007; for data on biomass, see Chaps. 13 and 14). Coal can be additionally enriched by these elements during the diagenesis process (when it is transformed from plant material into coal) and often by means of secondary inputs through formation water. This accumulation of elements is the reason why coal burning leads to a significantly higher emission of critical elements than wood burning if the emission is not reduced through filters. To compare the emissions, they have to be related to the produced energy, because the heating value of dry wood fuel is approximately half of that of hard coal. There has been no systematic comparison of the harmful organic compounds emitted from coal and biomass burning.

Since biofuels are often wet and contaminated and inefficiently burnt in small, traditional stoves – especially in poor households in developing countries – indoor air pollution (chiefly organic emissions, such as black carbon) is also a concern (Chum et al. 2011). In 2008, 2.7 billion people were estimated to be dependent on biomass for cooking. In total, indoor, air-pollution-related diseases cause 1.6 million additional deaths and casualties, including those of 900,000 children under five, and a loss of 38.6 million DALYs (Disability Adjusted Life Years) per year. Cleaner fuels and more effective and safer stoves that produce fewer emissions are beyond most of these households' reach. Advanced biomass cooking appliances include biomass gasifier-operated cooking stoves that run on solid biomass, such as wood chips and briquettes. These appliances have significantly lower emissions and are far more efficient than the traditional biomass cooking stoves (three-stone fires) that are still widely used in developing countries. Switching from traditional to modern bioenergy reduces the death and disease count from indoor air pollution significantly, frees women and children from having to collect fuel wood and reduces deforestation (GBEP 2011; Chum et al. 2011).

The local environmental and social impacts of activities (e.g., coal mining) and their effects on the surrounding areas (landscape destruction, water pollution through elutriated sulphuric acid and elements as well as emissions such as methane) should also be taken into account when comparing renewable and fossil energy sources (for coal, see Dones et al. 2007). These kinds of impacts are negligible in the case of regenerative bioenergy sources. Furthermore, the residues from coal and lignite burning (ashes) should be deposited in landfills, or, for example, used as an additive for cement production. The residues from bioenergy generation, however, can be recycled back into the areas from which the biomass was harvested: Fermentation plant residues can be recycled back into the farmland and grate ash residues can be recycled back into forests.

The emission situation regarding nuclear energy differs greatly from energy generation from organic materials. Energy from nuclear power is assumed to have very low emission rates of critical substances; however, there are important risks: The mining and processing of the uranium ore, as well as the reprocessing treatment of the spent nuclear fuel rods may release radioactive radiation (the Chernobyl accident, April 1986). Moreover, operating errors, equipment failure, natural catastrophes, such as earthquakes or tsunamis (the Fukushima accident in March 2011), as well as the consequences of landslides and floods may release radioactive material from nuclear reactors and should thus be taken into account. In addition to these risks, none of the 31 countries with nuclear power plants has as yet found an optimal solution for the ultimate storage of nuclear waste.

1.2.3.10 Financial Implications

While bioenergy production might contribute to increases in food prices, its impact is less severe than that of weather conditions, changes in food demand, production efficiency and increasing energy costs (Zeddies et al. 2012). While benefitting farmers, increases have adversely affected the poor, food security and nourishment in developing countries. On the other hand, bioenergy provides these countries with opportunities to progress with regard to rural developments and agricultural growth, both of which lead to job creation. If sustainability frameworks are properly designed, implemented, monitored and adhered to, they may help minimise negative socio-economic impacts and maximise the benefits of bioenergy production, particularly for local people (Chum et al. 2011; WBGU 2011). The real implications of bioenergy have to be investigated individually with regard to each form of bioenergy.

1.3 Bioenergy in Germany

In 2007, Germany's government established the German Integrated Energy and Climate Change Programme. Its main objectives are to ensure a secure, economically efficient and environmentally friendly energy supply. At the same time, the

Box 1.2 Important renewable energy implementations in Germany

1. The Renewable Energy Sources Act (EEG) 2012:
    - passed on 1 January 2012, replacing the previous acts of 2009 and 2004
    - most effective funding instrument at the German government's disposal
    - purpose: To further increase the share of renewable energies in electricity generation by 2020 as part of an integrated energy and climate protection programme
    - internationally observed as exemplary law.
2. The Renewable Energies Heat Act (EEWärmeG) 2008:
    - goal: to increase the percentage of renewable energies in heat supply to 14 % by 2020
    - purpose: to promote renewable energies in the heat sector
    - to achieve the sound management of fossil resources and decrease dependency on energy imports
    - to facilitate the sustainable development of energy supply and
    - further develop technologies to generate heat from renewable energy sources.
3. The Biomass Ordinance 2001 specifies:
    - which substances are recognised as biomass
    - which technical processes may be used for electricity generation
    - which environmental standards have to be met.
4. The Biomass Electricity Ordinance (BioSt-NachV) and Biofuel Sustainability Ordinance (Biokraft-NachV) 2009:
    - implements the EU's Renewable Energy Directive (RED) regarding the sustainability criteria for biomass and
    - specifies the legal and technical rules for recognising certification systems and certification bodies.

programme is supported by on-going legislative initiatives (e.g., the Renewable Energy Source Act, see Box 1.2) that aim to create more competition in the energy markets and offer new regulations for emissions trading.

One of the programme's initiatives is the National Biomass Action Plan, which provides a holistic solution for increasing bioenergy's contribution to Germany's total energy supply. Bioenergy is considered a means with which to mitigate the effects of climate change, to secure energy supply and enable the sustainable development of societies. It has also helped Germany increase its domestic value creation, especially in rural areas.

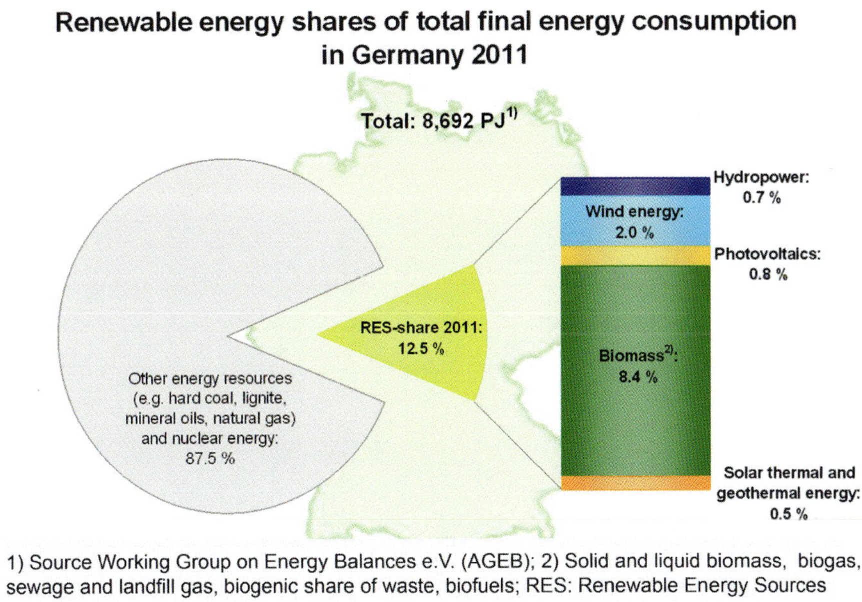

Fig. 1.2 Shares of renewable energies in relation to the total final energy consumption in 2011 (Source: BMU-KI III 1 based on working Group on Renewable Energy-Statics (AGEE-Stat) and Centre for Solar Energy and Hydrogen Research Baden-Württenberg (ZSW), according to AGEB (BMU 2012c); deviations in the totals are due to rounding; 1 PJ = 10^{15} J; as at: July 2012; all figures provisional)

The National Biomass Action Plan will be integrated into the German government's Renewable Energy Action Plan according to the requirements of the EU Renewable Energy Directive. The main goals are to:

- increase the percentage of renewable energy in electricity production to at least 30 % by 2020;
- use biofuels to further reduce greenhouse gas emissions in the transport sector; from 2015, biofuel quotas will be based on net greenhouse gas reductions rather than set relative to their energy content.
- increase the percentage of biofuels in the overall fuel consumption to reduce the net greenhouse gas emission by 7 % by 2020 (equivalent to approximately 12 % energy content).
- increase the percentage of renewables-generated heat from the current 12 % energy content.

Bioenergy met 8.2 % of Germany's final energy consumption needs in 2011 (BMU 2012c; Fig. 1.2). This amount will have increased by 2020 due to the targets stated in the EU Climate and Energy Package in April 2009 and in the German Integrated Energy and Climate Change Programme, which was launched in August 2007. Biomass is the most important contributor to the renewable energy mix (Fig. 1.2). Nuclear power's contribution to the energy mix has decreased by 30 % due to Germany's decision to abolish it. In contrast, owing to governmental subsidies, the total renewable energy supply (electrical and thermal power) grew by 9 % in 2011, which shows an

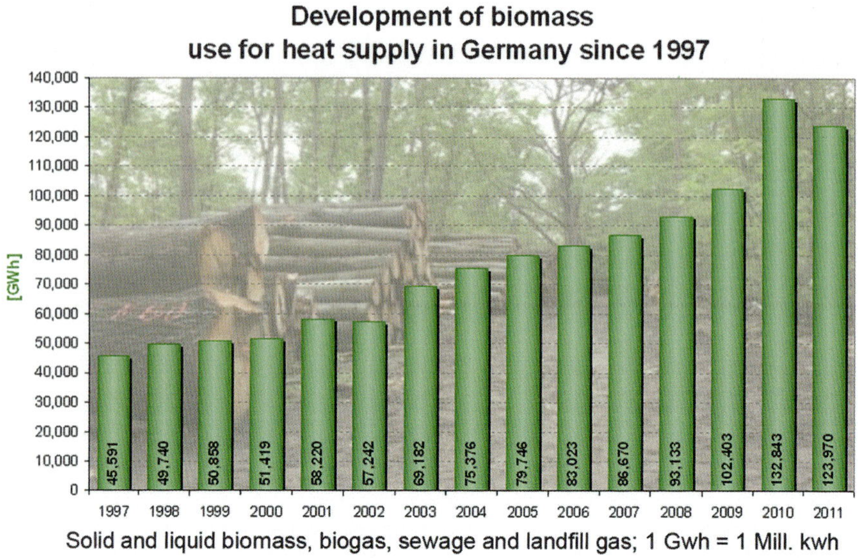

Fig. 1.3 Development of biomass use for heat supply in Germany since 2007 (Source: BMU-KI III 1 based on Working Group on Renewable Energy-Statics (AGEE-Stat); image BMU/ Brigitte Hiss; as at: July 2012; all figures provisional (Source BMU 2012c))

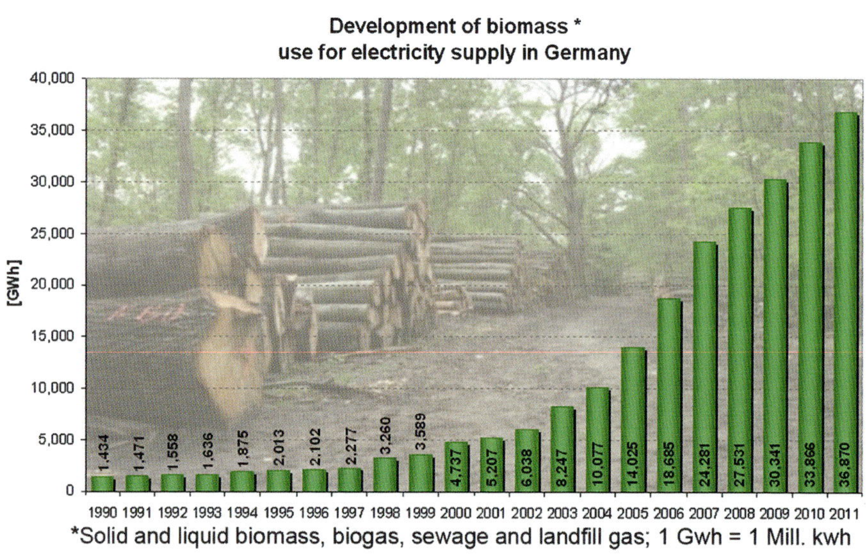

Fig. 1.4 Development of biomass use for electricity supply in Germany since 1990 (Source: BMU-KI III 1 based on Working Group on Renewable Energy-Statics (AGEE-Stat); image BMU/ Brigitte Hiss; as at: July 2012; all figures provisional (BMU 2012c))

impressive increase in the development of renewable energies to 12.2 % of the total final energy consumption. Figures 1.3 and 1.4 demonstrate the development of biomass use for the respective supply of heat and power over the last few years.

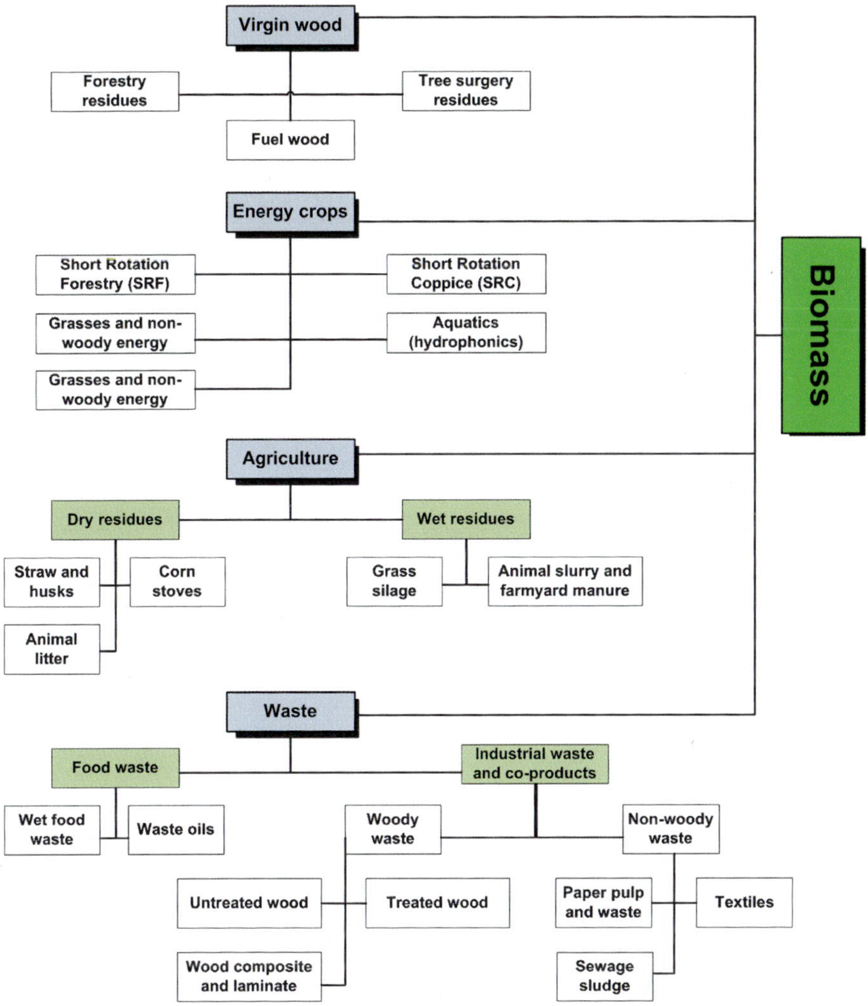

Fig. 1.5 Sources of biomass for the production of bioenergy (modified from Ladanai and Vinterbäck 2009)

The German Biomass Action Plan specifies the potential of biomass use in Germany and reveals Germany's strategies to promote bioenergy use in the heating, electricity and fuel sectors. Depending on the raw material used, biomass is a manifold energy source. Moreover, it can be used in many different technologies: wood is mainly used for heat production, biogas for both heat and power generation, while oil seed is used as biofuel in power stations.

According to Fig. 1.5, wood, agricultural sources (mostly energy crops) and wastes are the main sources of biomass (for an explanation of the biomass sources, also see Box 1.1). Figure 1.6 shows the ways in which heat, power and fuels are generated from biomass.

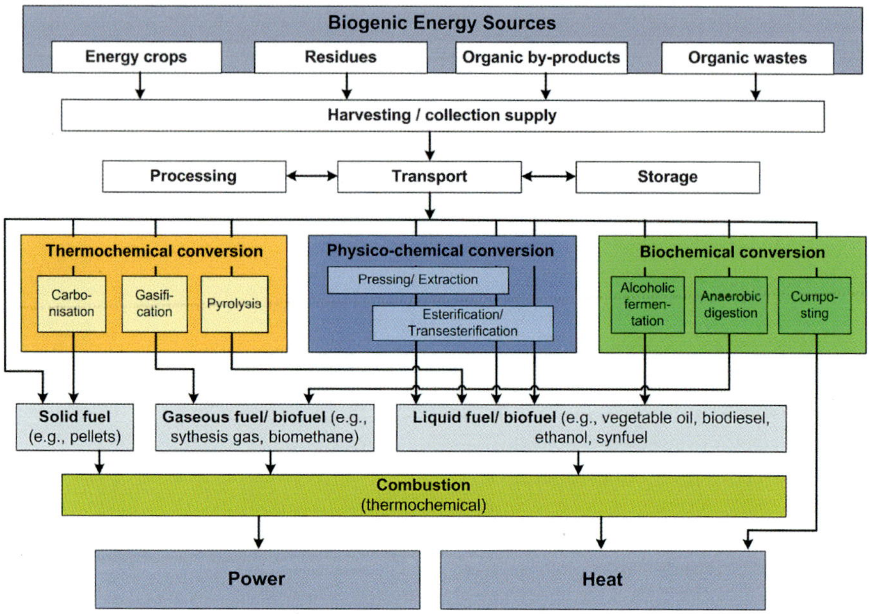

Fig. 1.6 Ways to provide heat, power and fuels from biomass (Modified from FNR/GIZ 2010)

1.4 The Promotion of Renewable Energies: Quota Systems and Feed-in Tariffs

According to the European Union's targets, renewable energy sources should comprise 20 % of the total energy supply by 2020 (European Commission 2012). At this stage, fossil energy production is still more cost-effective than renewable energy production. All countries should therefore have financial support schemes in place in order to "facilitate a sustainable development of energy supply, to reduce the costs of energy supply to the national economy, also by incorporating external long-term effects, to conserve fossil fuels and to promote the further development of technologies for the generation of electricity from renewable energy sources" (BMU 2012a).

All the individual countries in the EU have different goals and methods to achieve such schemes. Thus, each country also has its own means of promoting the development of its renewable energy sector. There are two general instruments that can be used to promote the use of renewable energies in the electrical power sector: the quota system and feed-in tariffs.

1.4.1 *Quota System*

Some European countries, including Great Britain, Sweden, Belgium and the Netherlands, have applied the quota system. The system requires energy operators and energy supply companies to share the quota of renewable energies to be contributed to the total electrical power supply. Therefore, each company is expected to contribute a specific percentage of electrical power generated from renewable energy sources to the power grid. The federal government defines this quota, which therefore varies from country to country. In some cases, the sharing quota depends on the type of renewable energy source, i.e. there might be different quotas for wind power and photovoltaic power.

The quota scheme rewards operators with certificates or penalises them, as this is known to stimulate the market and therefore reduce the total cost of changing the energy system. The reward and penalising approach is aimed at specifically promoting renewable energy systems with the lowest operating costs. In turn, this will facilitate access to markets without federal supply systems. Energy operators and energy supply companies can guarantee their quota if they produce the power themselves, or if they buy certificates from renewable installation companies. These certificates are therefore available on the market. Those energy supply companies that cannot guarantee a fixed quota, are penalised. The calculation of the certificates' prices and the penalties' amount differ significantly from country to country. These calculations are, however, the key factors for the success of the quota system and determine whether or not renewable technologies will be developed and will achieve their energy production targets.

The quota system not only applies to the electrical power sector, but also to the production of liquid biofuels. Fuel companies have to guarantee a certain percentage of biofuel in their gasoline. This percentage has been increased over the years to support the biofuel industry and to systematically reduce dependency on fossil fuel.

1.4.2 *Feed-in Tariffs*

Sijm (2002) defines feed-in tariffs as follows: "Usually, this term refers to the regulatory, minimum guaranteed price per kWh that an electricity utility has to pay to a private, independent producer of renewable power fed into the grid". Feed-in tariffs have been enacted in 50 countries, including Germany, Austria, Denmark, Canada and China. They have three key functions:

- To prioritise the connection to the grid system
- To appoint long-term contracts for electricity generation
- To base the purchase price on production costs.

1.4.2.1 Feed-in Tariffs in Germany

Feed-in tariffs were enforced in Germany in 1991 (BMU 2000). The Renewable Energy Sources Act (EEG), in which the feed-in tariffs are defined, was implemented in 2000. In order to address certain developments in the energy sector, including the federal decision to change the whole energy system from fossil-based to renewable-based production, the Act was amended in 2004, 2009 and 2012. The following key elements were added to the Renewable Energy Sources Act (BMU 2004):

- The prioritisation of the purchase and transmission of electricity
- The guarantee of a consistent fee – generally for a 20-year period – minus a degression rate
- The nation-wide equalisation of purchased electricity and the associated payable fee.

The main intention of the Act is to guarantee a high investment security for 20 years and to guarantee a tariff system that covers more or less all the installation and running costs over the 20-year period. This reduces installation companies' investment risks, because grid operators have to pay fixed tariffs for the feed-in of electricity from renewable energy sources. Moreover, the Act supports the installation of renewable energies and influences the quality of the production (e.g., of bioenergy) by adjusting the tariffs accordingly.

In principle, the fee that the grid operators have to pay depends on the energy source and on the installation size. Innovative technologies receive an additional bonus. Differentiated fee systems apply to bioenergy (see paragraph below). The fees generally depend on the materials used (bioenergy: variety of plants, wood, manure, etc.), the installation size and the installation type. The grid operators can take the additional expenses set by the Renewable Energy Sources Act into consideration in their charges for the use of the grid. Therefore, the total costs are apportioned to the electricity consumer.

A yearly degression rate, which depends on the energy source, reduces the costs. The reduction rate of all types of renewable energy varies if the costs of the technology decrease. The costs, for example, of installing photovoltaic systems have been reduced by about 65 % over the last few years (BSW 2012). Therefore, the degression rate was reduced several times between 2010 and 2012.

Since 2009, operators have had the opportunity to sell energy directly on the market at the price that the spot market determines plus a compensation payment. This firstly reduces the costs of the tariff-in scheme and facilitates entrance to the market. Furthermore, in 2012, a flexibility premium was established for bioenergy as it has the capacity to generate electricity on a demand basis. The aim of this is:

- to feed in electrical power when there is a high electricity demand, thus
- to reduce the amount of electricity that needs to be stored and
- to stabilise the grid by reducing the generation of power from biogas plants if too much electricity is generated by wind and solar power.

Figure 1.7 illustrates the share of the renewables according to different consumption schemes. It is noteworthy that electricity consumption is increasing continuously. In 2011, wind, photovoltaic, bioenergy and hydropower already provided 20.3 % of the gross electricity consumption. The percentage of wind and photovoltaic power also shows a rapid increase.

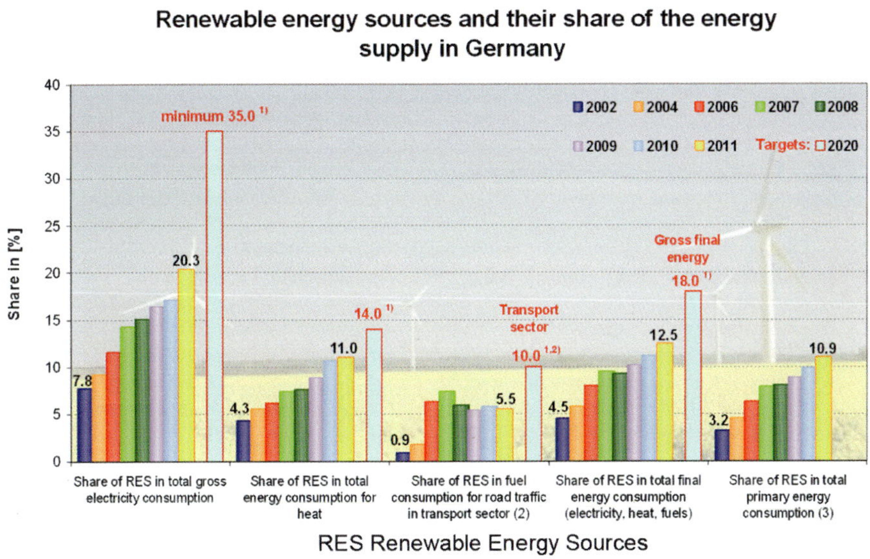

Fig. 1.7 Renewable energy sources and their share of the energy supply in Germany (Sources: Targets of the German Government, Renewable Energy Sources Act (EEG); Renewable Energy Sources Heat Act (EEWärmeG), EU-Directive 2009/28/EC, Total consumption of engine fuels, excluding fuel in air traffic, calculated using efficiency method; Source: Working Group on Energy Balances e.V. (AGEB)) (Source: BMU-KI III 1 based on working Group on Renewable Energy-Statics (AGEE-Stat); image BMU/Brigitte Hiss; as at: July 2012; all figures provisional (BMU 2012c))

1.4.2.2 Feed-in Tariffs in the Bioenergy Sector

The Renewable Energy Sources Act (EEG) has set different fees for the use of bioenergy, depending on the input material used. The biomass ordinance "regulates which substances are classed as biomass, the substances for which an additional substance-based tariff may be claimed, which energy-related reference values are to be used to calculate this tariff and how the substance-based tariff is to be calculated, which technical procedures for electricity generation from biomass fall within the scope of application of the Act and which environmental requirements must be met in generating electricity from biomass" (BMU 2012b). As mentioned above, the EEG promotes the quality of bioenergy production. Since 2012, bonuses are only paid if 60 % of the electricity is generated from combined heat and power plants, or if manure is used to generate 60 % of the electricity in biogas installations. Furthermore, maize, including corn-cob mixes, may not comprise more than 60 % of input substances in biogas installations (BMU 2012a).

References

Barona, E., Ramankutty, N., Hyman, G., & Coomes, O. T. (2010). The role of pasture and soybean in deforestation of the Brazilian Amazon. *Environmental Research Letters, 5*(2), 9 pp.

Bauen, A., Berndes, G., Junginger, M., Londo, M., & Vuille, F. (2009). *Bioenergy – A sustainable and reliable energy source – A review of status and prospects* (Main Report, 108 pp.), IEA Bioenergy. Retrieved November 27, 2012, from http://www.ieabioenergy.com/LibItem.aspx?id=6479

BMU. (2000). *Act on granting priority to renewable energy sources (Renewable Energy Sources Act – EEG)*. The Federal Ministry for the Environment, Nature Conservation and Nuclear Safety, 2000. Retrieved June 13, 2012, from http://www.erneuerbare-energien.de/files/pdfs/allgemein/application/pdf/res-act.pdf

BMU. (2004). *Act on granting priority to renewable energy sources (Renewable Energy Sources Act – EEG), Consolidated (non-binding) version of the Act in the version applicable as at 1 August 2004*. Berlin: The Federal Ministry for the Environment, Nature Conservation and Nuclear Safety. Retrieved June 13, 2012, from http://www.bmu.de/files/english/pdf/application/pdf/eeg_gesetz_merkmale_en.pdf

BMU. (2012a). *Act on granting priority to renewable energy sources (Renewable Energy Sources Act – EEG), Consolidated (non-binding) version of the Act in the version applicable as at 1 January 2012*. Berlin: The Federal Ministry for the Environment, Nature Conservation and Nuclear Safety. Retrieved November 25, 2012, from http://www.erneuerbare-energien.de/files/english/pdf/application/pdf/eeg_2012_en_bf.pdf

BMU. (2012b). *Ordinance on the generation of electricity from biomass. Consolidated (non-binding) version of the ordinance in the version applicable as of 1 January 2012*. Berlin: The Federal Ministry for the Environment, Nature Conservation and Nuclear Safety. Retrieved June 18, 2012, from http://www.bmu.de/files/english/pdf/application/pdf/biomasse_verordnung_en_bf.pdf

BMU. (2012c). *Development of renewable energy sources in Germany 2011. Graphics and tables.* Version: July 2012. Berlin: The Federal Ministry for the Environment, Nature Conservation and Nuclear Safety. Retrieved October 21, 2012, from http://www.erneuerbare-energien.de/files/english/pdf/application/pdf/ee_in_deutschland_graf_tab_en.pdf

Bringezu, S., Schütz, H., Schepelmann, P., Lange, U., von Geibler, J., Bienge, K., Kristof, K., Arnold, K., Merten, F., Ramesohl, S., Fischedick, M., Borelbach, P., Kabasci, S., Michels, C., Reinhardt, G. A., Gärtner, S., Rettenmaier, N., & Münch, J. (2009). Nachhaltige Flächennutzung und nachwachsende Rohstoffe – Optionen einer nachhaltigen Flächennutzung und Ressourcenschutzstrategien unter besonderer Berücksichtigung der nachhaltigen Versorgung mit nachwachsenden Rohstoffen. *Umweltbundesamt Texte 34*/2009, p. 26. Retrieved November 8, 2012, from http://www.umweltbundesamt.de/uba-info-medien/3861.html

Boßmann, T., Eichhammer, W., & Elsland, R. (2012). *Contribution of energy efficiency measures to climate protection within the European Union until 2050*. Policy Report of the Federal Ministry for the Environment, Nature Conservation and Nuclear Safety, Berlin, and Fraunhofer Institute for Systems and Innovation Research ISI, 42 pp. Retrieved November 19, 2012, from http://www.isi.fraunhofer.de/isi-en/e/projekte/bmu_eu-energy-roadmap_315192_ei.php

BSW. (2012). *Bundesverband Solarwirtschaft*. Retrieved June 17, 2012, from http://www.solarwirtschaft.de/preisindex

Chum, H., Faaij, A., Moreira, J., Berndes, G., Dhamija, P., Dong, H., Gabrielle, B., Goss Eng, A., Lucht, W., Mapako, M., Masera Cerutti, O., McIntyre, T., Minowa, T., & Pingoud, K. (2011). Bioenergy. In O. Edenhofer, R. Pichs-Madruga, Y. Sokona, K. Seyboth, P. Matschoss, S. Kadner, T. Zwickel, P. Eickemeier, G. Hansen, S. Schlömer, & C. von Stechow (Eds.), *IPCC special report on renewable energy sources and climate change mitigation* (pp. 209–331). Cambridge/New York: Cambridge University Press. Retrieved November 18, 2012, from http://srren.ipcc-wg3.de/report

Cordell, D., Drangert, J.-O., & White, S. (2009). The story of phosphorus: Global food security and food for thought. *Global Environmental Change, 19*(2), 292–305.

Deininger, K., & Byerlee, D. (2011). *Rising global interest in farmland: Can it yield sustainable and equitable benefits?* 214 pp., World Bank, Washington, ISBN 978-0-8213-8591-3. Retrieved November 1, 2012, from http://siteresources.worldbank.org/INTARD/Resources/ESW_Sept7_final_final.pdf

DERA. (2011). Deutsche Rohstoffagentur = German Mineral Resources Agency. *DERA Rohstoffinformationen. Annual Report. Reserves, Resources and Availability of Energy Resources 2011*. Federal Institute for Geosciences and Natural Resources (BGR, Ed.); 89 pp., Hannover (Germany); ISSN 2193–5319. Retrieved January 7, 2012, from http://www.bgr.bund.de/EN/Themen/Energie/Downloads/annual_report_2011_en.pdf?__blob=publicationFile&v=2

Dones, R., Bauer, C., & Röder, A. (2007). Kohle. In R. Dones, et al. (Eds.), *Sachbilanzen von Energiesystemen: Grundlagen für den ökologischen Vergleich von Energiesystemen und den Einbezug von Energiesystemen in Ökobilanzen für die Schweiz* (Final report ecoinvent No. 6-VI). Dübendorf, CH: Paul Scherrer Institut Villigen/Swiss Centre for Life Cycle Inventories, 346 pp.

Dornburg, V., Faaij, A., Langeveld, H., van de Ven, G., Wester, F., van Keulen, H., van Diepen, K., Ros, J., van Vuuren, D., van den Born, G.J., van Oorschot, M., Smout, F., Aiking, H., Londo, M., Mozaffarian, H., Smekens, K., Meeusen, M., Banse, M., Lysen E., & van Egmond. (2008). *Biomass assessment: Assessment of global biomass potentials and their links to food, water, biodiversity, energy demand and economy* (Report 500102 012), 107 pp. Retrieved November 25, 2012, from http://www.pbl.nl/sites/default/files/cms/publicaties/500102012.pdf

European Commission. (2012). Communication from the Commission to the European Parliament, The Council, The European Economic and Social Committee and The Committee of the Regions, Com (2012) 271final. Retrieved June 17, 2012, from http://ec.europa.eu/energy/renewables/doc/communication/2012/comm_en.pdf

EMPA (Swiss Federal Laboratories for Material Science and Technology). (2012). Harmonisation and extension of the bioenergy inventories and assessment. End report 31st August 2012; 97 pp. Retrieved October 7, 2012, from http://www.empa.ch/plugin/template/empa/*/125527 and http://www.bfe.admin.ch/php/modules/enet/streamfile.php?file=000000010923.pdf&name=000000290691

FAO (Food and Agriculture Organization). (2009). The state of food and agriculture 2009. Livestock in the balance. Retrieved November 9, 2012, from http://www.fao.org/docrep/012/i0680e/i0680e.pdf

FAO (Food and Agriculture Organization). (2012). FAO statistical yearbook 2012. Retrieved July 16, 2012, from http://www.fao.org/economic/ess/ess-publications/ess-yearbook/yearbook2012/en/

FNR (Fachagentur Nachwachsende Rohstoffe, Agency for Renewable Resources). (2009). Biogas – An introduction, 25 pp, 2nd ed. Retrieved November 27, 2012, from http://www.fnr-server.de/ftp/pdf/literatur/pdf_329-biogas—an-introduction_engl.pdf

FNR (Fachagentur Nachwachsende Rohstoffe, Agency for Renewable Resources). (2012). Bioenergy in Germany: Facts and figures. Retrieved August 29, 2012, from http://www.fnr-server.de/ftp/pdf/literatur/pdf_484-basisdaten_engl_web_neu.pdf

FNR/GIZ (Fachagentur Nachwachsende Rohstoffe and Deutsche Gesellschaft für Internationale Zusammenarbeit). (2010). Guide to biogas – From production to use, 232 pp. Retrieved November 27, 2012, from http://mediathek.fnr.de/media/downloadable/files/samples/g/u/guide_biogas_engl_2012.pdf

Foley, J. A., Ramankutty, N., Brauman, K. A., Cassidy, E. S., Gerber, J. S., Johnston, M., Mueller, N. D., O'Connell, C., Ray, D. K., West, P. C., Balzer, C., Bennett, E. M., Carpenter, S. R., Hill, J., Monfreda, C., Polasky, S., Rockström, J., Sheehan, J., Siebert, S., Tilman, D., & Zaks, D. P. M. (2011). Solutions for a cultivated planet. *Nature, 478*, 337–342.

Fritsche U.R., Fehrenbach, H., & Köppen, S. (2012). Nach Super E10: Welche Rolle für BiokraftStoffe? Fakten, Trends und Perspektiven, 42 pp. Retrieved November 2, 2012, from www.shell.de/biokraftstoffstudie

GBEP (Global Bioenergy Partnership). (2011). The Global Bioenergy Partnership sustainability indicators for bioenergy, 211 pp, ISBN 978-92-5-107249-3. Retrieved November 18, 2012, from http://www.globalbioenergy.org/fileadmin/user_upload/gbep/docs/Indicators/The_GBEP_Sustainability_Indicators_for_Bioenergy_FINAL.pdf

Gilbert, N. (2009). Environment: The disappearing nutrient. *Nature, 461*(7265), 716–718.

Grethe, H., Dembélé, A., & Duman, N. (2011). How to feed the world's growing billions. Heinrich Böll Stiftung and WWF Deutschland, 63 pp. Retrieved December 10, 2012, from http://www.wwf.de/fileadmin/fm-wwf/Publikationen-PDF/WWF_BOELL_How_to_feed.pdf

Gustavsson, J., Cederberg, C., Sonesson, U., van Otterdijk, R., & Meybeck, A. (2011). Global food losses and food waste section 3.2 (Study conducted for the international congress "Save Food!" at Interpack2011, Düsseldorf, Germany), FAO Rural Infrastructure and Agro-Industries Division, 38 pp. Retrieved September 23, 2012, from: http://www.fao.org/docrep/014/mb060e/mb060e00.pdf

IEA (International Energy Agency). (2011). *World energy outlook 2011*. Paris: IEA. 659 pp. ISBN 978-92-64-12413-4.

Jungbluth, N. (2007). Erdöl. In R. Dones (Ed.), *Sachbilanzen von Energiesystemen: Grundlagen für den ökologischen Vergleich von Energiesystemen und den Einbezug von Energiesystemen* (Ökobilanzen für die Schweiz, ecoinvent Report No. 6-IV). Duebendorf: Swiss Centre for Life Cycle Inventories, 327 pp.

Haberl, H., Erb, K.-H., Krausmann, F., Bondeau, A., Lauk, C., Müller, C., Plutzar, C., & Steinberger, J. K. (2011). Global bioenergy potentials from agricultural land in 2050: Sensitivity to climate change, diets and yields. *Biomass and Bioenergy, 35*, 4753–4769.

Haberl, H., Körner, C., Lauk, C., Schmid-Staiger, U., Smetacek, V., Schulze, E.-D., Thauer, R. K., Weiland, P., & Wilson, K. (2012). In Deutsche Akademie der Naturforscher Leopoldina – Nationale Akademie der Wissenschaften (Ed.), *The availability and sustainability of biomass as an energy source* (pp. 9–42). Retrieved July 30, 2012, form http://www.leopoldina.org/uploads/tx_leopublication/201207_Stellungnahme_Bioenergie_LAY_en_final.pdf

Kranert, M., Hafner, G., Barabosz, J., Schuller, H., Leverenz, D., Kölbig, A., Schneider, F., Lebersorger, S., & Scherhaufer, S. (2012). Ermittlung der weggeworfenen Lebensmittelmengen und Vorschläge zur Verminderung der Wegwerfrate bei Lebensmitteln in Deutschland, 483 pp. Retrieved November 11, 2012, from http://www.bmelv.de/SharedDocs/Downloads/Ernaehrung/WvL/Studie_Lebensmittelabfaelle_Langfassung.pdf?__blob=publicationFile

Ladanai, S., & Vinterbäck, J. (2009). Global potential of sustainable biomass for energy. SLU, Swedish University of agricultural sciences, Department of energy and technology (Report 13, Uppsala, ISSN 1654-9406). Retrieved November 27, 2012, from http://publikationer.slu.se/Filer/WBA_PP-1_100122final101.pdf

Liebetrau, J., Daniel-Gromke, J., Oehmichen, K., Weiland, P., Friehe, J., Clemens, J., & Hafermann, C. (2012). Emissionsanalyse und Quantifizierung von Stoffflüssen durch Biogasanlagen im Hinblick auf die ökologische Bewertung der landwirtschaftlichen Biogasgewinnung und Inventarisierung der deutschen Landwirtschaft (FKZ: 22023606). Deutsches BiomasseForschungsZentrum, 229 pp. Retrieved October 12, 2012, from http://www.fnr-server.de/ftp/pdf/berichte/22023606.pdf

MacDonald, M. (2012). Food security and equity in a climate-constrained world. In *Worldwatch institute's state of the world 2012: Moving toward sustainable prosperity,* Chapter 14 (pp. 161–168). Island Press. Washington. ISBN-13: 978-1610910378.

McMichael, A. J., Powles, J. W., Butler, C. D., & Uauy, R. (2007). Food, livestock production, energy, climate change, and health. *The Lancet, 370*, 1253–1263.

Mühlenhoff, J. (2010). Value creation for local communities through renewable energies. Renews special issue 46/December 2010. German Renewable Energy Agency (AEE), 22 pp. Retrieved September 6, 2012, from http://www.unendlich-viel-energie.de/uploads/media/46_Renews_Special_value-creation_for_local_communities.pdf

Noleppa, S. (2012). Klimawandel auf dem Teller. WWF Deutschland, 86 pp. Retrieved November 6, 2012, from http://www.wwf.de/fileadmin/fm-wwf/Publikationen-PDF/Klimawandel_auf_-dem_Teller.pdf

Noleppa, S., & von Witzke, H. (2012). Tonnen für die Tonne. WWF Deutschland, 52 pp. Retrieved November 11, 2012, from http://www.wwf.de/fileadmin/fm-wwf/Publikationen-PDF/studie_tonnen_fuer_die_tonne.pdf

OECD/FAO. (2012). *OECD-FAO agricultural outlook 2012–2021*. OECD Publishing/FAO. ISSN 1563-0447 (print). Retrieved November 10, 2012, from http://www.oecd-ilibrary.org/agriculture-and-food/oecd-fao-agricultural-outlook-2012_agr_outlook-2012-en

Parfitt, J., Barthel, M., & Macnaughton, S. (2010). Food waste within food supply chains: Quantification and potential for change to 2050. *Philosophical Transactions of the Royal Society B: Biological Sciences, 365*, 3065–3081.

Raschka, A., & Carus, M. (2012). Stoffliche Nutzung von Biomasse Basisdaten für Deutschland, Europa und die Welt. Retrieved November 7, 2012, from http://www.nova-institut.de/download/Stoffliche_Nutzung_von_Biomasse_nova

REN21 (Renewable Energy Policy Network for the 21st Century). (2012). Renewables 2012 (Global Status Report), 171 pp. Retrieved September 14, 2012, from http://www.ren21.net/Portals/97/documents/GSR/GSR2012_low%20res_FINAL.pdf

Reenberg, A., & Fenger, N. A. (2011). Globalizing land use transitions: The soybean acceleration. *Geografisk Tidsskrift-Danish Journal of Geography, 111*(1), 85–92.

Rees, W. E. (2006). Ecological footprints and biocapacity: Essential elements in sustainability assessment. In J. Dewulf & H. van Langenhove (Eds.), *Renewables-based technology: Sustainability assessment* (pp. 143–158). Hoboken: Wiley Series in Renewable Resource. ISBN 978-0-470-02241-2.

Sijm, J. P. M. (2002). *The performance of feed-in tariffs to promote renewable electricity in European countries*. Retrieved June 13, 2012, from ftp://ecn.nl/pub/www/library/report/2002/c02083.pdf

WBGU (German Advisory Council on Global Change). (2008). *Future bioenergy and sustainable land use* (Flagship Report 2008); ISBN 978-1-84407-841-7; 365 pp., London: Earthscan. Retrieved October 2012, from http://www.wbgu.de/fileadmin/templates/dateien/veroeffentlichungen/hauptgutachten/jg2008/wbgu_jg2008_en.pdf

WBGU (German Advisory Council on Global Change). (2011). *World in transition – A social contract for sustainability* (Flagship Report). Berlin: WBGU, 396 pp. Retrieved November 24, 2012, from http://www.wbgu.de/fileadmin/templates/dateien/veroeffentlichungen/hauptgutachten/jg2011/wbgu_jg2011_en.pdf

WHO (World Health Organization). (2012). *World health statistics 2012*, 175 pp, ISBN 978 92 4 156444 1. Retrieved December 1, 2012, from http://www.who.int/entity/gho/publications/world_health_statistics/2012/en/index.html

Witzke, H., von, Noleppa, S., & Zhirkova, I. (2011). *Fleisch frisst Land. WWF-study*, 73 pp. Retrieved November 8, 2012, from http://www.wwf.de/fileadmin/fm-wwf/Publikationen-PDF/WWF_Fleischkonsum_web.pdf

WWF (World Wide Fund for Nature). (2012). *Living planet report 2012*. WWF International, Gland, Switzerland, 164 pp. Retrieved November 21, 2012, from http://awsassets.panda.org/downloads/2_lpr_2012_online_single_pages_version2_final_120516.pdf

Zeddies, J., Bahrs, E., Schönleber, N., & Gamer, W. (2012). *Globale Analyse und Abschätzung des Biomasse-Flächennutzungspotentials* (Report for BMELV), 139 pp. Retrieved November 2, 2012, from http://www.fnr-server.de/cms35/index.php?id=6143&rid=P_1057&mid=649&aC=5ae0cc34&jumpurl=1

Chapter 2
Bioenergy Villages in Germany: Applying the Göttingen Approach of Sustainability Science to Promote Sustainable Bioenergy Projects

Peter Schmuck, Swantje Eigner-Thiel, Marianne Karpenstein-Machan, Benedikt Sauer, Hans Ruppert, Walter Girschner, and Folker Roland

Abstract This chapter describes the history of bioenergy villages in Germany between 2000 and 2008, providing an exemplifying introduction to the more detailed aspects of sustainable bioenergy use. Developed by a team of scientists at the University of Göttingen, the electricity and heat supply of an entire village was transformed from conventional to biomass energy sources between 2000 and 2005. This lighthouse project, the first "bioenergy village" in Germany, was realised through the active participation of the entire population of Jühnde, a village in Southern Lower Saxony (800 inhabitants). The technical concept comprises (1) an anaerobic digestion plant (fuelled by energy crops and liquid manure) with a

---

P. Schmuck (✉) • M. Karpenstein-Machan • W. Girschner
Interdisciplinary Centre for Sustainable Development (IZNE), University of Göttingen, Goldschmidtstr. 1, D-37077 Göttingen, Germany
e-mail: peterschmuck@gmx.de; mkarpen@gwdg.de

S. Eigner-Thiel
Interdisciplinary Centre for Sustainable Development (IZNE), University of Göttingen, Goldschmidtstr. 3, D-37077 Göttingen, Germany

HAWK, Büsgenweg 1a, D-37077 Göttingen, Germany
e-mail: eigner-thiel@hawk-hhg.de

B. Sauer • H. Ruppert
Interdisciplinary Centre for Sustainable Development (IZNE), University of Göttingen, Goldschmidtstr. 3, D-37077 Göttingen, Germany

Department of Sedimentology, University of Göttingen, Goldschmidtstr. 3, D-37077 Göttingen, Germany
e-mail: bsauer@gwdg.de; hrupper@gwdg.de

F. Roland
Interdisciplinary Centre for Sustainable Development (IZNE), University of Göttingen, Goldschmidtstr. 3, D-37077 Göttingen, Germany

Department of Business, University of Applied Sciences Hochschule Harz, Friedrichstraße 57-59, D-38855 Wenigerode, Germany
e-mail: froland@hs-harz.de

H. Ruppert et al. (eds.), *Sustainable Bioenergy Production - An Integrated Approach*, DOI 10.1007/978-94-007-6642-6_2, © Springer Science+Business Media Dordrecht 2013

combined heat and power (CHP) generator producing electricity and heat, (2) a central heating plant fired by locally produced wood chips to satisfy the additional heat demand during the winter as well as (3) a hot water pipeline delivering the heat energy to the connected households. The chapter explains the history of the project, its social implementations and the results thereof regarding the ecological, economic and social changes in the village. Furthermore, this chapter describes the successful transfer of the model to dozens of other villages in Germany. The process of developing bioenergy villages is embedded in the methodological framework of sustainability science, which is based on the principles of inter- and transdisciplinary collaboration and on participatory action research aimed at sustainable development.

Keywords Sustainability science • Action research • Bioenergy village

Within a broader sustainability framework, this chapter focuses the methodological background of our scientific approach to replace our heat and electricity supply with renewable bioenergy. The authors of this chapter, most of whom are founding members of the Interdisciplinary Centre for Sustainable Development at the University of Göttingen, initiated the complete conversion of the heat and electricity supply of Jühnde from fossil to biomass fuels. We follow an elaborated approach, which we call the "Göttingen Approach of Sustainability Science". We start with a short introduction on our understanding of "sustainable development" and "sustainability science". Thereafter, we describe how sustainability science emerged in the village of Jühnde, focussing mainly on the processes leading to the first success. The methodological basis for the ongoing project "Sustainable use of bioenergy: bridging climate protection, nature conservation and society" was partly provided by the systematic research that complemented the bioenergy village project from 2000 to 2008.

2.1 Sustainable Development

We share the view that the current environmental, social and economic problems require all societal groups of all countries to cooperate closely if the problems are to be solved (Cervinka and Schmuck 2010). If we want to find (1) alternatives to fossil and nuclear fuels with their known impact on the environment, (2) alternatives to the disparities in the distribution of resources between countries as well as within countries and (3) alternatives to the economy-driven and ever-increasing consumption of meat-based diets and automobile-centred transportation, we have to bundle our efforts as scientists and, with the cooperation of other societal groups, create new and sustainable ways of life. The concept of "sustainable development" is explained in Box 2.1

Box 2.1 The Sustainable Development Concept

Sustainable development embedded in intra and intergenerational justice may serve as a guideline (despite some limitations of the concept; see Schmuck and Schulz 2002) if it is based on at least five principles:

1. The **respect principle** maintains that all forms of life have an equal right to live (Schweitzer 1991; Gorke 1999).
2. The **precautionary principle** is aimed at avoiding irreversible human-caused changes in the balance of our biosphere/ecosphere (Komiyama and Takeuchi 2006, p. 5): "The primary objective is [...] to achieve, as soon as possible, substantial improvements in [...] the interaction between the sciences and decision-making, using the precautionary approach, where appropriate, to change the existing patterns of production and consumption and to gain time for reducing uncertainty with respect to the selection of policy options".
3. The **principle of participation** encourages the population to take part in searching for, evaluating and implementing sustainable ways of life. Many chapters in Agenda 21 emphasise this principle, i.e.: "The primary objective is [...] to achieve, as soon as possible, substantial improvements in [...] participation of people in setting priorities and in decision-making relating to sustainable development" (UNO 1992, Chapter 35.6). "The objective is to promote broad public awareness as an essential part of a global education effort to strengthen attitudes, values and actions which are compatible with sustainable development. It is important to stress the principle of devolving authority, accountability and resources to the most appropriate level with preference given to local responsibility and control over awareness-building activities" (UNO 1992, Chapter 36.9). "Governments at the appropriate level, with the support of the relevant [...] regional organizations, should [...] launch applied research on participatory methodologies, management strategies and local organizations" (UNO 1992, Chapter 14.22). "The public should be assisted in communicating their sentiments to the scientific and technological community concerning how science and technology might be better managed to affect their lives in a beneficial way" (UNO 1992, Chapter 31.1).
4. The goal of the **efficiency principle** is to avoid wasting limited resources.
5. The **consistency principle** is aimed at replacing the use of finite resources (the actual main base of our economy) with renewable resources without any waste products, thereby following naturally occurring biospheric cycles. The input of harmful substances and nutrient matter into the ecosystem should be minimised. The state of our landscapes has to be improved to increase future generations' living conditions.

2.2 Sustainability Science

Sustainability science, which was initially formulated by Kates et al. (2001), is a new approach to tackling today's global problems with scientific tools. The methodological principles of traditional science have to be complemented by additional principles. The classical view regards scientific activities as value-free endeavours to mainly develop and test hypotheses in laboratories in a monodisciplinary, analytical and linear way by means of basic research and with a strict division between research and application as the ideal. The main motivation to include new approaches lies in the nature of today's global problems: They are based on non-linear, highly interwoven complex processes and there are often long time lags between actions and their consequences. Therefore, the advocates of sustainability science believe that the chances of contributing substantially to solving the current global problems are greater if science (1) acts explicitly to support sustainable development, (2) tries an interdisciplinary approach and (3) if science is transdisciplinary in terms of undertaking action-oriented research. In action-oriented research, scientists apply ideas for sustainable development to a society and simultaneously investigate the interactions that occur between the members of this society once they have adopted a more sustainable approach. The following sections summarise some of the most convincing arguments for the proposed new approach within science.

2.2.1 Science for Sustainable Development

Agenda 21, an environmental plan of action drawn up by global political representatives, clearly mentions scientists as co-responsible for creating sustainable life patterns; for instance, Chapter 35, entitled "Science for sustainable development" states: "The sciences should continue to play an increasing role in providing for an improvement in the efficiency of resource utilisation and in finding new development practices, resources, and alternatives. [. . .] Thus, the sciences are increasingly being understood as an essential component in the search for feasible pathways towards sustainable development" (UNO 1992, Chapter 35.2). This new role of science is confirmed in several later scientific documents. For instance, Kates et al. (2001, p. 642) emphasise that "research itself must be focused on the character of nature-society interactions, on our ability to guide those interactions along sustainable trajectories, and on ways of promoting the social learning that will be necessary to navigate the transition to sustainability. Science must be connected to the political agenda for sustainable development". According to Clark and Dickson (2003, p. 8059), we need "international consensus on goals and targets for targeting problem-driven research in support of a sustainability transition"; Komiyama and Takeuchi (2006, p. 3) state that "sustainability science must therefore adopt a comprehensive, holistic approach to identification of problems and perspectives involving the sustainability of these global, social, and human systems. [...] The ultimate purpose of sustainability science is to contribute to the preservation and improvement of the sustainability of these three systems". To conclude, we see a growing consensus within the scientific community that science should direct its efforts explicitly to supporting sustainable development (for more details see Schmuck and Vlek 2003; Sheldon et al. 2000).

2.2.2 Interdisciplinary Approach

Komiyama and Takeuchi (2006, pp. 4–5) believe that sustainability science "can help resolve one of the fundamental dilemmas of contemporary scholarship – the inability of our overly specialised disciplines to offer comprehensive solutions to the conditions that threaten the sustainability of global, social, and human systems" by replacing "the current piecemeal approach with one that can develop and apply comprehensive solutions to these problems". Likewise, Kates et al. (2001, p. 641) see the success of the new approach as dependent on close collaboration between scientists: "Progress in sustainability science will require fostering problem-driven, interdisciplinary research". In Agenda 21, the "Science for Sustainability" chapter also stresses the interdisciplinarity of research as a precondition for solving global problems. Specifically, the social sciences are seen as an indispensable part of interdisciplinary teams: "The primary objective is [...] to achieve, as soon as possible, substantial improvements in [...] cooperation between scientists by promoting interdisciplinary research programmes and activities" (UNO 1992, Chapter 35.6). "The scientific and technological means include [...] supporting new scientific research programmes, including their socio-economic and human aspects, at the community, national, subregional, regional and global levels, to complement and encourage synergies between traditional and conventional scientific knowledge and practices and strengthening interdisciplinary research related to environmental degradation and rehabilitation" (UNO 1992, Chapter 35.9).

"Social processes are subject to multiple variations across time and space, regions and culture. They both affect and are influenced by changing environmental conditions. Human factors are key driving forces in these intricate sets of relationships and exert their influence directly on global change. Therefore, the study of the human dimensions of the causes and consequences of environmental change and of more sustainable development paths is essential" (UNO 1992, Chapter 35.10).

2.2.3 Transdisciplinary Approach

In broad terms, transdisciplinarity means the close collaboration between (interdisciplinary interconnected) groups of scientists on the one hand and the broad public on the other. The necessity of such a collaboration is cogently expressed by Kates et al.: “In a world put at risk by the unintended consequences of scientific progress, participatory procedures involving scientists, stakeholders, advocates, active citizens, and users of knowledge are critically needed” (2001, p. 641). Similarly, Clark and Dickson (2003, p. 8059) argue that “the multiple movements [...] with the goal of creating and applying knowledge in support of decision making for sustainable development [...] are grounded in the belief that for such knowledge to be truly useful it generally needs to be “coproduced” through close collaboration between scholars and practitioners”. Moreover, Komiyama and Takeuchi (2006, p. 5) conclude that “[i]f sustainability science is to contribute practical solutions to the problems we face, cooperation among researchers, industry, and the general public is imperative”. In Chapters 31 and 35 of Agenda 21, we find that such argumentations are particularly relevant when “the cooperative relationship existing between the scientific and technological community and the general public should be extended and deepened into a full partnership” (UNO 1992, Chapter 31); or when the “participation of people in setting priorities and in decision-making relating to sustainable development” is required (UNO 1992, Chapter 35).

The transdisciplinary approach implies that scientists following this new approach have a double role. In addition to the classical role of the analyser of objective data patterns, scientists today also form part of social groups that conjointly create and apply demonstration models for new production, distribution and consumption patterns. Research and application take place simultaneously. Kates et al. argue that “pertinent actions are not ordered linearly in the familiar sequence of scientific inquiry, where action lies outside the research domain. In areas like climate change, scientific exploration, and practical application must occur simultaneously. They tend to influence and become entangled with each other” (2001, p. 641).

According to Clark and Dickson (2003, pp. 8059–8060), scientists have new roles. They argue that “perhaps the strongest message to emerge from dialogues induced by the Johannesburg Summit was that the research community needs to complement its historic role in identifying problems of sustainability with a greater willingness to join with the development and other communities to work on practical solutions to those problems. This means bringing our science and technology to bear on the highest-priority goals of a sustainability transition, with those goals defined not by scientists alone but rather through a dialogue between scientists and the people engaged in the practice of meeting human needs while conserving the earth’s life support systems and reducing hunger and poverty. [...] The commitment of sustainability science to problem-driven agenda setting does not mean that it has been confined to ‘applied’ research. Indeed, the pursuit of practical solutions to the pressing challenges of sustainability has driven the field to tackle an array of fundamental questions”.

This new kind of close interconnectedness of basic and applied research seems to be an important and unavoidable characteristic of sustainability science, as Komiyama and Takeuchi (2006, p. 5) explicate: “One problem unique to sustainability science lies in the process of shifting from the stage of phenomena identification and analysis to that of problem solving. For sustainability science this process necessarily differs from the conventional transition from basic to applied research, because solutions to problems may have to be sought before those problems have been sufficiently analysed or even identified. Global warming is the prime example of this dilemma. Future scenarios predicted by various models of global warming remain unverifiable, yet the search for solutions cannot wait. [...] What is demanded of sustainability science is not only the development of scientifically sound models for predicting future scenarios and evaluating the effects of different countermeasures and solutions but also effective management of the process by which these forecasts and evaluations are accepted by society, to generate the social reforms necessary to ensure global sustainability".

To summarise this section, the advocates of sustainability science call for science and scientists to accept a double role within society: Instead of restricting their role to producing scientific knowledge (Role A), scientists are *additionally* invited to apply that knowledge in transdisciplinary teams to solve urgent global problems (Role B). This does not mean that science's traditional role, which lies in its objective methodology (Role A), is abandoned: The new scientist does not fill either the one *or* the other role, but can apply, combine and balance both roles.

2.3 The Göttingen Approach of Sustainability Science

In this section, we describe how we integrated the defining characteristics of sustainable development and sustainability science into our approach. It consists of seven elements comprising the specific tasks scientists have to fulfil during the research cycle. The approach requires a group of scientists willing to cooperate and who share an intrinsic sustainability motivation. The first task is defined as the traditional scientist's role (traditional research producing scientific knowledge, Role A). The other six tasks comprise different practical problem-solving activities (the application of scientific knowledge in inter- and transdisciplinary teams, Role B) that occur consecutively (Fig. 2.1).

The research activities are distributed over the whole cycle, whereas the problem-solving activities are modelled consecutively. The detailed description starts with the latter.

2.3.1 Problem-Solving Activities

2.3.1.1 Select a Critical Global Problem

Problem-solving activities start with the selection of a problem. If the global level is taken into consideration in this early phase, the more serious problems will be given priority. When the urgency of certain global problems, such as climate change, water crises, etc. is examined, one concludes that the world scientists should focus

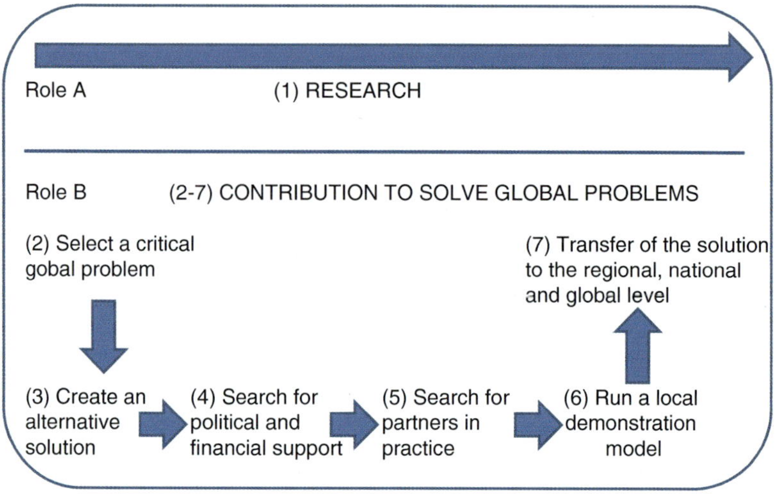

Fig. 2.1 Seven elements of the Göttingen approach to sustainability science

their energy on the most pressing problems if they want to prevent other catastrophes, like the oil disaster in the Gulf of Mexico in 2010, the nuclear disaster in Fukushima in 2011, or the increased melting of the Arctic ice.

2.3.1.2 Formulate Alternative Solutions Starting at a Regional Level

When formulating possible solutions to a global problem, the regional level seems to be an appropriate place to start, because scientists usually have neither the power nor the will to change world politics directly. Therefore, the creative process could be started in the area in which an active group of scientists live and work.

2.3.1.3 Find Political and Financial Support

The vast majority of scientists are mostly specialists in specific science subjects and are not explicitly assigned or have the financial means to pursue inter- and transdisciplinary sustainability science. Therefore, political and financial support is needed for sustainability projects. In order to obtain this support, it is helpful to refer to international and, where applicable, national political agreements regarding the promotion of sustainable development. Here, Agenda 21 again serves as an example as it contains many paragraphs on the energy sector; For instance, "governments [. . .] with the cooperation of [. . .] non-governmental organizations, should [. . .] promote the research, development, transfer and use of technologies and practices for environmentally sound energy systems, including new and renewable energy systems" (UNO 1992, Chapter 9.12). In Article 20a of the

German constitution, entitled "Protection of the natural bases of life", the following formulation is found: "Mindful also of its responsibility toward future generations, the state shall protect the natural bases of life through legislation and, in accordance with law and justice, through executive and judicial action, all within the framework of the constitutional order" (Federal Ministry of the Interior 1998).

There are two ways for scientists to become active in sustainability science: The one is to wait until governments or funding agencies create funding programmes for sustainability research. However, it is also possible for scientists to take the first step, meaning they need to share their sustainability research ideas with political authorities, which is what happened in the bioenergy village project under discussion. This is described in more detail in the next section.

2.3.1.4 Search for Practice Partners

The next step comprises motivating practice partners outside the research community to collaborate on the sustainability project.

2.3.1.5 Run a Pilot Project on the Local Level

During a project's implementation, scientists are focused on providing practitioners with scientifically based advice. Clark and Dickson (2003, p. 8059) express this idea as follows: "The transcendent challenge is to help promote the relatively 'local' (place or enterprise-based) dialogues from which meaningful priorities can emerge, and to put in place the local support systems that will allow those priorities to be implemented".

2.3.1.6 Transfer to the Regional, National and Global Level

After realising the pilot project successfully, an additional task could be to actively support the transfer of the model to other regions and, where applicable, to other countries.

2.3.2 Research Activities

The results of traditional research are, if available, a more or less suitable base for problem-solving activities. Thus, when selecting a critical problem to investigate, researchers should consider which global problems are the most harmful (the group of scientists' competence fields will, of course, limit this) to ensure they tackle only very relevant problems. The researchers' actual scientific knowledge of the -problem fields should then be assessed. These fields include, among others, water, energy, health, agriculture and biodiversity (the WEHAB priority targets as defined at the Johannesburg Summit; Clark and Dickson 2003, p. 8060). During the later problem-solving process, scientific and technological knowledge should guide all the individual steps. Before the first demonstration of the alternative solution models are held, hypotheses regarding the consequences – in, for instance, longitudinal designs – should be posited and tested if possible. This would mean that new scientific knowledge can be produced from such alternative demonstration models.

2.4 Application of the Göttingen Approach in the Bioenergy Field

In the following sections, we describe the implementation of our approach within a specific problem field. Following the notion of the two roles of those scientists who accept the challenge of sustainability science, we start with the problem-solving activities to provide some background to the research activities and results that follow. However, when implementing a project in practice, the problem-solving and the research activities are closely interwoven and sometimes occur simultaneously. However, the linear sequence of the text requires us to discuss these two aspects separately.

2.4.1 Specific Problem-Solving Activities

At the University of Göttingen, scientists from seven disciplines (sociologists, psychologists, political scientists, economists, agronomists, agrarian economists, biologists and geologists), who share the intention to contribute actively to sustainable development, came together for two days during the spring of 1997 for a "future workshop" (Zukunftswerkstatt). The goal of this workshop was to initiate a model project in the field of sustainable development, demonstrating that it is possible to change our ways of life and enable future generations to have a good life. Robert Jungk developed the "future workshop" concept in the 1970s (Jungk and Müllert 1991) in order to exploit modern societies' democratic potential and creativity to solve their problems. This workshop concept is often used in communal processes in Europe, but has not been widely used in scientific settings, probably because many members of the scientific community still undervalue the systematic inclusion of emotions and intuitions. Such a workshop mainly comprises three phases: the criticism phase, the phantasy phase and the realisation phase.

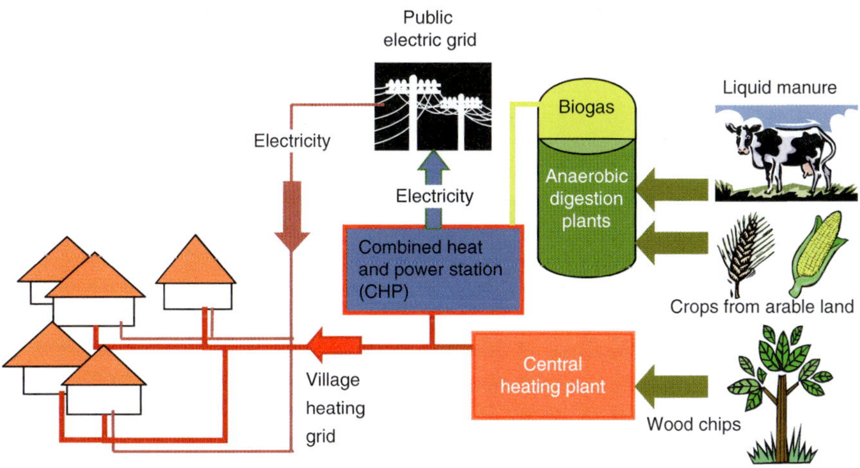

Fig. 2.2 Heat and electricity production and distribution in a bioenergy village

2.4.1.1 Select a Critical Global Problem: The Side Effects of Exploiting Fossil and Nuclear Energy Resources

During the criticism phase, actual problems and challenges are outlined and one problem field, which combines the interests and competencies of the group of persons present, is selected. In our case, we decided to focus on energy production and distribution questions, because we agreed that there are unsolved problems of energy production based on fossil and nuclear resources (mainly their finite nature and the side effects of their waste products such as carbon dioxide and nuclear waste). Furthermore, they are causally interconnected with many other adverse effects (e.g., climate change, decreased biodiversity, and socially unfair distribution patterns).

2.4.1.2 Formulate an Alternative Solution at a Regional Level

The second phase of the future workshop is a phantasy and brainstorming process enriched by creativity-evoking activities, like game-playing, listening to music, meditation, dreaming, or drawing pictures of one's visions for the future. Here, the goal is to foster the participants' creative processes to find alternative solutions to the specified problems. The method was successful: During the first day, the idea of a "bioenergy village" emerged: Motivating an entire village to participate in a collective effort to convert the village's energy supply – based on non-renewable sources – into one that uses locally available biomass to provide electricity and heat (see Fig. 2.2); to plan the necessary processes and help the villagers implement them.

In the concluding phase of the workshop – the realisation phase – the goal is to formulate the concrete steps required to put the idea into practice. Here, we agreed that the most important step would be to obtain political and financial support from the authorities outside the University.

2.4.1.3 Obtain Political and Financial Support

In 1998, after many further in-depth discussions on the very complex problems and their interconnectedness, a research project was formulated. Since there was no viable funding programme for our idea, we sent the project proposal to ten funding agencies and German ministries. All of them dismissed the proposal as unrealistic: It was considered too unlikely that a whole village would accept such a transformation. However, refusing to give up, we contacted leading people in the German Ministry for Food, Agriculture and Consumer Protection (BMELV) to convince them of our cause.

In 2000 – the industrialisation of the German agriculture over the previous decades had already led to a dramatic decrease in rural employment – the BMELV decided to back the project financially as it appreciated the project's potential to provide sustainable employment in the countryside. The Ministry wanted us to first choose a model village to demonstrate that our idea would work both economically and socially. If this succeeded, we would subsequently be allowed to apply the idea to other villages to revitalise the role of agriculture in Germany's labour market. The project kicked off in October 2000.

2.4.1.4 Search for Practice Partners: Village Competition

From 2000 to 2002, the first project phase was focussed on identifying a suitable village in the Göttingen rural district that would possibly participate in the project. A kick-off meeting with local politicians and some press publicity resulted (unexpectedly) in several villages showing a great interest in participating in the project. The project team then presented the idea to 17 interested villages; four of these, which had particularly suitable criteria, such as a broad agricultural base and social coherence, were formally invited to apply to be partner villages for the model project. This led to a competition – which we had not foreseen – between the four villages, indicating the villagers' strong motivation to transform their villages into ones with renewable energy sources with our support. In these four villages, an engineering company developed concepts for the technical implementation. On the basis of these technical concepts and the suitability criteria developed by the group of scientists, the village Jühnde – located 12 km southwest of Göttingen and with a population of 800 inhabitants at that time – was selected as our model village as it had the best prerequisites for the transformation into a bioenergy village.

2.4.1.5 Run a Pilot Project on the Local Level: The Transformation Process in the Bioenergy Village Jühnde

Between 2002 and 2004, preparations were undertaken to technically install a new infrastructure in Jühnde. During this phase, the scientists' main role was to develop and offer technical, economic and social support. Furthermore, the project not only required the villagers to install the technical equipment in the village themselves, but also to plan the details of the conversion project. Consequently, the residents were involved in the planning process and worked on site from the very beginning. After the initial general meetings with all the villagers, eight working groups were formed. In these working groups, several relevant project aspects, which the university's team proposed and initially moderated, were discussed: agricultural resources, electricity production, heat production, the heat distribution grid, the form that the company to be founded would take, the housing technique, public relations and the energy crop cultivation.

The results of the groups' work had to be communicated to the villagers. The university team suggested establishing a central planning group comprising the heads of the specific planning groups and the local authorities, for example, the mayor, members of the district council, the chairpersons of village clubs, etc. When formed, the inhabitants would legitimise the group by public acclamation. During the subsequent planning phase, the central planning group made important decisions; for example it decided on the location and the power capacity of the energy plants as well as determined the prices of the biomass and heat energy. The combination of planning processes at different levels within (1) the specific planning groups and (2) the central planning group, as well as (3) the regular inhabitants meetings led to a transparent and very powerful participatory process. By implementing a planning procedure based on intensive village participation, the scientists ensured that the project would become the villagers' venture, although they had conceived the idea. The plan worked: The villagers accepted responsibility for the project and required less and less support from the university team.

After the green energy plants (see more information on energy plants in Chap. 4) have been harvested, they are chopped and stored on three concrete plates, where the plant material, due to its compaction and the subsequent lack of air, is transformed into silage. If properly stored, silage is stable for many months. The technical equipment responsible for using silage to ultimately produce electricity and heat in Jühnde, was installed between 2004 and 2005 and consists of three main components:

1. A combined heat and power (CHP) generator with an electric capacity of 680 kW that produces electricity and space heat by burning biogas. The capacity is adapted to the required electricity and heat output to run the plant economically. Biogas is generated from biodegradable organic matter in an anaerobic digestion plant. The plant contains two fermentation units with a combined capacity of 7,800 m^3. Over the course of two months, micro-organisms enzymatically digest liquid manure (about 10,000 m^3/year) and crops cultivated on approximately 220 ha of arable land around Jühnde under anaerobic conditions

and transform these into biogas. The CHP unit converts the energy content of the biogas into roughly 35 % electricity and 50 % usable heat energy. The electricity is fed into the national electricity grid. German law guaranteed a feeding-in price of about 17 Eurocent/kWh in 2004 for 20 years (BMU 2004), thus promoting energy production from biomass. The CHP station's heat output is partly used for the digestion process. However, most of the heat can be used for space heating and to meet about 75 % of the village households' hot water demand. In summer, surplus heat is used to dry wood chips and cereals. Consequently, renewable fuels replace fossil fuels, like oil, gas, coal and nuclear power, as sources of heat and electricity.

2. In winter, a central combustion furnace with a thermal capacity of 550 kW, fired by locally produced wood chips, provides the additional heat energy required in the Central European climatic conditions. The capacity of the wood chip plant covers the peak heat demand in winter. Furthermore, an oil-fired peak load boiler with a capacity of 1,600 kW has been installed to provide heat for the peak load in winter and if the biomass plants were to fail and during their routine maintenances. Less than 5 % of the heat demand is covered by oil. The whole system is therefore highly reliable.
3. The heat energy from the plant is fed into a 5.5 km long hot water grid, which delivers the heat energy to the connected households in the village. The heat transfer in the houses occurs through heat exchangers (with a heat meter included), which have replaced the individual heating systems.

2.4.1.6 Publicising the Project on a Regional, National and Global Level

The successful outcome of the model project, which was completed in 2005, has been widely communicated via public relations activities (mass media, scientific publications and practical guides for formulating the generalised principles for the conversion process from fossil fuels to bioenergy). This has attracted the interest of many of Germany's rural population, especially farmers and local politicians, such as mayors and district administrators. Consequently, inspired by the successful implementation of the first bioenergy village in Germany, several other activities were initiated:

Between 2006 and 2009, again with the university team's support, four other villages in the Göttingen district (Reiffenhausen, Wollbrandshausen, Krebeck and Barlissen) followed the Jühnde model and installed similar communally organised bioenergy systems (for details see Wüste et al. 2011). In 2010, a process was started to initiate bioenergy villages in the biosphere sanctuary region Schorfheide in the federal state of Brandenburg. Five villages in the region showed interest in the conversion. The governments of the federal states Baden-Württemberg, Mecklenburg-Vorpommern and Brandenburg decided to support the development of bioenergy villages financially. In 2008, following the success of the bioenergy villages, the German government started a grant programme to support bioenergy regions: 210 regions in Germany applied for support. From 2009 to 2012, networking activities in 25 bioenergy regions in Germany were supported financially.

In two federal competitions held in 2010 and 2012, 35 and 41 individual villages respectively competed for the prize that the German government offered for the "most innovative bioenergy village" in Germany. This is indicative of the many German villages following our, or a similar, project model.

2.4.2 *Selected Research Activities and Results*

Between 2000 and 2008, before, during and after the communal transformation process in Jühnde, scientific analyses were undertaken of the ecological, economical, and social changes in the village. The essential research results are outlined in the following sections (for more details see Karpenstein-Machan and Schmuck 2007, 2010):

2.4.2.1 Natural Science: Reduction of Greenhouse Gas Emissions

The programme "Globales Emissions-Modell Integrated Systems (GEMIS)" Version 4.5 (Ökoinstitut 2008) was used to calculate the decrease in the greenhouse gas emissions in Jühnde after it changed to bioenergy supply. With this programme, it is possible to calculate the greenhouse gas emissions of various energy production models. The energy used (a) for the construction of the biogas plant and the other structures such as the silage plates (e.g., concrete, PVC granulate, rock wool and steel), (b) for the production and transport of the energy crops and manure to the biogas plant and to recycle the digestion residues on the fields and (c) for the maintenance of the processes in the fermentation plant (electricity and heat) is transformed into comparable accumulated CO_2 equivalents. For example, the production of corn silage needs energy to provide the seed, to transport it, to till the cropland, to sow the grains, to fertilise (including the energy required to produce and supply the fertiliser), to apply pesticides, to harvest, to transport it to the silage plate, etc. The cumulated energy demand can be converted into CO_2 equivalents and can be compared with emissions from conventional power stations that deliver the same amount of electricity.

The 2007 electricity and heat production data were used to calculate the decrease in greenhouse gas emissions in Jühnde (Sauer 2009). In 2007, 4,933 MWh of electricity and 3,956 MWh of heat were generated (Friehe 2007). Subsequently, 3,379 MWh of waste heat from the CHP was turned into useful heat, while the wood chip heating plant produced and an additional 577 MWh of heat. Only the amount of heat that was actually used to heat the households and the digester was included in the calculation. The total amount of generated electricity was included because it is fed into the public power grid and fully consumed completely. The less heat is wasted – especially during summer – the more CO_2 equivalents can be saved.

Table 2.1 shows a comparison between the greenhouse gas emissions from Jühnde's bioenergy facilities and those of other power plants.

Table 2.1 Comparison between the CO_2-equivalent emissions of the Jühnde bioenergy facilities and those of other power plants (From GEMIS; Öko-Institute 2008)

Electricity generation	Emissions of 4,933 MWh electricity (CO_2 equivalents in tons)
Coal-fired power plant 2005	5,396
Gas-fired power plant 2005	2,116
Brown-coal-fired power plant (Rhenish) 2005	6,158
Nuclear power plant 2000[a]	158
Electricity mix Germany 2005	3,213
Bioenergy facility Jühnde	267
Avoidance in Jühnde compared to the German electricity mix	−2,946
Heat generation	**Emissions of 3,956 MWh heat (CO_2 equivalents in tons)**
Oil heating system 2005	1,486
	Emissions of 577 MWh heat (CO_2 equivalents in tons) (heat from CHP plant already subtracted)
Chip wood heating plant Jühnde and heating grid	20
Avoidance in Jühnde compared to oil heating	−1,467
Avoidance in Jühnde regarding electricity and heat	−4,413

[a]The low value of nuclear power plants is misleading, because the storage/processing of spent nuclear fuel and the decommissioning of the plant are NOT included as there are no reliable data on these aspects. There is currently no final storage space for nuclear waste in Germany. Just the auxiliary energy used to store nuclear waste for at least 100,000 years would counteract the good CO_2-emission value of nuclear power

The Jühnde CO_2 emission data for electricity production were compared with the whole of Germany's 2005 electricity emissions data. As 3,379 MWh of heat from the CHP are used at Jühnde, these emissions are already calculated at the electricity side. With regard to heat production, Jühnde only emits CO_2 equivalents of 577 MWh, which the wood chip heating plant and the heating grid generate. However, in comparison, a village the size of Jühnde and mainly using fossil fuel heating systems would consume 3,956 MWh of heat.

In sum, the conversion of Jühnde into a bioenergy village prevents about 4,400 t of CO_2 equivalents every year. Approximately, 440 persons in Jühnde are connected to the heat grid. If we attribute the decrease in greenhouse gas emissions to these people, everyone has saved 10 t of CO_2 equivalents annually. In 2007, the average total greenhouse gas emission in Germany was around 11.9 t of CO_2 equivalents per capita and year (data from the Umweltbundesamt 2012). Compared to the average German, Jühnde showed an 84 % decrease in greenhouse gas emissions per capita. An ecologically acceptable worldwide annual average lies around 2.5 t per capita. The balance for Jühnde is very favourable, because approximately 2.5 times more electricity is generated than the village uses. Therefore, it also prevents others from emitting greenhouse gas emissions.

2.4.2.2 Agriculture

Since 2005, three types of organic substances have been used to generate enough power to satisfy Jühnde's electricity and heat energy needs: (1) energy crops, cultivated on arable land to produce electricity and heat energy in a biogas plant, (2) liquid manure from husbandry farms and (3) wood chips mainly burned in a central heating plant during winter. About 80 % of the total produced energy is generated from annually cultivated crops fermented in the biogas plant. This means that energy crops and their sustainable cultivation are very important for the village's energy concept. Therefore, this section is mainly focussed on sustainable energy crop cultivation and the relevant advising of the farmers.

Energy crop cultivation can contribute positively to achieving climate goals. However, if not implemented carefully, it could exacerbate the degradation of land, water bodies and ecosystems as well as increase the greenhouse gas emissions, leading to the citizens' rejection of the initiative.

The energy cultivation concepts regarding biogas use differ from cultivation concepts regarding food crops (Karpenstein-Machan 1997, 2002, 2005). The selection of crops, varieties, seed densities, harvest time and fertilisation have to be managed to gain a high fermentable biomass yield. To sustainably manage these, the following criteria were included in the energy crop cultivation concept implemented in Jühnde:

- A high diversity of crops – no monoculture
- Reduce agricultural pesticides
- Avoid nitrate and pesticide leaching to groundwater
- Avoid soil erosion and humus degradation
- Optimise nutrient recycling
- Optimise crop yields
- Optimise the energy input–output ratio of energy crop cultivation.

Locally adapted and environmentally friendly concepts for energy crop production were developed and tested over many years at the University of Kassel-Witzenhausen (Scheffer and Stülpnagel 1993; Karpenstein-Machan 2003, 2005, Karpenstein-Machan and Stülpnagel 2000). These new concepts were implemented in the crop rotations of the food and feed crops in the Jühnde district.

Furthermore, part of the energy crop farmland is located in the water protection area of the Jühnde district. Scheffer and Stülpnagel's (1993) "double-cropping system" with its more balanced nutrient extraction was tested on different soils to investigate whether the ground water quality could be improved by decreasing the leaching of nutrients.

Another goal was to integrate all the available liquid manure from husbandry into the fermentation process to avoid further climate-change-relevant gas emissions from the husbandry farms. Nutrient recycling was thus optimised and the consumption of mineral fertiliser reduced.

Energy balances of the crop cultivation and the operation of the biogas plant had been made to get information about efficiency of energy production.

The following sections describe selected results, starting with crop rotation.

Crop Rotation

Before the implementation of the biogas plant, the Jühnde farmers' produce consisted of 72 % winter cereals – mainly winter wheat and winter barley –, 20 % winter rape and 8 % maize. After the implementation of the energy plants, the wheat and barley cultivation for the market was reduced to 11 % and replaced with triticale and rye cultivation for the biogas plant. The maize cultivation area in the district was expanded to 11 % and the winter rape cultivation area to 22 %.

On fairly fertile soils that have a German soil fertility number higher than 40 (the best fertility number is 100), the farmers changed their crop rotation from winter rape – winter wheat – winter wheat – winter barley to a more diverse rotation of winter rape – winter wheat – energy winter triticale – green manures (mustard) – maize. Owing to the early harvest of winter triticale for biogas production, a second crop is feasible in the same year. In Jühnde, mustard or other green manure crops were sown to cover the soil during winter, thus preventing soil erosion and nitrate leaching. In the following year, energy maize could be sown between the stubbles of the dead green manure (which is killed by frost) with minimum tillage. On less fertile soils (with a soil fertility number lower than 40), the crop rotation winter rape – winter wheat – winter barley was changed to winter rape – energy winter triticale – energy winter rye – winter barley.

On both soil types, winter wheat and winter barley were replaced with energy crops. The replacement of wheat and barley in the crop rotation improves the environment. The replaced crops, which were extensively cultivated in the district, required several pesticide and herbicide applications as well as treatments against diseases. Replacing these with triticale and rye, two rarely cultivated and healthier crops, improved the crop rotation in Jühnde.

Optimal Harvest Time for Digestion

Biogas is the final product of an anaerobic transformation process caused by bacteria in the fermenter. The anaerobic bacteria only develop stable life communities under ideal environmental conditions, i.e. an optimal temperature, pH value and nutrient composition in the fermenter. Such conditions are a prerequisite for high gas yields. Given these requirements, the feeding of the biogas plant with energy plants plays a central role. In order to supply the biogas plant with easily degradable substrates rich in energy, annual crops should be harvested at the milk-ripe stage or early dough ripeness when the whole plant contains 25–35 % dry matter (Karpenstein-Machan 2005). This ensures that the bacteria can easily

Table 2.2 Nitrogen fertilisation in kg N/ha regarding the percentage of area treated with pesticides and the number of pesticide treatments applied to the energy crop cultivation (triticale and energy maize) compared with that applied to winter wheat for grain production and fodder maize

	N fertilisation (digestate and mineral N)	Growth regulator	Herbicides	Fungicides	Insecticides	Treatments
Crops	kg N ha^{-1}	% of area				Numbers
Energy triticale	152	58	68	58	17	2–3
Winter wheat grain production	196	100	100	100	88	6–7
Energy maize	146		100	0	20	1–2
Fodder maize	170		100	0	20	1–2

degrade the green plants' organic substance. At this stage of their development, the plants' lignification is not yet very advanced.

To meet these requirements for digestion, the optimal harvest time for triticale, wheat, rye and oats was explored in Jühnde area. We looked for a high dry matter yield in combination with a dry matter content of about 30 %. Samples of winter crops were taken from the beginning of June until the end of July at different stages of their development. All tested cereals still showed high dry matter increments in June, which lasted until the beginning of July.

With 16 t of dry matter per hectare, the highest biomass yields were reached at the end of June with a dry matter content of 32 %, which is still optimal for digestion. After this time, the dry matter yield declined in triticale, rye and wheat, while the dry matter content increased to 40 %, which is suboptimal for digestion. The younger oat plants reached the highest dry matter yield later – in the middle of July – amounting to 17 t of dry matter per hectare.

Regarding both parameters – high dry matter yield and optimal dry matter content – we can conclude that, under the specific climatic conditions of the hilly areas of southern Lower Saxony, the optimal harvest time for winter cereals is from the end of June until mid-July.

Regarding maize cultivation, the development of the crop is limited by the vegetation time in autumn. Location-adapted varieties, which reach the milk-ripe/dough stage of development in autumn, should be chosen for cultivation. These varieties can be harvested at the end of their vegetation time – which is normally mid-October for maize in the climatic conditions of southern Lower Saxony.

Pesticide Use and Fertilisation of Energy Crop Cultivation

Table 2.2 shows a comparison between the nitrogen fertilisation and pesticide use in conventional crops – like winter wheat for grain production, or maize for fodder production – and in crops for energy production (triticale and maize). These data were provided by Jühnde farmers who produce biomass for the biogas plant. In the energy triticale, the nitrogen fertilisation was 44 kg N/ha lower than in the wheat production for grain use. Only two fertilisation treatments were applied to energy crops while three were applied to wheat grain production. Digestate from the biogas plant was used for the first intensive application. The second nitrogen application was less intensive and was applied by means of mineral fertiliser. A comparison of the pesticide use shows that fewer pesticides were applied to energy triticale production than to wheat grain production. The use of insecticides, fungicides and growth regulators was specifically reduced. This result indicates that far fewer pesticide treatments were applied to energy triticale. Many tests have shown that, in winter, the application of herbicides, fungicides and growth regulators to energy producing winter cereals rarely increases the biomass yield and is mostly not economically feasible (Sodikin 1994; Karpenstein-Machan 1997, 2002; FNR 2008).

The pesticide treatments of maize for fodder and energy production are very similar. Compared to the winter cereal production, the treatments are generally on a lower level as the plant health of the maize is still good. The amount of nitrogen fertiliser applied to energy maize is lower than that applied to maize for fodder production.

Concluding our analysis of the pesticide and nitrogen applications, we point out the positive aspects of energy crop production in the Jühnde district's water protection area. In the long term, this means that the quality of the drinking water from the water protection area can be improved by the cultivation of energy crops. The area's water protection administration is aware of these reduced applications and promotes the cultivation of winter crops for biomass energy. Further ecological and economical improvements could be realised in the district if the farmers were to eliminate growth regulators and reduce herbicide input. The application of growth regulators on marginal soils is critical and can lead to a biomass yield decrease, especially under drought conditions in early summer (Karpenstein-Machan 1994). Furthermore, a shorter culm leads to lower biomass yields (von Buttlar 1996). However, farmers fear crop lodging and therefore apply growth regulators. The use of varieties with stable culms and an adapted nitrogen fertiliser input are preferred means of fertilisation and prevent crop lodging.

Yield and Yield Stability

The energy crops triticale and maize have been cultivated in the Jühnde district as a fodder for the biogas plant since 2004. Within three years, the yearly average yield of the triticale biomass was 11.1 t of dry matter per hectare (1.8 t/ha standard deviation). The maize yields were 12.6 t of dry matter per hectare and year, but with a much higher standard deviation (4.7 t/ha).

Triticale cultivation was mainly planted in soils with lower fertility (fertility numbers 30–50) and the maize was cultivated in soils with higher fertility (fertility numbers above 50). The correlation between the yield and the soil fertility was low with regard to triticale and high regarding maize. This shows that triticale is adaptable to a wider range of soils. More than 40 % of the soils in the Jühnde district have soil fertility numbers below 40. Therefore, the cultivation of winter triticale as an energy crop is a good option with many ecological benefits. Furthermore, triticale's high yield stability is an important safeguard for the adequate supply of biomass for the biogas plant.

Double-Cropping System

To optimise the ecological effect of energy crop rotation, an ecologically-oriented cultivation system was developed at the University of Kassel (Scheffer and Stuelpnagel 1993; Karpenstein-Machan 2001, 2005). It is based on a diverse crop rotation system, with several winter and summer crops. In moderate climates with a growing period of six months or more (days with mean temperatures of over 10 °C), two crops (C3 and C4 crops) per year are feasible, as both crops are harvested in the milk-ripe stage of development. This double cropping system can reach high annual biomass yields per hectare (Schuette 1991; Scheffer and Stuelpnagel 1993; Karpenstein-Machan 1997; Graß and Scheffer 2003). However, the climatic conditions and soil quality should be sufficiently adequate to realise a high annual biomass yield of more than 20 t/ha.

The double-cropping system was tested under Jühnde's climatic conditions (elevation: 270–375 m above sea level; a mean yearly temperature of 7.9 °C; a yearly precipitation of 800 mm) with a shorter vegetation time (155–160 days). In contrast to the original double-cropping system with a C3 crop (winter rye) and a C4 crop (maize) (Scheffer and Stuelpnagel 1993), the double-cropping system in the Jühnde district was tested with two C3 crops due to the shorter vegetation time. The first crop was winter triticale, the most yield-stable crop, while sunflowers, summer rye and mustard were tested as a second crop.

To realise a high biomass yield from the first crop, triticale was harvested in the beginning of July when it was at its highest biomass yield during its milk-ripe stage with a dry matter content of 34–36 %. Sunflowers, summer rye and mustard were sown with minimum tillage immediately after the triticale harvest. These crops were harvested at the beginning of October. Triticale had a dry matter yield of 13 t, while the second crops yielded between 6 and 7 t of dry matter per hectare and year. Consequently, two crops per year realised nearly 20 t of dry matter per hectare. Whereas a satisfactory dry matter content of 30 % was achieved with the summer rye, the sunflower and mustard only reached a dry matter content of 20 %, which is insufficient for silaging. To avoid plant juices percolating through the harvest and silage, the dry matter contents in biomass should be at least 28 %. Under Jühnde's climatic conditions, the double-cropping system with two C3 crops can be recommended on fertile soils with a water storage capacity of 200 l/m^3 or more. In addition to its ecological advantages, the double-cropping system can contribute to a more efficient use of arable land and help prevent strong competition for land.

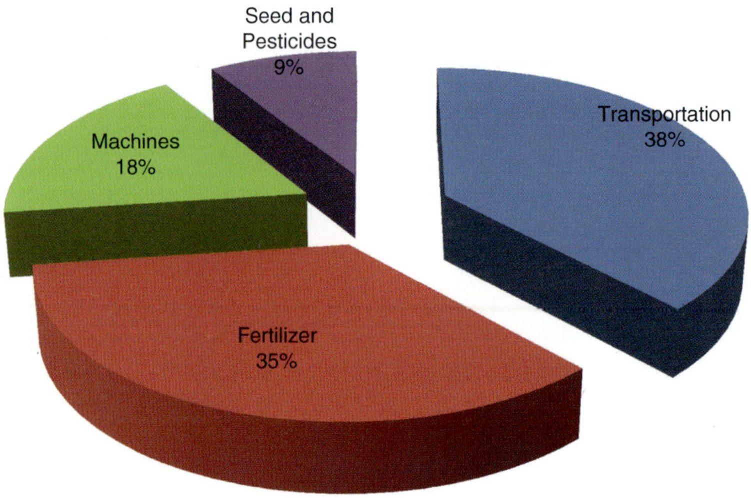

Fig. 2.3 Distribution of fossil energy input (in %) for cultivation of crops, transportation and silaging of biomass

Energy Balances of Energy Crop Production

An energy balance was undertaken regarding the energy crops' cultivation, transportation and silaging. Three years' cultivation (2005–2007) data on the Jühnde district were taken into account. All the supply chain data regarding farm energy inputs – fuels for field work and transportation, lubricants, machines, fertilisers, seeds and pesticides as well as silaging – were taken into account (see Fig. 2.3). According to these calculations, the energy input/output ratio was 1:19 for triticale and 1:18 for maize. Transportation and fertiliser are the main energy inputs. Farmers fertilise energy crops with digestate and mineral fertiliser, therefore fertiliser is still a main input factor. The production of mineral fertiliser is very energy intensive. In spite of higher yields in maize, the input/output ratio is better in triticale as it requires a lower energy input, especially of phosphate fertiliser. This "under root fertilisation" with mineral phosphate leads to a higher energy input in maize cultivation. This ratio shows that the cultivation system can replace fossil energy with renewable energy on a remarkable scale. Owing to the higher mean yields in maize, the net energy output was 230 GJ/ha for maize and 200 GJ/ha for triticale.

Energy Balance of the Biogas Plant

A further calculation was done to estimate the total fossil energy input necessary to operate the energy plant and its production (operating energy) as well as to deliver the crops and liquid manure the energy plant. This calculation is called the cumulated energy input (CEI). The calculation estimates an economic lifetime of 20 years (see Fig. 2.4).

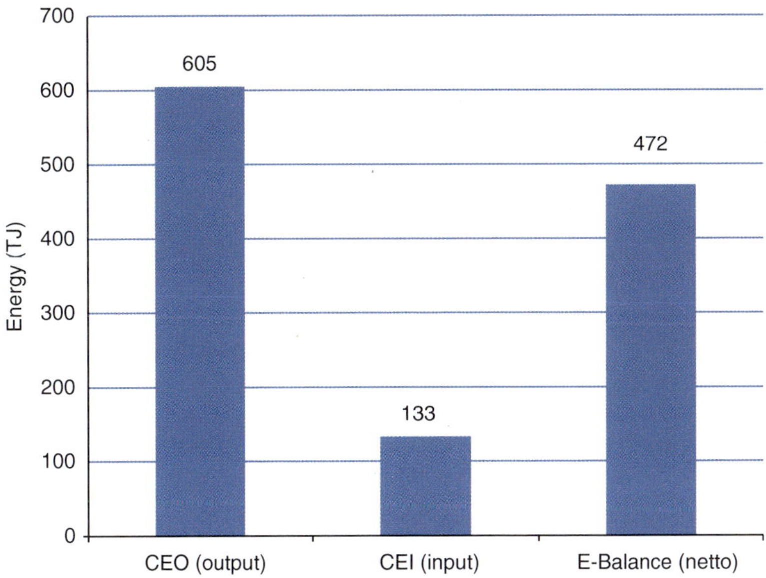

Fig. 2.4 Energy balance of the biogas plant in Jühnde (2005–2007) calculated for an economic life-time of 20 years

To calculate the cumulated energy output (CEO), we took the produced electricity and the used heat energy over a period of twenty years into account. The cumulated energy production, which is divided by the cumulated energy input, is called the harvest index. The harvest index for the biogas plant in Jühnde is 4.5. This means we need 1 kWh of fossil energy to produce 4.5 kWh of bioenergy (electricity and heat energy). After running for 21 weeks, all the fossil energy input for the biogas plant's production is amortised and, after 5 years, all the fossil energy input is amortised for a period of 20 years.

2.4.2.3 Psychology

The following section focuses on selected psychological aspects of the project. The main question was how to successfully motivate the inhabitants of rural areas to participate in such a conversion process. Consequently, the social success factors established in similar projects were analysed and then applied to the own project. Furthermore, psychological hypotheses were tested regarding the changes in the psychological variables – for example, social support, self-reported environmental behaviour, self-efficacy and well-being – as a result of the villagers' activities. Below, the results are reported of a longitudinal study (both before and after the conversion) of a broad sample of villagers – who answered a questionnaire – and a subgroup of the villagers – with whom we had a semi-structured interview – who were extraordinarily engaged in the project over a longer period (see also Schmuck 2013; Eigner-Thiel 2005).

Social Success Factors to Motivate People for a Collective Climate Protection Project

The social success factors established in climate protection projects similar to that of the Jühnde project were analysed with regard to motivating the inhabitants of rural areas to participate collectively (for details, see Eigner and Schmuck 2002) and then successfully applied in Jühnde.

Visiting model sites. Firstly, by visiting model sites fears can be decreased and prejudices eliminated regarding the technical equipment that needs to be installed. Success stories disseminated by important, accepted and favoured people in a particular village may also contribute to this process. In Jühnde, the villagers' interest was piqued after they had visited the first well-functioning bioenergy site. These experiences are congruent with the empirical findings of Mosler (1998), Aronson and O'Leary (1983) and Schuster and Marx (1998).

Being for something, not against it. Moreover, it is important that the aim of the project is formulated positively and constructively. A project's objective should be directed *for* something, not *against* persons or corporations. In Jühnde, for example, the active group called itself the "initiative for a bioenergy village" and not the "initiative against nuclear energy". This positive view is also advocated by Csikszentmihalyi (1993) and Richter (see Schmuck et al. 1997, p. 11), because this "pro-attitude" motivates people to solve conflicts, to love and help others and preserve nature, whereas a "contra-attitude" often has a destructive outcome.

Setting realistic goals. Another suggestion is to not set lofty goals; for example, instead of trying to change the energy politics in Germany, rather focus on a smaller region or a village, as was done in Jühnde. This is in line with self-efficacy research findings (Bandura 1982). Achieving smaller goals from time to time and experiencing success engender feelings of internal control, which motivate people to continue pursuing a distant goal.

Well-established advocates. It is important to ensure that prominent villagers (like the local bank director) support the project, at least ideologically. If well-established people with broad recognition and respect in the local population support the project, it will be considered more important and will be taken more seriously. In Jühnde, the most popular major advocated the bioenergy idea, which is one of the reasons why so many people participated in the conversion process.

Good contact with the local media. Having good contact with the local newspapers is of great benefit, because these are usually read and the contents discussed by most people in the district. In the district of Göttingen, the local newspapers regularly reported on the search for a suitable village. This motivated several villages to compete to become the first bioenergy village. Moreover, using plausible, easily understood terms or symbols for initiatives or projects is good for publicity.

Having a person in charge of a village district. The village should be divided into several districts. In Jühnde, for example, this was done according to the streets, with a person was in charge of disseminating information and providing an overview of the households' willingness to participate in the common heat supply. Having such a person in charge of a small district can increase a feeling of unity.

"Neutral" approach. In projects where many persons have to be motivated, it is beneficial to get the politicians of different parties involved; a "neutral" approach has been shown to be effective. This was also confirmed in our experience of mobilising the villages in the district of Göttingen: The University's neutral stance was a good basis for persuading and motivating people.

Spreading information orally. Initiators should provide informational stalls at markets or festivals attended by many people, and where they can talk face-to-face and provide additional written material. In Jühnde, such stalls were set up at nearly each festivity attended by many people and where doubts and concerns could be minimised through personal communication. Furthermore, the involvement of the local clubs and societies as well as the involvement of the council and municipality are important success factors (see also Mieneke and Midden 1991; Scherhorn et al. 1999).

Festivities. Public festivities should also be used to transfer ideas and stimulate others to participate in a project. Herzog (1997) found this type of participation to be a critical factor. In the villages around Göttingen, the inhabitants decorated wagons for the parish fair very creatively with elements pertaining "bioenergy", such as a little wood-fired oven, etc.

Personal contact. Another successful way to mobilise individuals is to contact them *personally*; for example, by going from house to house and informing them. In Jühnde, the initiators elaborated this strategy by selecting a particular person from the initiative group to speak to the residents of each house. In some cases, listening to the daily events may help create a trusting atmosphere that increases understanding and willingness to participate in an energy project. Individuals' doubts and scruples should always be taken seriously and should receive careful consideration. The personal approach is one of the most important ways to motivate people. This is consistent with research findings on face-to-face contact, which is considered more effective than written material (Ammann et al. 1997; Gonzales et al. 1988; Burn and Oskamp 1986). Furthermore, best-practice analyses found that personal contact is more efficient than impersonal contact, such as sending out mail (Fischer and Kallen 1995; Hennicke et al. 1997; Schuster and Marx 1998). If impersonal forms of information are used (e.g., posters and mail), it is more convincing if specific persons write about their experiences and state their names and addresses than if only technical or financial information is given (Schmittknecht 1998).

Authenticity and conviction. Technical details are often unimportant when one wishes to persuade people. Personal conviction and authenticity, as well as plausible arguments for engagement in the project are often more important.

One of the residents in Jühnde mentioned: “We also live in this village, and we would not plead for an electricity supply that is not reliable”. Furthermore, it is important to point out how the project will benefit the region. Another successful approach could be to get the children involved, which will in turn lead to more parents being involved (Herzog 1997). In Jühnde, this was realised by means of a drawing competition for children with the “bioenergy village” as the topic.

Humour. Whatever strategy is used, it is good to make people laugh in order to open to new ideas; fantasy and humour also promote open-mindedness toward ideas. The inhabitants of Jühnde, for example, learned about bioenergy villages through a few theatre projects. Emphasising the positive aspects of a particular project’s consequences can also be helpful. In the case of Jühnde, natural, economic and social scientists gave introductory presentations that pointed out the benefits (see also Csikszentmihalyi 1993, and Richter, cited in Schmuck et al. 1997).

The Impact of the Collective Engagement in a Bioenergy Village on Different Psychological Levels: Results from a Questionnaire Study

Schmuck and Sheldon (2001) collected data from several research groups all over the world that demonstrated that self-transcending life goals directed at social and environmental thriving tend to serve individual well-being. Furthermore, empirical findings show that social belongingness contributes to health and well-being (Baumeister and Leary 1995) and that high rates of self-efficacy are positively related to health (Bengel et al. 1998).

Given that many of the Jühnde inhabitants participated in the planning and conversion process and were engaged in different working groups – for example, one for “public relations”, one for “technique” and one for “biomass production” –, we expected positive changes in the mentioned psychological variables (for details see Eigner-Thiel 2005).

1. To examine these questions, a 14-pages questionnaire, which included the mentioned variables’ and the environmental behaviour’s scales, was distributed (a) to the 238 households in Jühnde and (b) to 240 households in a comparable control village. The design was a longitudinal study of the two villages. Data were collected before the conversion in 2001, and after the conversion in 2007. The following differences were found: Self-efficacy was higher in the converted village (both temporal measurements) and the self-reported environmental behaviour had increased over the period (both villages). Neither the other variables, nor the expected interactions (villages and time) showed significant effects.
2. Satisfaction with the heat supply from biomass. The people of Jühnde linked to the heat supply system were also asked to what extent they were satisfied with this system. On the whole, 89 % said they were “very satisfied”, 11 % were “satisfied”, while nobody was “dissatisfied”.
3. Furthermore, the people in Jühnde were asked how they felt about the large number of visitors they had received (in 2007 around 8,000). A total of 78 %

chose the answer option: "This makes me proud, it stimulates me"; the visitors did not bother 16 %, 4 % did chose not to answer, and 2 % had been bothered by the visitors.

The Impact of the Collective Engagement in a Bioenergy Village on Different Psychological Variables: Results from an Interview Study

In a semi-structured interview study (according to Witzel 2000) with 11 persons belonging to the subgroup of bioenergy villagers who had been particularly engaged in the project over a longer period (e.g., as a representative of a working group), evidence was found of an increase in social support and well-being during the project implementation. For details on the interview manual and the analysis of the interviews, see Eigner-Thiel (2005) and Eigner-Thiel et al. (2004). The results of the interviews are described in the following paragraphs:

Group-feeling: Most of the interviewees (10 out of 11) said that they got to know and value many others in the village through the project. Prejudices concerning neighbours were partly diminished. Especially people who had only recently moved to the village valued this outcome: They felt better integrated into the village community after the project. Even long-time residents, who had already known many people in the village before the project, said that the contents of their discussions within the village were more profound after the bioenergy project and that their conversations were no longer merely small-talk. The village community was described as "more interesting" since the project had started. A greater feeling of oneness also became evident in statements like "*We* were on the TV last week" or "*We* have indeed realised the project".

Environmental behaviour: Most of the interviewees said that their environmental behaviour had already been very proactive before the project had started (9 out of 11). Examples of their behaviour were: "not tossing anything out of the car window", "not leaving old refrigerators in the forest" and "not wasting electricity". Only a few people said that they had further changed their behaviour (2 out of 11): One person, for example, reported that since the project, he obtained electricity from a more expensive eco-provider and had also bought a gas-driven car.

Self-efficacy: Concerning the question of whether an individual can do anything about climate protection, most of the interviewed persons (8) answered that they alone could not do anything. However, their experience of being a tourist guide in their biogas plants was very positive and gave them the feeling that they had sparked something in others. Only two persons felt that new developments should only be driven by politicians.

Well-Being: "If it had not been fun, I would not have engaged in this project". This, or a similar statement, was the answer most of the interviewees (10 out of 11) gave regarding the question of whether or not they had considered the project fun. Their reasons for enjoying it were, for example, that they could act from conviction; this was described as an intrinsic motivation during the processes' good and less good times. Some of them felt that taking responsibility for the project and getting to know their personal boundaries were very fulfilling. For many of the interviewees, it was very liberating to see the village community implement plans. Moreover, it was heartening to learn more about the different functions of the working groups (e.g., techniques for running households, operating companies and cultivating energy crops). The interviewees felt that finding solutions to difficult problems (e.g., where to place the biogas plant in the village) was exciting. Some of those involved also found observing and participating in different forms of learning and the presentation that the university group members shared with the inhabitants fascinating.

Some of the interviewees also referred to the negative consequences of the involvement; for example, "having less time for the family". However, even those who had negative experiences felt that the positive aspects had had a greater impact. Guiding tourists through the energy plants allowed them to share their acquired knowledge and was reported to be fun. Today, the interviewees are proud to see their village and the news about its pioneering activities on the Internet or on German television. Helping other interested villages become a bioenergy village was also considered fun. When asked how the project impacted their life satisfaction, one group (five people) responded: "Yes, this project has totally affected my contentment with life", which means that the experience had given their lives additional meaning. Persons from this group described the project as one of the highlights of their life; they feel as if they are part of something really big and important, which makes them proud. These experiences are expected to have a lifelong impact. The other group (six persons) was pleased that the biogas plants were built, that they are functioning and that the project was implemented successfully. However, they stated that, in their life, there are still matters that are more important than the project, for example, their family. Interestingly, one person said that he felt physically quite exhausted throughout the project, but that he nevertheless felt a mental or spiritual contentment as a result of his engagement in the project.

On the whole, the mentioned positive psychological consequences (more details in Eigner-Thiel et al. 2004; Eigner-Thiel and Schmuck 2010) can serve as a driver to transfer the idea to other villages. If the inhabitants of other villages see the potential psychological gains from this collective action, it could be a strong motivation to spread the idea of bioenergy villages, thereby supporting ecological and economical movements. We focus on the economical movements on in the next section.

2.4.2.4 Financial and Economic Aspects

One of the aims of this project was for all the stakeholders (e.g., households/heat customers, farmers, the operating company and the region) to benefit from the bioenergy village project. This meant that no one would suffer economic disadvantages from participating in the project.

Perspective of the Households/Heat Customers

The ways in which electricity was supplied to the houses have remained the same. To calculate the heating costs (of heating the rooms and water in the houses), three components have to be taken into account: the costs of the heat supply (which depends on the amount of heat required), the connection fee (in Jühnde 1,000 EUR per heat customer) and individual conversion costs (a one-time payment of approximately 2,600 EUR per household to install the new heating system). The operating company guaranteed a fixed buying price for energy until 2008, which refers to the price of heating oil at the time of contracting (0.35 EUR/l). Since the oil price rose at that time (e.g., 0.95 EUR/l in August 2008), an average household saved 1,800 EUR in heating costs annually.

Perspective of the Farmers

Cultivating crops for energy production is an alternative way for farmers to generate income besides the traditional markets for foods and animal feed. This can be an advantage for the farmers because these markets' prices fluctuate heavily over time. Producing biomass for energy will therefore lead to a constant basis income.

The operating company and the farmers agreed upon a price for the biomass that equalled the farmers' winter wheat profit. A potential problem could therefore be that the Jühnde bioenergy plant can only be run profitably if the operating company pays a price that is comparable with a price for winter wheat of 185 EUR/t. The average production costs of a ton of winter wheat amount to approximately 130 EUR. The associated market price fluctuated between 120 EUR/t and 290 EUR/t from 2005 to 2008. In this situation, it would be reasonable to agree on long-term supply contracts that set the prices for wheat in a range between 130 EUR/t and 185 EUR/t. This would smooth out the volatility of prices in the world markets for both the farmers and the operating company. Unfortunately, in real life, it is not so easy to close long-term supply contracts.

Perspective of the Operating Company

The owners of the operating company in Jühnde are farmers, villagers and (a few) external shareholders. Consequently, the profits remain in the region. If only external investors held the shares, the price of the heat provision would have been much higher and the price of the biomass would have been lower to allow the operating company to maximise its profits. This would have meant high payouts to the investors with the money lost to the region.

In Jühnde, the operating company invested a total of 5.4 million EUR: 2.9 million EUR in biogas and electric power production, 0.9 million EUR in the central heating plant and 1.6 million EUR in the hot water grid. This sum was financed by means of equity capital (0.5 million EUR), government grants (1.5 million EUR) and loans (about 3.4 million EUR).

Perspective of the Region

In the Jühnde project, 58 % of the invested sum was given to regional companies, and most of the annual turnover (80 %) remained in the region too. This clarifies that the installation of a bioenergy village supports local economic cycles. On the whole, the stakeholders of the Jühnde project have gained their expected economic benefits.

2.5 Conclusions

This chapter showed that sustainability science principles can be successfully applied to initiate renewable energy solutions in German communities. In the Göttingen Approach, sustainability science is not only based on interdisciplinary research, but preferably on transdisciplinary research. This kind of science should not be an end in itself, but should initiate, and contribute to, the solution of actual practical problems in cooperation with active practice partners from outside the scientific community. Typical consecutive steps for such activities can be: (1) Select a critical global problem; (2) formulate alternative solutions starting at a regional level; (3) find political and financial support; (4) search for practice partners; (5) run a pilot project on a local or regional level; and (6) transfer the successfully accomplished pilot project to other regions or to national or international levels, if applicable. The scientists should accompany and investigate all the individual steps during the project realization. New scientific knowledge can be produced from such alternative demonstration models. The double role of scientists within sustainability research is one approach to cope with the challenges of the global ecological crisis.

The application of our sustainability approach in the bioenergy field comprised the following elements: (a) Communicate the side effects of exploiting and applying fossil and nuclear energy resources within the scientist group; (b) find and elaborate an attractive alternative energy supply at a regional level (bioenergy village concept); (c) convince political and financial supporters; (d) search for partners in the region and in villages; (e) run a pilot project to transform a village's conventional heat and electricity supply into a renewable energy basis with the villagers as the main actors (the bioenergy village is born); and (f) bring the successful lighthouse project to the media's attention on a regional, national and global level to motivate other villages or regions to attempt similar projects.

The important research results can be summarised as follows:

- The transformation of the heat and electricity supply of the bioenergy village Jühnde by means of crops, manure and wood decreased the greenhouse gas emissions by 84 % when compared to Germany's average total emission in 2007. The energy balances of the crop production show that their energy input/output ratio was 1:19 for triticale and 1:18 for maize. The harvest index for heat and electricity production in the Jühnde biogas plant is convincing: 1 kWh of fossil energy input produces 4.5 kWh of bioenergy.

- New, locally adapted and environmentally friendly concepts for the parallel cultivation of food, feed and energy crops, such as crop rotation and double-cropping systems, create a high crop diversity (no monoculture), which may enrich the landscape and increase the population's acceptance.
- The reduction of pesticide use during the energy crop cultivation and the more balanced nutrient cycles lessen the translocation of pesticides and nitrate to surface and ground water, which is especially important in water protection areas.
- The new concept also decreases soil erosion and humus degradation; even the build-up of soil humus and the corresponding carbon fixation are possible.
- By harvesting energy crops several weeks earlier than food crops, the farmers' workload becomes more balanced.
- Energy production is an additional way for farmers to generate income. If the owners of the operating company are mainly local farmers and villagers, the profits remain in the region. In Jühnde, 58 % of the invested sum was given to local companies and 80 % of the annual turnover has remained in the region, thus supporting the local economic cycles.

All these positive aspects help to convince and motivate citizens – especially environmentalist, farmers, etc. – to follow the bioenergetic pathway. The following social and psychological motivating actions were used in Jühnde: (a) Visit model sites with successful installations together with the citizens; (b) formulate positive and constructive, but objective, arguments in favour of the project; (c) inspire well-established people with broad recognition and who are respected for the project; (d) involve local clubs and societies and the parish, as well as the council and the municipality; (e) use festivities and other similar events to transfer the ideas and stimulate people to participate in the project; humour, authenticity and objectivity are important ingredients to convict people; (f) have someone in charge of the dissemination of information; (g) also spread the information orally and contact individuals personally; and (h) establish and cultivate good contact with the local media.

The evaluation of a questionnaire shows that the self-efficacy and self-reported environmental behaviour in the bioenergy village increased and that all the inhabitants of Jühnde who are linked to the hot water grid are very satisfied (89 %) or satisfied (11 %) with the heat supply by means of biomass.

On average, 11 Jühnde interviewees engaged in the project perceived a better group feeling and integration into the village, more profound communication, a greater feeling of unity and well-being.

Altogether, the implementation of the bioenergy village was a success story not only for the villagers and farmers in Jühnde, but it also formed the basis for hundreds of other communities in Germany that realized similar or other renewable energy projects decentrally (Schmuck et al. 2006; Ruppert et al. 2008). The trans-regional, national and international interest in Jühnde was very high. In 2007, more than 8,000 visitors (mostly in the form of groups) arrived to familiarise themselves with Jühnde, the first bioenergy village in Germany.

References

Ammann, F., Mosler, H.-J., & Gutscher, H. (1997). Wie wirkt Überzeugungsarbeit in einer Population entsprechend Umweltbetroffenheit, Umweltwissen und vorhandener Voreingenommenheit? In R. Kaufmann-Hayoz (Ed.), *Allgemeine Ökologie zur Diskussion gestellt. Bedingungen umweltverantwortlichen Handelns von Individuen* (pp. 91–99). Bern: IKAOE.

Aronson, E., & O'Leary, M. (1983). The relative effectiveness of models and prompts on energy conservation: A field experiment in a shower room. *Journal of Environmental Systems, 12*, 219–224.

Bandura, A. (1982). Self-efficacy mechanism in human agency. *AmericanPsychologist, 37*, 122–147.

Baumeister, R. F., & Leary, M. R. (1995). The need to belong: Desire for interpersonal attachments as a fundamental human motivation. *Psychological Bulletin, 117*, 497–529.

Bengel, J., Strittmatter, R., & Willmann, H. (1998). Was erhält Menschen gesund? Antonovskys Modell der Salutogenese – Diskussionsstand und Stellenwert. In *Forschung und Praxis der Gesundheitsförderung* (Vol. 6). Köln: Bundeszentrale für gesundheitliche Aufklärung.

BMU. (2004). *The renewable energy source act*. Berlin. Retrieved November 29, 2012, from: http://www.bmu.de/files/english/pdf/application/pdf/eeg_en.pdf

Burn, S. M., & Oskamp, S. (1986). Increasing community recycling with persuasive communication and public commitment. *Journal of Applied Social Psychology, 16*, 29–41.

Cervinka, R., & Schmuck, P. (2010). Umweltpsychologie und Nachhaltigkeit. In V. Linneweber, E. Lantermann, & E. Klas (Eds.), *Enzyklopädie der Psychologie, Band Umweltpsychologie* (pp. 595–641). Göttingen: Hogrefe.

Clark, W., & Dickson, N. (2003). Sustainability science: The emerging research program. *Proceedings of the National Academy of Sciences of the United States of America, 100*, 8059–8061.

Csikszentmihalyi, M. (1993). *The evolving self. A psychology for the third millennium*. New York: Harper Collins Publishers.

Eigner, S., & Schmuck, P. (2002). Motivating collective action: Converting to sustainable energy sources in a German community. In P. Schmuck & W. Schultz (Eds.), *Psychology of sustainable development* (pp. 241–257). Boston: Kluwer Academic Publishers.

Eigner-Thiel, S. (2005). *Kollektives Engagement für die Nutzung erneuerbarer Energieträger – Motive, Mobilisierung und Auswirkungen am Beispiel des Aktionsforschungsprojekts "Das Bioenergiedorf"*, Studien zur Umweltpsychologie, Band 1. Kovac: Hamburg.

Eigner-Thiel, S., & Schmuck, P. (2010). Gemeinschaftliches Engagement für das Bioenergiedorf Jühnde – Ergebnisse einer Längsschnittstudie zu psychologischen Auswirkungen auf die Dorfbevölkerung. *Zeitschrift für Umweltpsychologie, 14*, 98–120.

Eigner-Thiel, S., Schmuck, P., & Lackschewitz, H. (2004). Kommunales Engagement für die energetische Nutzung von Biomasse: Auswirkungen auf Umweltverhalten, soziale Unterstützung, Selbstwirksamkeitserwartung und seelische Gesundheit. *Umweltpsychologie, 8*, 146–167.

Federal Ministry of the Interior. (1998). *Basic law for the Federal Republic of Germany (Grundgesetz, GG)*. Retrieved October 25, 2012, from: http://www.iuscomp.org/gla/statutes/GG.htm#20

Fischer, A., & Kallen, C. (Eds.). (1995). *Erfolgsbedingungen für kommunale Energie- und Klimaschutzkonzepte*. Berlin: Deutsches Institut für Urbanistik.

FNR (Ed.). (2008). *Standortangepasste Anbausysteme für Energiepflanzen*. Hürth: Nova Institut GmbH.

Friehe, J. (2007). *Wissenschaftliche Begleitung der Strom- und Wärmeversorgung des Bioenergiedorfes Jühnde*. 2. Zwischenbericht 2007, Fachagentur Nachwachsende Rohstoffe, Förderkennzeichen 22006205, Gülzow.

Gonzales, M. H., Aronson, E., & Costanzo, M. A. (1988). Using social cognition and persuasion to promote energy conservation: A quasi-experiment. *Journal of Applied Social Psychology, 18*, 1049–1066.

Gorke, M. (1999). *Artensterben. Von der ökologischen Theorie zum Eigenwert der Natur.* Stuttgart: Klett-Cotta.

Graß, R., & Scheffer, K. (2003). Direkt- und Spätsaat von Silomais nach Wintererbsenvorfrucht – Erfahrungen aus Forschung und Praxis. *Beiträge zur 7. Wissenschaftstagung zum Ökologischen Landbau,* (pp. 45–48), Wien.

Hennicke, P., Jochem, E., & Prose, F. (1997). *Interdisziplinäre Analyse der Umsetzungschancen einer Energiespar- und Klimaschutzpolitik.* Karlsruhe/Kiel/Wuppertal: Forschungsbericht.

Herzog, M. (1997). *Psychologische und ökonomische Aspekte bei der Förderung von regenerativen Energieträgern.* Frankfurt: Lang.

Jungk, R., & Müllert, N. (1991). *Zukunftswerkstätten. Mit Phantasie gegen Routine und Resignation* [Future workshops. How to create desirable futures]. München: Heyne.

Karpenstein, M., & Scheffer, K. (1994). Die ertrags- und wachstumsbeeinflussenden Wirkungen eines Wachstumsregulators auf der Basis von Mepiquatchlorid und Ethephon auf zwei Sommergerstensorten bei differenzierter N-Düngung. *Landwirtschaftl. Forschung, 37*(3–4), 277–285.

Karpenstein-Machan, M. (1997). *Konzepte für den Energiepflanzenbau; Perspektiven eines pestizidfreien Anbaus von Energiepflanzen zur thermischen Verwertung im System der Zweikulturnutzung.* Frankfurt: DLG Verlag.

Karpenstein-Machan, M. (2001). Sustainable cultivation concepts for domestic energy production from biomass. *Critical reviews of plant science, special issue on bioenergy;* New York, *20*(1).

Karpenstein-Machan, M. (2002). Low input energy crop rotations without herbicides. *Pflanzenbauwissenschaften, 6*(1), 36–46.

Karpenstein-Machan, M. (2003). Low input energy crop rotations without herbicides. *Pflanzenbauwissenschaften, 6*(1), 36–46.

Karpenstein-Machan, M. (2005). *Energiepflanzenbau für Biogasanlagenbetreiber.* Frankfurt: DLG-Verlag.

Karpenstein-Machan, M., & Schmuck, P. (2007). The bioenergy village. Ecological and social aspects in implementation of a sustainability project. *Journal of Biobased Materials and Bioenergy, 1*, 148–154.

Karpenstein-Machan, M., & Schmuck, P. (2010). The bioenergy village in Germany: A lighthouse project for sustainable energy production in rural areas. In M. Osaki, A. Braimoh, & K. Nakagami (Eds.), *Local perspectives on bioproduction, ecosystems and humanity* (pp. 184–194). Tokyo: United Nations University Press.

Karpenstein-Machan, M., & Stülpnagel, R. (2000). Biomass yield and nitrogen fixation of legumes monocropped and intercropped with rye and rotation effects on a subsequent maize crop. *Plant and Soil, 28*(1/2), 215–232.

Kates, R., Clark, W., Corell, R., Hall, J., Jaeger, C., Lowe, I., McCarthy, J., Schellnhuber, H., Bolin, B., Dickson, N., Faucheux, S., Gallopin, G., Grubler, A., Huntley, B., Jager, J., Jodha, N., Kasperson, R., Mabogunje, A., Matson, P., Mooney, H., Moore, B., O'Riordan, T., & Svedin, U. (2001). Environment and development: Sustainability science. *Science, 292*, 641–642.

Komiyama, H., & Takeushi, K. (2006). Sustainability science: Building a new discipline. *Sustainability Science, 1*, 1–6.

Mienecke, W. H. W., & Midden, H. J. (1991). Communication network influences on information diffusion and persuasion. *Journal of Personality and Social Psychology, 61*, 734–742.

Mosler, H.-J. (1998). Umweltpsychologische Interventionsformen für die Praxis. *Umweltpsychologie, 2*(2), 64–79.

Öko-Institut (Institut für angewandte Ökologie e.V). (2008). *Globales Emissions-Modell Integrierter Systeme (GEMIS).* Version 4.5; Internet-release: www.gemis.de. October 2009.

Ruppert, H., Eigner-Thiel, S., Girschner, W., Karpenstein-Machan, M., Roland, V., Ruwisch, V., Sauer, B., & Schmuck, P. (2008). *Wege zum Bioenergiedorf – Leitfaden für eine eigenständige Strom- und Wärmeversorgung auf Basis von Biomasse im ländlichen Raum* 120 pp., with DVD. Gülzow: Fachagentur für Nachwachsende Rohstoffe; ISBN 978-3-9803927-3-0.

Sauer, B. (2009). *Elementgehalte und Stoffströme bei der Strom- und Wärmegewinnung im Bioenergiedorf Jühnde, Schriftenreihe Fortschritt neu denken* (124 pp), Heft 2, ISSN 2190–7722, Göttingen.

Scheffer, K., & Stülpnagel, R. (1993). Wege und Chancen bei der Bereitstellung des CO_2-neutralen Energieträgers Biomasse – Grundgedanken zu einem Forschungskonzept. *Der Tropenlandwirt, Beiheft Nr., 49*, 147–161.

Scherhorn, G., Hellenthal, F., & Schroedl, S. (1999). Umweltschonung als kollektive Aktion. *Umweltpsychologie, 3*, 48–55.

Schmittknecht, I. (1998). *Klimaschutz-Forum für kleine Kommunen. Dokumentation der Workshops in Osnabrück und Wernigerode, 1998*. Deutsche Umwelthilfe/Klimabündnis Alianza del Clima, Krammer.

Schmuck, P. (2013). The Göttingen approach of sustainability science: Creating renewable energy communities in Germany and testing a psychological hypothesis. Umweltpsychologie (in press).

Schmuck, P., & Schultz, W. (2002). Sustainable development as a challenge for psychology. In P. Schmuck & W. Schultz (Eds.), *Psychology of sustainable development* (pp. 3–19). Boston: Kluwer Academic Publishers.

Schmuck, P., & Sheldon, K. (Eds.). (2001). *Life goals and well-being. Towards a positive psychology of human striving*. Seattle: Hogrefe & Huber Publishers.

Schmuck, P., & Vlek, C. (2003). Psychologists can do much to support sustainable development. *European Psychologist, 8*, 66–76.

Schmuck, P., Eigner, S., Krapoth, S., & Kaufhold, A. (1997). *Wie kommen Menschen zu ganzheitlichem Denken und Handeln? Ein psychologischer Annäherungsversuch über biographische Analysen und ein Interview mit dem Nobelpreisträger Ilya Prigogine* (Vol. 9). Göttingen: Schriftenreihe des Koordinations- und Studienzentrums Frieden und Umwelt.

Schmuck, P., Eigner-Thiel, S., Karpenstein-Machan, M., Ruwisch, V., Sauer, B., Girschner, W., Roland, V., Ruppert, H., & Scheffer, K. (2006). Das Bioenergiedorf Jühnde. In G. Altner, H. Leitschuh-Fecht, G. Michelsen, U. Simonis, & E. U. von Weizsäcker (Eds.), *Jahrbuch Ökologie 200* (pp. 104–112). München: C.H. Beck Verlag.

Schuster, K., & Marx, R. (1998). *Sozialwissenschaftliche Evaluation des Förderprogramms Biomasse*. Kiel: Endbericht der Energiestiftung Schleswig-Holstein.

Schütte, A. (1991). *Anbauversuche mit Zuckerhirse (Sorghum bicolor (L.) Moench), Teil I: Biomasseerträge bei verschiedenen Anbauverfahren unter besonderer Berücksichtigung minimalen Pflanzenschutzaufwandes, Teil II: Bewertung der Stickstoffdynamik und Abschätzung von Nitratausträgen (anhand von Nmin-Daten) bei verschiedenen Varianten des Zuckerhirseanbaus*. Dissertation, Universität Kassel-Witzenhausen.

Schweitzer, A. (1991). *Menschlichkeit und Friede. Kleine philosophisch-ethische Texte*. Berlin: Verlags-Anstalt Union.

Sheldon, K., Schmuck, P., & Kasser, T. (2000). Is value-free science possible? *American Psychologist, 55*, 1152–1153.

Sodikin, E. (1994). *Die Ertragsbildung der Wintergetreidearten in Abhängigkeit von Herbizid- und Fungizidmaßnahmen bei Körner- und Ganzpflanzenutzung zur Energiegewinnung*. Dissertation, Universität Kassel-Witzenhausen.

Umweltbundesamt (2012): *Less greenhouse gases with less nuclear energy*. Press Release No. 17/2012. Retrieved October 30, 2012, from: http://www.umweltbundesamt.de/uba-info-presse-e/2012/pdf/pe12-017_less_greenhouse_gases_with_less_nuclear_energy.pdf

UNO, United Nations Organization (1992). *Agenda 21. The Rio declaration on environment and development*. Retrieved May 3, 2011, from: http://www.un.org/esa/dsd/agenda21/

von Buttlar, C. (1996). *Erhaltung pflanzengenetischer Ressourcen über den Weg der energetischen Nutzung von Ganzpflanzen am Beispiel Wintergerste*. Dissertation, Kassel-Witzenhausen.

Witzel, A. (2000). Das problemzentrierte Interview/The problem-centered Interview. *Forum Qualitative Sozialforschung/Forum Qualitative Social Research [Online-Journal], 1*(1). Retrieved May 4, 2011, from: http://www.qualitative-research.net/fqs-texte/1-00/1-00witzel-d.htm

Wüste, A., Schmuck, P., Eigner-Thiel, S., Ruppert, H., Karpenstein-Machan, M., & Sauer, B. (2011). Gesellschaftliche Akzeptanz von kommunalen Bioenergieprojekten im ländlichen Raum am Beispiel potenzieller Bioenergiedörfer im Landkreis Göttingen. *Zeitschrift für Umweltpsychologie, 15*, 135–151.

Part II
Do We Have Enough? – Biomass Potentials for Energy Generation

Chapter 3
Estimation of Global Bioenergy Potentials and Their Contribution to the World's Future Energy Demand – A Short Review

Martin Kappas

Abstract The global energy question is currently dominated by three concerns that strongly affect decisions on energy development priorities, i.e. the security of the energy supply, the security of the food supply and climate change. A very challenging question in this context is the estimation of global bioenergy potentials and their possible contribution to the world's future energy demand. The sustainability potential of global biomass for energy is widely recognised and thus a primary concern of the book. The annual global primary production (GPP) of biomass is equivalent to the 4,500 EJ (EJ = 1 Exajoule = 10^{18} J = 1,000 Petajoule; 14.0 EJ = Germany's primary energy consumption in 2008, while 508 EJ = the primary energy consumption of mankind in 2009) of solar energy captured each year. Around 5 % of that energy (225 EJ) could deliver 50 % of the world's total energy use today. This approximation is in accordance with other estimates that show a sustainable annual bioenergy production of around 270 EJ. The 50 EJ that biomass contributed to the global energy supply in 2006 (the approximate energy demand was 490 EJ) was mainly used in the form of traditional non-commercial biomass fuels and contributed only 10 % to global energy use. This chapter provides a synthesis of analyses of the longer term potential of biomass resource availability on a global scale. Various studies have assessed global biomass potentials and have arrived at widely varying results. These studies highlight the reasons for these uncertainties and explain the factors that can affect biomass availability. Estimates, for instance, are sensitive to assumptions about crop yields and the amount of land that could be made available for the production of biomass for energy usage.

The sustainable use of biomass as an energy source requires comprehensive management of specific landscapes and their natural resources, which are subject to restrictions (e.g., nature protection, contaminated land, priority for food production, etc.). Knowledge of the regional landscape's potential to provide biomass and

---

M. Kappas (✉)
Institute of Geography, Department of Cartography, GIS and Remote Sensing, University of Göttingen, Goldschmidtstrasse 5, D-37077 Göttingen, Germany
e-mail: mkappas@gwdg.de; mkappas@uni-goettingen.de

H. Ruppert et al. (eds.), *Sustainable Bioenergy Production - An Integrated Approach*,
DOI 10.1007/978-94-007-6642-6_3,

hence bioenergy, is urgently needed and best provided by bottom-up approaches, because unsustainable biomass production would diminish the climate-related environmental advantage of bioenergy.

Therefore, based on a review of currently available studies on the subject, this chapter discusses the role of sustainable biomass in the future global energy supply.

Keywords Sustainable biomass • Bioenergy • Global biomass potential • Bioenergy potential • Bottom-up approaches

3.1 Introduction

The world's energy demand is growing rapidly. Estimations of commercial energy use increased from 467 EJ in 2004 (reported by IEA 2006a, b) to approximately 490 EJ in 2006 (see Fig. 3.1). Around 88–90 % of this demand is provided by fossil fuels. Fossil fuel-derived CO_2 emissions are the most important contributor to atmospheric concentrations of greenhouse gases (GHGs), which reached a historical high in 2010 and 2011 (Le Quéré et al. 2012).

There is consensus in the scientific and political communities that GHG emissions should be reduced to mitigate related global warming and climate change impacts. Both these communities believe that GHG emissions should be reduced to less than half of the 1990 global emission levels (IEA 2007a). The world's current energy supplies are dominated by fossil fuels (388 EJ per year) with much smaller contributions from nuclear power (26 EJ) and hydropower (28 EJ).

In the face of the nuclear accident at Fukushima (Japan), many countries have rethought their nuclear power strategy or, in Germany's case, have decided to completely phase out the use of nuclear power. Today, biomass delivers about 45EJ (±10 %) of energy, making it the most important renewable energy source.

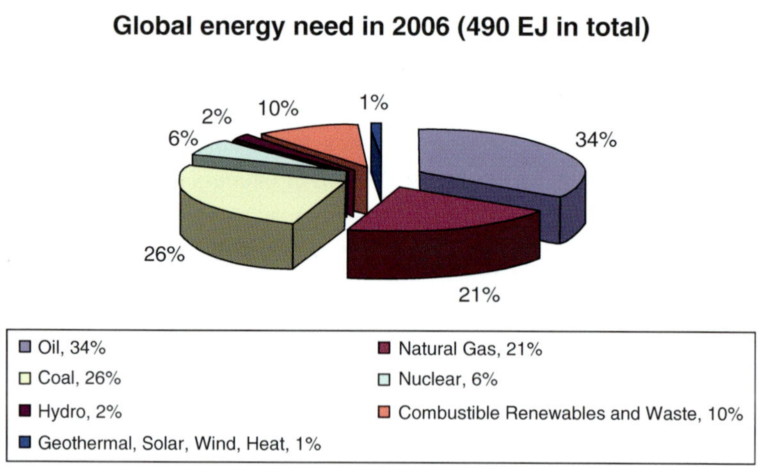

Fig. 3.1 Global energy needs in 2006 (numbers after IEA 2008; Source: Ladanai and Vinterbäck 2009)

Box 3.1 Flowchart – Possibilities to Provide Heat and/or Power as Well as Fuels from Biomass (Source: FAO United Bioenergy Terminology)

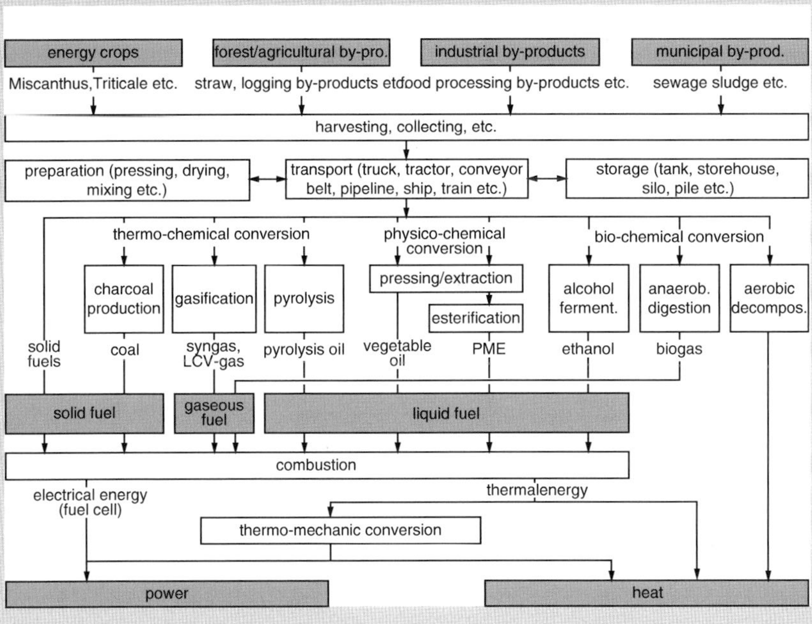

However, in industrialised countries, biomass currently accounts for less than 10 % of the total energy supply. In developing countries, on the other hand, the contribution is as high as 20–30 % and, in some of these poorer countries, biomass supplies 50–90 % of their total energy need (IEA 2007a). The vast bulk of biomass energy is used non-commercially. It is often used by poorer people for cooking and room heating. The recently introduced commercial production of biogas for power, heat generation or transport fuels contributes to a lower, but very significant, portion of the total energy supply (around 7 EJ/year in 2000 after WEA 2000). The use of modern bioenergy is growing. Ten years ago, 40 GW of biomass-based electricity production capacity and 200 GW of heat production capacity had been installed worldwide (producing 0.6 EJ electricity per year and 2.5 EJ heat per year; WEA 2000). Biomass potentials also depend on land availability. At present, only 0.19 % (approximately 0.025 billion hectares) of the world's total land area (13.2 billion hectares) and only 0.5 % of global agricultural land are used for growing energy crops for biofuels (Ladanai and Vinterbäck 2009). The significant potential of using algae to generate biomass energy, as illustrated in a number of studies (e.g., Christi 2007; Beer et al. 2009), is not taken into account in this review. In the context of this book, we understand bioenergy as energy from biomass sources, including energy crops, residues and wastes from agriculture, forestry, food production and waste management (see also Box 3.1: Biomass sources for energy production).

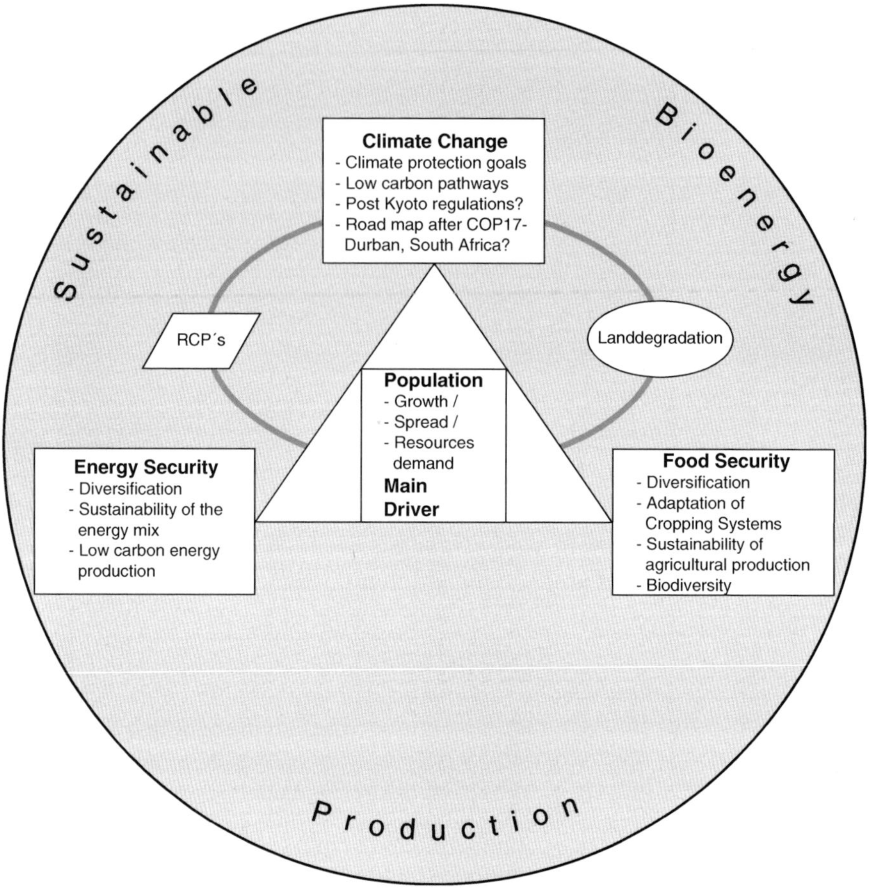

Fig. 3.2 The future role of bioenergy within the triangle of climate change/protection – energy security/diversification and sustainability – food security (food and fodder)

Thus, this review is based on existing literature and tries to point out the main trends and shifts in bioenergy issues related to global sustainable biomass potentials. The discussion on the future role of bioenergy is currently dominated by three issues that strongly affect decisions on energy development priorities: the security and sustainability of the energy and food supply as well as climate change (see Fig. 3.2). The main driver in the relation triangle is the world population and its dynamics in space and time. Population growth, the per capita demand for resources and the population's life style (consumption pattern) directly influence climate change, energy demand and food security. Estimates about population development and its future influence on the climate as well as on energy and food consumption are described in various scenarios, such as the IPCC SRES scenarios (A1, A2, A1B, B1, B2) and the newer Representative Concentration Pathways (RCPs) scenarios, which will be used in the IPCC AR5 report (Moss et al. 2010). Both estimations (population and future bioenergy potentials) have a high degree of inaccuracy. Therefore, depending on the chosen scenario, the results of the potential estimates can vary significantly. This chapter makes assumptions about biomass's future potential on the basis of a comparison of the different scenarios' predictions.

3.2 Types of Biomass Potential

Most of the analysed studies assessing future biomass potentials differ in their definitions of biomass potentials. Offermann et al. (2011) differentiate between demand-driven assessments and supply-driven assessments. Demand-driven assessments focus on bioenergy's contribution to the entire energy system. In general, supply-driven assessments are best suited for showing the resource availability and are better capable of handling sustainability criteria. Therefore, the chapter's focus is on supply-driven assessments. Within the reviewed literature, different types of biomass potential are mentioned, such as:

Theoretical potential: All biomass produced by plant photosynthesis (from land, oceans, lakes, etc.). This process is only limited by physical and biological constraints (e.g., latitude, light use efficiency, plant type C3/C4, etc.).

Geographical potential: The biomass potential of all land areas. Marine biomass is excluded.

Technical potential: The part of the geographical potential that is restricted to the need for land for food production, infrastructure and housing, and the protection of areas (e.g., forest protection, nature reserves, etc.). Moreover, the technical potential is determined by the level of agricultural technology in a specific geographical area.

Economic potential: The part of the technical potential that is used to achieve the most economically profitable outcome.

Implementation potential: The part of economic potential that can be achieved within a certain time. However, the time required is constrained by policy decisions as well as institutional and social behaviour.

The above-mentioned descriptions are an adapted summary of various definitions found in the literature review, which includes studies by Hoogwijk (2004), the World Energy Council (2004), Smeets et al. (2007) and Offermann et al. (2011).

Most of the reviewed studies on biomass potential focus on technical potentials. The principal reason for this is that the available data allow a significant comparison between single findings and the applied methods.

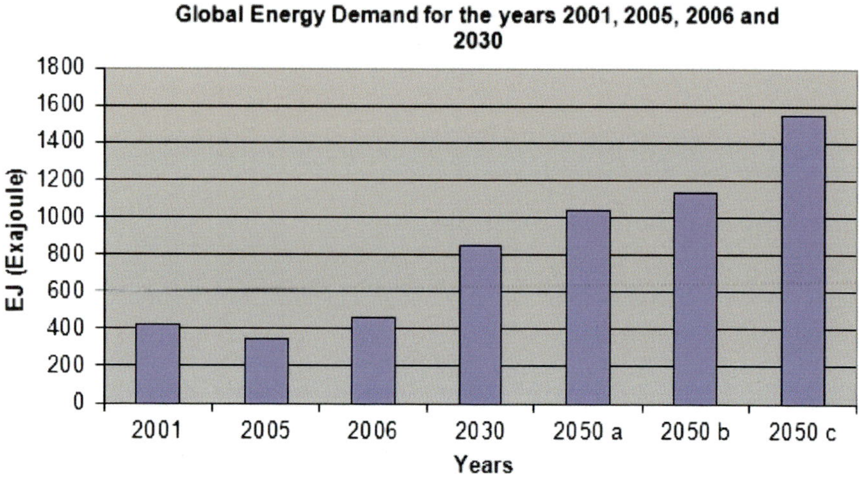

Fig. 3.3 Global Energy Demand in 2001, 2005, 2006 and 2030 and estimations of the total global bioenergy production in 2050
Sources:

- The numbers for 2001 (420 EJ), 2005 (343 EJ) and 2006 (464 EJ) are respectively taken from IEA 2003, 2005 and 2007a, b, c
- The estimate of the energy use in 2030 is calculated by adding the total power generation to the total power consumption, and does not include heat generated from heat pumps or electricity (After IEA 2008)
- 2050 a (1,041 EJ) is based on a high consumption scenario described by Smeets et al. (2004)
- 2050 b (1,135 EJ) is the total world potential of biomass energy based on the upper limit of the amount of biomass that may be available as a primary energy source without affecting food security (After Hoogwijk et al. 2003)
- 2050 c (1,548 EJ) is the potential global bioenergy production based on a scenario in which, in all corners of the globe, a type of agricultural management is applied that uses the best available techniques, such as those used in industrialised areas (After Smeets et al. 2006)

3.3 Global Biomass Potential: Current Scenarios

As mentioned before, the global primary production (GPP) of biomass is comparable to 4,500 EJ of captured solar energy per year. Only 5 % of this large energy amount would have been required to cover 50 % of the world's energy needs in 2006. Figure 3.3 illustrates the global energy use for each year since 2000 as well as the total global bioenergy production potentials for 2050, which were derived from different scenarios (after Ladanai and Vinterbäck 2009; EC 2005).

Another prerequisite for determining the future potential of biomass energy is the availability of land. The world's total land area is 13.2 billion hectares, of which 0.19 % is used for growing crops for biofuels, which accounts for 0.5 % of the global agricultural land (Ladanai and Vinterbäck 2009). However, we have to keep in mind that many land use statistics are not exact because the specific types of land use are not included in the different land use categories (i.e. in the FAO land use categories), which are the basis for modelling and scenario building. Therefore, there are many assumptions about the distribution of land use type with regard to the world's total land area and, specifically, its agricultural land area. Berndes et al. (2003) offer a comprehensive overview of studies on global biomass production potential, including studies that estimate the future demand and supply of bioenergy (e.g., studies by Smeets and Faaji 2007; Lashof and Tirpak 1990; Hall et al. 1993; WEC 2004, 2007; Fujino et al. 1999; IPCC 2000; Rogner 2000; Fischer and Schrattenholzer 2001).

Berndes et al.'s (2003) comparison of studies on global biomass production potentials includes previously published numbers on the total global bioenergy production potential in 2050. These numbers range from 33 to 1,135 EJ per year (Hoogwijk et al. 2003), of which 0–358 EJ per year are from woody biomass (Hoogwijk et al. 2003). Energy crops from surplus agricultural land have the largest potential, contributing 0–988 EJ per year (Hoogwijk et al. 2003, 2005). However, only a few publications have assessed the global potentials of the different world regions. In addition to Berndes et al. (2003), we examined the work of Offermann et al. (2011), who reviewed 19 publications on global biomass production potentials and strengthened the land availability aspect. Offermann et al. point out that future energy crop potential will probably range from 200 to 600 EJ per year in 2050, whereby the residue potentials are estimated to vary between 62 and 325 EJ per year. The areas with the highest potential are Asia, Africa and South America, whereas Europe, North America and the Pacific regions have a weaker potential. The most optimistic scenario offers a biomass potential of 1,548 EJ per year in 2050, which is three times higher than the current global energy supply. If the biomass potential is divided into the biomass potential from energy crops and residues, the relationship between the potentials will be as follows: The energy crop potentials for 2050 (based on more than 50 % of all the reviewed studies) will range from 0 to 1,272 EJ per year (again based on the optimum quantity; 75 % of the studies reveal an energy crop potential of below 657 EJ per year), while the residues potentials for 2050 will range from 62 to 325 EJ per year.

Despite this high estimate of 1,548 EJ per year (optimum value without restrictions) for 2050, over 50 % of the literature studies assume that the future energy crop potential will vary between 200 and 600 EJ per year in 2050. From a sustainability point of view, the sustainable energy crop potentials will realistically be towards the lower end of the mentioned range of 200–600 EJ per year. The latest studies of energy crop potentials on abandoned land and studies committed to sustainable biomass production (Teske et al. 2008) show lower bioenergy potentials. These studies estimate that biomass could contribute anything from below 100 EJ per year to above 400 EJ per year to the future global energy supply in 2050. This shortage of future bioenergy potential is mainly caused by uncertainty regarding future land availability and future yield levels in energy crop production. Both these parameters vary widely, causing the current assessments of the bioenergy potential of surplus agricultural land (plantation supply) in 2050 to range from below 50 EJ per year to almost 240 EJ per year. Moreover, most of the literature studies only provide preliminary estimations of the future availability of forest wood and of residues from agriculture and forestry. In sum, owing to this imponderability, only rough estimations are possible of the sustainable bioenergy potential in the future global energy supply.

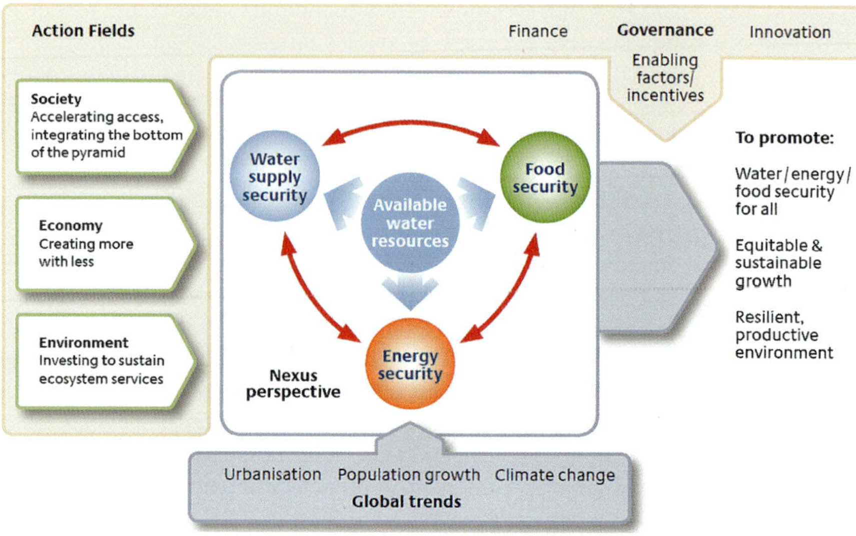

Fig. 3.4 The water, energy and food security nexus (Source: Hoff 2011)

3.3.1 The Imponderability of Future Bioenergy Potential – Possible Constraints

According to the water, energy and food security nexus approach (see Fig. 3.4, Hoff 2011), the expansion of the bioenergy sector should interact sensibly with other land uses and ecosystem services, such as food and fodder production, soil and nature conservation, biodiversity and carbon sequestration. The assortment of energy plants, accompanied by certain crop sequences, can alter the carbon sequestration potential of an area, which is an important factor in the climate protection chain. In addition to the production of energy crops, the alteration of the soil and the associated ecosystem services may also influence the carbon sequestration balance of an area.

In general, the following constrain bioenergy potentials:

1. Population growth, consumption per capita and lifestyle (decisive for the amount of area reserved for food production)
2. Land availability (areas used for reforestation, areas lost by soil degradation, areas reserved for nature protection)
3. Future crop yield increases under changing edaphic situations (climate and soil moisture, soil temperature)
4. The preservation of biodiversity and the need to expand nature reservation areas
5. Land degradation (the most important driver is soil degradation and reduced land availability)

6. Severe water scarcity (a limiting factor in terms of quality and quantity).

Owing to these six constraints, it is difficult to estimate the impact of bioenergy expansion as an advantageous option for climate change mitigation within the energy sector.

3.3.1.1 Population Growth, Consumption Per Capita and Lifestyle

According to the latest projections, the world population will increase from 7 billion in 2011 to around 8.4 billion in 2030 (about 20 %). Developing countries will contribute most to this increase, with their total population increasing from 4.7 (2011) to 6.9 billion (2030).

During the last few decades, agricultural yields have developed faster than the earth's population (UNEP 2009). This means more food has been produced on already existing croplands. In the near to mid future, this trend might develop less favourably, as average crop yields may balance out the population growth but not the increasing demand for animal-based food. The global population is estimated to grow by 36 % from 2000 to 2030 (UN/FAO medium projection, UNEP 2009). This is comparable to the expected increase in crop yields during this time (UNEP 2009). However, the demand for food is simultaneously changing towards a higher share of animal-based food, with developing countries showing a particularly high growth rate.

The FAO has forecasted that the world's meat consumption will increase by about 22 % per person from 2000 to 2030, milk consumption will increase by 11 % and the need for vegetable oils will rise by 45 % (UNEP 2009). In contrast to this trend, agricultural commodities with lower land requirements (cereals, root and tuber crops) will rise at lower rates per person.

On average, the world's population is predicted to grow about as fast as the cereal yields. If the forecasted yield increases are not realized, the world would suffer significantly from an insufficient food supply. Consequently, any additional demand for biomass production can only be generated by expanding cropland at the expense of other land uses. To date, there is no clear assessment of changes in global land use due to changing food demand (especially with regard to a stronger shift towards animal-based food). A more recent RFA report by Gallagher (2008) reveals that an additional 144–334 Mha of global cropland will be required for food production in 2020 (see 'The Gallagher Review of the indirect effects of biofuels production'). In general, the FAO (2008) statistics show that the production of beef, pork, poultry, mutton and milk have increased, and many developing countries show a growth rate of more than 10 %. In the EU, meat production is more or less stagnant and EU milk production has decreased. The increase in meat production in certain key regions (the USA and Brazil) is expected to decline, but meat production and consumption will increase in developing countries (India and China). If the change in lifestyle and its influence on food production per area is kept in mind, the question of future land availability for bioenergy production arises, providing a new perspective.

3.3.1.2 Land Availability

The available land area is the most important constraint when assessing the potential of energy crops, which are estimated to deliver the greatest amount of future bioenergy potential. Land categories suitable for energy crop cultivation include surplus agricultural land and abandoned agricultural land (also called degraded or marginal land). Surplus agricultural land is defined as the land that remains after subtracting the land required for food and feed production from the total amount of agricultural land. Most studies in the literature review focus primarily on energy crop potentials from surplus agricultural land. The latest studies also examined the potentials of degraded land or abandoned land, a topic that has received increasing attention in studies assessing the potential of bioenergy (e.g. Offermann et al. 2011). A major problem is, however, that the different land categories are not clearly defined and the data on these land categories (global land use classifications) are weak (the statistics are inconsistent). Scientists also need to determine the variety of produce that will be yielded in relation to the land availability to ensure that balanced nutrition will be available in future. In doing so, the following challenging question must be answered: What happens when people move away from meat as a food source and what impact would this scenario have on future land availability? If people were to eat less meat, a lot of land would become available for food (a more vegetarian variety of produce or mostly vegetarian produce) and bioenergy production.

Land scarcity is escalating rapidly due to the increasing demand for meat as a result of population growth, changing lifestyles, resource degradation and climate change. Even if more sustainable crop production and consumption patterns and, simultaneously, a declining population growth rate could be achieved, agricultural production would have to grow by 70 % and agricultural land would have to expand by about 10 % globally (by 20 % in developing countries and by 30 % in Latin America; de Fraiture et al. 2007; Bruinsma 2009; FAO 2011) by 2050 if all people are to receive balanced nutrition. Even the most optimistic scenarios of improvements in productivity through technological development still assume that the demand for agricultural water will increase by at least 20 % by 2050 (de Fraiture et al. 2008).

In Germany, the area under cultivation has declined over the last years. A report by the German Statistic Bundesamt (Federal Statistical Office 2012) at the World Food Day 2011 documented that the agricultural area used for farming in Germany regressed from 17.3 Mha (1995) to 16.8 Mha (2011). Over the last 16 years, the area under cultivation has diminished by 3 %. In 2011, 47 % of Germany's total area was agricultural land, of which most was used as cropland (11.9 Mha ~ 70 %). Germany's cropland, which is the country's most important basis for food production, has been relatively constant over the last 20 years.

The cultivation of cereals, which required 6.5 Mha in 2011, has remained relatively stable over the last 15 years. However, the acreage has diminished by about 78,000 ha since 2010 and the total harvest was 37 million tonnes in 2011,

Table 3.1 Germany's cultivated land in 1,000 ha (from 1995 to 2011 preliminary results); (Source: Federal statistical office 2012)

	1995	1999	2005	2010	2011
Area under agricultural use	17,344.3	17,151.6	17,035.2	16,704.0	16,757.7
Cropland	11,834.5	11,821.5	11,903.3	11,846.7	11,909.6
Subdivided into:					
Cereals	6,526.7	6,634.7	6,839.0	6,595.4	6,517.5
Root crops	856.9	813.5	705.4	624.3	664.8
Subdivided into:					
Potatoes	315.2	308.5	276.9	254.4	259.4
Forage crops	1,792.5	1,708.9	1,805.0	2,571.0	2,824.1
Subdivided into:					
Silage maize	1,251.8	1,202.8	1,262.5	1,828.9	2,042.0

which is 2.8 million tonnes less than in 2010. On the other hand, potato cultivation in Germany has regressed since 1995. The acreage for potato cultivation regressed by 56,000 ha from 1995 to 2011. 259,000 ha were still under potato cultivation in 2011. Nevertheless, the potato production has increased over the last years due to favourable weather conditions (11.9 million tonnes of potato in 2011, which is 2 million tonnes more than in 1995).

In general, the area under cultivation for forage crops, silage maize, field grass and clover has increased over the last years. In 2011, 2.8 Mha was used for forage crops (an increase of 10 % since 2010 and 58 % since 1995).

In 2011, the areas for silage maize were expanded by about 2 Mha, which represents an increase of 63 % since 1995. This expansion was due to the rising importance of silage maize to fuel bioenergy plants. From 2010 to 2011, the area for silage maize increased by 12 % (for a comparison, see Table 3.1). In terms of the amendment to the new German Renewable Energy Source Act (BMU 2012), 60 % of maize will be specifically restricted for use in bioenergy plants. Moreover, small bioenergy plants (75 KWh) will be promoted (Figs. 3.1, 3.5, 3.6 and 3.7).

3.3.1.3 Future Crop Yields

Parameters, such as water availability, within climate change and the evolution of agricultural markets (commodity prices) obfuscate the development of future crop yields (Kappas 2009). In general, it is doubtful that the growth rates of global agricultural yields over the past 60 years can be continued. The evolution of global agricultural yields will be decisive for the degree to which biomass for food and non-food use can be delivered from existing cultivated land. The increase in commodity prices is and will be a function of future yield changes. The latest statistics from the FAO (data from 1961 to 2005) show reduced average annual yield developments with regard to six field crops (Fig. 3.8, six field crops after Lobell and Field 2007).

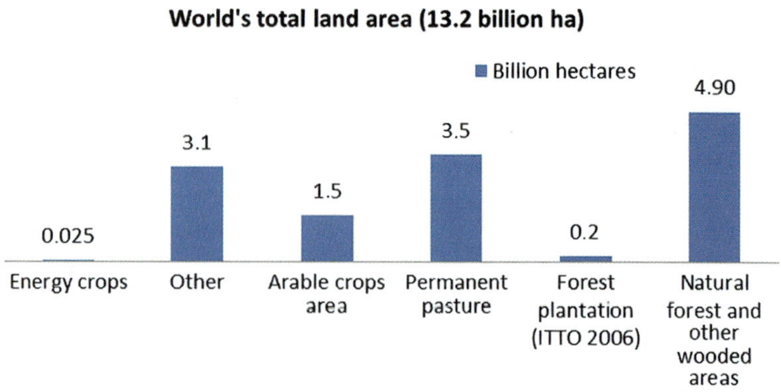

Fig. 3.5 World's total land area and its subdivision into major land use types (Figures from Faaij 2008; ITTO 2006a, b; Smeets et al. 2004)

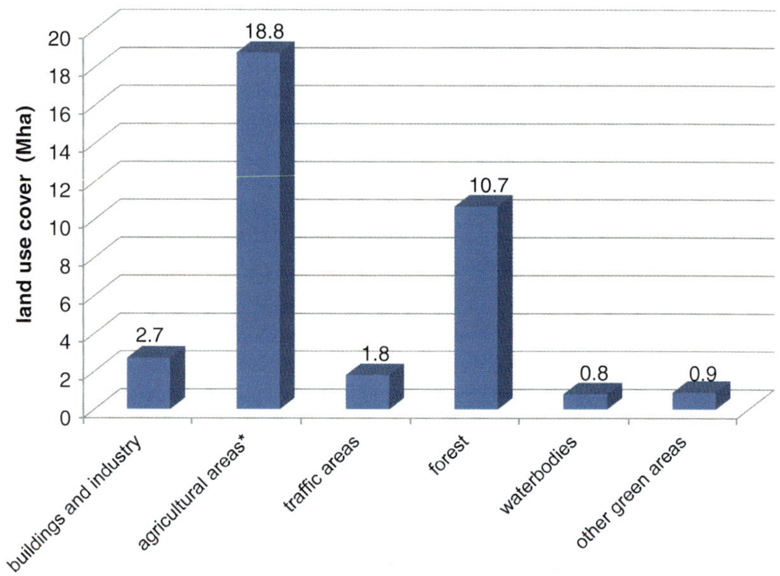

Fig. 3.6 Germany's land use cover in Mha in 2008 (entire country's territory: 35.7 Mha, Source: Statistisches Bundesamt Germany 2012)

Figure 3.8 illustrates the range of yield development in six crops. In general, there has been a decrease in yield changes that differs from crop to crop. Soybeans show the strongest decrease over the last decades. This trend is reflected in many agricultural areas. Most international organizations, like the FAO and the International Food Policy Research Institute (IFPRI) see a potential for future yield increases in developing countries (especially those in Africa). The OECD-FAO (2004) *Agricultural Outlook* has estimated an increase rate of 1.0–1.1 % per year

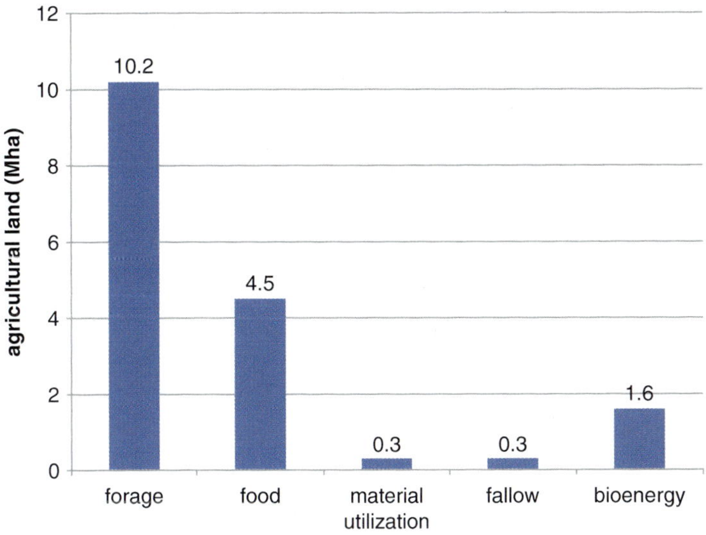

Fig. 3.7 How Germany's agricultural land is used in 2008 (Mha) (Source: BMELV 2012)

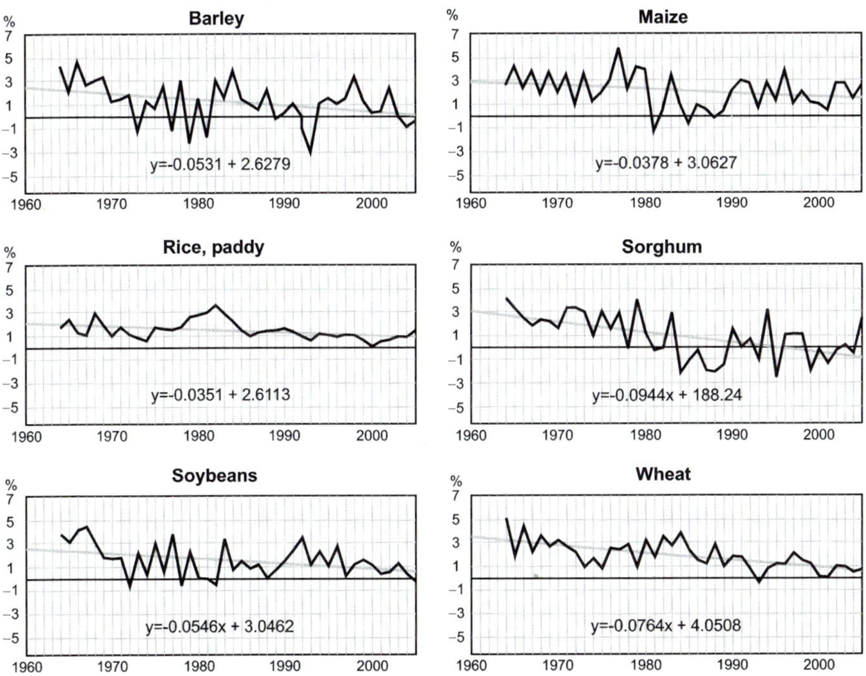

Fig. 3.8 Five-year running average of changes in global crop yields in % (Source: Adapted from UNEP 2009, Assessing biofuels, p. 43)

(1.3 % for roots and tubers). Although more recent calculations offer a somewhat better average increase in yields per year, the future rates of increase are still below the global rates of the past decades.

Multiple factors influence future crop yields. Among these are water availability, climate (temperature) change and environmental restrictions. Different IFPRI scenarios show that under "water stress" conditions, the global cereal yields would increase by only 0.9 % per year instead of by 1.2 % for the business as usual (BAU) scenario (IFPRI 2012). On a global scale, climate trends have had a negative impact on crop yields since the early 1980s. Lobell and Field (2007) estimate that, due to rising temperatures, maize and sorghum yields will decrease, with an average yield loss of about 8 % for each 1 °C increase in temperature. This assumption needs to be re-evaluated since the global average temperature is expected to increase to more than 2 °C above pre-industrial levels (Lobell and Field 2007).

Other plants like switchgrass, which is grown in North America, will increase in yields by up to 50 % for a 3.0–8.0 °C increase in temperature. Extreme weather events due to climate change, such as droughts and heavy rains or flooding, may have a negative impact on future yields (IPCC 2007). Therefore, it can be concluded that the future progression of yields will be accompanied by a higher level of uncertainty than today's assessment.

3.3.1.4 Biodiversity and Nature Reserves

Land conversion for bioenergy crops can lead to negative environmental impacts, such as reduced biodiversity and increased GHG emissions. Replacing natural vegetation and nature reserves with crops affects the carbon balance when the previously stocked carbon is mobilised. The results of biodiversity science's large-scale international programmes, such as DIVERSITAS, have shown that increased biofuel production could have a severe impact on biological diversity (Jackson et al. 2012). Different crops' biodiversity to GHG reduction balance showed that GHG reductions due to biofuel production is often too weak to compensate for the biodiversity losses due to the land use conversion process and its ecological side effects. Therefore, the identification of detrimental changes in biodiversity and ecosystem services is important to provide knowledge to avoid such changes when producing bioenergy. Positive effects regarding biodiversity production has only been found in terms of formerly abandoned or intensively used agricultural land or for degraded/contaminated land. On such land, bioenergy production could even lead to biodiversity benefits.

Rethinking the linkages between bioenergy, climate change (limiting global temperature change to 2 °C), land use and water requires an integrated assessment of the energy, land and water nexus.

3.3.1.5 Land Degradation

Land degradation and land desertification due to human activities are the main drivers of the decline in available cropland and make future assessments uncertain. The decrease in cropland is the result of many factors, such as deforestation, overgrazing, intensive agricultural activities, industrialization and urbanization. The degradation of land includes soil erosion, salinization, nutrient depletion and desertification. The rate of degradation in space and time has increased with population growth and technology. Serious land damage results from large-scale agricultural activities and restoration is very problematic. The continued loss of cropland will jeopardise our ability to feed the world and produce bioenergy.

Land degradation is a worldwide phenomenon and appears in developed and developing countries. To date, there are no reliable data on land degradation's future influence on the projected yield. Just in Germany, sealing of the landscape is responsible for the loss of 78 ha per day (Federal Statistical Office 2012).

3.3.1.6 Severe Water Scarcity

Water plays a major role in the water, energy and food nexus, because water is non-substitutable for food and bioenergy production. Although water is a renewable resource, we find many areas with water stress or water scarcity. Water scarcity is defined as the ratio between green water and blue water availability and the water need to produce a daily diet of 3,000 kcal (20 % animal product included).

Agriculture currently uses about 70 % of fresh water globally and bioenergy production would add to this. Food production is still the biggest user of blue water (around 80 %) and also requires a large percentage of green water (Hoff 2011). Water-related food productivity varies among crops, cropping systems and agricultural management systems. Therefore, water consumption depends on the crop types used as feedstock as well as the production methods and conversion technologies. Feedstock production for bioenergy in water-scarce regions requires irrigation, which may lead to competition with food production as well as pressure

Box 3.2 Green and Blue Water

Green water: Refers to groundwater generated directly from precipitation. Green water is available to plants and maintains the agricultural system. It is managed by choosing a typical land use type and agricultural practice (cropping system).

Blue water: Refers to water in aquifers (rivers and lakes) and is used for irrigation as well as municipal (e.g., sanitation) and industrial purposes. It is managed by the underlying water infrastructure. In contrast to green water, which is only used by plants, blue water can be locally allocated and recycled.

Table 3.2 Water productivity of different crops per kcal/m^3 in comparison to the average water productivity of selected livestock products per kcal/m^3

	Water productivity per kcal/m^3
Wheat[a]	660–4,000
Potato[a]	3,000–7,000
Tomato[a]	1,000–4,000
Apple[a]	520–2,600
Meat from beef cattle[b]	34
Meat from sheep and goats[b]	30
Milk from dairy cattle[b]	332
Meat from pigs[b]	666
Meat from poultry[b]	371
Eggs from poultry[b]	578

[a]Data from Molden et al. (2010)
[b]Data from Gerten et al. (2011)

on water resources beyond their restoration capacity (classic over-exploitation). Extreme weather events (inundation, droughts and heat waves) due to climate change might increase the uncertainty regarding available water resources. Table 3.2 shows different crops' water productivity per kcal/m^3 in comparison to the average water productivity of selected livestock products. According to Foley et al. (2011), more than one-third of global crop production is used for animal feed rather than for direct human consumption. The future development of the human diet (especially meat consumption) will have a severe impact on food and water security. Even the most optimistic estimations of improvements in agricultural productivity (mainly due to technological innovation) will demand an increase in agricultural water (an increase in blue water demand) of at least 20 % by 2050 (Hoff 2011). This demand for agricultural water could increase significantly if bioenergy strategies are fully implemented without taking the water, energy and food security nexus into consideration.

3.4 Lessons Learnt from Reviewing Global Bioenergy Potentials

All these constraints and different perspectives allow us to formulate a simple conclusion: While biomass has a promising potential for energy production, the mentioned constraints, which might reduce the future potential and sustainability of biomass production, should also be taken into account. In the near and mid-term future, bioenergy will be the most important renewable energy type. Impact factors like land availability, water scarcity, biodiversity preservation and land degradation have often been excluded from potential estimates of global trends. Therefore, potential estimates of future bioenergy have varied widely to date. On the other hand, bioenergy from residual material and waste promises an energy source based on biomass that does not compete with food and fodder production.

The future use of biomass as an energy type needs a comprehensive assessment and will require careful management of natural land resources, such as soil and water. Unsustainable biomass use would disturb the climate-related advantage of bioenergy. The advantages of sustainable bioenergy production use should always outweigh the effect of its possible environmental damage. This book thus develops a new research concept that focuses on interactions between different land uses, biodiversity and food production. This research concept provides new insights into the competition for resources and the synergies between different land uses. Its aim is to improve people's understanding of the potential of large-scale bioenergy production and future land use management (and biomass management).

So far, the transition towards renewable energy has been misunderstood as an economic demand and not as a means to gain diverse societal and economic advantages. Today, biomass could be produced to generate energy and renewable raw materials while simultaneously securing soil and water resources. The transition to a greener economy is an important precondition to achieve the sustainable development of societies. Environmental and social costs are part of the economic system (including external environmental costs) and external diseconomies have to be taken into account when pursuing sustainable consumption and production. The significance of energy alternatives based on locally available renewable resources, such as bioenergy, is an important aspect of creating a new energy mix. At the same time, increasing the energy efficiency of the whole economy and all energy alternatives is an indispensable precondition for transitioning towards a renewable energy system and a society oriented towards sustainability.

In order to develop a modern and forward-looking energy supply from biomass, such as biomass for heat and power generation and liquid biofuels for transport, there has to be a balance between the amount of biomass required for food production and for material purposes. Crop types, production methods and conversion technologies need to be matched with local conditions within the different landscapes to establish a national transformation plan and to reduce the increasing land use competition between food/fodder production, bioenergy and forests.

3.5 Take-Home Messages

- According to "Agenda 1" of "Agenda 21", renewable energy is the key to sustainable development and replacing fossil fuel energy supplies with renewable supplies can be achieved with existing technology.
- There are many estimations of the future biomass potential. However, if we compare the average global bioenergy production potential in 2050 with the highest predictions regarding global primary energy consumption in 2050, we have to conclude that bioenergy will play an important part in the future energy supply.
- From a sustainability point of view, the sustainable energy crop potentials will realistically be at the lower end of the mentioned range of 200–600 EJ per year.

Taking possible imponderability into account, biomass's estimated contribution to the future global energy supply will vary from below 100 EJ per year to above 400 EJ per year in 2050. The annual sustainable bioenergy market is estimated to generate 270 EJ per year (Hall and Rosillo-Calle 1998).

- Despite energy crop potentials, residue potential is seen as an important bioenergy source that could contribute between 60 and 325 EJ per year.
- Land availability, water scarcity and biodiversity concerns are the most important factors influencing bioenergy potential estimates. More realistic studies should focus on these factors.
- Bioenergy production is largely regulated by land availability. Today, only 0.19 % or 25 million hectares of the world's total land area are used for bioenergy production.
- Confirmation is needed of bioenergy production's sustainability. The World Bioenergy Association (WBA) is preparing a report called "Certification Criteria for Sustainable Biomass for Energy". The current book is a best practice example to illustrate a comprehensive pathway to sustainable biomass production in Germany.
- The sustainable use of bioenergy needs a comprehensive assessment and management of natural resources, such as land and water. Each landscape is unique and requires a specific assessment of its inherent biomass potential (see Sects. 4 and 5: Biomass assessment via BETHY in Sect. 4 and BIOSTAR in Sect. 5). Tools for the assessment of an area's biomass potential are already available (e.g., BIOSTAR) as well as comprehensive decision support methods, such as MCDA.

References

Beer, L. L., Boyd, E. S., Peters, J. W., & Posewitz, M. C. (2009). Engineering algae for biohydrogen and biofuel production. *Current Opinion in Biotechnology, 20*(3), 264–271.

Berndes, G., Hoogwijk, M., & van den Broek, R. (2003). The contribution of biomass in the future global energy supply: A review of 17 studies. *Biomass and Bioenergy, 25*(1), 1–28.

BMU. (2012). *Renewable Energy Source Act. Federal Ministry for the Environment, Nature Conservation and Nuclear Safety*. Berlin. Retrieved December 4, 2012, from http://www.bmu.de/files/english/pdf/application/pdf/eeg_2012_en_bf.pdf

Bruinsma, J. (2009, June). *The resource outlook to 2050*. Paper presented at the expert meeting on "How to feed the world in 2050", Rome.

Bundesministerium für Ernährung, Landwirtschaft und Verbraucherschutz (BMELV). (2012). *Ausgewählte Daten und Fakten der Agrarwirtschaft 2012*. Retrieved December 7, 2012, from http://www.bmelv.de/DE/Service/Publikationen/PublikationenLandwirtschaft/publikationen-Landwirtschaft_node.html

Christi, Y. (2007). Biodiesel from micro algae. *Biotechnology Advances, 25*, 294–306.

de Fraiture, C., et al. (2007). Looking ahead to 2050: Scenarios of alternative investment approaches. In D. Molden (Ed.), *Water for food, water for live*. London: Earthscan.

de Fraiture, C., Giordano, M., & Liao, Y. (2008). Biofuels and implications for agricultural water use: Blue impacts of green energy. *Water Policy, 10*, 67–81.

EC. (2005). *Report on the green paper on energy. Four years of European initiatives*. Directorate-General for Energy and Transport, European Commission. Office for Official Publications of the European Communities, Luxembourg. Retrieved December 7, 2012, from www.jet.efda.org/documents/ad-hoc/2005-green-paperreport-en.pdf

Faaij, A. (2008). *"Sustainable Biofuels"*. Paper presented at Rockefeller Bellagio Conference on North-south Biopact, Bellagio, Italy.

FAO. (2008). *Global forest products consumption, production, trade and prices: Global forest products model projections to 2010* (p. 345). Rome: United Nations Food Agricultural Organisation.

FAO. (2011). *The state of land and water resources for food and agriculture (SOLAW) – Managing systems at risk*. London: Earthscan.

Federal Statistical Office. (2012). *Sustainable development in Germany. Indicator report 2012*. Wiesbaden. Retrieved December 7, 2012, from https://www.destatis.de/EN/Publications/Specialized/EnvironmentalEconomicAccounting/Indicators2012.pdf?__blob=publicationFile

Fischer, G., & Schrattenholzer, L. (2001). Global bioenergy potentials through 2050. *Biomass and Bioenergy, 20*, 151–159.

Foley, J. A., Ramankutty, N., Brauman, K. A., Cassidy, E. S., Gerber, J. S., Johnston, M., Mueller, N. D., O'Connell, C., Ray, D. K., West, P. C., Balzer, C., Bennett, E. M., Carpenter, S. R., Hill, J., Monfreda, C., Polasky, S., Rockstroem, J., Sheehan, J., Siebert, S., Tilman, D., & Zaks, D. P. M. (2011). Solutions for a cultivated planet. *Nature, 478*, 337–342.

Fujino, J., Yamaji, K., & Yamamoto, H. (1999). Biomass-balance table for evaluating bioenergy resources. *Applied Energy, 63*(2), 75–89.

Gallagher, E. (2008). *The Gallagher Review of the indirect effects of biofuels productions*. Brighton: Renewable Fuels Agency (RFA).

Gerten, D., Heinke, J., Hoff, H., Biemans, H., Fader, M., & Waha, K. (2011). Global water availability and requirements for future food production. *Journal of Hydrometeorology, 12*, 885–899. http://dx.doi.org/10.1175/2011JHM1328.1.

Hall, D., Rosillo-Calle, F., Williams, R., & Woods, J. (1993). Biomass for energy: Supply prospects. In T. Johansson, H. Kelly, A. Reddy, & R. Williams (Eds.), *Renewable energy: Sources for fuels and electricity*. Washington, D.C: Island.

Hall, D. O., & Rosillo-Calle, F. (1998). Biomass – Other than wood. In *Survey of Energy Resources 1998* (18th ed., pp. 227–241). London: World Energy Council.

Hoff, H. (2011). *Understanding the nexus*. Background paper for the Bonn 2011 Conference: The water, energy and food security nexus. Stockholm Environment Institute, Stockholm.

Hoogwijk, M. (2004). *On the global and regional potential of renewable energy sources*. Dissertation, Utrecht University.

Hoogwijk, M., Faaij, A., Van den Broek, R., Berndes, G., Gielen, D., & Turkenburg, W. (2003). Exploration of the ranges of the global potential of biomass for energy. *Biomass and Bioenergy, 25*(2), 119–133.

Hoogwijk, M., Faaij, A., Eickhout, B., de Vries, B., & Turkenburg, W. (2005). Potential of biomass energy out to 2100, for four IPCC SRES land-use scenarios. *Biomass and Bioenergy, 29*(4), 225–257.

IEA. (2003). *Key world energy statistics*. Paris: International Energy Agency/Energy Statistics Division.

IEA. (2005). *Key world energy statistics*. Paris: International Energy Agency/Energy Statistics Division.

IEA. (2006a). *World energy outlook*. Paris: OECD/IEA.

IEA. (2006b). *Energy technology perspectives – Scenario's and strategies to 2050*. Paris: OECD/IEA.

IEA. (2007a). *Potential contribution of bioenergy to the world's future energy demand*. ExCo: 2007:02, IEA Bioenergy.

IEA. (2007b). *Key world energy statistic*. Retrieved from http://www.iea.org/textbase/nppdf/free/2007/key_stats_2007.pdf

IEA. (2007c). *Renewables in global energy supply*. Retrieved from http://www.iea.org/textbase/papers/2006/renewable_factsheet.pdf

IEA. (2008). *World energy outlook*. Paris: OECD/IEA/IEA, Head of Communication and Information Office.

IFPRI (International Food Policy Research Institute). (2012). *Global hunger index. The challenge of hunger: Ensuring sustainable food security under land, water, and energy stresses*. Retrieved from http://www.ifpri.org/sites/default/files/publications/ghi12.pdf

IPCC. (2000). *Special report on emissions scenarios*. Cambridge: Intergovernmental panel on climate change/Cambridge University Press.

IPCC. (2007). *IPCC fourth assessment report*, Working group III. Retrieved from http://www.ipcc.ch

ITTO. (2006a). *Global study on forest plantations. Private sector investments in industrial plantation in the tropics*. Yokohama: International Tropical Timber Organization (ITTO).

ITTO. (2006b). *Global study on forest plantations: Market study on tropical plantation timber*. Yokohama: International Tropical Timber Organization (ITTO).

Jackson, L. E., Pulleman, M. M., Brussaard, L., Bawa, K. S., Brown, G. G., Cardoso, I. M., de Ruiter, P. C., Garcia-Barrios, L., Hollander, A. D., Lavelle, P., Ouedraogo, E., Pascual, U., Setty, S., Smukler, S. M., Tscharntke, T., & Van Noordwijk, M. (2012). Social-ecological and regional adaptation of agrobiodiversity management across a global set of research regions. *Global Environmental Change, 22*(3), 623–639.

Kappas, M. (2009). *Klimatologie. Klimaforschung im 21. Jahrhundert – Herausforderung für Natur- und Sozialwissenschaft*. Heidelberg: Spektrum Akad. Verlag, Springer.

Ladanai, S., & Vinterbäck, J. (2009). *Global potential of sustainable biomass for energy* (Report 013). Uppsala: SLU, Institutionen för energi och teknik, Swedish University of Agricultural Sciences.

Lashof, D. A., & Tirpak, D. A. (1990). *Policy options for stabilizing global climate*. New York: United States Environmental Protection Agency, Hemisphere.

Le Quéré, C., Andres, R. J., Boden, T., Conway, T., Houghton, R. A., House, J. I., Marland, G., Peters, G. P., van der Werf, G., Ahlström, A., Andrew, R. M., Bopp, L., Canadell, J. G., Ciais, P., Doney, S. C., Enright, C., Friedlingstein, P., Huntingford, C., Jain, A. K., Jourdain, C., Kato, E., Keeling, R. F., Klein Goldewijk, K., Levis, S., Levy, P., Lomas, M., Poulter, B., Raupach, M. R., Schwinger, J., Sitch, S., Stocker, B. D., Viovy, N., Zaehle, S., & Zeng, N. (2012). The global carbon budget 1959–2011. *Earth System Science Data Discussions, 5*, 1107–1157. doi:10.5194/essdd-5-1107-2012.

Lobell, D. B., & Field, C. B. (2007). Global scale climate crop yield relationships and the impacts of recent warming. *Environmental Research Letters, 2*, 014002. 7 pp.

Molden, D., Oweis, T., Steduto, P., Bindraban, P., Hanjra, M. A., & Kijne, J. (2010). Improving agricultural water productivity: Between optimism and caution. *Agricultural Water Management, 97*, 528–535.

Moss, R. H., Edmonds, J. A., Hibbard, K. A., Manning, M. R., Rose, S. K., van Vuuren, D. P., Carter, T. R., Emori, S., Kainuma, M., Kram, T., Meehl, G. A., Mitchell, J. F., Nakicenovic, N., Riahi, K., Smith, S. J., Stouffer, R. J., Thomson, A. M., Weyant, J. P., & Wilbanks, T. J. (2010). The next generation of scenarios for climate change research and assessment. *Nature, 463* (7282), 747–756.

OECD-FAO. (2004). *OECD agricultural outlook: 2004/2013*. ISBN-92-64-02008-X © OECD 2004. Retrieved from http://www.oecd.org/tad/agriculturaltrade/33669817.pdf

Offermann, R., Seidenberger, T., Thrän, D., Kaltschmitt, M., Zinoviev, S., & Miertus, S. (2011). Assessment of global bioenergy potentials. *Mitigation and Adaptation Strategies for Global Change, 16*(1), 103–115.

Rogner, H-H. (2000). Chapter 5, Energy resources. In J. Goldemberg (Ed.), *World energy assessment: energy and the challenge of sustainability* (pp. 135–171). New York: United Nations Development Programme (UNDP)/United Nations Department of Economic and Social Affairs (UN-DESA)/World Energy Council (WEC).

Smeets, E., & Faaij, A. (2007). Bioenergy potentials from forestry in 2050 – An assessment of the drivers that determine the potentials. *Climatic Change, 8*, 353–390.

Smeets, E., Faaij, A., & Lewandowski, I. (2004). *A quick scan of global bioenergy potentials to 2050. An analysis of the regional availability of biomass resources for export in relation to the underlying factors* (Report NWS-E-2004-109). Utrecht: Utrecht University.

Smeets, E., Faaij, A., Lewandowski, I. M., & Turkenburg, W. (2006). A bottom-up assessment and review of global bioenergy potentials to 2050. *Progress in Energy and Combustion Science, 33*, 56–106.

Smeets, E. M. W., Faaij, A., Lewandowski, I. M., & Turkenburg, W. (2007). A quick scan of global bio-energy potentials to 2050. *Progress in Energy and Combustion Science, 33*(1), 56–106.

Statistisches Bundesamt. (2012). *Land- und Forstwirtschaft, Fischerei Bodennutzung der Betriebe (Landwirtschaftlich genutzte Flächen)*, Fachserie 3 Reihe 3.1.2. Retrieved from https://www.destatis.de/DE/Publikationen/Thematisch/LandForstwirtschaft/Bodennutzung/LandwirtschaftlicheNutzflaeche2030312127004.pdf?__blob=publicationFile

Teske, S., Schäfer, O., Zervos, A., Krewitt, W., Simon, S., Pregger, T., & Schmid, S. (2008). *Energy [r]evolution: A sustainable global energy outlook. Greenpeace International, European Renewable Energy Council (EREC)*. Retrieved December 4, 2012, from http://www.energyblueprint.info/fileadmin/media/documents/energy_revolution2009.pdf?PHPSESSID=19ca1307fc8bddb35e37437e7bc7dcac. Cited 14 April 2010-

UNEP. (2009). *Towards sustainable production and use of resources: Assessing biofuels*. UNEP Earthprint, ISBN number: 978-92-807-3052-4

WEA. (2000). *World energy assessment of the United Nations, UNDP, UNDESA/WEC*. New York: UNDP.

WEC (World Energy Council). (2007). Survey of Energy Resources 2007. http://www.worldenergy.org/publications/survey_of_energy_resources_2007/default.asp

WEC. (2004). *New renewable energy sources. A guide to the future*. London: World Energy Council/Kogan Page Limited.

Chapter 4
A Process-Based Vegetation Model for Estimating Agricultural Bioenergy Potentials

Markus Tum, Kurt P. Günther, and Martin Kappas

Abstract We present an approach to estimate sustainable straw energy potentials by means of a modelled net primary productivity (NPP) product validated against empirical data on the managed area and mean yields of the main crops in Germany. We used the Biosphere Energy Transfer Hydrology Model (BETHY/DLR) as a theoretical framework for estimating the NPP of agricultural areas in Germany. The BETHY/DLR was driven by remote sensing data from SPOT-VEGETATION, meteorological data from the European Centre for Medium-Range Weather Forecast (ECMWF) and additional static datasets such as land cover information (GLC2000), a soil map (ISRIC-WISE) and an elevation model (ETOP05). The output of the BETHY/DLR, i.e. the yearly accumulated NPP, was first converted into straw potentials through simple allocation rules (root-to-shoot and yield-to-straw ratios). Thereafter it was converted into energy potentials through species-specific lower heating values. The 2006 and 2007 results were compared with data from the literature. Using this method for estimating sustainable bioenergy potentials, we found good compatibility between the established approaches with only little overestimations (up to 12 %) and high correlations with the R^2 of up to 0.78. Our analysis shows that the presented approach fills an important gap in estimating energy potentials from the modelled NPP. The estimated straw biomass energy potentials play an important role in the sustainable energy debate.

Keywords Bioenergy potentials • Biosphere Energy Transfer Hydrology Model (BETHY/DLR) • NPP • Agriculture • Remote sensing

M. Tum (✉) • K.P. Günther
German Aerospace Center (DLR), German Remote Sensing Data Center (DFD), Oberpfaffenhofen, D-82234 Weßling, Germany
e-mail: markus.tum@dlr.de; kurt.guenther@dlr.de

M. Kappas
Institute of Geography, Department of Cartography, GIS and Remote Sensing, University of Göttingen, Goldschmidtstr. 5, D-37077 Göttingen, Germany
e-mail: mkappas@uni-goettingen.de; mkappas@gwdg.de

H. Ruppert et al. (eds.), *Sustainable Bioenergy Production - An Integrated Approach*,
DOI 10.1007/978-94-007-6642-6_4,

4.1 Introduction

Over the last 60 years, the treatment of straw as a side product of cereal production has changed considerably in many developed countries. This is mainly due to a reduction in field straw burning, which used to be done to fertilise the soil, control pests and avoid nitrogen immobilisation (Børresen 1999). The decreasing demand for straw as bedding litter in feeding lots due to changes in housing systems (Jorden et al. 2008) also resulted in many regions experiencing an increase in the available straw on the fields. However, while leaving straw residues on fields has certain positive effects, like the stabilization of the topsoil, specific crop rotations require their removal (Zebarth et al. 2009). Cropping systems with a high straw supply rate thus offer the possibility of straw removal without changing the soil conditions.

The focus of the current – politically motivated – energy discussion has shifted toward renewable energy sources and the energetic use of agricultural by-products, such as straw. Since no competition with human food is related to its use, it has considerable potential, but is limited by several factors. Two major limiting factors apply to central Europe: animal husbandry and the demand for organic material for the humus balance. Over the last 10 years, several studies have been conducted to assess Germany's total and regional straw potentials (e.g., Gauder et al. 2011; Zeller et al. 2011; Pacan and Dröge 2010; Thrän et al. 2009; Fritsche et al. 2004).

These studies' general approach is to use empirical data on land use and mean yields to estimate the theoretically informed straw and sustainable energy potentials after considering the use competitions. Thus, there is always a spatial limitation of the area on which the empirical data source is based.

Besides these empirical approaches, remote sensing-driven vegetation models, established to assess the carbon uptake of plants, can also provide information on the straw potential, but at a significantly higher – raster-based – resolution. Vegetation models have become an important tool for answering questions on the mechanisms that drive the carbon cycle and the roles of terrestrial carbon sinks and sources (Cox et al. 1999). Models, such as the Biosphere Energy Transfer Hydrology (BETHY/DLR)[1] model, have already been tested to estimate the sustainable energy potentials of forests in Germany (Tum et al. 2011) and have shown reasonably good results. More detailed information on the local availability of straw potentials is needed if a sustainable and cost-efficient use is to be achieved in terms of the current political discussion on renewable energy sources.

The primary objective of this study is to investigate an approach to estimate straw potentials by using the modelled net primary productivity (NPP) from the BETHY/DLR in a 1 km^2 area. Statistical data on the land use and the main crops' yields, which are at Level 3 of the 'Nomenclature des Unités Territoriales Statistiques' (NUTS), are used to calibrate the estimated straw potentials. Specific allocation schemes, such as the root-to-shoot and yield-to-straw ratios, are used to

[1] The BETHY/DLR was originally designed for global applications (Knorr and Heimann 2001), after which Wisskrichen (2005) adapted it for regional use.

estimate the straw potentials. Germany was selected as the test area due to its data availability. Computing time and hard disk storage issues restricted our modelling to 2006 and 2007.

4.2 Model Description

The BETHY/DLR is a special soil-vegetation-atmosphere-transfer (SVAT) model of photosynthesis that takes environmental conditions that affect it into account. SVAT models track the plant-mediated transformation of atmospheric carbon dioxide into energy-storing hydrocarbons, such as sugars; this process is called carbon fixation.

The process of photosynthesis is parameterised following a combined approach based on methodologies by Farquhar et al. (1980) and Collatz et al. (1992). Photosynthetic reactions to dark and light are calculated on the leaf level and treated separately. With this approach, the photosynthesis rate can be limited either by light availability or the carboxylation enzyme Rubisco – the key player in the Calvin cycle, which fixes carbon. Owing to the significant differences between their carbon fixation physiologies, a distinction is made between C3 and C4 plants. –In the BETHY/DLR, C4 plants, such as sugar beet and corn, can fix more atmospheric carbon dioxide at higher temperatures than C3 plants, such as barley and wheat.

To extrapolate photosynthesis from the leaf to the canopy level, the canopy structure and the soil-atmosphere-vegetation interaction is taken into account. The photosynthetic rate of closed and open canopies (forests, shrubs, grassland and crops) depends on the Leaf Area Index (LAI). Self-shading is considered by reducing the photosynthetic rate from the canopy top to the soil by using Sellers's (1985) 'two-flux scheme' with three canopy layers.

In addition to photosynthesis, other energy transfers, such as heat fluxes between vegetation and the atmosphere as well as the cooling effect of evapotranspiration, are taken into account. Furthermore, we consider the soil heat flux and the storage of heat in the canopy. The coupling of these processes is of great importance, since temperature-dependent photosynthesis transforms light energy into chemical energy and finally into carbohydrates by using water and CO_2.

The water cycle is also modelled and included in the interaction scheme. Three reservoirs are considered: soil water, snow, and 'skin' or 'intercepted' water on leaves and other parts of the vegetation, which change in space and time. Soil water is available for vegetation, while evapotranspiration from vegetation and evaporation from soil determine the water loss to the atmosphere. Water limitation is modelled by calculating the demand for evapotranspiration. We do so by using Monteith's (1965) approach, to which we have applied criteria by Federer (1979) which assume that evapotranspiration cannot be greater than the limit determined by the soil water supply and the water uptake of a plant's roots. Thus, when the dynamic interaction of, for instance, the soil water balance and photosynthesis is examined, this reflects the natural behaviour of vegetation, which motivated us to use the SVAT approach.

Using the BETHY/DLR, autotrophic respiration is modelled as the sum of the maintenance respiration and the growth respiration. Maintenance respiration is limited by vegetation-specific dark respiration rates. Growth respiration is assumed to be a constant fraction of the NPP.

The model output of the BETHY/DLR is given as a time series of the NPP in daily steps with a spatial resolution and a projection of the land cover classification. For this study, we used the Global Landcover Classification 2000 (GLC2000) with an area size of 1 km^2.

4.3 Input Data

The inputs for the BETHY/DLR model include two remote sensing datasets derived from SPOT-VEGETATION, meteorological time series data provided by the ECMWF and two static datasets describing the soil type and land elevation.

4.3.1 Meteorological Data

The BETHY/DLR requires a meteorological time series with a temporal resolution of at least once per day. The European Centre for Medium-Range Weather Forecasts (ECMWF) provides the data, which indicate a spatial resolution of $0.25° \times 0.25°$ and a temporal resolution of up to four times per day. The ECMWF INTERIM dataset contains a broad variety of parameters of which air temperature (at 2 m height), wind speed (at 10 m height), soil water content (in the four uppermost layers), cloud cover and precipitation are used. The INTERIM reanalysis combines the meteorological station, satellite and airborne-based measurements. We used these data to calculate the daily mean minimum and maximum temperatures as well as the daily mean cloud cover at three heights. The daily temperatures were adjusted to the 1 km^2 resolution of the model output to compensate for the elevation difference between the ECMWF data and the elevation of each model pixel. We did this by using a 1 km^2 elevation map and the temperature gradient of the international standard atmosphere (−0.65 K per 100 m).

Using Burridge and Gadd's method (1974), we calculated the daily average photosynthetically active radiation (PAR) from the global radiation. We estimated the PAR by using the incident sunlight for the given day and year, limited by atmospheric transmissions, which depend on the degree of cloudiness. The daily average cloud cover was calculated using a weighted sum of each cloud layer. The advantage of this approach is that it produces more accurate results than the direct use of radiation forecast data (Wisskirchen 2005).

Daily volumetric soil water content data were needed to calculate the model's soil water budget. The soil type information was taken from the International Soil Reference and Information Centre-World Inventory of Soil Emission Potentials

(ISRIC-WISE) dataset, which is a harmonization of the global FAO-UNESCO Soil Map of the World (FAO-UNESCO 1974) and is available with a 5 × 5 arc-minutes resolution.

4.3.2 *Remote Sensing Data*

In addition to meteorological data, the BETHY/DLR is driven by two remote-sensing-based datasets. These consist of the LAI time series and a detailed and homogenous land cover/land use product. The LAI time series are used to indicate the phenology of vegetation and are based on the CYCLOPES 10-day composites dataset that the POSTEL (Pole d'Observation des Surfaces continentales par Teledetection) database provides.

For each 1 km^2 pixel, a time series analysis, namely the harmonic analysis, was applied to fill the data gaps and eliminate outliers. The harmonic analysis decomposes a time series into a linear combination of suitable trigonometric functions (sine and cosine oscillations) of particular periodicities. In principal, the power spectrum is deconvolved by iteratively finding and subtracting the highest peak of the time series power spectrum.

The CYCLOPES database also provides land cover and land use information, indicated as the GLC2000 (Fritz et al. 2003; Bartholome and Belward 2005). To derive the GLC2000 land cover classes, the FAO's Land Cover Classification System (LCCS) (DiGregorio and Jansen 2001) was used. The GLC2000 dataset represents the year 2000 and includes 22 different land cover classes. The CYCLOPES dataset was chosen because it is thought to be the most accurate dataset for agricultural areas (Garrigues et al. 2008).

4.4 Energy Potentials

The main objective of this study is to derive sustainable straw energy potentials from the modelled and validated NPP (Tum and Günther 2011) of agricultural areas in Germany, and to compare these with recently published estimates. Straw energy potentials are of considerable importance for the sustainable energy discussion and the development of a sustainable energy policy.

Before the energy content of straw is estimated, the modelled NPP needs to be transferred to dry above-ground biomass. This can be done by using simple crop-specific allocation schemes. Since the GLC2000 only contains information about general land use, an additional dataset, describing the area use and yields of the main crops, had to be implemented in order to differentiate between straw crops, such as wheat and barley, and non-straw crops, such as sugar beet and potatoes. We used empirical data from the German Federal Statistical Office, which conducts a yearly farm structure survey. It contains yield and area use information on the main crops grown in each NUTS 3 region. The NUTS hierarchical spatial classification starts with the member states of the European Community (EU) (NUTS 0), followed by the regions of the EU (NUTS 1), which are separated into basic administrative units (NUTS 2), and ends with the subdivisions of these basic administrative units (NUTS 3). However, a criterion was needed to fill the gaps in the dataset. We thus assumed that the gaps in a given crop could be filled by using the mean yield of the given crop from the German NUTS 3 units.

In a first step, the modelled NPP of the BETHY/DLR was aggregated into NUTS 3 units and compared to the NPP values of each NUTS 3 unit, which, as described by Tum and Günther (2011) were calculated from the empirical data. To calculate the NPP of straw-providing crops (NPP_s) the NPP of non-straw-providing crops (NPP_{ns}) was subtracted from the aggregated modelled NPP per NUTS area (NPP_N). The percentage of land use was taken into consideration as described in the empirical dataset.

$$NPP_s = NPP_N - NPP_{ns} \tag{4.1}$$

The remaining NPP_S was then transferred to above-ground NPP (NPP_a) by subtracting the below-ground NPP part (NPP_b), using crop-specific root-to-shoot ratios:

$$NPP_a = NPP_s - NPP_b \tag{4.2}$$

In a next step, the straw content (NPP_{st}) of NPP_a was calculated by subtracting the yield content (NPP_{yi}), using crop-specific yield to straw ratios:

$$NPP_{st} = NPP_a - NPP_{yi} \tag{4.3}$$

The final sustainable straw potential (S_{pot}) was then calculated by adding non-carbon ($_{non}C$) and water (H_2O) contents to NPP_{st}. We used Gauder et al.'s (2011) empirical factor of 0.29 in respect of the use competitions of the harvested straw.

$$S_{pot} = (NPP_{st} + H_2O + {}_{non}C) \times 0.29 \tag{4.4}$$

In addition, we applied Tum and Günther's (2011) crop-specific root-to-shoot and yield-to-straw ratios, water and non-carbon contents.

The recently available S_{pot} values per NUTS 3 region can be used directly to estimate sustainable straw energy potentials. To do so, species-specific lower heating values (H) are needed to convert dry above-ground biomass into energy. The heating values represent the maximum energy output from burning biogenic solid fuels and are measured in megajoules per kilogram. Since the GLC2000 does not provide any information on crop species, we calculated a mean heating value per NUTS 3 unit (<*H*>). Kaltschmidt and Hartmann's (2001) heating values for rye, wheat, barley and rapeseed straw were used in this study.

We calculated the energy potential (J_n) of each NUTS 3 unit, as shown in Eq. 4.5.

$$J_n = S_{pot} * \langle H \rangle \tag{4.5}$$

In a last step, the energy potentials per NUTS 3 unit were spatially reallocated, using the modelled NPP values. To do so, we assumed that the high NPP values of the model output represented high energy potentials and vice versa. We calculated the energy content (J_i) of each pixel (i), as presented in Eq. 4.6.

$$J_i = \frac{NPP_i}{NPP_N} \times J_n \tag{4.6}$$

With this approach, we assumed that each pixel's percentage of straw-providing crops is similar to that of the full NUTS 3 region.

4.5 Results and Discussion

Figure 4.1 depicts the estimated 2006 and 2007 annual straw energy potentials in Germany in accordance with the study's spatial resolution of 1 km^2. In both years, central Germany was identified as the area with the highest energy potential values. This area is located in the Magdeburger Börde, which is well known for its extensive agricultural use. Other areas, such as the Münsterland in northwest Germany and parts of southeast Germany, show significant variability over the 2 years of observation. Overall, the calculated energy potentials in 2006 were lower than in 2007, which we assume was caused by climate conditions. The mean annual energy potential in 2006 was 0.52 [TJ km^{-2} year^{-1}] with a maximum of 2.85 [TJ km^{-2} year^{-1}]. In 2007, the mean annual energy potential was 0.70 [TJ km^{-2} year^{-1}] with a maximum of 2.75 [TJ km^{-2} year^{-1}]. The total annual estimated energy potential was thus 156 PJ in 2006 and 217 PJ in 2007.

Our estimates agree well with the values of the mean straw potentials reported in the literature (Zeller et al. 2011). Three methods were used to estimate the annual straw potentials in Germany and its 16 federal states. These methods consider the humus balance, which is required for sustainable crop and soil management as well as forming the basis of the direct payment obligation, i.e. the accounting regulation. Depending on the method of estimation, the mean annual energy potentials of 112–186 [PJ year^{-1}] are calculated for Germany by applying a mean heating value H of 14.05 MJ kg^{-1}. The heating value represents dry matter with 14 % moisture.

In addition to the annual sum, we analysed the correlation between the modelled sustainable energy potential of both years and the mean sustainable energy straw potential of each Federal State in Germany as presented by Zeller et al. (2011).

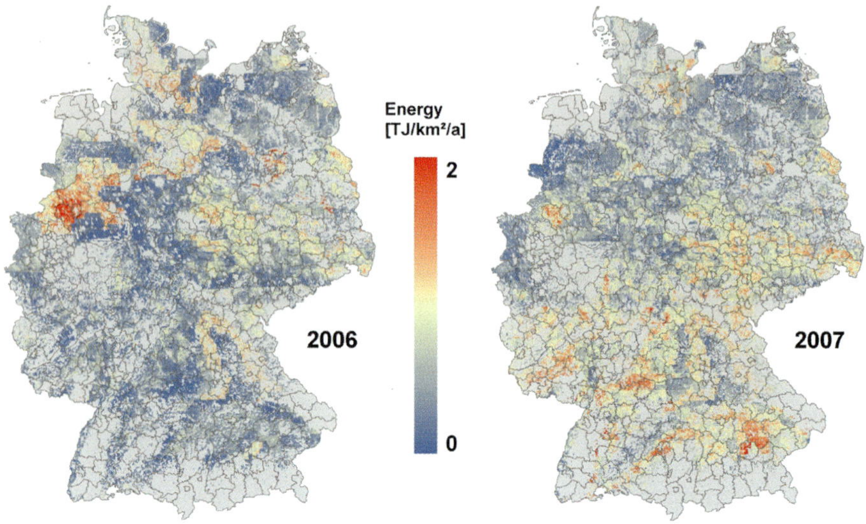

Fig. 4.1 Sustainable energy potential in terajoules per 1 km^2 pixels of agricultural areas in Germany in 2006 and 2007 modelled by the BETHY/DLR. Low energy potentials are indicated in *blue*, intermediate in *beige*, and high energy potentials in *red*. *Grey* represents areas that the GLC2000 has not designated as managed regions

Therefore, the biomass potential was converted into energy potentials, again using a mean *H* value of 14.05 MJ kg^{-1}. The results are presented as linear regressions in Fig. 4.2.

Figure 4.2 shows that, in both years, the sustainable energy potential, which was calculated using the BETHY/DLR model, tended to be slightly overestimated on the Federal State (FS) level. The R^2 values of 0.78 and 0.70 indicate a high degree of correlation between 2006 and 2007. In order to quantify the correlation between our estimations and the literature data, the root mean square error (RMSE) was calculated for both years; the RMSE for 2006 is 3.9 PJ FS^{-1} $year^{-1}$ and, for 2007, it is 6.5 PJ FS^{-1} $year^{-1}$. Figure 4.1 clearly indicates that, in 2006, the sustainable energy potentials for most regions in Germany were lower than in 2007. Our assumption that this finding is related to meteorological conditions is supported by a note in the agro-meteorological bulletin posted by the MARS (Monitoring Agriculture with Remote Sensing) project (MARS 2006). The MARS project characterised 2006 as 'a below-average cereal season explained by hot and dry summer followed by over-wet conditions at harvest'. A mean wheat yield (including soft and durum wheat) of 6.6 tons per hectare was reported for Germany, while the 5-year moving average was 7.4 tons per hectare. This indicates a reduction of about 11 %. On the other hand, barley and grain maize show the same yields as in previous years. Thus, in 2006, Germany's total cereal yield was slightly lower than the 5-year average. In 2007, the cereal production in "Germany was again limited by wet conditions at harvest (winter cereals) but not on the same amplitude as in 2006" (MARS 2007). In 2007, the wheat yield was more or less on par with the

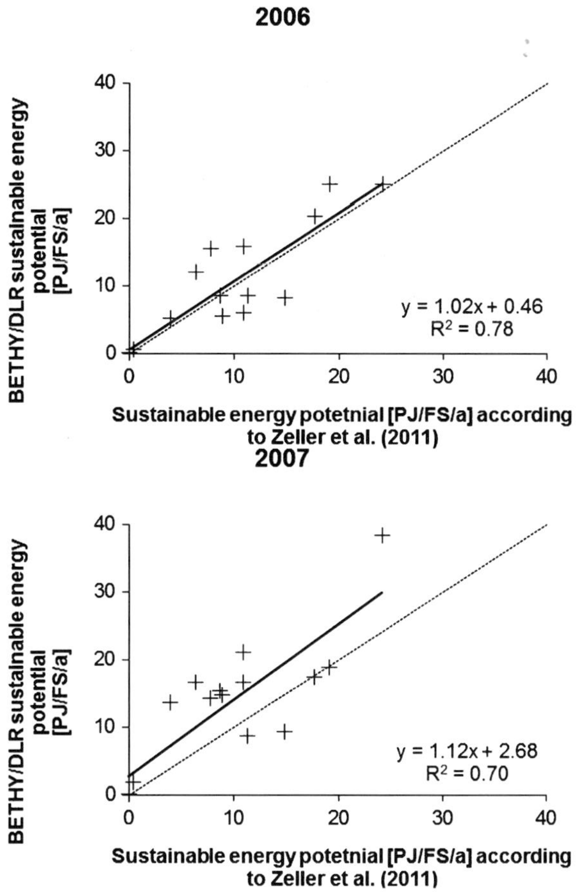

Fig. 4.2 Correlation between sustainable energy potentials of the 16 Federal States of Germany, derived from the modelled NPP with data from Zeller et al. (2011). The 2006 and 2007 energy potentials are modelled. Data points indicate the Federal States' energy potentials. *Dotted lines* indicate a perfect correlation while *solid lines* indicate the correlation found by means of a linear regression. Energy potentials are given in PJ per Federal State (FS) and year

5-year average. The barley yield estimates were about 3 % lower, while that of grain maize was about 8 % higher than the average yields. According to the MARS bulletins, 2007 was overall a more productive year than 2006. Our results support this finding.

Revisiting Fig. 4.1, the Magdeburger Börde can clearly be identified as an area in central Germany with high energy potentials. This is also an area with extensive agricultural use. The ISRIC-WISE dataset reports large amounts of cambisols and chernozems, which are very fertile soils. Areas rich in chernozems, which are among the most fertile soils, are especially sought after as agricultural land. Thus, a constantly high straw potential is expected for areas with these types of soils, as seen in both years.

In northwest Germany, namely the Münsterland, high energy potentials are observed for 2006 and, to a lesser degree, 2007. In the Münsterland, the total

Table 4.1 The precipitation rates of the Münsterland and Landshut areas in millimetres and the 2006 and 2007 mean temperatures in °C

	Münsterland		Landshut area	
10-year mean precipitation sum [mm]	781		779	
10-year mean precipitation sum [mm] for the growing season (15 March–30 September)	432		422	
10-year mean temperature (15 March–30 September) [°C]	15.2		14.7	
	2006	**2007**	**2006**	**2007**
Precipitation sum (1 January–31 December) [mm]	732	831	736	971
Mean temperature (1 January–31 December) [°C]	11.4	11.4	9.4	10.1
Precipitation (15 March–30 September) [mm]	475	496	427	558
Mean temperature (15 March– 30 September) [°C]	14.9	15.4	13.9	15.0

amount of precipitation in 2006 (732 mm) was considerably lower than in 2007 (831 mm) – even lower than the 10-year average (781 mm). A similar precipitation pattern is seen in 2006 and 2007 in the area around Landshut in southeast Germany. In the Landshut area, the precipitation was about 736 mm in 2006 and 971 mm in 2007, as shown in Table 4.1. When discussing the energy potential of straw and biomass development, the most important meteorological parameters are the precipitation and the mean temperature during the growing season. For our analysis, we defined the growing season as the period between 15 April and the end of September. During the growing season, the precipitation in both years was higher than the 10-year average of both regions.

An analysis of the monthly mean temperatures, which we calculated using the daily values taken from the ECMWF data, was performed for both regions to investigate the potential warming or cooling effects on the plant growth and, ultimately, on the straw energy potential. Figure 4.3 presents the time series of the mean monthly temperature of the Münsterland and Landshut areas in both years.

The 2006 and 2007 monthly mean temperatures differed significantly in the non-productive time period (from January to mid-March and from October to December) in both regions. However, there were slight differences in the temperature between the growing seasons (mid-March to September) in the two areas. The mean temperature during the growing period from mid-March to September was lower in the Landshut region in 2006 than in 2007 and equal in the Münsterland region. Compared to the 10-year average, the 2006 mean temperature was lower in both regions. An explanation for the high energy potentials in the Münsterland in 2006 and in the Landshut region in 2007 is found when examining the scatterplot of the mean temperature and precipitation in the growing season from 1999 to 2010, as shown in Fig. 4.4. All the mean values of the meteorological parameters are based on the daily ECMWF data. It is evident that the 2006 growing season was relatively cold and wet in the Landshut region, while the 2007 growing season was relative warm and wet (compared to the 10-year average – shown in Fig. 4.4 as an open

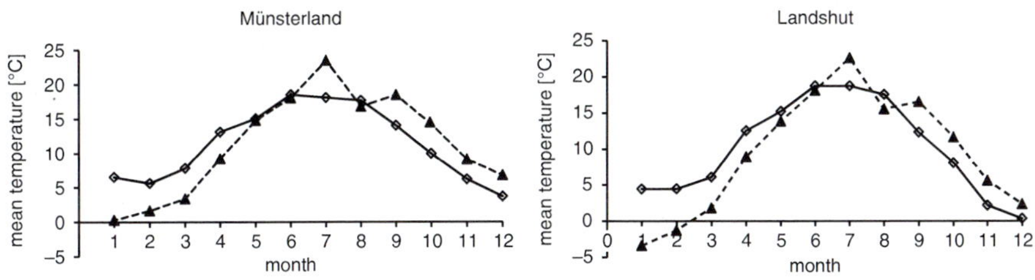

Fig. 4.3 The 2006 (*triangles*) and 2007 (*diamonds*) monthly mean temperatures of the Münsterland (*left*) and Landshut areas (*right*) in °C

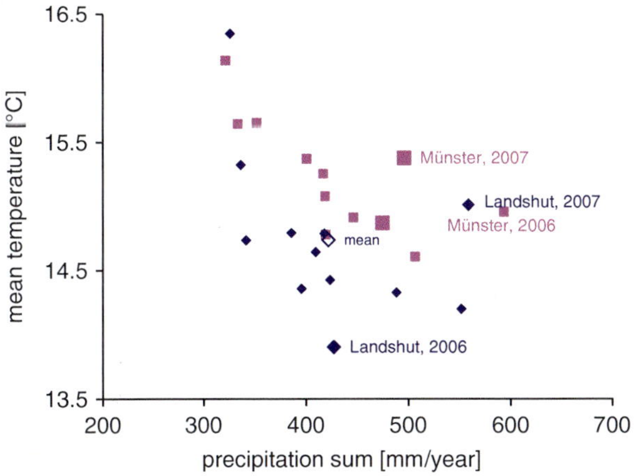

Fig. 4.4 Scatterplot of the precipitation sum [mm] and the mean temperature [°C] of the growing season (15 March–30 September) over 11 years (1999–2010). The Münsterland data are presented as *squares* (*magenta*) while the Landshut region data are indicated by *diamonds* (*blue*). The average values of both regions is presented as *open symbol*

diamond). The 2006 growing season in the Münsterland region was only a little colder and wetter than the 10-year average, but the 2007 growing season was a little warmer and significantly wetter than the 10-year average. In sum, in 2006, the meteorological conditions in the Münsterland region were more favourable than in 2007. In 2006, the cold conditions in the Landshut region reduced the biomass growth and, thus, the energy potential of straw.

The mean yields of the two NUTS 3 units we investigated were derived from an agricultural statistical survey. When the mean yields of the two NUTS 3 units, which are representative of the two described regions, are studies, it becomes evident that our modelled NPP data show lower yields and, thus, lower straw potentials (Table 4.2) for the Landshut region in 2006 and for the Münsterland in 2007.

Table 4.2 Mean 2006 and 2007 yields of important straw-providing crops in two NUTS 3 units in Germany

	Winter-wheat	Rye	Winter-barley	Summer-barley	Oats	Triticale	Other cereals
Landshut							
2006	70.4	54.3	56.8	46.7	49.6	68.8	57.8
2007	78.6	62.1	65.3	51.0	50.0	76.2	63.9
Steinfurt							
2006	68.4	60.4	59.2	44.5	39.7	54.7	54.5
2007	60.7	40.7	48.5	34.3	38.3	47.3	45.5

Steinfurt is representative of the Münsterland area, while Landshut represents the area surrounding Landshut. The values are given in dt ha^{-1}

4.6 Conclusion

Germany's sustainable straw energy potentials in 2006 and 2007 were calculated using the modelled NPP from the BETHY/DLR vegetation model. Inputs for the model were the LAI time series from the VEGETATION satellite, meteorological data from the ECMWF and land cover/land use data from the GLC2000. In this chapter, we presented an approach to estimate sustainable energy potentials by using empirical data on average grain yields and on the acreage of main crops on the NUTS 3 level. Using conversion factors (root-to-shoot and corn-to-straw ratios), the modelled NPP data were converted into harvested straw per NUTS unit. Thus, the NUTS's specific land use practices were taken into account. Compared to recently published straw potential values (Zeller et al. 2011), this method yielded reasonably high coefficients of determination (R^2 up to 0.78), combined with a slight overestimation (up to 12 %), therefore allowing strong conclusions to be drawn about the usability of the presented method.

We indicated the differences between two areas' rate of precipitation and mean annual temperature. We furthermore proved that lower mean temperatures and wet conditions, especially during the growing season, correspond to lower mean grain yields. We hypothesised that significantly cooler mean temperatures during the growing seasons, combined with high precipitation rates, cause yield losses. This phenomenon also corresponds to our calculated sustainable energy potential, which is a good indicator of our method's usefulness.

This study illustrated an approach to calculate sustainable straw energy potentials that we believe will be useful in estimating the energy potentials of the modelled NPP products with a medium resolution. This method could also be used as a downscaling approach to empirically derived straw potential data on a NUTS level, as the model's results could help to spatially represent the NUTS information.

Acknowledgements This study was conducted under the EU FP7 projects EnerGEO (Grant agreement no.: 226364). We would like thank the ECMWF, MediasFrance and the IIASA for providing us with data.

References

Bartholome, E., & Belward, A. S. (2005). GLC2000: A new approach to global land cover mapping from earth observation data. *International Journal of Remote Sensing, 26*, 1959–1977.

Børresen, T. (1999). The effect of straw management and reduced tillage on soil properties and crop yields of spring-sown cereals on two loam soils in Norway. *Soil and Tillage Research, 51*, 91–102.

Burridge, D. M., & Gadd, A. J. (1974). The meteorological office operational 10-level numerical weather prediction model. *British Meteorological Office Tech. Notes Nos. 12 and 48.*

Collatz, G. J., Ribas-Carbo, M., & Berry, J. A. (1992). Coupled photosynthesis – Stomatal conductance model for leaves of C4 plants. *Australian Journal of Plant Physiology, 19*, 519–538.

Cox, P. M., Betty, R. A., Bunton, C. B., Essery, R. L. H., Rowntree, P. R., & Smith, J. (1999). The impact of new land surface physics on the GCM simulation of climate and climate sensitivity. *Climate Dynamics, 15*, 183–203.

DiGregorio, A., & Jansen, L. J. M. (2001). *Land cover classification system (LCCS): Classification concepts and user manual for software version 1.0.* Rome: United Nations Food and Agricultural Organization.

Farquhar, G. D., von Caemmerer, S., & Berry, J. A. (1980). A biochemical model of photosynthesis in leaves of C3 species. *Planta, 149*, 58–90.

Federer, C. A. (1979). A soil-plant-atmosphere model for transpiration and availability of soil water. *Water Resources Research, 15*, 555–562.

Food and Agriculture Organization – United Nations Educational, Scientific and Cultural Organization (FAO-UNESCO). (1974). *Soil map of the world. Volume I: Legend.* Paris: UNESCO.

Fritsche, U. R., Dehoust, G., Jenseit, W., Hünecke, K., Rausch, L., Schüler, D., Wiegmann, K., Heinz, A., Hiebel, M., Ising, M., Kabasci, S., Unger, C., Thrän, D., Fröhlich, N., Scholwin, F., Reinhardt, G., Gärtner, S., Patyk, A., Baur, F., Bemmann, U., Groß, B., Heib, M., Ziegler, C., Flake, M., Schmehl, M., & Simon, S. (2004). *Stoffstromanalyse zur nachhaltigen energetischen Nutzung von Biomasse – Endbericht.* Berlin: BMU.

Fritz, S., Bartholomé, E., Belward, A., Hartley, A., Stibig, H., Eva, H., Mayaux, P., Bartalev, S., Latifovic, R., Kolmert, S., Roy, P., Agrawal, S., Bingfang, W., Wenting, X., Ledwith, M., Pekel, F., Giri, C., Mücher, S., de Badts, E., Tateishi, R., Champeaux, J.-L., & Defourny, P. (2003). *Harmonisation, mosaicing and production of the Global Land Cover 2000 database (beta version).* Luxembourg: EUR 20849 EN/European Commission (EC)/Joint Research Centre (JRC).

Garrigues, S., Lacaze, R., Baret, F., Morisette, J. T., Weiss, M., Nickeson, J. E., Fernandes, R., Plummer, S., Shabanov, N. V., Myneni, R. B., Knyazikhin, Y., & Yang, W. (2008). Validation and intercomparison of global leaf area index products derived from remote sensing data. *Journal of Geophysical Research, 113*, G02028.

Gauder, M., Graeff-Hönninger, S., & Claupein, W. (2011). Identifying the regional straw potential for energetic use on the basis of statistical information. *Biomass and Bioenergy, 35*, 1646–1654.

Jorden, D., Žgur, S., Gorjanc, G., & Štuhec, I. (2008). Straw or hay as environmental improvement and its effect on behaviour and production traits of fattening pigs. *Archiv Tierzucht/Archives Animal Breeding, 51*, 1–11.

Kaltschmitt, M., & Hartmann, H. (Eds.). (2001). *Energie aus Biomasse – Grundlagen, Techniken und Verfahren.* Berlin/Heidelberg: Springer.

Knorr, W., & Heimann, M. (2001). Uncertainties in global terrestrial biosphere modeling, Part I: A comprehensive sensitivity analysis with a new photosynthesis and energy balance scheme. *Global Biogeochemical Cycles, 15*, 207–225.

MARS. (2006). *Agrometeorological bulletin of the MARS project, 14*(6). Retrieved December 7, 2012, from ftp://mars.jrc.ec.europa.eu/Bulletin/Europe/2006/TR2006_V14_6_PRINT.pdf

MARS. (2007). *Agrometeorological bulletin of the MARS project, 15*(6). Retrieved December 7, 2012, from ftp://mars.jrc.ec.europa.eu/Bulletin/Europe/2007/TR2007_15_6.pdf

Monteith, J. L. (1965). Evaporation and environment in the state and movement of water in living organisms. *Symposium of the Society of Experimental Biology, 19*, 205–234.

Pacan, B., & Dröge, S. (2010). *Abschlussbericht zum Forschungsprojekt: Stoffliche und energetische Nutzung von Stroh*. Pirmasens: PFI.

Sellers, P. J. (1985). Canopy reflectance, photosynthesis and transpiration. *International Journal of Remote Sensing, 6*, 1335–1372.

Thrän, D., Bunzel, K., Seyfert, U., Zeller, V., Buchhorn, M., Müller, K., Matzdorf, B., Gaasch, N., Klöckner, K., Möller, I., Starick, A., Brandes, J., Günther, K. P., Tum, M., Zeddies, J., Schönleber, N., Gamer, W., Schweinle, J., & Weimar, H. (2009). *Regionale und globale räumliche Verteilung von Biomassepotenzialen*. Berlin: BBSR.

Tum, M., & Günther, K. P. (2011). Validating modelled NPP using statistical yield data. *Biomass and Bioenergy, 35*, 4665–4674.

Tum, M., Buchhorn, M., Günther, K. P., & Haller, B. C. (2011). Validation of modelled forest biomass in Germany using BETHY/DLR. *Geoscientific Model Development, 4*, 1019–1034.

Wisskirchen, K. (2005). *Modellierung der regionalen CO2-Aufnahme durch Vegetation*. Dissertation, Friedrich – Wilhelm –Universität Bonn.

Zebarth, B. J., Scott, P., & Sharifi, M. (2009). Effect of straw and fertilizer nitrogen management for spring barley on soil nitrogen supply to a subsequent potato crop. *American Journal of Potato Research, 86*, 209–217.

Zeller, V., Weiser, C., Hennenberg, K., Reinicke, F., Schaubach, K., Thrän, D., Vetter, A., & Wagner, B. (2011). *Basisinformationen für eine nachhaltige Nutzung landwirtschaftlicher Reststoffe zur Bioenergiebereitstellung. Schriftreihe des BMU- Förderprogramms "Energetische Biomassenutzung" Band 2*. Berlin: BMU.

Chapter 5
Modelling Site-Specific Biomass Potentials

Roland Bauböck

Abstract During the past few years, increasing energy prices and climate protection policies have boosted the use of renewable energy sources in Germany. In particular, the conversion of biomass from agricultural land into liquid or gaseous fuels is estimated to make up nearly 40 % of the country's future renewable energies mix (BMU 2009 and BMVBS, Globale und regionale Verteilung von Biomassepotentialen. Status-que und Möglichkeiten der Präzisierung. BMVBS-Online Publikationen 27/2010, Ministerium für Verkehr (BRD), 2010). Different methodologies and modelling approaches can be used to estimate or calculate biomass potentials. In this chapter, we describe a method for estimating agricultural biomass potentials, namely the carbon-based crop modelling approach (Azam Ali et al. Perspectives in modelling resource capture by crops. In *Resource capture by crops. Proceedings of the 52nd University of Nottingham Easter School* (pp. 125–148). Nottingham: Nottingham University Press, 1994), and briefly compare it with other methods.

Scientists at the University of Göttingen and the LBEG (Lower Saxony state office of mining, energy and geology) in Hanover developed a carbon-based crop model (BioSTAR) with which to assess site-specific and larger area biomass potentials in Lower Saxony. Using measured agricultural harvest data from a farm in the Wolfenbuettel district (from 2005 to 2008), the first validations of the model have rendered satisfactory results. In respect of sugar beet, winter wheat and maize, the stability index of the modelled yields spans from $R^2 = 0.72$ (s-beet) and $R^2 = 0.82$ (w-wheat) to $R^2 = 0.88$ (maize). In order to further expand agricultural biomass's use in biogas facilities in the administration district of Göttingen, the Jühnde district's biomass potential was calculated using the BioSTAR tool. Depending on the intensity and crop rotation, the Jühnde district's (≈5–7.5 km radius) annual biomass potentials are between 12,935 and 46,306 t of total dry mass.

R. Bauböck (✉)
Institute of Geography, Department of Cartography, GIS and Remote Sensing, University of Göttingen, Goldschmidtstr. 5, D-37077 Göttingen, Germany
e-mail: roland.bauboeck-1@geo.uni-goettingen.de; rbauboe1@gwdg.de

H. Ruppert et al. (eds.), *Sustainable Bioenergy Production - An Integrated Approach*,
DOI 10.1007/978-94-007-6642-6_5,

Keywords Biomass potentials • Agricultural biomass potentials • Crop model • BioSTAR • Energy crops

5.1 Introduction

Owing to an increasing international demand for energy and the rising prices of fossil fuels, biomass fuels (solid and liquid) have been experiencing a renaissance in Germany the last few years. Since 2000, when the German government released the first version of the Renewable Energy Source Act (EEG), the use of non-traditional biomass – i.e. forms of biomass usage other than burning wood in household fireplaces – has increased considerably. Particularly, the use of agricultural biomass for energy production has risen continually since then. The number of biogas production facilities has increased from 1,043 in 2000 to about 5,800 in 2010. Installed electrical power increased from 450 MW in 2000 to 2,400 MW in 2010 (FNR 2011). However, the use of traditional biomass (for small furnaces in private homes) has also increased by 60–80 % (BMU 2009). This development is probably due to the increase in oil prices.

Together with the rapid development of bioenergy facilities, the agricultural area used for renewable resources (energetic and material usage) increased from about 600,000 ha in 2000 to over 2,000,000 ha in 2009. The production of energy crops (mainly maize) for use in biogas facilities covered about 30 % of this area. Another 44 % was used to grow crops for fuel production (mainly oilseed rape) (FNR 2011).

The German government aims to reduce the country's dependence on fossil fuels and to reduce greenhouse gas emissions by 40 % by 2020 (compared to the 1990 emissions). It therefore established the Integrated Energy and Climate Programme (IEKP) and the National Biomass Action Plan, according to which bioenergy's contribution to electrical energy consumption and to the total heat demand should respectively reach 8 and 9.7 % by 2020 (BMU 2009). The share of biofuels is meant to increase to 12 %. An increase in energy efficiency is part of the IEKP in order to achieve these goals. The current primary energy consumption (13.5 EJ in 2009) is projected to drop to about 10.8 EJ in 2020 (an approximate reduction of 2 % p.a., BMU 2007)

Two recent studies (BMU 2009 and BMVBS 2010) have come to similar conclusions – depending on the type of bioenergy production and the efficiency of the resource capture – regarding the potentials of biomass and its contribution to the future total energy production (Table 5.1). If, however, the amount of energy produced from biomass stems from less efficient technologies and the usage thereof (e.g., first generation biofuels or electricity generation without heat utilisation), these potentials will be lower. An energy yield of 180–230 GJ * ha^{-1} * p.a.$^{-1}$ has been assumed for the BMU scenario (Table 5.1). If less efficient bioenergy production strategies are dominant, the yield potential can drop as low as 100 GJ * ha^{-1} * p.a.$^{-1}$.

Table 5.1 Contribution potentials of biomass to the energy demand of Germany

	BMU			
	Peta Joule	**%**	**Peta Joule**	**%**
Forestry	200	16.5	250	14.7
Agr.: Crops	360	29.8	800	47.1
Agr.: Pasture	100	8.3	100	5.9
Waste Materials	550	45.5	550	32.4
Total	**1210**	**100**	**1700**	**100**
	BMVBS			
	Peta Joule	**%**	**Peta Joule**	**%**
Forestry	511	34.1	511	28.4
Agr.: Crops	501	33.4	860	47.8
Agr.: Pasture	100	6.7	100	5.6
Waste Materials	388	25.9	329	18.3
Total	**1500**	**100**	**1800**	**100**

Source: BMU (2009), BMVBS (2010)
Shaded columns are minimum scenarios, non-shaded columns are maximum scenarios

5.2 Modelling Agricultural Biomass Potentials

The history of crop modelling dates back to the late 1960s with the pioneering work of C.T. de Wit and others (Bouman et al. 1996). At that time, computer technology, knowledge of plant physiology as well as the understanding of plant-environment interactions (Monsi and Saeki 2005; Monteith 1977; Mc Cree 1970 to name just a few) had reached a point where the construction of crop models became feasible.

The objective of these first modelling attempts was to attain a better understanding of the underlying physiological processes on a crop scale and to describe agro-ecological systems (van Ittersum et al. 2003). Since its start, the modelling of agricultural crops has become an important tool in the fields of research, education and farming. The application of models can lead to the more effective use of existing knowledge and thus support plant breeding, the training of agriculturalists at universities and more efficient crop production in general (Penning de Vries et al. 1989).

In the scientific community – mainly in the US, Europe and Australia (Bouman et al. 1996) –several teams have developed crop models (Whistler et al. 1986). Moreover, new models appear regularly in the literature (Brisson et al. 2003). Whistler et al. (1986) and van Ittersum et al. (2003) provide a broad overview of existing crop models.

With so many models available, one might question why there is a need for yet another model. However, as Sinclair and Seligman (1996) argue, no single model can be comprehensive enough or sufficiently adequate to serve a modeller's every need. Different demands and the lack of data on the spatial scale, input/output variables and the user group for a model, often make building user and scale-adapted models necessary.

As a general rule, crop models can be classified as empirical, semi-empirical or mechanistic. The mechanistic, simulation-type model is also called an explanatory model (Penning de Vries et al. 1989). Models can be further distinguished in accordance with their main features, i.e. whether they are dynamic or static as well as whether they are deterministic or stochastic (Azam-Ali et al. 1994).

As the empiricism in a model decreases and the number of mechanistic functions increases, the model becomes increasingly comprehensive and its capabilities broaden from a simple summary of data to the interpretation of experimental results. Depending on its number of mechanisms, a model is ranked hierarchically according to the working-level of detail (Whistler et al. 1986). This may entail, for instance, the spatial level of analysis (ecosystem, crop, cell or molecule), the time frame (years, months, days or hours) and the database (field experiments or laboratory experiments).

Penning de Vries et al. (1989) note that comprehensive models are often unwieldy and unsuitable for use outside the research group that developed them. Moreover, they argue that summary models (abstracts of comprehensive models) are a better choice as far as educational use and applicability are concerned.

5.2.1 General Crop Model Functions and Processes

All crop models aim to simulate the resource capture of the crops or individual plants and to generate one or more output variables, which are generally the harvestable yield of the crop or components thereof. At the heart of any model's functions is a "growth engine", which calculates the assimilate turnover through the process of photosynthesis (Azam-Ali et al. 1994). Any plant or crop not growing in a controlled or unnatural environment is part of the soil-plant-atmosphere continuum (SPAC) (Philip 1966) and is therefore subject to soil or atmospheric resource limitations.

De Wit proposed a classification of crop production systems based on growth-limiting factors to account for the limitations in modelling plants' resource capture (Penning de Vries et al. 1989). Crop production at level one implies that ample water and nutrients are available for the crops and that these crops produce a higher yield than at any other level. At the second production level, growth is limited by water shortages during at least a part of the growing season. At production level three, nitrogen becomes a limiting factor for at least part of the growing season and water shortage or poor weather for the remainder thereof. At the fourth production level, there is a shortage of phosphorus and other mineral nutrients and the growing season often lasts less than 100 days.

The atmospheric limitations to crop growth are primarily the availability of water (precipitation), the temperature and radiation, and, secondarily, the prevailing wind speeds and the vapour pressure deficits. Crop models also take another important factor into account, namely that green plants reduce CO_2 and other substrates during the photosynthesis process, incorporate the assimilated substance into new plant structures and maintain themselves as a living unit (Loomis and Amthor 1999).

5.2.2 *The Solar Engine*

A central function of every crop model is to simulate the process of photosynthesis in one way or another. Leaf photosynthesis is simulated by using genotype-specific coefficients for the maximum photosynthetic rate (A_{max} in mmol $CO_2{}^{*}$ m^{-2} s) and the quantum efficiency of the species (measured in mmol CO_2 per mol radiation received).

The fraction of radiation that is photosynthetically active or usable for plants (~400–700 nm) is called photosynthetically active radiation (PAR) and amounts to about 50 % of the total incoming shortwave radiation (Sinclair and Muchow 1999). Leaf photosynthesis can be calculated with these two base parameters (coefficients for the maximum photosynthetic rate and quantum efficiency) and knowledge of the incident radiation. The leaf area index of the crop and the crop-specific light attenuation coefficient have to be known to extrapolate the leaf photosynthetic rate to the whole canopy. As a last step, the gross canopy assimilation can be predicted by means of mathematical functions (Boote and Loomis 1991).

To accurately model CO_2 assimilation, respiration (maintenance and growth) has to be accounted for. Different species' respiration rates vary and are influenced by temperature and the development stage of the crop (Penning de Vries et al. 1989). Amthor (1989) provides a listing of the respiration coefficients of several crop species. This approach to modelling resource capture is called carbon-based modelling.

A simpler way of simulating a crop's dry matter accumulation is to employ the radiation use efficiency (RUE) coefficient of the species. This way of accounting for biomass accumulation is called the Monteith approach (Monteith 1977). Crop modellers often use these coefficients to demonstrate dry matter accumulation's dependence on incoming solar radiation by means of the linear regression method.

Since the accumulation of crops' dry matter is also dependant on climatic variables other than radiation, for example, vapour pressure, and on water availability, the transfer of RUE values to different locations can be problematic and can lead to a false simulation of dry matter accumulation. The problems associated with using crop-specific RUE values have been extensively discussed in the literature (Demitriades-Shaw et al. 1992, 1994; Arkebauer et al. 1992; Monteith 1994; Loomis and Amthor 1999) and have, to date, not been entirely rejected.

5.2.3 *The Water Engine*

The transpiration function is closely related to the photosynthesis process. Under production level one, a crop's water consumption would not need to be modelled at all. Owing to soil water retention, water can be a limiting factor for growth even in humid or sub-humid climates.

To assess the soil-water balance during the growing period and especially in those months with high evaporative demand, crop models need to take evapotranspiration (evaporation plus transpiration) into account. The Penman-Monteith method is a widely used approach to calculate potential evapotranspiration (Allen et al. 1998). Furthermore, in order to account for water loss due to runoff and interception evaporation, a crop's daily water balance and that of the soil profile underneath it need to be calculated.

Depending on the availability of water through precipitation and the soil reservoir, the actual rate of evapotranspiration can fall below its calculated potential. To account for this situation, a layered soil sub-model is needed to calculate the water extraction by the root system.

The pf curve and the water balance of each soil layer represent the input data for water movement in the soil. Model-generated values of root depth and root distribution as well as information on root-length and root-weight ratios are needed to calculate the actual water extraction by the roots.

Another approach to modelling crops' biomass accumulation uses a water productivity (WP) coefficient. Like the RUE-based models, these models avoid the more complex algorithms for converting radiation interception and CO_2 assimilation and dissimilation; they may therefore require a smaller number of crop-specific input variables. Water productivity models are based on the observation that biomass accumulation in a crop is linearly proportional to the water transpired over a given time period (Todorovic et al. 2009). Another advantage of WP-based models is that they can be normalised according to the climate (vapour pressure and CO_2) and, unlike radiation-driven models, can therefore be transferred to different climate environments more easily (Steduto et al. 2007). CropSyst (Stöckle et al. 2003) and AquaCrop (Steduto et al. 2009) are examples of models that use WP coefficients for substance accumulation.

5.3 The Crop Model BioSTAR

5.3.1 *Model Development Objectives*

The model presented here was designed to be application oriented and to serve as a decision support tool for the economic and ecological optimisation of bioenergy cropping systems in Lower Saxony. Lower Saxony yields are modelled with the current, long-term and projected future climate data. Changes in the atmospheric CO_2 concentration are considered for model runs with the projected climate data.

The model was developed as part of the research project "Sustainable use of Bioenergy – bridging climate protection, nature conservation and society". The yield potentials that the model generates are used as a database for economic and ecological planning in three test districts. In each district, modelling is supported by close collaboration with a cooperating farm.

On the cooperating farms, new cropping rotations and crops were tested as part of the field research trials (see Chap. 6). Weather data from the adjacent climate stations (Liebenburg, Seesen, Celle, Goslar and Pabstorf) and from a weather station at the Goslar test site (the administration district) were used to calibrate the model. The model's soil input data were taken from the Lower Saxony Soil Information System (NIBIS®) on a scale of 1:5,000.

In the first development phase, the nitrogen dynamics and vertical water movement in the soil were ignored. The model was designed to simulate crop growth at the first or second production level (see above), depending on the user specifications. Since the regionalised input climate data are an aggregation of data obtained over more than 40 years, the administration districts' modelled yields are average potentials within the climate and soil's given production limitations. Weather station data (point data) were used to calculate the yield generation on the cooperating farms and to calibrate the model, thus allowing individual years to be modelled. The location-specific attainable yields from various crops and cropping systems could subsequently be used as a basis for economic cost calculations as well as regional or ecological planning.

The modelling objective of the BioSTAR development stage is to generate site-specific maximum attainable yields reflecting the variations in soil and climate and given the best farming practices with the current agricultural technology.

5.3.2 Description of the Model

The algorithms on which the model is built can be roughly divided into three main groups. The first set of algorithms keeps track of CO_2 flows and radiation utilisation during the assimilation and respiration processes. The second set of algorithms is linked to the photosynthetic process and the crops' water usage, which – depending on the water availability (soil moisture and precipitation) and water demand (evapotranspiration) – it regulates. The third set contains those equations required to define the crop's development speed and development stage, the development of the leaf area, the expansion of root growth and the distribution of plant biomass in roots, grain and straw.

The BioSTAR model's prototype, which was used to calculate triticale and maize yields in the Göttingen district (Bauböck 2009), used the RUE coefficient approach for the crop-specific computation of photosynthesis and respiration. Using RUE coefficients in a crop model reduces the number of crop-specific input variables. However, these coefficients might have to be adjusted when the model is used in climates that differ from the one in which the RUE values were originally measured. Especially the influence of temperature on photosynthesis and the effect of water

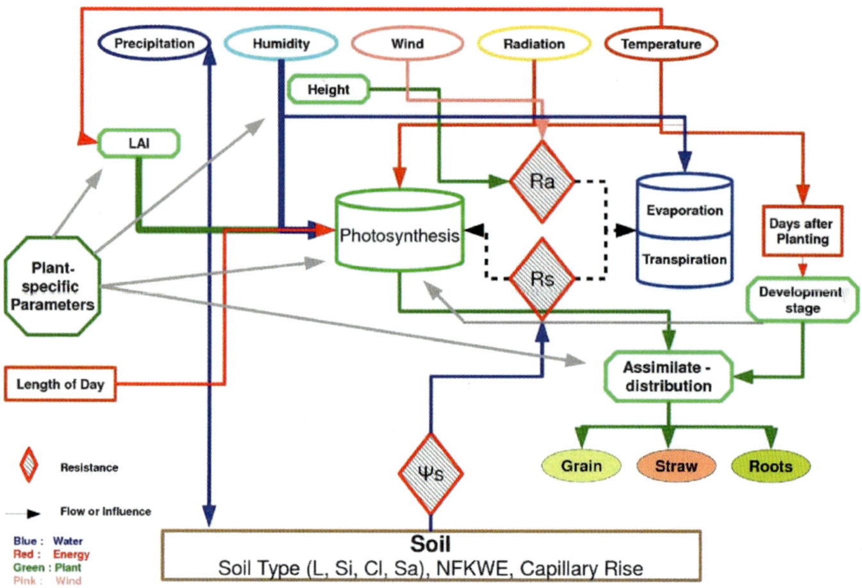

Fig. 5.1 Flow chart of the BioSTAR model. Abbreviations: *L* loam, *Si* silt, *Cl* clay, *Sa* sand, *NFKWE* usable field capacity in the rooted zone, *Rs* stomata resistance, *Ra* aerodynamic resistance, ψs soil matric potential, *LAI* leaf area index

vapour pressure (atmospheric water demand) and soil moisture content (water supply) can be difficult to model with a RUE coefficient (Todorovic et al. 2009).

With these problems in mind, the successor of the prototype model was developed as a carbon-based model (Azam-Ali et al. 1994). This more complex approach to modelling photosynthesis made it is possible to incorporate several functions into the model. These functions are used to either lower or increase the photosynthetic rate in response to the limiting factors in the SPAC. Figure 5.1 presents a flow chart of the model, thus providing an overview of the different components and their interconnections.

5.3.3 *Modelling Photosynthesis, Leaf Area and Transpiration*

To keep track of a crop's photosynthetic rate (abbreviated to PR in the following), the amount of the PAR present per square meter is fed into a crop-specific radiation utilization curve. The output value of the curve is the gross CO_2 assimilation rate.

By using temperature-dependent input parameters for this curve (initial light use efficiency and the maximum point of the PR), it is possible to continually adjust the PR to the temperature in the model. Using additional functions, we can model

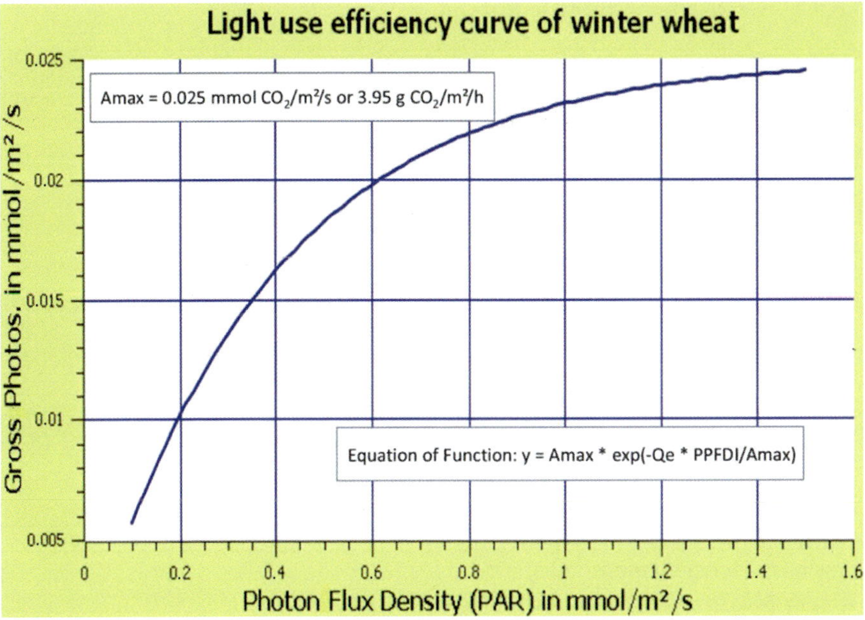

Fig. 5.2 Light use efficiency curve of winter wheat (Source: Data used in the BioSTAR model)

the influences of water stress, vapour pressure and nitrogen availability on the PR. The functions used in the BioSTAR model are either linear or exponential, depending on the effect they are mimicking.

Since this chapter only intends to provide an introductory description of the model, the following equations offer a brief overview of its important functions. The first three equations are common to many crop models, whereas Eqs. 5.4, 5.5, 5.6, 5.7, 5.8, 5.9 and 5.10 were specifically chosen for the BioSTAR model. In Fig. 5.2, the light use efficiency curve for winter wheat is shown. The x-vector plots the PAR in mmol $^* \mathrm{m}^{2-1} \, ^* \mathrm{s}^{-1}$ and the y-vector displays the corresponding gross CO_2 uptake rate in mmol $^* \mathrm{m}^{2-1} \, ^* \mathrm{s}^{-1}$. The flattening of the curve at 1.4 mmol $^* \mathrm{m}^{2-1} \, ^* \mathrm{s}^{-1}$ indicates the species' light saturation point.

5.3.3.1 Light Interception of the Crop

$$\mathrm{PPFDI} = \mathrm{PPFD} \, ^* \left(1 - \exp^{(-k^* \mathrm{LAI})}\right) \tag{5.1}$$

Where:
PPFDI = Intercepted photosynthetic photon flux density
PPFD = Photosynthetic photon flux density
k = Crop-specific light attenuation coefficient
LAI = Leaf area index

5.3.3.2 Gross Photosynthetic Rate (A_g)

$$A_g = A_{max}{}^* \ (1 - exp^{(-Qe^*PPFDI/Amax)} \tag{5.2}$$

Where:
Qe = Initial light use efficiency (quantum efficiency)
PPFDI = Intercepted photosynthetic photon flux density
A_{max} = Maximum photosynthetic rate (crop specific)

5.3.3.3 Temperature Influence on A_{max} (e.g., on Triticale)

$$A_{maxT} = \left(10^{-5} + (0.13332 \ ^* \ Tcel)\right) \ ^* \ A_{max} \tag{5.3}$$

Where:
A_{maxT} = Temperature dependent A_{max} value
Tcel = Air temperature in °C
A_{max} = Maximum photosynthetic rate (crop-specific)

5.3.3.4 Stomata Resistance

$$Rs = 2{,}815 \ ^* \ exp^{(-3.13^*RSMP)} \tag{5.4}$$

Where:
Rs = Stomata resistance in s m^{-1}
RSMP = Relative soil matrix potential (value from 0 to 1, corresponding to 1.5 to 0.01 MPa)

5.3.4 Development of Leaf Area in the Course of the Vegetation Period

The leaf area indices (LAIs) of different crops and stages of development can vary greatly (Lindquist et al. 2005; Steduto and Albrizio 2005; Wang 2001; Garcia et al. 1988). According to Gardener et al. (1985), an optimum LAI (>95 % light interception) is achieved at approximately or slightly below five, which means that, for each square meter of ground area, there is a corresponding leaf surface (upper side) of 5 m^2.

The LAI development of a crop is governed by several factors that influence the temperature. Therefore, leaf area expansion and water availability can easily be

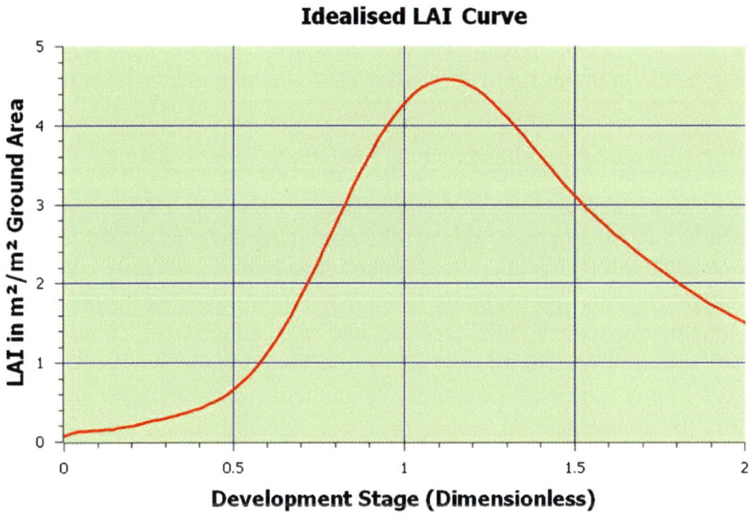

Fig. 5.3 Idealized leaf area index curve

modelled in a crop model (Mailhol et al. 1997; Teruel et al. 1997). Row spacing can also influence the LAI of a crop, but, in order to reduce complexity, it is always assumed to be optimal in the BioSTAR model.

If the temperatures are below a crop-specific optimum, crop development and leaf area growth are delayed. The same can be said for water stressed crops, at least up to the point of anthesis. After anthesis, water stress can induce a faster senescence of the leaves and a premature deposition of assimilates in the reproductive organs (e.g., grains). An optimum LAI curve (polynomial, 5th degree) is used (Fig. 5.3) to mimic the leaf area expansion and decline processes.

In the model, this optimum LAI curve is adjusted with each calculation step using two further influencing variables. The first variable is the dimensionless development stage of the crop (Penning de Vries et al. 1989), which is, in turn, governed by the ambient air temperature and is controlled by crop-specific coefficients for the period before and after the point of anthesis. The second variable is the summarised quotient of the possible-to-actual photosynthetic rate of the crop (the mean PR), which is influenced by the availability of water (Eq. 5.5).

$$PR_{mean} = \Sigma\left\{ PR_{stcon}/\ PR_{pot}\right\}/NR_{days} \tag{5.5}$$

Where:

PR_{mean}: mean photosynthetic rate over the course of the crop development
PR_{stcon}: stomata-conductance-regulated photosynthetic rate
PR_{pot}: potential photosynthetic rate (non-regulated)
NR_{days}: number of growing days

The function PR_{mean} renders the value in percentage, which is then used to regulate the LAI development. Before anthesis, low PR_{mean} values lead to a lowering of the optimum LAI curve; after anthesis, they induce a faster decline of the LAI.

An important variable for any crop model is the rate of transpiration, mainly via the leaves. Ground evaporation is usually high when the crop is at the beginning of its growing cycle and declines exponentially as the leaf area increases (Merta et al. 2006; Ritchie 1972). Merta et al.'s (2006) equations have been used to model the LAI-dependent share of evaporation. Since transpiration is directly connected to a crop's photosynthesis (gas exchange through the same stomata), the computation of water vapour movement from leaves into the atmosphere (transpiration) is modelled with a direct link to crop's CO_2 exchange rate in the BioSTAR model. In doing so, BioSTAR uses the respective gradients of H_2O vapour and CO_2 from the leaf to the atmosphere.

In a first step, the crop-specific internal-to-external ratio of CO_2 in the leaves (Ci) is adjusted with the quotient of the possible-to-actual PR (stomata-induced reduction) in order to link the photosynthetic rate to transpiration (Eq. 5.6). This mimics an effect that has been observed in plants. When the photosynthetic rate drops, the internal or external CO_2 ratio of the leaf is lowered accordingly and the CO_2 gradient rises.

In a second step, the H_2O gradient from the leaf to the atmosphere is calculated using the vapour pressure deficit of the air (demand in grams), the mole fraction of 1 g of dry air and the molecular weight of water (Eq. 5.7).

Thirdly, a water use coefficient is calculated by multiplying the quotient of the H_2O gradient and the CO_2 gradient (Eq. 5.8) with the factor 1.56 (relative diameter of a water molecule to the diameter of a CO_2 molecule) (Eq. 5.9). In this equation, two processes can be mimicked. The first process is an increasing transpiration rate with an increasing vapour pressure deficit of the air. The second is a lowering of the transpiration rate with an increasing CO_2 gradient, caused by a lower internal/external CO_2 ratio. In this way, water stress will reduce the modelled crop's water demand and the water usage of species that, in general, require less water for dry matter production – such as plants with a C4 metabolism – can be modelled adequately.

In a final step, the PR-dependent water usage is calculated by multiplying the water use coefficient with the gross CO_2 assimilation rate (Eq. 5.10).

5.3.4.1 Internal/External CO_2 Ratio

$$\mathrm{IntExt} = Ci^{*}\ (\mathrm{PR}_{\mathrm{stcon}}/\mathrm{PR}_{\mathrm{pot}}) \tag{5.6}$$

Where:

IntExt: The PR-adjusted internal/external CO_2 ratio (linear modulation)
Ci: The crop-specific internal/external CO_2 ratio
PR_{stcon}: The stomata-conductance-regulated photosynthetic rate
PR_{pot}: The potential photosynthetic rate (non-regulated)

5.3.4.2 H_2O Gradient

$$H_2O_{grad} = ((H_2O_{dem} {}^* Mol_{air})/18)^* 1,000 \quad (5.7)$$

Where:
H_2O_{grad}: water gradient from leaf to air
H_2O_{dem}: water demand of the air (saturation deficit) in $g^* \ m^{-3}$
Factor 18: molecular weight of water
Factor 1,000: conversion from mol to mmol

5.3.4.3 CO_2 Gradient

$$CO_{2grad} = (390 - (390^* IntExt))/1,000 \quad (5.8)$$

Where:
CO_{2grad}: CO_2 gradient from leaf to atmosphere
Factor 390: CO_2 content of air (ppm)
Factor 1,000: conversion from μmol to mmol
$Mol_{air})$: mole fraction of 1 g of dry air

5.3.4.4 Water Use Coefficient

$$Use_{coeff} = (H_2O_{grad}/CO_{2grad})^* 1.56 \quad (5.9)$$

Where:
CO_{2grad}: CO_2 gradient from leaf to atmosphere
H_2O_{grad}: water gradient from leaf to atmosphere
Factor 1.56: relative diameter of a water molecule to the diameter of a CO_2 molecule

5.3.4.5 Actual Water Use

$$H_2O_{use} = \left(PR_{gross} {}^* 44^* 3,600/1,000^* L_{day}/1,000\right)^* Use_{coeff} \quad (5.10)$$

Where:
H_2O_{use}: actual water usage in $l\ m^{-2}$
PR_{gross}: gross CO_2 assimilation rate

L_{day}: length of day (daylight)
Use_{coeff}: water use coefficient (ratio of transpiration to CO_2 assimilation)
Factor 44: mol weight of CO_2
Factor 3,600: conversion of seconds to hours
Factors 1,000 (2×): conversion of mmol to mol and mg to litres

5.3.4.6 Respiration

Following Amthor's (1989) and Choudhury's (2000 and 2001) approach, the respiration of assimilates in the processes of maintenance and growth respiration as well as the net primary production (NPP) is computed using the crop-specific coefficients Rm and Y_G (see Eqs. 5.11, 5.12, 5.13 and 5.14).

$$Rm = 2.1^{*}10^{-4*}(nf + ns + nst + 2nr) \tag{5.11}$$

$$Rm(T) = Rm(T = 20)^{*}2^{((T-20)/10))} \tag{5.12}$$

$$R_{TOT} = (1 - Y_G)^{*}A_g + Y_G{}^{*}Rm(T) \tag{5.13}$$

$$NPP = A_g - R_{TOT} \tag{5.14}$$

Where:
Y_G: growth conversion efficiency
A_g: gross CO_2 assimilation in mol * m^{-2} * day^{-1}
Rm: maintenance respiration coefficient
Rm(T): temperature-dependent maintenance respiration coefficient
R_{TOT}: total respiration (maintenance and growth)
nf, ns, nst, nr: nitrogen contents of foliage, stem, storage and roots in mmol * m^{-2}

The value attained after the respiration costs have been subtracted corresponds to the net primary production (dry mass NPP) measured in gram per square meter. The above-described equations can be used to estimate two important processes, namely the gross and net uptake of CO_2 and the loss of water to the atmosphere in the process of transpiration.

5.3.5 Model Calibration

Model calibration requires good datasets of yield and weather data. Without accurate and site-specific measurements of grain or fresh weight yields, the modeller can easily mistake these measurement errors for mistakes in the model and will try to adjust the model to fit the false input data. The data best suited for the calibration of a model stem from controlled field trials, including weather measurements over the course of (at least) the vegetation period.

One of the datasets used for the initial calibration of BioSTAR was taken from site-specific biomass (fresh weight and grain) yields of a farm in the administration district of Wolfenbuettel in Lower Saxony, Germany between 2005 and 2008. The crops cultivated on these sites included winter wheat ($n = 51$), maize ($n = 22$) and sugar beet ($n = 30$). The agricultural plots on which the crops were grown consist of a relatively wide spectrum of soil surface types, including clays, loams, sandy loams, silt loams and peat sites (lowland moor).

The usable field capacity of each site's rooted zone is the key variable of the model (nFKWe → nutzbare Feldkapazität im effektiven Wurzelraum). These data are available on Lower Saxony's soil information system (NIBIS® → Niedersächsisches Bodeninformationssystem). The sites' nFKWe values, used for model calibration, range from $100\ mm\ ^{*}\ m^{-2}$ to $270\ mm\ ^{*}\ m^{-2}$ of water content at field capacity. In addition to the nFKWe, some of the sites' ground water tables reach less than 2 m from the ground surface. Thus, capillary rise into the rooted zone is possible at these sites.

The climate data used to feed the model were taken from two different sources. The first is a regionalised (areal data) dataset of long-term (1960–2000) monthly mean climate values of the area surrounding the farm sites (NIBIS® data). The second dataset is a record of daily weather data from the DWD station in Hildesheim (point data). To adjust the regionalised dataset to the individual years from 2005 to 2008, its mean values (1960–2000) were compared with the long-term means of the Hildesheim station. The discrepancy between Hildesheim and the areal data was then used to modify the Hildesheim datasets of the individual years in order to adapt them to the farm site's climatic situation. Figures 5.4, 5.5 and 5.6 show the radiation, precipitation and temperature curves for the farm in Wolfenbüttel from 2005 to 2008.

5.3.6 *Interpretation of Weather Data*

5.3.6.1 Radiation

When examining the radiation curve for 2008 (Fig. 5.4), there was clearly a good surplus of incident solar radiation from April until July. From June 2005 until July 2005, the radiation level fell below that of the other 3 years.

5.3.6.2 Precipitation

When comparing all four precipitation curves (Fig. 5.5), 2005 and 2006 show a sharp drop from May until mid-June. In 2008, a similar drop can be seen reaching a minimum in May. This drop was also prevalent in 2007, but again shifted one

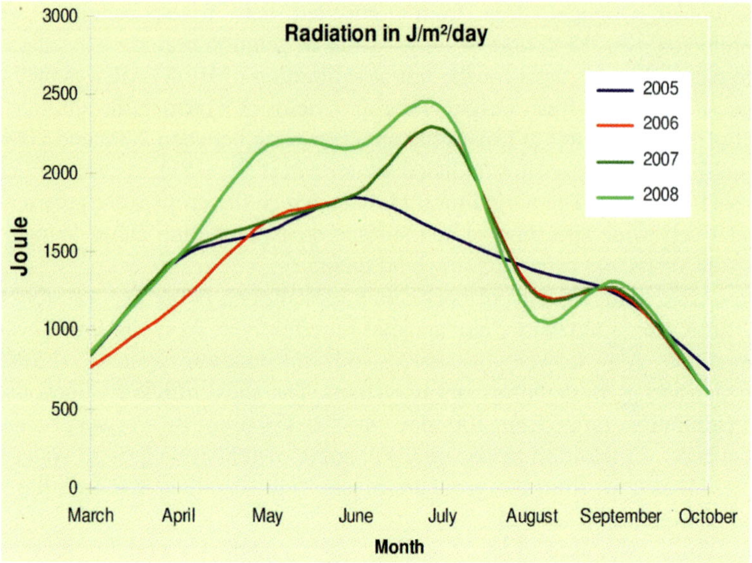

Fig. 5.4 Radiation curve (global) in J * m^{-2} *day^{-1} of farm plots in Wolfenbuettel during 2005–2008

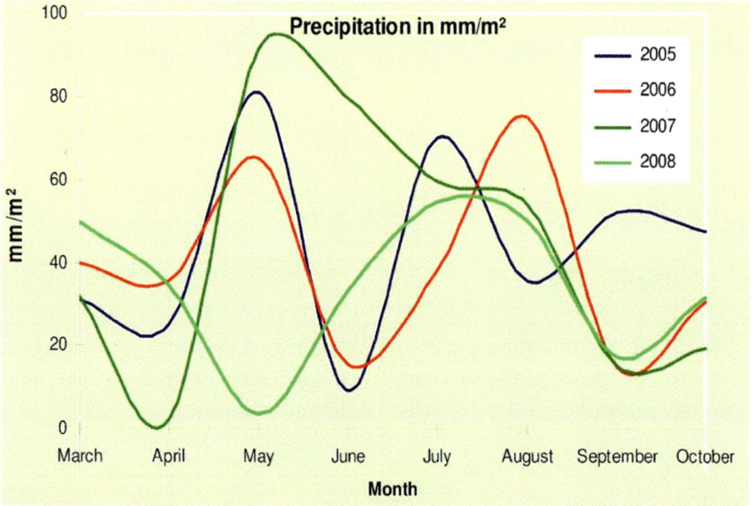

Fig. 5.5 Precipitation curve in mm of farm plots in Wolfenbuettel during 2005–2008

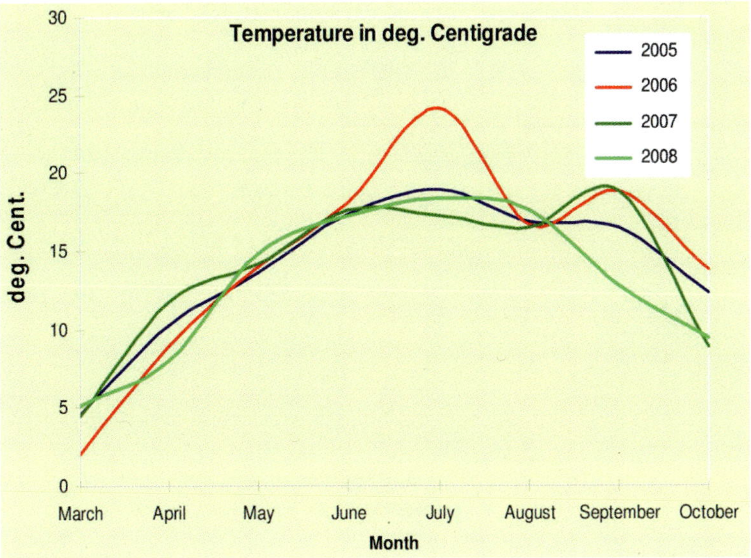

Fig. 5.6 Temperature curve in °C of farm plots in Wolfenbuettel during 2005–2008

month earlier to April. Apart from the drop in precipitation, 2007 had a good surplus of precipitation from May to mid-June.

5.3.6.3 Temperature

When comparing all four temperature curves (Fig. 5.6), 2006 displays elevated early to mid-summer temperatures (June–July), which are several degrees higher than those in the other 3 years. Both 2007 and 2008 had lower than average autumn temperatures, dropping below 10 °C in October in both months.

5.3.7 Interpretation of Yield Data

5.3.7.1 Interpreting Maize Yields in the Context of Climate and Field Capacity

In 2005 and 2006, the maize yields (modelled and actual) were at a high level with fresh weight yields of 610–621 dt ha^{-1} (deci-tonnes per hectare) for soils with average field capacities of about 200 mm (Fig. 5.7). In 2007 and 2008, the maize yields (modelled and actual) dropped below 600 dt ha^{-1} (587–594 dt ha^{-1}) for soils with average field capacities of 217 mm.

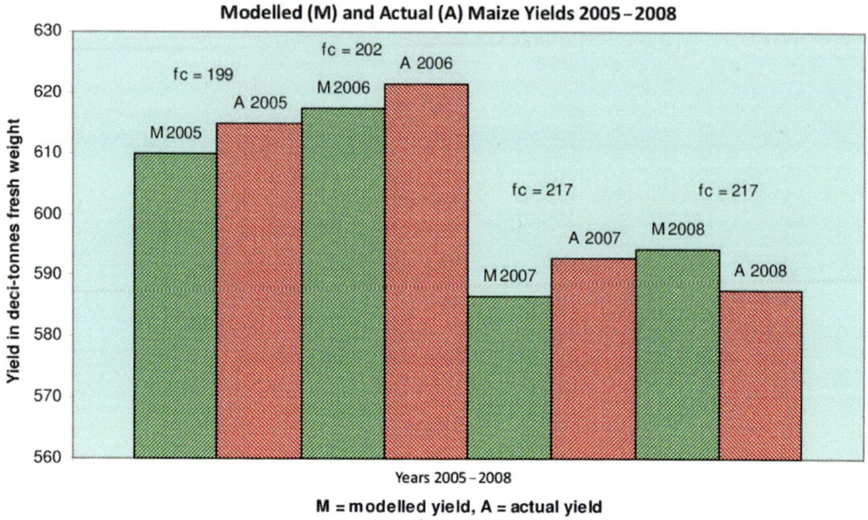

Fig. 5.7 Modelled and actual maize yields of farm plots in Wolfenbuettel during 2005–2008. fc = average field capacity in mm m^{-2}

With regard to the precipitation and radiation curves, all 4 years display good growing conditions for maize. Even the sharp drop in June (2005/2006) probably did not affect the maize, since there was plenty of rain in May to saturate the soils.

The lower yields in 2007 and 2008 were probably caused by the drop in temperature to below 10 °C from September to October in both years. The shortened maize vegetation period, in this case by one to 2 weeks, causes the dry matter content of the harvested biomass to be below the optimum 32–35 %.

5.3.7.2 Interpretation of Sugar Beet Yields in the Context of Climate and Field Capacity

In 2005 and 2006, the sugar beet yields (modelled and actual) were at a lower level than in 2007 and 2008 (555–621 vs. 709–748 dt ha^{-1}) (Fig. 5.8). The precipitation curves for 2005 and 2006 indicate a dry period in June, which might have played a role in lowering the yields. Another reason for the lower yields could be the overall higher radiation level in 2007 and 2008 as well as the increase in precipitation in 2007 and the field capacity in 2008.

5.3.7.3 Interpretation of Winter Wheat Yields in the Context of Climate and Field Capacity

In 2005, 2007 and 2008, the winter wheat yields were at a high level with grain yields in the range of 90–104 dt ha^{-1} (modelled and actual) (Fig. 5.9). In 2006,

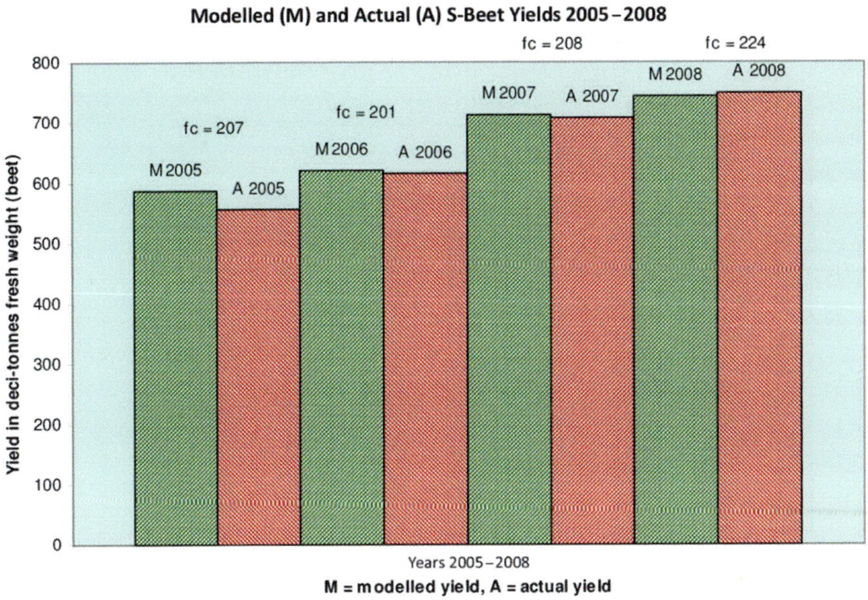

Fig. 5.8 Modelled and actual sugar beet yields of farm plots in Wolfenbuettel during 2005–2008 fc = average field capacity in mm * m^{-2}

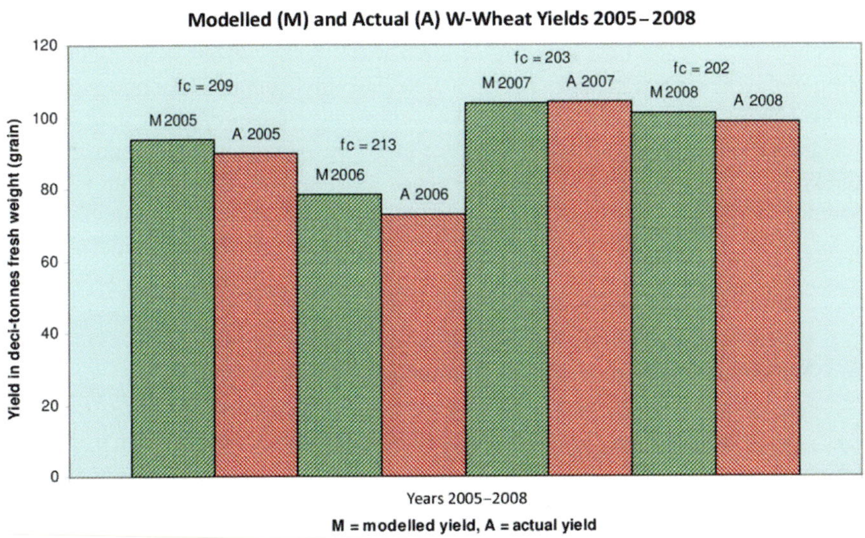

Fig. 5.9 Modelled and actual Winter Wheat yields of farm plots in Wolfenbuettel during 2005–2008 fc = average field capacity in mm * m^{-2}

Table 5.2 Weather data of farm plots from 2005 to 2008

	March	April	May	June	July	August	Septem.	October	04-10	03-07
R 2005	846	1,452	1,627	1,845	1,622	1,388	1,223	759	9,916	7,393
P 2005	31.0	25.2	80.9	9.2	70.1	35.6	52.2	47.5	320.7	216
T 2005	4.6	10.3	13.6	17.3	18.9	16.9	16.5	12.3	15.1	12.9
R 2006	773	1170	1,695	1,863	2,282	1,262	1,233	610	10,115	7,783
P 2006	39.9	35.6	65.0	15.6	38.1	75.0	14.9	30.7	274.9	194
T 2006	2.0	8.9	14.0	18.1	24.1	16.7	18.8	13.8	16.3	13.4
R 2007	845	1,470	1,695	1,863	2,282	1,247	1,252	610	10,419	8,155
P 2007	31.9	2.0	91.6	79.5	59.6	54.0	15.2	19.4	321.3	265
T 2007	4.5	11.8	14.2	17.6	17.2	16.5	18.8	8.8	15.0	13.1
R 2008	862	1470	2,176	2,166	2,413	1,100	1,300	610	11,235	9,087
P 2008	49.69	34.54	3.64	33.08	54.60	49.9	17.4	31.6	224.8	176
T 2008	5.1	8.0	14.8	17.2	18.3	17.6	12.8	9.5	14.0	12.7

R = radiation in J * m^{-2} * day^{-1}, P = precipitation in mm m^{-2}, T = mean air temperature in °C. Columns 04–10 and 03–07: R, P, T = sum of mean for periods April–October and March–July

there was a depression in the modelled and actual yields, reaching respectively 78 and 73 dt ha^{-1}.

Since the average field capacity in 2006 was slightly higher than in 2005, the drop in the precipitation curve between June and July cannot be the reason for this drop. The most obvious explanation is the temperature curve of 2006, with its peak reaching 24.1 °C in July (Table 5.2). Since these are average monthly temperatures, the peaks on individual days must have been even higher. The optimum temperature for the photosynthesis of wheat is about 20 °C and, therefore, an inhibition of photosynthesis in the June-July period probably led to lower yields in 2006.

5.3.8 Stability Index of the Modelled Data

We can conclude that the model renders good results for the three modelled crops. The best results were achieved with maize ($R^2 = 0.88$, Fig. 5.10), followed by winter wheat ($R^2 = 0.82$, Fig. 5.11) and then sugar beet ($R^2 = 0.72$, Fig. 5.12).

Sugar beet proved the most difficult to model. Therefore, the model needs to be further adjusted and calibrated. The model rendered exceedingly high yields at three sites with clay-type soils, which were therefore omitted from the sample.

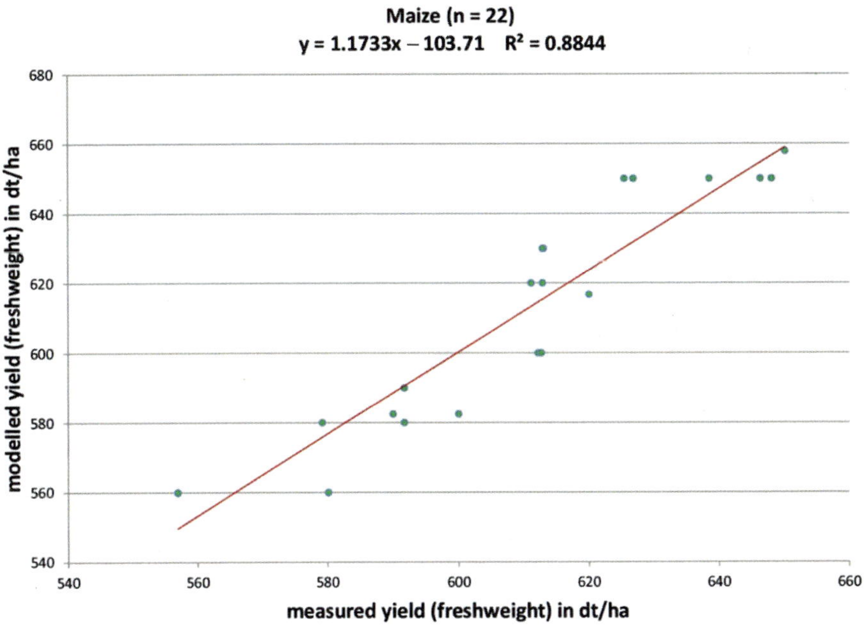

Fig. 5.10 Regression analysis and stability index for maize

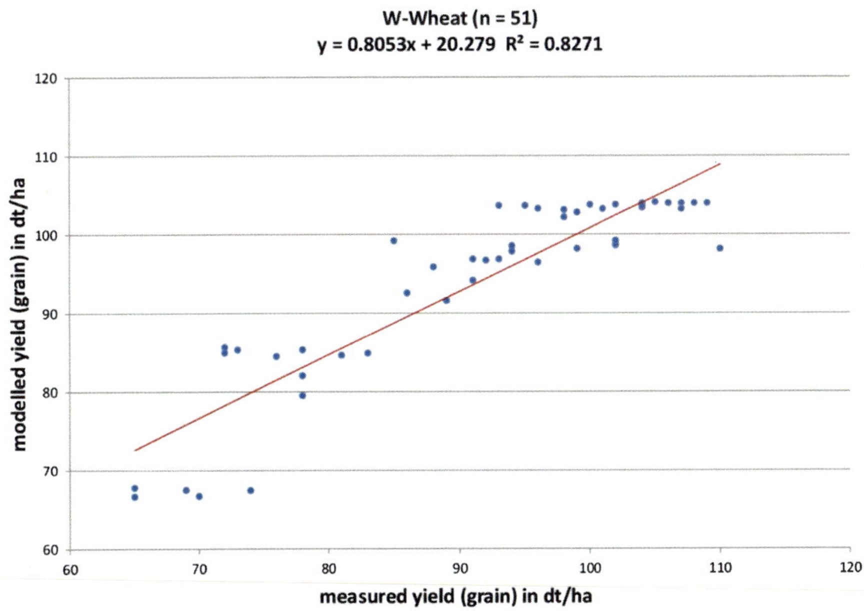

Fig. 5.11 Regression analysis and stability index for winter wheat

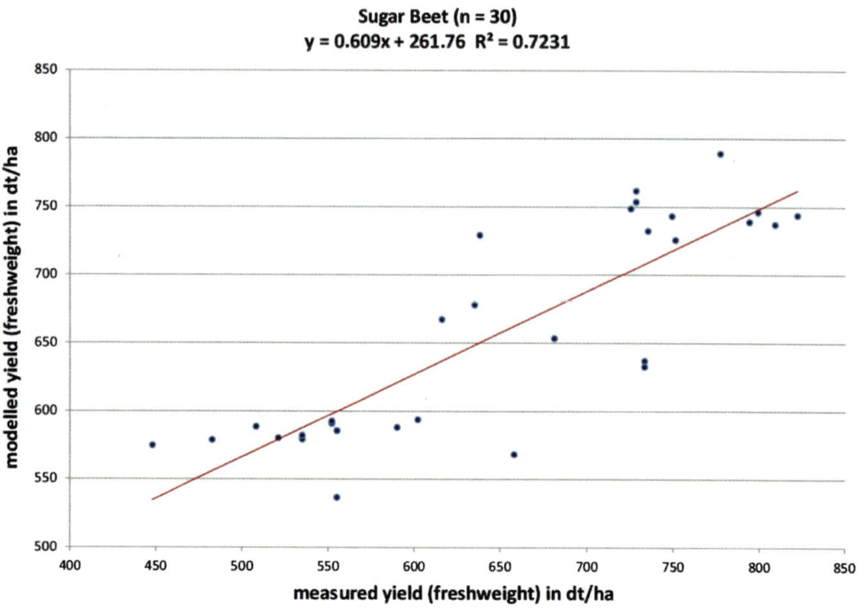

Fig. 5.12 Regression analysis and stability index for sugar beet

5.4 Modelling Biomass Potentials in the Jühnde Vicinity

The administrative district of Göttingen, situated in southern Lower Saxony, is the home of Germany's first bioenergy village, Jühnde. Since the facility in Jühnde became operational in 2005, three more bioenergy villages (Raiffenhausen, Wollbrandshausen and Krebeck) have been established in the Göttingen district.

To supply a bioenergy facility or a bioenergy village with biomass requires a certain percentage of agricultural area within a radius of the facility. This percentage and the associated radius are further defined by the size of the facility (installed electrical power) and the type and intensity of the biomass grown to fuel it.

In this context, "intensity" is the percentage of agricultural area used solely for biomass production in the course of a year, while the type of biomass is defined by the biomass crops and the crop rotations used. If, for instance, the energy crop rotation cycle is rather narrow and consists of only maize the radius needs to be bigger, since European cross-compliance measures require a 2-year break between two maize cultivations on one site. If the energy crop rotation cycle comprises two or more cultures, the radius can be smaller (see also Chap. 6).

The radius around the facility is limited to a certain distance (≈5–10 km) to keep transport costs of the biomass within a certain range. For a small facility, 10 % of the agricultural area around a facility might suffice for energy crop cultivation. For a larger facility, or to keep transport costs low, 30 % might be needed. Therefore, the first step in the assessment of the biomass potential available for a planned facility is to define the desired type of facility and size (electrical power in kW or MW) thereof. As a rule of thumb, for each 500 kW of installed electrical power, 250 ha of agricultural land are needed for biomass production. This value may vary, depending on the sites' productivity (soil and climate).

This process can also be reversed when the maximum potential for bioenergy production in an area needs to be established in the process of regional planning. In this case, the result of the analysis will identify how many bioenergy facilities can be installed in an area and how much electrical power they will generate.

In the following example, the six boroughs (Dransfeld, Niemetal, Rosdorf, Buehren, Scheden, and Jühnde) in the Jühnde district, situated in the western part of Göttingen, were analysed with regard to their biomass potential for energy production.

In a first step, all the agricultural sites (excluding pastures) were identified and soil and climate data from the general soil map 1:50,000 (Buek50, NIBIS®) were processed for usage in the BioSTAR model.

For reasons of practicality and with an open option as to where one or more bioenergy facilities could be located in the area, the agricultural sites were divided into eastern, central and western areas. The eastern part (Rosdorf) is characterised by mostly silt-type soils (river deposits) with high water retention capacities. The central part (Jühnde) has a combination of loam and clay type soils with lower water retention capacities than the Rosdorf area. The western part (Dransfeld) has an increased soil water retention capacity with loam and silt-loam types of soils.

The annual long-term (1960–2000) precipitation was the lowest in the eastern part (735 mm), higher in Dransfeld (766 mm) and the highest in Jühnde (794 mm) due to its higher elevation. The annual long-term (1960–2000) mean temperature was the highest in Rosdorf (8.4 °C), followed by Dransfeld (8.3 °C) and the lowest in Jühnde (8.1 °C).

Two different scenarios with three different intensities of use were computed to calculate the potential biomass available for the production of energy. The first scenario applies to a bioenergy village with 500 kW of electrical power (the same dimension as Jühnde) while the second is for a bigger biogas facility with 1 MW or more installed electrical power. The first scenario would require an energy crop mix of 50 % maize, 25 % triticale (or a similar winter cereal) and 25 % rye-grass. The second scenario would require a larger share of maize (80 %) and only 20 % cereal silage.

In total, 9,995 ha of agricultural land are available in all six boroughs. As a rule of thumb, 2, 4, or 6 MW of electrical power could be respectively installed for each of the three different intensities (10, 20 or 30 % energy crop cultivation), assuming an area of 250 ha for each 500 kW (Table 5.3).

Table 5.3 Total agricultural area in hectares and installable electrical power

Total area (agr.)	99,950,000	Square Meters
	9,995	Hectare
Theoretically available area for energy crops cultivation		
	Hectare	Installable electric power in kW
10 % energy crops	1,000	2,000
20 % energy crops	1,999	4,000
30 % energy crops	2,999	6,000

Table 5.4 Hectares and tonnes of dry mass (total and per hectare) for three areas as well as sums for total area and averages per hectare

	Rosdorf (east)	Jühnde (centre)	Dransfeld (west)	Sum/avg.
Area in hectare	3,326	2,377	4,292	9,995
Maize tonnes	53,013	35,905	68,131	157,049
Triticale tonnes	48,254	33,043	62,282	143,578
Maize (t ha^{-1})	15.9	15.1	15.9	15.6
Triticale (t ha^{-1})	14.5	13.9	14.5	14.3
Rye-grass (t ha^{-1})	6.0	5.0	6.5	5.8

Table 5.5 Hectares and tonnes of dry mass (total and per hectare) for the three areas with 10, 20 and 30 % of agricultural area used for energy crop cultivation scenarios

	Rosdorf (east)			Jühnde (centre)			Dransfeld (west)		
	10 %	20 %	30 %	10 %	20 %	30 %	10 %	20 %	30 %
Area in hectare	333	665	998	238	475	713	429	858	1,288
Maize tonnes	5,301	10,603	15,904	3,591	7,181	10,772	6,813	13,626	20,439
Triticale tonnes	4,825	9,651	14,476	3,304	6,609	9,913	6,228	12,456	18,685
Maize (t ha^{-1})	1.59	3.19	4.78	1.51	3.02	4.53	1.59	3.17	4.76
Triticale (t ha^{-1})	1.45	2.9	4.35	1.39	2.78	4.17	1.45	2.9	4.35
Rye-grass (t ha^{-1})	0.6	1.2	1.8	0.5	1	1.5	0.65	1.3	1.95

The modelled results – measured in total tonnes and tonnes per hectare dry mass for maize, triticale and rye-grass – of the three areas are summarised in Tables 5.4, 5.5 and 5.6. Table 5.7 renders the total biomass (dry matter per year) that can be harvested with the crops, intensities and crop rotations mentioned above. Figures 5.13 and 5.14 display the triticale and maize yield data generated by the model. Figure 5.15 depicts the division of the test sites into the three areas.

Table 5.6 Hectares and tonnes of dry mass (total and per hectare) for the three areas with 10, 20 and 30 % of agricultural area used for energy crop cultivation scenarios (100 % is not a scenario but only displayed for comparison)

	Three areas combined			
	Tonnes of dry mass per year			
	10 %	20 %	30 %	100 %
Hectare	1,000	1,999	2,999	9,995
Maize tonnes	15,705	31,410	47,115	157,050
Triticale tonnes	14,358	28,716	43,073	143,576
R-grass tonnes	5,974	11,948	17,922	59,740
Maize (t ha^{-1})	1.56	3.13	4.69	15.63
Triticale (t ha^{-1})	1.43	2.86	4.29	14.3
Ryegrass (t ha^{-1})	0.58	1.17	1.75	5.8

Table 5.7 Potential of annual biomass production in tonnes for two options of bioenergy facilities (500 kW bioenergy village or big facility with 1 MW or more electrical power) and three intensity scenarios (10, 20, 30 %)

	Small facility 500 kW			Big facility 1 MW or more		
	Tonnes of dry mass per year			Tonnes of dry mass per year		
Percentage	10 %	20 %	30 %	10 %	20 %	30 %
Maize tonnes	7,852	15,705	23,557	12,564	25,128	37,692
Triticale tonnes	3,589	7,179	10,768	2,872	5,743	8,615
R-grass tonnes	1,493	2,987	4,480			
Comb. total tonnes	12,935	25,871	38,806	15,435	30,871	46,306

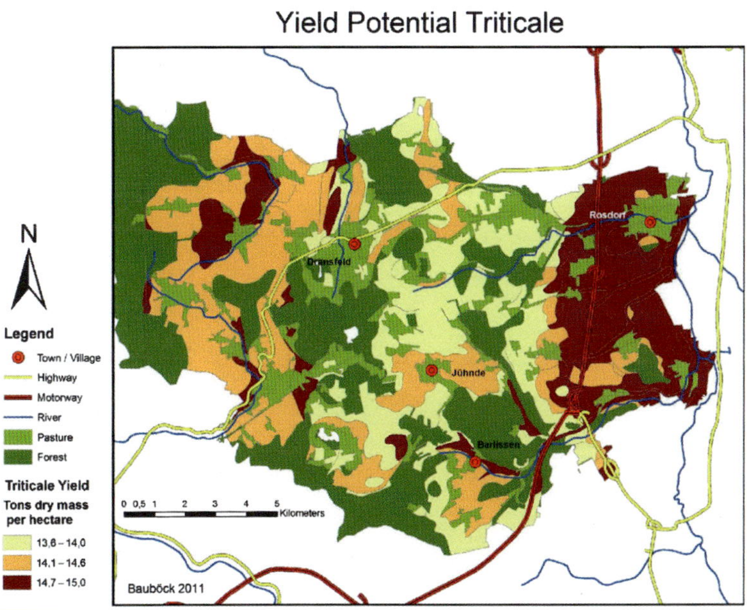

Fig. 5.13 Yield potential of triticale in the Juehnde vicinity (climate data: 1960–2000) (Source: BioSTAR)

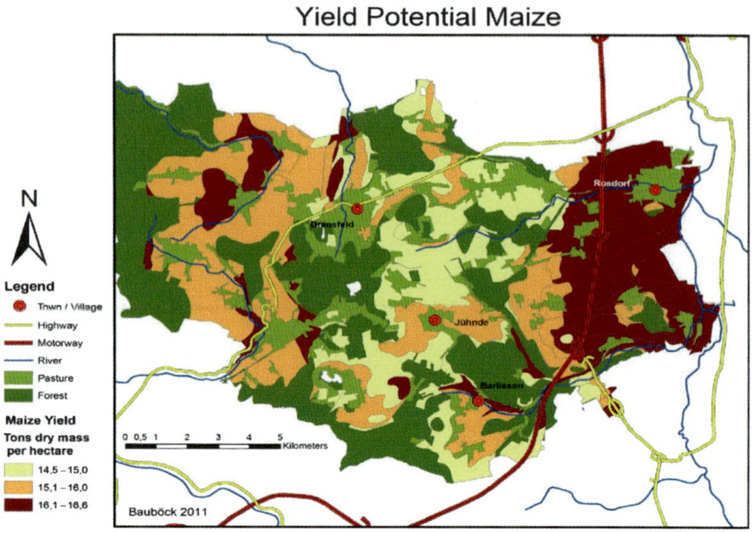

Fig. 5.14 Yield potential of maize in the Juehnde vicinity (climate data: 1960–2000) (Source: BioSTAR)

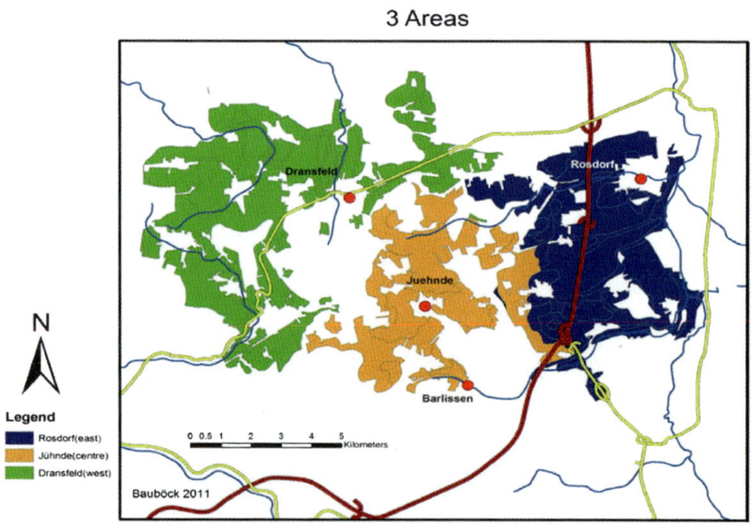

Fig. 5.15 Division of agricultural sites in the vicinity of Juehnde into three areas

5.5 Conclusion

The biomass potentials calculated by using the BioSTAR model (or any other crop model) can now be used to approximate the number or size of possible future biogas facilities in the Jühnde vicinity. These potentials can subsequently be compared with those calculated using the rule of thumb.

According to the first scenario, an annual potential of 12,935 t of dry mass can be achieved with an intensity of 10 % energy crops in the crop rotation and a combination of 50 % maize, 25 % triticale (or other winter cereal) and 25 % rye grass. With an average annual demand of 3,000 t of dry mass for a 500 kW facility, four of these (Jühnde-type) biogas plants could be installed.

In the second scenario (the other extreme), an annual biomass potential of 46,306 t could be achieved with a 30 % intensity of energy crops and a combination of 80 % maize and 20 % triticale (or other winter cereal). Again assuming a demand of 3,000 t of annual dry mass for each 500 kW, seven 1 MW biogas facilities could theoretically be built in the Jühnde vicinity without having to import biomass from outside the area.

However, it should be stressed that these are only theoretical potentials and whether or not they can actually be achieved will depend on various social, political and economic factors as well as on farming practices and future climate changes.

References

Allen, R. G., Pereira, L. S., Raes, D., & Smith, M. (1998). FAO irrigation and drainage papers no.56, *Crop Evapotranspiration,* Rome.

Amthor, J. S. (1989). *Respiration and crop productivity*. New York/Berlin/Heidelberg: Springer. ISBN 0-387-96938-1.

Arkebauer, T. J., Weiss, A., Sinclair, T. R., & Blum, A. (1992). In defence of radiation use efficiency: A response to Demetriades-Shah et al. *Agricultural and Forest Meteorology, 68* (3–4), 221–227.

Azam-Ali, S. N., Crout, N. M. J., & Bradley, R. G. (1994). Perspectives in modelling resource capture by crops. In *Some perspectives in modelling resource capture by crops. Proceedings of 52nd Easter School in Agricultural Science*. Sutton Bonington, UK.

Bauböck, R. (2009). *Bioenergie im Landkreis Göttingen – GIS-gestützte Biomassepotentialabschätzung anhand ausgewählter Kulturen, Triticale und Mais*. Stuttgart: Ibidem.

BMU. (2007). The integrated energy and climate programme of the German government. Berlin. Retrieved December 8, 2012, from http://www.bmu.de/files/english/pdf/application/pdf/hintergrund_meseberg_en.pdf

BMU (2009). Nationaler Biomasseaktionsplan für Deutschland [Online]. Retrieved February 16, 2012, from http://www.bmu.de/files/english/pdf/application/pdf/broschuere_biomasseaktionsplan_en_bf.pdf

BMVBS. (2010). Globale und regionale Verteilung von Biomassepotentialen. Status-que und Möglichkeiten der Präzisierung. BMVBS-Online Publikationen 27/2010. Ministerium für Verkehr (BRD), 2010. ISSN 1869–9324.

Boote, K. J., & Loomis, R. S. (1991). The prediction of canopy assimilation. *Modeling photosynthesis – From biochemistry to canopy*. CSSA Special Publication no.19.

Bouman, B. A. M., van Keulen, H., van Laar, H. H., & Rabbinge, R. (1996). The "School of de Wit" crop growth simulation models: A pedigree and historical overview. *Agricultural Systems, 52*(2/3), 171–198.

Brisson, N., Gary, C., Justes, E., Roche, R., et al. (2003). An overview of the crop model STICS. *European Journal of Agronomy, 18*, 309–332.

Choudhury, B. J. (2000). A sensitivity analysis of the radiation use efficiency for gross photosynthesis and net carbon accumulation by wheat. *Agricultural and Forest Meteorology, 101*, 217–234.

Choudhury, B. J. (2001). Modeling radiation- and carbon-use efficiencies of maize, sorghum, and rice. *Agricultural and Forest Meteorology, 106*, 317–330.

Demetriades-Shaw, T. H., Fuchs, M., Kanemasu, E. T., & Flitcroft, I. (1992). A note of caution concerning the relationship between cumulated intercepted solar radiation and crop growth. *Agricultural and Forest Meteorology, 58*, 193–207.

Demetriades-Shaw, T. H., Fuchs, M., Kanemasu, E. T., & Flitcroft, I. (1994). Further discussions on the relationship between cumulated intercepted solar radiation and crop growth. *Agricultural and Forest Meteorology, 68*, 231–242.

FNR. (2011). Fachagentur für nachwachsende Rohstoffe: *Basisdaten Bioenergie*, Gülzow. Retrieved March 25, 2013, from http://mediathek.fnr.de/media/downloadable/files/samples/f/n/fnr_basisdaten_2012_web_neu.pdf

Garcia, R., Kanemasu, E. T., Blad, B. L., Bauer, A., Hatfield, J. L., Major, D. J., Reginato, R. J., & Hubbard, K. G. (1988). Interception and use efficiency of light in winter wheat under different nitrogen regimes. *Agricultural and Forest Meteorology, 44*, 175–186.

Gardener, F., Pearce, R. B., & Mitchell, R. L. (1985). *Plant physiology*. Ames: Iowa State University Press.

Lindquis, J. L., Arkebauer, T. J., Walters, D. T., Cassman, K. G., & Dobermann, A. (2005). Maize radiation use efficiency under optimal growth conditions. *Agronomy Journal, 97*(1), 2–78.

Loomis, R. S., & Amthor, J. S. (1999). Yield potential, plant assimilatory capacity, and metabolic efficiencies. *Crop Science, 39*, 1584–1596.

Mailhol, J. C., Olufayo, A. A., & Ruelle, P. (1997). Sorghum and Sunflower evapotranspiration and yield from simulated leaf area index. *Agricultural Water Management, 35*, 167–182.

Mc Cree, K. J. (1970). An equation for the rate of respiration of white clover plants grown under controlled conditions. In I. Setlik (Ed.), *Prediction and measurement of photosynthetic productivity* (Proceedings of IBP/PP Technical Meeting, Trebon, Czechoslovakia), (pp. 221–229). Wageningen: PUDOC.

Merta, M., Seidler, C., & Fjodorowa, T. (2006). Estimation of evaporation components in agricultural crops. *Biologia Bratislava, 61*(Suppl. 19), 280–283.

Monsi, M., & Saeki, T. (2005). On the factor of light in plant communities and its importance for matter production. *Annals of Botany, 95*, 549–567.

Monteith, J. L. (1977). Climate and the efficiency of crop production in Britain. *Philosophical Transactions of the Royal Society Series B, 281*, 277–294.

Monteith, J. L. (1994). Validity of the correlation between intercepted radiation and biomass. *Agricultural and Forest Meteorology, 68*, 213–220.

Penning de-Vries, F.W.T., Jansen, D. M., ten Berge, H. F. M., & Bakema, A. (1989). *Simulation of ecophysiological processes of growth in several annual crops. Simulation monographs*. Wageningen: Pudoc.

Philip, J. R. (1966). Plant water relations: Some physical aspects. *Annual Review of Plant Physiology, 17*, 245–268.

Ritchie, J. T. (1972). Model for predicting evaporation from a row crop with incomplete cover. *Water Resources Research, 8*(5), 1204–1213.

Sinclair, T. R., & Muchow, R. C. (1999). Occam's Razor, radiation-use efficiency, and vapor pressure deficit. *Field Crops Research, 62*(2), 239–243.

Sinclair, T. R., & Seligman, N. G. (1996). Crop modeling: From infancy to maturity. *Agronomy Journal, 88*, 698–704.

Steduto, P., & Albrizio, R. (2005). Resource use efficiency of field grown sunflower, sorghum, wheat and chickpea. II. Water use efficiency and comparison with radiation use efficiency. *Agricultural and Forest Meteorology, 130*, 269–281.

Steduto, P., Hsiao, T. C., & Fereres, E. (2007). On the conservative behaviour of biomass water productivity. *Irrigation Science, 25*, 189–207.

Steduto, P., Raes, D., Hsiao, T. C., & Fereres, E. (2009). AquaCrop—The FAO crop model to simulate yield response to water: I. Concepts and underlying principles. *Agronomy Journal, 101*, 426–437.

Stöckle, C., Donatelli, M., & Nelson, R. (2003). CropSyst, a cropping systems simulation model. *European Journal of Agronomy, 18*, 289–307.

Teruel, D. A., Baarbieri, V., & Ferraro, L. A. (1997). Sugarcane leaf area index modeling under different soil water conditions. *Science Agriculture Piracicaba, 54*, 39–44.

Todorovic, M., Albrizio, R., Zivotic, L., Abi SAAB, M. T., Stöckle, C., & Steduto, P. (2009). Assesment of Aqua Crop, CropSyst, and WOFOST models in the simulation of sunflower growth under different water regimes. *Agronomy Journal, 101*, 508–521.

van Ittersum, M. K., Leffelaar, P. A., van Keulen, H., Kropff, M. J., Bastiaans, L., & Goudriaan, J. (2003). On approaches and applications of the Wageningen crop models. *European Journal of Agronomy, 18*, 201–234.

Wang, S. H. (2001). *Einfluss von Blattstellung und Bestandesdichte auf Ertrag, Qualität, Lichtaufnahme und Blattflächenindex bei Silomaissorten verschiedenen Wuchstyps*. Dissertation, Humbold Universität Berlin.

Whistler, F. D., Acock, B., Baker, D. N., Fye, R. E., Hodges, H. F., Lambert, J. R., Lemmon, H. E., McKinion, J. M., & Reddy. V. R. (1986). Crop simulation models in agronomic systems. *Advances in Agronomy, 40*, 141–208.

Part III
Can Bioenergy Production Be Environmentally Sound?

Chapter 6
Integrative Energy Crop Cultivation as a Way to a More Nature-Orientated Agriculture

Marianne Karpenstein-Machan

Abstract The vision of integrative energy cultivation concepts is to contribute to a more diverse and sustainable rural landscape, keep nature in balance and conserve ecosystems. Integrative cultivation concepts also harmonise utilisation/production with the protection of landscapes. An overview is given of the status quo of energy crop cultivation management on farms in Lower Saxony, Germany. This overview explains the opportunities, but also the many risks associated with current bioenergy cultivation practices. Examples are presented of ecological and economical optimisation of farmland use for the production of food, feed and energy. In addition, sustainable cultivation concepts are presented, which include several winter annuals, summer annuals, perennials and wild herbs found in cultivation concepts adapted to local climate and soil conditions. In the model farms, the ecological challenges regarding the current cultivation concepts are described and farm-specific examples of more sustainable concepts are described. Subsequently, the opportunities to implement integrative energy cultivation concepts in agricultural practice are evaluated.

Keywords Biogas • crop rotation • integrative cultivation concept • energy crops

6.1 Bioenergy Production in the Contradictory Contexts of Nature, Environment and Society

Landscapes provide many services, offering agriculture, forestry, biodiversity, local recreation, buildings and streets. In industrial societies, an increasing amount of land is used for homes, buildings, industry and mobility (streets). In Germany,

M. Karpenstein-Machan (✉)
Interdisciplinary Centre for Sustainable Development, University of Göttingen,
Goldschmidtstrasse 1, 37077 Göttingen, Germany
e-mail: mkarpen@gwdg.de

H. Ruppert et al. (eds.), *Sustainable Bioenergy Production - An Integrated Approach*,
DOI 10.1007/978-94-007-6642-6_6,

approximately 87 ha of soil are sealed for a settlement area and infrastructure implementation (Federal Statistic office 2012) every day. However, land is limited, especially for agriculture, which is needed for food, fodder and bioenergy production. Furthermore, with increasing intensity in agricultural production, more space is necessary for biodiversity protection to reduce intensive agriculture's negative effects. Over the past 5 years, bioenergy has gained importance in Germany. Triggered by the 2004 Renewable Energy Sources Act (EEG), the total area for energy crop cultivation has increased rapidly. In Lower Saxony, the total area for bioenergy has increased by 5 % since 2004 to a total area of 10 % of arable land (ML 2010). This can bc considered a positive development, because bioenergy reduces the CO_2 output and contributes to climate protection (BMU/AGEE 2010); however, more conflicts have arisen between farmers, locals and nature conservation organisations due to their differing opinions of bioenergy (see the Chap. 10). The most frequent misgivings voiced by opponents of bioenergy are the increasing monoculture associated with maize and winter oilseed rape (Brassica napus L.) cultivation, the increasing use of pesticide, soil degradation and the reduction in fauna and flora.

The problems associated with energy crop cultivation could be avoided in sustainable bioenergy projects. Our work seeks to establish sustainable and integrative cultivation concepts for food, fodder and energy. The synergy effects between different utilisation options should therefore be identified and used. In the following sections, the bioenergy status quo in Lower Saxony is summarised, integrative concepts are described and examples are given. I start off by defining integrative cultivation.

6.2 Integrative Cultivation Concepts for Food, Fodder, Energy and Wildlife

Integrative cultivation can be defined as a scientific approach in which scientists working on concepts combine different landscape utilisation options to produce food, fodder and energy, as well as support wildlife (Karpenstein-Machan 1997, 2001, 2004, 2009a; Rode and Kanning 2010). Integrative cultivation concepts harmonise utilisation/production and landscape protection. The agricultural utilisation of farmland and landscape protection should no longer be seen as mutually exclusive. In the long term, only sustainable concepts are economically sound for society, due to the external costs of unsustainable systems.

Energy crop cultivation can act as a bridge between different landscape utilisation systems, such as grassland, cropland and forest, as well as between ecological and conventional agricultural systems. Furthermore, water and nature protection areas, as well as problematic locations (e.g., contaminated soils), do have a place in integrative concepts.

Integrative cultivation's vision is to contribute to a more diverse agricultural landscape, to keep nature in balance and conserve rare ecosystems. Integrative cultivation's vision for bioenergy includes the cultivation of locally adapted biomasses and the transformation of energy into locally scaled energy plants (decentralised concepts).

6.3 Examples of Integrative Cultivation

Figure 6.1 shows an integrative cultivation model with food, feed and energy crops. This can be a model for a farm, but also for a greater area, for instance a community area. Annual crops for food, feed and energy are cultivated in conventional ways. They are rotated in crop rotations (the minimum crop rotation length should be 3 years) and form the basis of high agricultural production. These annual crops are surrounded by herbicide-free buffer strips (flower strips). Such flower strips should increase the flora and fauna biodiversity and stabilise the agro-ecosystem, both of which should reduce the pesticide use on the annual crops. Flower strips can be harvested after flowering and utilised in the biogas plant, or remain there until the grain harvest. Maize and winter triticale/winter rye mixtures, which are typical biogas production crops, undergo the food and feed crop rotation together with winter cereals, sugar beet and field grass. Contrary to food production, pesticide use

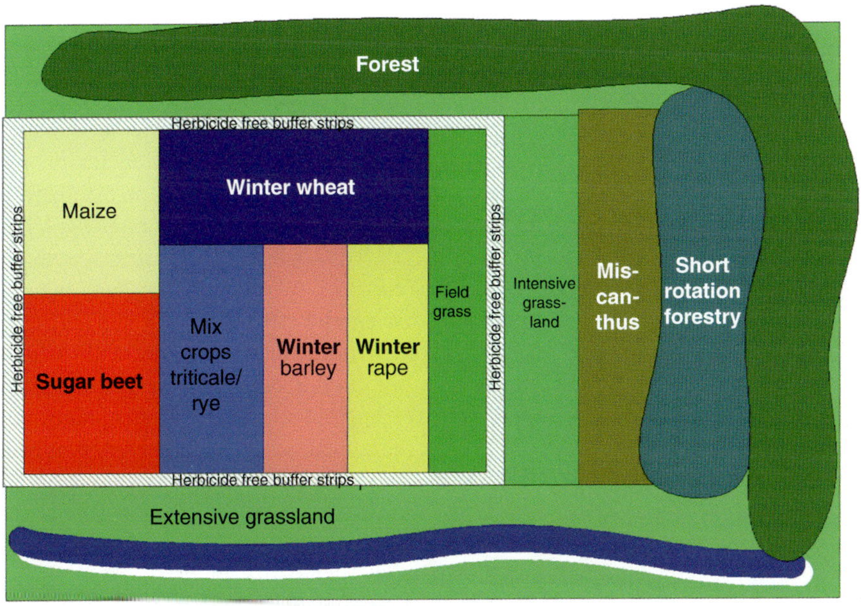

Fig. 6.1 Model of an integrated cultivation concept with food, feed and energy (Modified from Karpenstein-Machan 2004)

on biogas crops can be reduced, due to their lower sensitivity to diseases and premature harvest time (Karpenstein-Machan 2000a, 2002). In this way, the mixture of energy crops and food crops in one crop rotation leads to higher diversity and reduces pesticide use.

In this example, intensive grassland culture forms the transition between annual and perennial crops. Ecologically sensitive soils, which tend to leach nitrate and erosive hills, are better suited to perennial crops. These crops cover the soil all year round, which prevents leaching problems. In addition, extensive grassland builds a buffer that prevents erosion as well as pesticide and nutrient contamination of the river. Furthermore, permanent grassland can absorb the water from floods and does not form a barrier to run-off water. However, extensive grassland with nature protection status must fulfil certain harvest time and frequency requirements. A late harvest after flowering ensures wild flower reproduction. The removal of chopped biomasses from grassland is important since it prevents nutrient accumulation. Biomass from extensive grassland has a low fodder quality, due to advanced plant lignification. In special biogas plants, this biomass can be utilised for biogas production (dry fermentation). Low-input woody perennials, such as miscanthus or the newly discovered perennial for biogas use (see Sect. 6.6.6), and short-rotation forestry act as transition zone between open landscape and dense forest.

6.4 Bioenergy Status Quo in Lower Saxony

In Lower Saxony, the energy crop cultivation reached 7.3 % of the total agricultural area (arable land and grassland) and 10.6 % of the arable land in 2008 (ML 2010). Energy crop production includes production for biodiesel (share: 22 %), ethanol (share: 12 %) and biogas (share: 66 %). Generally, crop cultivation for biogas enjoys high priority in Lower Saxony, but this does differ depending on the district. In some districts, energy crops are cultivated on a 20 % share of the arable land, but provide 90 % of the biogas (i.e. the Celle district), while other districts have an energy crop share of under 5 % (district Göttingen). As energy crops, they mainly produce biodiesel with an 80 % share of the energy crop area. Problems arise in those districts where a high concentration of husbandry coincides with a high concentration of biogas plants. In districts high in husbandry, maize was the main crop even before the biogas boom. After the implementation of biogas plants, the farmers cultivated additional maize for these plants; consequently, the maize concentration in some districts comprises a 60 % share of the arable land (Karpenstein-Machan 2010). Figure 6.2 shows the maize cultivation shares (in percentage) of the arable land in the Lower Saxony districts.

The Ministry of Agriculture (ML 2010) calculates that, in 2012, 1,480 biogas plants with a 783 MW_{el} capacity will have been implemented in Lower Saxony. The produced electricity could cover the demand from approximately one million households. Further biogas plants, especially in critical districts, can create

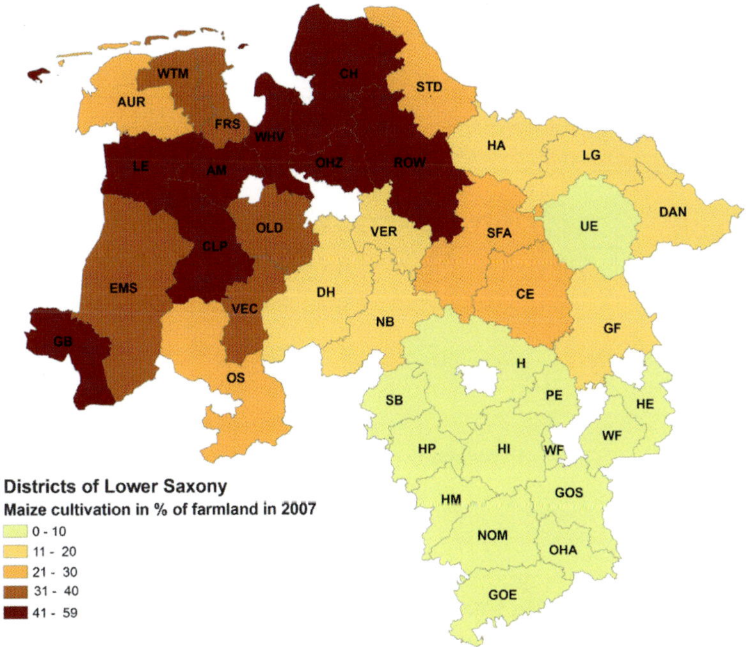

Fig. 6.2 Maize cultivation share (%) on arable land in districts of Lower Saxony (Status 2007)

environmental problems and problems with the local people. Therefore, scientists and experts from different disciplines should formulate new ecological standards for energy crop production, especially for biogas. Our work seeks to address antagonism to the energetical use of biomass and promote sustainable bioenergy development in Lower Saxony. More information about the cultivation situation on biogas farms was necessary to optimise existing cultivation concepts. A survey (a questionnaire and interviews) was designed to obtain information from the farmers on how they cultivate energy crops (fertilisation, pesticide treatments, crop rotation) and how they integrate crops into their crop rotation.

6.4.1 Results of Survey of Farmers

The results are based on the questionnaire and interviews with 76 farmers from six different districts in Lower Saxony. All the interviewees cultivate energy crops for a biogas plant. Approximately 50 % operate husbandry farms (n = 39) and 50 % cultivate field crops (n = 37). The share of farmers with an own biogas plant and farmers without one is also balanced. Figure 6.3 shows that, in most cases, small farms produce energy crops for foreign biogas plants. Biogas plant owners have more agricultural land. According to the study, most biogas plant operators have

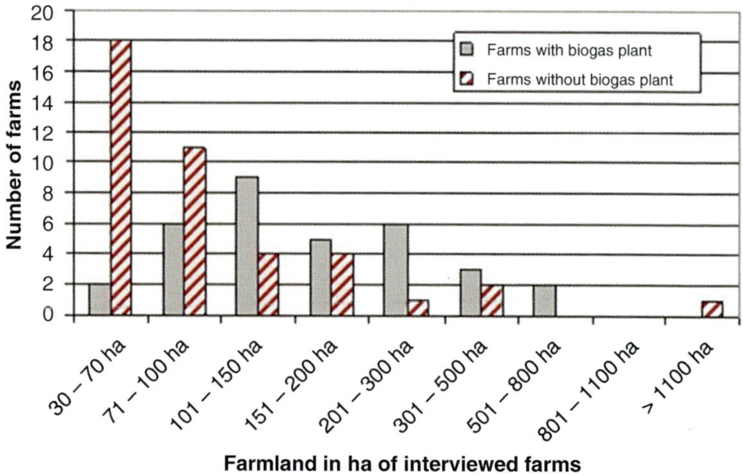

Fig. 6.3 Relationship between farmland size of interviewed farms and number of farms with biogas plants and without biogas plants

100–150 ha of farmland. For small farms, energy crop production for foreign plants offers an opportunity to stabilise their farm income if the biomass prices are acceptable and fixed over the long term. In most cases, the biomass prices are adapted, within a price corridor, to the wheat prices.

The farmers also cultivate other crops besides energy crops. Most farms (65 %) set aside 20 % of their farmland for energy crops. Only 10 % of farms cultivate energy crops on more than 50 % of their arable land. The farms high in energy crops have their own biogas plants, with one exception.

Nearly 50 % of energy crops are cultivated on fertile soils, with fertility numbers above 60. Soils with middle fertility numbers (40–60) have a 28 % share and soils with low fertility numbers (<40) 26 %.

The farmers were also asked which crops they cultivate and the shares of these crops. Maize was highest at 74 %, followed by winter rye (10 %), winter triticale (4 %), grassland (4 %), field grass (3 %), sugar beet (3 %) and diverse other crops (2 %).

A problem can arise for the environment, because 26 % of the energy crops, mainly maize, were cultivated on slopes. Furthermore, alluvial soils (7 %) and boggy soils (4 %) can create environmental problems if the cultivation concepts are not adapted. New sustainable energy concepts should specifically be tested for these soils in practice (see Sect. 6.8.4).

Through energy crop cultivation, other crops were replaced. Winter wheat was replaced the most at 62 %, followed by winter rape (17 %), winter barley (8 %), sugar beet (5 %), winter rye (3 %), triticale (2 %), potatoes (2 %) and diverse other crops (1 %).

The replacement of winter wheat can be viewed as an improvement for the environment. It has already reached a high concentration in many districts and needs several pesticide treatments against diseases and weeds. On fertile soils, winter wheat and sugar beet are often the only crops in the rotation (e.g., winter wheat, winter wheat, sugar beet). Owing to its self-incompatibility, winter rape should be cultivated with a 3-year break after cultivation. Since farmers do not always follow these rules, a reduction in the rape cultivation on locations with high concentrations can impact crop health positively.

In the next section, I analyse the situation before and after the restructuring of energy crop rotations. An example of a typical crop rotation before the cultivation of energy crops for a biogas plant: winter wheat, winter wheat, winter barley, winter rape. In fertile soils, sugar beets were cultivated instead of winter rape. However, very often, crop rotations were undertaken with only two crops (winter wheat, winter wheat, sugar beet). Since the restructuring, wheat-dominated crop rotations have been enhanced with maize (e.g., winter rape, winter wheat, winter barley, maize). Some farms lower in energy crops have integrated their energy crops very positively, which results in a more diverse crop rotation. Examples are winter rye, field grass, maize, triticale, potatoes; or winter rye, sorghum, maize, triticale, field grass, winter wheat.

Some farms high in energy crops run partial crop rotations on fertile soils with only two crops (e.g., maize, winter wheat; maize, sugar beet; or maize in monoculture). These one-sided crop rotations or monocultures can create many problems, for instance, humus degradation, plant diseases, soil erosion and nitrate leaching. Nonetheless, if all the farms are taken into consideration, the crop rotation changes that include energy crops have positive results. On average, across all the farms, the number of crops increased significantly from 3.5 crops to 4.0 crops in the crop rotation. About 50 % of the farms have had a more diverse crop rotation since the restructuring. Only 18 % of the farms have reduced the number of crops in their crop rotation.

With a crop rotation change, the humus reproduction demand changes, too. Maize is involved in most of the new crop rotations. Owing to its low soil covering in the spring and early summer and long vegetation time until October, maize is a humus-degrading crop. To retain the soil humus content, additional treatments are necessary, such as higher organic fertilisation and the cultivation of catch crops, cover crops or undersown crops.

Since the restructuring, 80 % of the farms have had a higher humus reproduction demand (on average, 91 kg C/ha^1/a^1) than before. However, 20 % of the farms have improved their crop rotations with humus-increasing crops such as field grass or mixtures of alfalfa and field grass, which they use as energy crops.

In energy crop cultivation, the crops requiring pesticide usage have been clearly reduced compared to the replaced crops. Table 6.1 provides an overview of these results. Seventy-seven percent of the farms use far fewer pesticides on energy crops (mainly maize). Fungicides and insecticides have been specifically reduced compared to the replaced crops. These findings can be attributed to maize diseases and pests currently not occurring in Lower Saxony; in addition, the two main maize pests (Ostrinia nubilalis, Dabrotica virgifera) have not reached Lower Saxony. This may change with a higher maize concentration in the crop rotations. In southern Germany (e.g., Baden-Wurttemberg), major problems arose due to the many years

Table 6.1 Pesticide applications in energy crops compared with the replaced crops

Pesticide applications		Energy crop maize n = 66		Energy crops winter cereals n = 15	
		Number	in %	Number	in %
Pesticides in general	No application	1	2	1	7
	Significant fewer applications	51	77	2	13
	Fewer applications	6	9	8	53
	No change	7	11	3	20
	Significant more applications	1	2	1	7
Fungicides	No application	58	88	3	20
	One application	1	2	7	47
	More than one application	0	0	3	20
Herbicides	No application	2	3	0	0
	Application before leaves emerge	9	14	1	7
	Application after leaves emerge	60	91	11	73
Insecticides	No application	58	88	3	20
	One application	0	0	3	20
	More than one application	0	0	1	7

of maize monoculture and the appearance of Dabrotica virgifera. Since no efficient insecticides are available against this pest, the government has forbidden maize cultivation in some districts to prevent the pest from spreading. This situation should be avoided in Lower Saxony through forward-looking sustainable crop rotation.

Main Cultivation Concept Changes in Lower Saxony Due to Energy Crops

- Energy maize mainly displaced winter wheat
- Number of crops on farms increased significantly from 3.5 to 4.0
- On 80 % of the farms, the humus reproduction demand increased
- Compared to reference crops, nitrogen fertilisation of and pesticide application to energy crops were reduced significantly

6.5 Optimisation of Farm Land Use for Energy, Food and Feed Production

Farmers – especially owners of bigger biogas plants – must consider how to optimise their land use, because biogas plants based on energy crops require much farmland to produce these crops. Table 6.2 provides an overview of how much land is needed to feed a plant depending on the biogas plant's size and the availability of liquid manure.

Table 6.2 Necessary farmland in ha to run a combined heat and power station as a function of increasing electric capacity of power station and livestock units (Karpenstein-Machan 2005)

	Necessary farmland area in ha			
	CHP-electricity capacity			
Livestock units (LU)	100 KW	150 KW	500 KW	1,000 KW
No LU	52	78	260	520
100 LU (=100 cows)	45	71	253	513
200 LU	37	63	245	505
500 LU	15	41	223	483
1,000 LU	0	4	186	446
2,000 LU	0	0	111	371
8,000 LU	0	0	0	24
10,000 LU	0	0	0	0

Fifty-seven percent of Lower Saxony's biogas plants have an electricity capacity of between 200 and 500 kW, while Lower Saxony's average plant size is 520 kW (ML 2010). As can be seen in Table 6.2, about 260 ha of farmland is needed to feed a 500 kW biogas plant. If liquid manure is available and used as a substrate, less farmland is necessary. However, due to liquid manure's low energy concentration, the savings are low. The manure of 8,000 live stock (manure of 8,000 cows/year) is necessary to operate a 500 kW_{el} biogas plant without energy crops. Fairly large quantities of manure are necessary to reduce the energy crop input from farmland.

Biogas production from farmland is still very land-use intensive. According to Table 6.2, 1 m^2 land can produce about 1.5 kWh electricity and 3 kWh heat energy. Approximately 50 % of the produced heat energy is needed to heat the fermentation tank, which means the usable heat energy is reduced to 1.5 $kWh_{thermal}$. Compared to photovoltaics (PV), biogas's efficiency is relatively low (Pimentel 2008). In northern Germany's climate, PV can produce approximately 100 kWh/m^2. In terms of land-use efficiency, that of PV is 33 times higher than that of biogas. However, it should be kept in mind that biogas is a renewable resource in rural areas. All types of wet biomass, crop residuals, manure and organic waste materials can be used for biogas production. Nevertheless, land use and biogas production need to be optimised to prevent negative effects – such as competition for land and unfavourable conditions for other production lines (food, husbandry, renewables for industrial uses) and to protect nature. Sustainable projects such as Jühnde's bioenergy village show that 83 % of the energy produced by the biogas plant is utilised for electricity and space heating. The energy input/output ratio of the biogas plant is high due to the village households largely using biomass heat energy for space heating (see Chap. 2).

Figure 6.4 shows possible pathways to optimise energy crop production. All the items provide optimisation without using more fertilisers and pesticides. Through those agricultural treatments (e.g., higher biodiversity through multi-cropping), crops are adapted to the location, which should increase the crop yield.

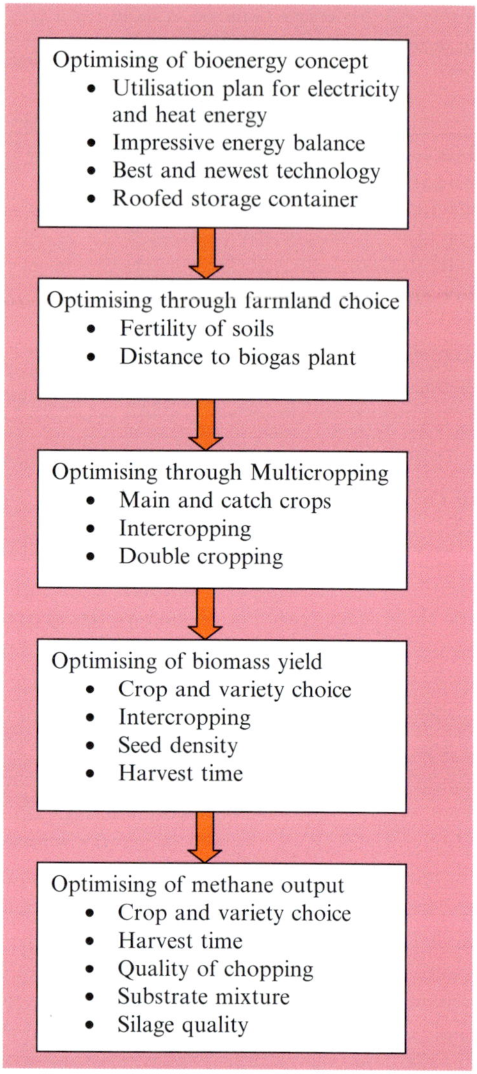

Fig. 6.4 Pathways to optimize energy crop production for biogas

6.5.1 Optimising Farm Land Usage for Biogas

To avoid long transportation routes, energy fields should be located close to the biogas or combustion plant. According to the questionnaires (Sect. 6.4), the farmers prefer fertile soils for annual crops for biogas production, as they mostly cultivate maize. However, other crops well suited to poorer soils are also suitable. Winter rye, which is highly drought resistant and is harvested as a total plant before maturity, may be a better option for poor soils than grain production, due to its shorter vegetation time and its lower risk of having to endure early summer drought.

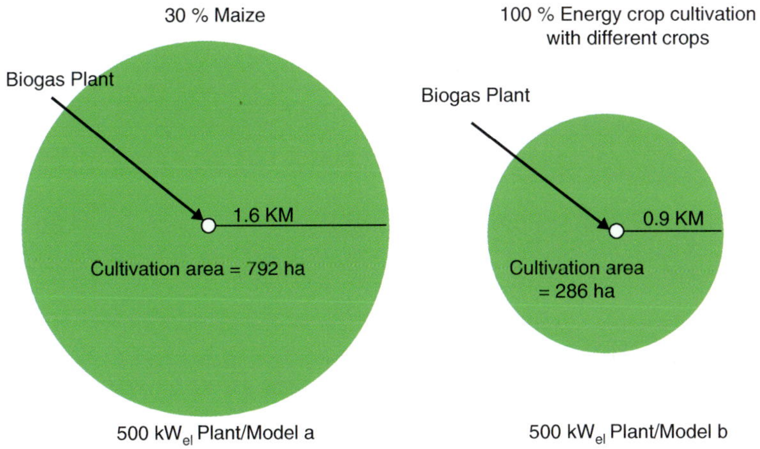

Fig. 6.5 Exemplary catchment area for biomass substrate, calculated for a 500 kW electricity power plant. Model a: only maize is used as substrate, Model b: catchment area for a pure energy crop rotation with three different energy crops

The crops cultivated around the biogas plant determine the size of the biomass catchment area. If farmers intend to feed their biogas plants only with maize, they need a catchment area that is three times larger than if they were to additionally also produce other crops. This is due to these farmers having to fulfil cross-compliance regulations concerning crop rotation diversity. The minimum is a 3-year crop rotation with maize being rotated with two other crops, mostly food market crops. With a pure energy crop rotation (e.g., winter rye/field grass, maize, winter triticale) cross-compliance can be fulfilled and the catchment area is much smaller (Fig. 6.5). The logistic concept can be optimised due to the shorter transportation routes for the harvested biomasses and the fertilisation of the fields with digestive material. In addition, perennial crops (Sect. 6.6.6) can be integratively cultivated in the biogas plant catchment area.

6.5.2 *Optimising Fields Through Multi-cropping*

The current climate conditions in Central Europe allows the cultivation of just one grain crop during vegetation time, while in forage production systems, two or three harvests per year are usual. Catch crops, which were established in the first half of the twentieth century, extended the fodder period in summer, which meant more farmland area could be used for market crops:

- Catch crops could supply high quantity and quality forage
- Catch crops delivered silage, which could be utilised during winter.

Biogas farmers can benefit from these experiences with fodder production concepts. Catch crops and multi-cropping concepts increase productivity and soil fertility simultaneously (Finckh and Karpenstein-Machan 2002).

Catch Crops

Catch crops are fast growing crops sown between regular crops grown in consecutive seasons. A great number of different catch crops are suitable for feed and energy production, for example, different species and varieties of cabbage, field grass, beans, feed pea, winter rape and Phacelia (see Sect. 6.6.4.)

6.5.3 Optimising Biomass Yield Through Intercropping

Beside optimisation through multi-cropping, location-adapted crops as well as crop and variety mixtures also increase the biomass yield. Several field trials show that mixtures yield better and have a better yield stability than pure stands; they are also healthier and need fewer pesticides than pure stands due to their higher genetic diversity (Aufhammer 1999; Finckh and Karpenstein-Machan 2002; Karpenstein-Machan and Finckh 2002). At this point, economic and ecological goals converge. Furthermore, to gain a high biomass yield and utilise biomass crops' potential, it is important to determine the optimal harvest time (see Sect. 6.5.4).

6.5.4 Optimal Harvest Time

The annual biomass crop yield for biogas use is dependent on the optimal harvest time. Plant development follows a growth curve, with diminishing yield increase and increasing lignification towards maturity. To ensure the best harvest time, a high biomass yield (dry matter yield) and the best conditions for bacteria to digest the biomass, plants should reach a dry matter content of 25–35 % (Karpenstein-Machan 1997; Herrmann et al. 2009). In maize and winter cereals, this dry matter content corresponds with the milky to doughy development stage. Furthermore, the dry matter content range in plants is a precondition for high-quality silage. In southern Lower Saxony's climate conditions, winter cereals (rye, triticale, wheat) reach a milky maturity stage between mid-June and mid-July. It is important to choose locally adapted maize varieties that reach the milky maturity stage before the first autumn freeze.

6.5.5 Optimising Methane Output

Biogas farmers are interested in the methane yield per hectare. The biogas plant power station is fuelled by biogas. However, only methane is a burnable gas that can be transformed into electricity and heat energy in a combined heat and power station. Methane (CH_4) and carbon dioxide (CO_2) are the dominant gases in biogas. The CH_4 to CO_2 ratio determines the biogas quality. The methane contents in biogas vary between 50 and 75 %, depending on the input substrates. The fermentation of fatty crops and substrates leads to higher methane contents in biogas. Furthermore, compared to the mono-fermentation of maize, the co-fermentation of liquid manure and dung with crops leads to more stable fermentation and a higher specific methane content (Leonhartsberger et al. 2008). If crops and manure are fermented together, this results in higher digestion rates and higher methane contents in biogas. Anaerobic digestion trials with different crops show the same effect: The fermentation supplies the bacteria with diverse foods that have all the necessary micro-nutrients. This causes higher specific biogas and methane outputs (Leonhartsberger et al. 2008). The diverse energy crop cultivation concept therefore has a strong economic basis. We can thus conclude that:

- crop and variety mixtures lead to a higher biomass yield
- the anaerobic digestion of crop mixtures and manure lead to a higher methane yield.

Furthermore, the methane yield per hectare is influenced by the optimal harvest time, as well as the chopping and silage quality.

6.5.6 Chopping and Silage Quality

Good chopping quality is associated with the harvested material's short and constant chopping length. The shorter the harvested material, the better the biomass can be compacted in the silo and the quicker the lactic acid fermentation can start. Furthermore, short chopping length improves the anaerobic digestion rate in the fermenter. However, more diesel fuel must be spent when harvesting to obtain a short chopping length. Therefore, farmers seek to balance the optimal chopping length and the energy input. In practice, a chopping length between 4 and 40 mm is common.

The right harvest time at the milky to doughy stage of crop development and a short chopping length are the best preconditions for good silage quality. Heiermann et al. (2009) show that ensiled biomass has positive effects on biomethanation, producing higher biogas yields and methane contents than fresh material. They also show that ensiling can be considered a pre-treatment with the potential to also improve methane production from plant matter. To achieve high-quality silage, crops should be harvested, rapidly and well compressed and, as soon as possible after the silo has been filled, sealed tightly with a plastic cover. The plastic cover prevents oxygen from entering the stored material and minimises further biomass decomposition.

6.6 Adapting Cultivation Concepts to a Location

The area requirement for a biogas plant is considerable. Locally adapted cultivation concepts, which enable farmers to exploit income possibilities from biogas under different climatic conditions, are of great importance. For the best methane output, annual crops should be harvested when the kernels are milky to doughy. The biomass is harvested with a fodder harvester. Compared to grain crops, which are harvested about 4–6 weeks later, biomass crop production shortens the vegetation time. The early harvest of bioenergy crops allows additional cropping on the same land, which, in turn, means that, depending on the climate conditions, new cultivation concepts can be introduced:

- The winter main crop in cool and dry locations, for instance on the foothills of low mountains ranges.
- The winter main crop and summer catch crop in cool and moderate wet locations, for instance on the foothills of low mountains ranges.
- The winter catch crops and summer main crops in moderately dry and more temperate locations.
- Two main crops when the climate is very favourable, has sufficient summer precipitation or irrigation and a long summer vegetation period.
- The summer main crop when climatic conditions are dry but the temperatures favourable.
- The perennial crops in moderately dry and moderately wet locations.
- Permanent grassland or perennial forage mixtures in moist and cool regions with a short summer vegetation period.

Figure 6.6 shows energy cultivation concepts adapted to climate conditions with different combinations of winter and summer main crops as well as winter and summer catch crops. All the crops in Fig. 6.6 are energy crops but for different utilisation purposes. Grain crops can also utilised for human nutrition and bioenergy crops as fodder for cattle.

Climate conditions are defined by means of the soil moisture level (SML) and summer vegetation period length. The SML characterises a location's moisture situation. Pedological, hydrological, morphological and climatic parameters influence the SML (LBEG 2011). The summer vegetation period length is defined as the number of months with a daily average temperature of more than 10 °C. In dry locations with short summer vegetation periods, winter annuals (e.g., rye, barley, triticale, rape) generally reach grain maturity. In moderate dry regions with a short summer vegetation period, winter triticale and winter rye yield well. A perennial crop, such as the undemanding Silphie, is also possible (see Sect. 6.5). A further moisture increase allows double-crop farming with winter and summer cereals for biomass or grain use and in keeping with each crop's vegetation time length. In wet and very wet soils, annual or perennial grass-legume mixtures yield well. In locations with higher summer temperatures (5 and 6 months of daily average temperatures of more than 10 °C), more thermopile crops with good dry resistance can reach high

Water supply in the vegetation time	Summer vegetation time in month > 10° C								
	3 month° C	4 month° C		5 month > 10° C		6 month > 10° C		7 month > 10° C	
		Winter crop	Summer crop	Winter crop	Summer crop	Winter crop	Summer crop	Winter crop	Summer crop
Dry (SM 3)		Rye, Barley, Triticale, Rape (all grain prod.)			Maize BM, Sunflowers BM Sugar beet		Maize BM, Sugar Millet BM Amaranth BM		Maize -CCM Maize grain, Sugar Millet BM Amaranth BM
Moderately dry (SM 4)		Triticale BM, Rye BM; Ethanol wheat; Rape seed		Barley BM	Maize BM, Sunflowers BM	Barley Grain, Triticale BM, Rye BM	Maize BM, Sugar Millet BM Amaranth BM	Rye grain	Maize BM, Sugar Millet BM Amaranth BM
				Winter catch crop (e.g., Phacelia, Mustard)	Ethanol-potatoes	Winter catch crop (e.g., Phacelia, Brassicaceae)	Ethanol potatoes		
		Perennial Silphie		Perennial Silphie		Perennial Silphie			
Moderately wet (SM 5)		Triticale BM, Rye BM; Ethanol wheat; Rape seed	Rye BM; Oat BM	Early Rye BM	Maize BM, Sugar Millet BM	Rape Seed	Maize BM, Sugar Millet BM Amaranth BM		
		Perennials Fast-growing trees: willow, poplar		Perennials e.g., Miscanthus, fast-growing trees: willow, poplar		Perennials e.g., Miscanthus, fast-growing trees: willow, poplar			
Wet (SM 6)	Permanent grass land	Wheat BM Triticale BM, Rape Seed	Rye grass, Landsberger "Gemenge"						
Very wet (SM 7)	Permanent grass land	Permanent grass land or perennial forage, grass-legume mixtures							

Fig. 6.6 Climate conditions as a function of possible crop combinations of winter and summer annuals and cultivation concepts. BM = Biomass

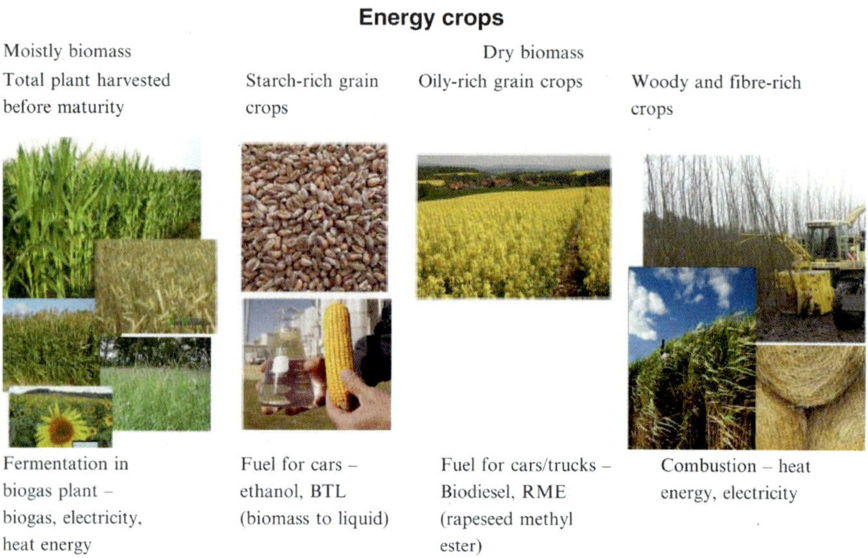

Fig. 6.7 Energy crops and their utilisation lines

yields. In moderately dry and moderately wet conditions, double-cropping systems with many different crops as well as perennials are feasible. Under very favourable summer temperatures (7 months of daily average temperatures above 10 °C), but in dry locations, thermopile subtropical crops such as maize – for corn cobs or grain production –, sugar millet and amaranth yield well.

6.6.1 Characterisation of Energy Crops

Energy crops can be defined as crops utilised for the production of electricity, space heating energy, cooling energy and fuel energy for mobility. Figure 6.7 shows the different utilisations of energy crops according to the maturity stage at which the biomass is harvested and the part of the biomass used for energy production (the total plant or just the grain). Many different crops are suitable for fermentation in a biogas plant. Anaerobic digestion depends on a high moisture content (about 70 %) in the biomass. Therefore, the biomass for anaerobic digestion must be harvested before maturity. The product of the fermentation process is biogas, which can be transformed into electricity, heat and cooling energy. Biogas can be used as fuel for cars with gas engines. Starch-rich grain crops, such as maize, cereals and potatoes, are the raw materials for ethanol production. In Germany, ethanol is mixed with other fossil fuels (gasoline) and utilised as a car fuel. Oil-rich grain crops, such as rape seed or sunflower seed, are used for biodiesel production

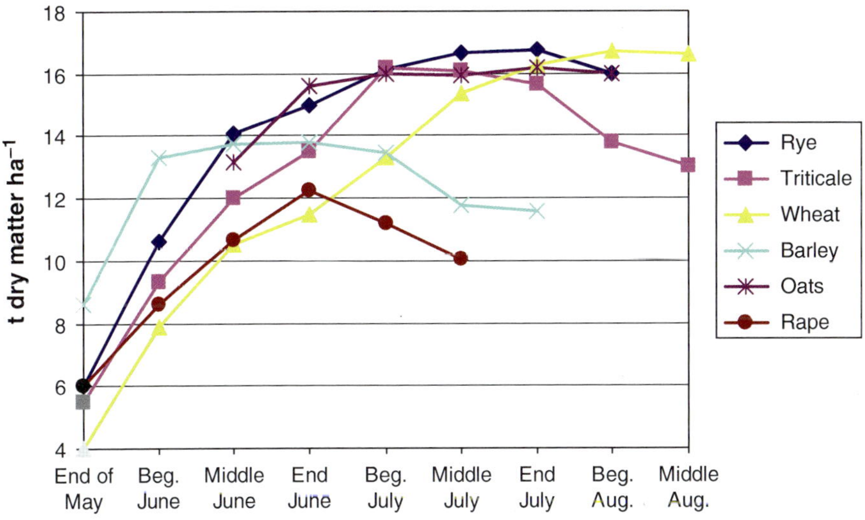

Fig. 6.8 Development of biomass dry matter yield in winter annuals (Karpenstein-Machan 2005)

to fuel diesel cars and trucks. In Germany, nearly one million ha of rape seed are cultivated for this purpose (FNR 2010). Woody and fibre-rich crops, such as fast-growing trees, hemp, miscanthus and straw, which is a by-product of grain production, are suitable for direct use as an energy carrier in a special biomass combustion plant, or, together with fossil energy carriers (e.g., coal), as an energy carrier in a co-firing plant (biomass and fossil fuels are burned together).

6.6.2 Winter Annuals

As energy crops, winter annuals are suitable for locations with cool and moderate climates and locations that lack a high summer precipitation. Winter annuals utilise winter soil moisture to produce biomass in the spring. They already reach the maximum biomass yield in the first half of the year. They are therefore hardly affected by summer dryness. Figure 6.8 shows biomass dry matter yield development and Fig. 6.9 the dry matter content of winter annuals harvested at different times between early May and mid-August. Triticale, rye, wheat, oats, barley and rape have different biomass yield curves that depend on the length of their vegetation time, their development rate and productivity. Cereals and rape reach their maximum biomass yields between end-June and mid-July. Dry matter yields range between 12 and 16 t/ha. Triticale and rye grow in valued locations in southern Lower Saxony; they are the most productive winter energy crops for biogas production. Even on poorer soils, rye and triticale are very productive biomass

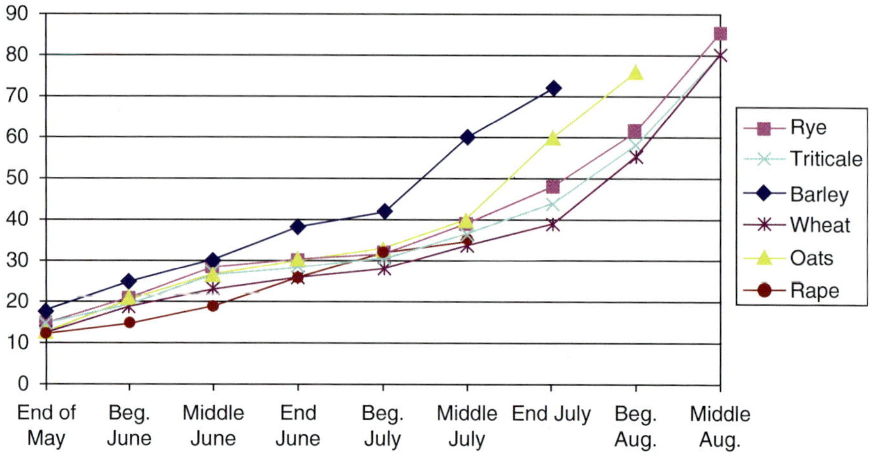

Fig. 6.9 Development of dry matter content in winter cereals (Karpenstein-Machan 2005)

producers (Karpenstein-Machan 2005). As is shown in Sect. 6.5.4 dry matter content is essential for anaerobic digestion, with a content of between 25 and 35 % optimal for anaerobic digestion. Depending on the crop development and climate conditions, the optimal harvest time is between mid-June and early July. A high dry matter yield and optimal dry matter content occur between mid-June and early July in most crops besides winter wheat, which reaches its highest dry matter yield too late for optimal digestion. For digestion, winter wheat has to be harvested before the maximum dry matter yield is reached.

The grain of winter annuals such as winter wheat and winter triticale is very suitable for ethanol production, due to its high starch content. If they are to be used for this purpose, cereals should be harvested at full maturity.

Winter rye and winter triticale are the most suitable of the winter cereals for mixing with winter legumes such as winter vetch (Vicia villosa L.), winter pea (Pisum sativum L.) and winter crimson clover (Trifolium incarnatum L.) (Karpenstein-Machan and Stülpnagel 2000; Aufhammer 1999).

6.6.3 Summer Annuals

Maize (Zea mays L.) is the most important summer annual for biogas production. Currently, alternative crops like sunflower (Helianthus annuus L.), sorghum (Sorghum ssp.) (see Fig. 6.10) and amaranth (Amaranthus spp.) receive attention and have been tested in field trials and in practice. While maize breeding is advanced, breeding work must still be done on sorghum spp. and amaranth to adapt these crops to mid-European climates. However, they have the potential for very high biomass yields in favourable climate conditions.

Fig. 6.10 Different varieties of Sorghum subspecies

Sunflowers are better adapted to mid-European climates because they originated in Middle and North America. They have shown a high biomass yield potential in many field trials (see Fig. 6.11). Sunflowers' high genetic diversity (Khoshbakht and Hammer 2008) offer many possibilities for breeding optimal varieties for biomass use. Furthermore, as a substrate for digestion, sugar and fodder beets are an option to increase the diversity in biogas crop rotations. Farmers have cultivated beets for many years as crops for sugar and fodder production. For biogas production, soil must be removed from the beets after harvesting and they must be chaffed before fermentation. Summer cereals are also suitable as a biomass source, but due to their limited vegetation time, the biomass yield is lower than that of winter cereals. Summer cereals can be used in double-cropping systems as catch crops after a winter annual main crop (see Sect. 6.6.4).

6.6.4 Catch Crops

Catch crops like Brassica napus L., Phacelia tanacetifolia L., Sinapsis alba L., Trifolium incarnatum L., Raphanus sativus var. oleiformis L., Fagopyrum esculentum L., Lolium multiflorum L. as well as summer cereals and sunflowers can be used in double-cropping systems as a complement crop after the main crop. Photoperiod-insensitive varieties can utilise residual vegetation time after the main crop for biomass production. For biogas production, photoperiod-insensitive varieties are sown in June to early July and harvested in the autumn before the first frost. They have a vegetation time of approximately 12 weeks. As green manure, they are not

Fig. 6.11 Herb free buffer strip with sunflowers on maize field

harvested but last through winter and are killed by frost. In field trials, the biomass yields of different catch crops, which are harvested in October, range between 4 and 8 t of dry matter per hectare when cultivated after a winter annual (see Fig. 6.12) (Karpenstein-Machan 2009b).

6.6.5 *Undersown Crops*

Winter main crops and summer main crops can be undersown with other crops. If main crops and undersown crops are sown together in one operation to save time, energy and costs, the undersown varieties in the winter main crops must be winter hardy. The following crops are suitable for this purpose in winter main crops: winter crimson clover (Trifolium incarnatum L.), winter vetch (Vicia villosa L.), winter pea (Pisum sativum L.), ryegrass (Lolium sp. L.) and red fescue (Festuca rubra L.) Different varieties of ryegrasses, red fescue (see Fig. 6.13), or white clover, can be utilised as undersown crops with summer main crops (e.g., maize). Especially with maize, crop competition must be considered. Since young maize plants compete poorly against weeds and other crops, different undersowing concepts have been developed for maize:

1. A very slow-growing grass (e.g., red fescue) is sown before maize seeding.
2. A faster-growing grass (e.g., Italian ryegrass) is sown after maize seeding when the maize has developed four to six leaves.

Fig. 6.12 Fagopyrum esculentum (Buckwheat) and Sinapsis alba (white mustard) mixtures as catch crops in a double cropping system

Fig. 6.13 Maize with undersown red fescue

Concept 1 is very suitable as protection against erosion and nitrate leaching with maize, while, with concept 2, an additional biomass yield can be realised the following spring with ryegrass. Both of these grasses survive winter and protect the soil against erosion and ensure a balanced humus content in soil.

6.6.6 Perennials

Perennial forage crops (e.g., red and white clover, alfalfa, ryegrass) are suitable energy crops. They are mostly cultivated with legumes and grasses and can be utilised for 2–3 years. Depending on the climate conditions and soil fertility, they can deliver several harvests per year.

A very long useful life is anticipated for Silphie (Silphium perfoliatum L.), also known as the cup plant. With its cupped leaves, Silphie can collect air moisture and is therefore relatively resistant to dry conditions. It is adapted to the moderate climate conditions of eastern North America and can be cultivated 400 m above sea level (Conrad and Biertümpfel 2010). Silphie has been cultivated as fodder for cattle in North America and in the former GDR. It was tested as an alternative biogas crop in field trials in Germany from 2005 onward (FNR 2010). In 2010, farmers cultivated Silphie on about 20 ha of farmland. The best results have been obtained when the seeds are sown and nursed in greenhouses and transplanted as young plants with three or four leaves into the fields in May or June (Biertümpfel and Conrad 2013). In the first year, the crop should establish itself in the soil and the plants should only build a leaf rosette before winter (see Fig. 6.14). In the following spring, the plants grow very quickly and can deliver their first harvest in the autumn. The first results show that Silphie has a very high yield, similar to that of maize (FNR 2010). Its advantage is that, after the first year, the crop needs no further weed control and no additional pesticides. However, the seed quality and cultivation concepts must be improved to help broaden Silphie's use as a crop.

6.6.7 Wild Herbs as Biogas Substrate

Some breeders, together with nature protection organisations and seed producers, try to select productive wild herbs as mixtures for biogas (Vollrath et al. 2011). The idea is to combine ecological (a low input of fertiliser and agricultural treatments) and economic aims (a high yield, high methane output, good silage quality). In conventional agriculture, there is a lack of flowerings plants, especially in summer. Bees need flowering herbs' pollen and nectar (bee bread) to survive and reproduce. The newly bred mixtures are perennials, which flower for long owing to herbs' different development rhythms. They change their composition from year to year. In the first year, annuals are dominant in the mixture, but in the following years, high-yielding perennials (shrubs) form the canopy (Vollrath et al. 2011). Further research is necessary to stabilise the yield and other economically important parameters of wild herb mixtures, as well as to multiply the seed mixtures before these concepts can be optimally utilised in practice.

Fig. 6.14 Silphie (Silphium perfoliatum L.) in the first year of development

6.7 Energy Crop Rotation Design

Energy cultivation concepts can be designed as pure energy crop rotations, or as mixed rotations with food, feed and energy crops. Many crops can be used as energy, food or forage crops. However, the cultivation concepts must fulfil the cross-compliance regulations regarding crop diversity and humus balance. At least three crops should be combined in a rotation. The advantage of pure energy rotations is that the plant catchment area for biomass production for the biogas plant is much smaller (see Fig. 6.5) than for mixed rotations with food, feed and energy crop rotations.

Figure 6.15 shows an example of pure energy crop rotations designed as a 3-year rotation with five different crops. In the first year, winter rye is cultivated, followed by Italian ryegrass. The field grass can be harvested twice – in autumn and in spring – before maize is planted in May. Maize is sown in early May with a conventional corn seed drill machine; approximately 2 weeks later, fescue is sown with a pneumatic seed drill between the rows of maize. The later fescue seed gives maize a head start and the grass develops very slowly under the maize canopy and does not compete with the maize (see Fig. 6.13). After the maize harvest in October, the fescue continues to grow and builds a stable green cover against soil erosion over the winter. The vegetative growth of fescue ends with the ploughing at the end of April. While red fescue is generally not harvested owing to its low yield, the grass adds much subsoil and root biomass and is inserted to protect the soil over winter, increases the soil's carrying capacity and supports humus reproduction (see Fig. 6.16). The last crop in the rotation is sunflowers. Sunflowers interrupt cereals' cultivation sequence before the rotations restarts with winter rye as a biomass crop. With the recycling of digestate, the soil's humus content can be kept in balance for biogas production with this pure energy crop rotation.

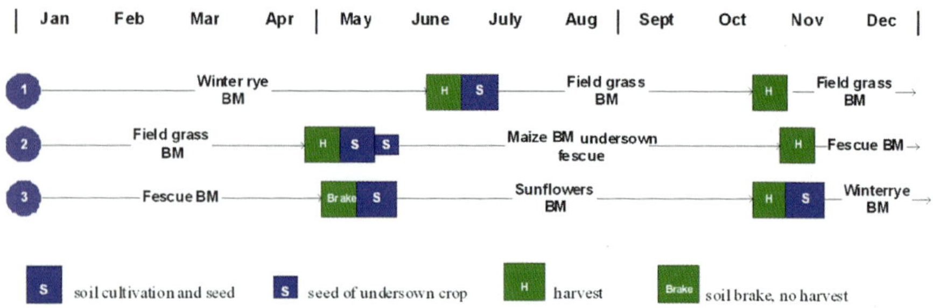

Fig. 6.15 Example for pure energy crop rotation

Fig. 6.16 Subsoil root biomass of red fescue delivers humus reproduction material

Figure 6.17 shows a mixed rotation with food, feed and energy crops. Eight different crops are involved in this exemplary crop rotation, with a 33 % share of energy crops and a 66 % share of market crops (winter wheat, sugar beet, winter rape). These market crops for food or fodder can also be used as energy crops: Winter rape for biodiesel and winter wheat and sugar beet for ethanol production. Furthermore, sugar beets are currently also utilised for biogas production.

The 6-year rotation starts with winter triticale, winter vetch and a field grass mixture. The development of field grass is reduced by the fast-growing mixture of winter rye and winter vetch. After the harvest of the biomass mixture winter rye/ winter vetch at the end of June, field grass grows swiftly and provides biomass for the biogas plant in autumn. The following crop – winter wheat – can be used as a market crop for food or animal fodder. The wheat crop's straw remains on the field to keep the soil's humus content balanced. The following catch crop helps turn the straw into

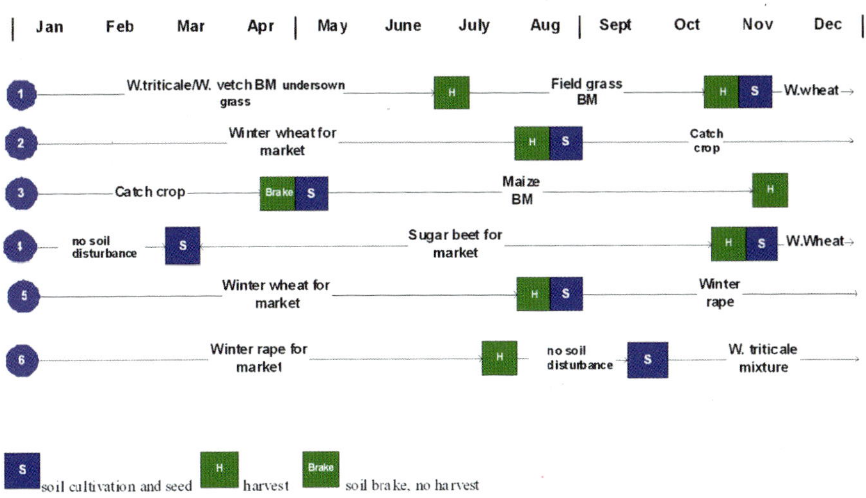

Fig. 6.17 Example for mixed crop rotation with energy and food/fodder crops for market Karpenstein-Machan in Schmuck et al. (2012)

humus and prepares the soil for conservation tillage systems. Maize is then sown by mulch till or strip till. After the maize harvest, the soil is not disturbed until March or April for sugar beet drilling. Sugar beets are used for sugar production and are a market crop. Two further market crops follow with winter wheat and winter rape. These two grain crops, with their straw residuals, are necessary for humus reproduction because maize and sugar beets are very humus-draining crops. After the rape has been harvested, the soil remains undisturbed until September, when the rotation restarts with a mixture of winter triticale, winter vetch and field grass for biogas production.

6.8 Model Farms as Lighthouse Projects

6.8.1 Why Model Farms?

During the district partner selection process (see Chap. 11), we also looked for suitable partners to research agricultural questions. The farmers were given an opportunity to express their willingness to cooperate with the university team and answer a questionnaire. Representatives of agricultural organisations distributed the questionnaire to farmers with energy crop production. The questionnaires also gave the farmers the opportunity to have their farms recognised as model farms. In three selected districts, we began to cooperate with three interested farmers, who were keen to try out new approaches to improve their cultivation concepts. The model farm initiative sought to develop new ecologically and economically optimal cultivation concepts for bioenergy, food and forage crops with the help of the farmers. They would act as new project leaders to motivate other farmers to change their cultivation systems to obtain increased sustainability and productivity.

6.8.2 *Characterisation of Model Farms*

6.8.2.1 Farm Types and Biogas Plant Operation

Table 6.3 provides an overview of the model farm types and biogas plant specifics. Farm A uses most of its farmland for energy production (80 %). Maize, rye kernels and a part of the sugar beet cultivation are supplied to the biogas plant. Farm B uses only 10 % of its farmland for energy production, mainly for maize production; the market crops winter wheat and sugar beets are produced on 80 % of its farmland. Farm C produces fodder for dairy cattle on farmland (50 %) and grassland (25 %). On the remaining 50 % farmland, energy crops for biogas (mainly maize) are produced. Liquid manure is only used as an energy carrier in farm C's plant, while the other biogas plants are based on renewable resources from farmland only.

Farm A has a contract with the other farmers to produce energy crops for the biogas plant with an 800 kW electrical capacity. Together, Farms B and C, which operate biogas plants in cooperation with other farms, own enough farmland to operate the biogas plant with their biomass. A combined heat and power station (CHP) is attached to every biogas plant and this, in turn, produces electricity and heat. Electricity is fed into the public grid. However, only farm C's plant has a sufficient heat utilisation concept. Communal industry buildings and private homes are heated with the heat from the combined heat and power plant. At Farm B, the CHP's heat output is used to run an organic rankine cycle (ORC) plant. The ORC process converts heat output from CHP into electricity. Owing to the low temperatures generated, ORC heat output can no longer be used for heating. Much of the heat energy is therefore still unused. In farm B's biogas plant, the biogas process takes place in two fermentation tanks. The hydrolysis process takes place in fermenter 1 and is separated from the biogas production, which takes place in fermenter 2. With these facilities, the different demands of bacteria on temperature, pH-value and nutrient ratio should be fulfilled better. Another technical upgrading is a press that separates the digestive material in a liquid phase and a solid phase. The liquid, water-rich digestate is used as fertiliser on soils near the plant, while the solid phase can be transported over longer distances and is especially used on low humus content soils.

6.8.2.2 Climate, Soil Specifics and Crop Rotations

Table 6.4 show the three model farms' climatic conditions and soil characteristics. The mean annual air temperature increases from farm A, farm B to farm C. Farm A is located close to the Harz Mountains, while the other two farms have a more

Table 6.3 Specifics of the model farms

	Farm type and % area for crops	Capacity of biogas plant	Type of energy plant	Owner/ Operator of the plant	Substrates for biogas plant	Specific feature of the plant
Farm A						
213 ha farm land	Field (20%) and energy crops (80%)	800 kW_{el}	Renewable resources plant with separated hydrolysis	1 farmer, supply contracts with other farmers	Maize; sugar beet; rye cornels	Part of the residuals is separated in the liquid and solid phase, no heat concept
Farm B						
253 ha farm land	Field (90%) and energy crops (10%)	500 kW_{el}	Renewable resources plant	6 farmers in a cooperation	Maize, field grass, winter rye, sunflowers catch crops	Heat output from the combined heat and power plant (CHP) is used in the ORC plant
Farm C						
90 ha farm land, 90 ha grassland	Dairy cattle (75%) and energy crops (25%)	500 kW_{el}	Co-fermentation plant	4 farmers in a cooperation	Maize, cattle manure	Heat output from the combined heat and power plant (CHP) is used for space heating (industry and communal buildings)

favourable lowland climate. Mean annual precipitation is between 600 and 720 mm/ year, which is typical for moderate dry to moderate wet climates. The soil types range from sand, loam and less loam to organic soils. The soil heterogeneity is reflected in the soil fertility code, which ranges from 30 to 100.

On farm A, 35 % of the soil has developed from karst, is rich in limestone and has a low rooting depth. Much of the soil has been irrigated with sewage for more than 50 years; one can therefore assume that it is contaminated with toxic substances. For the past 10 years, the farm's sandy soil has been part of a water protection area. In water protection areas, land utilisation is regulated by water protection guidelines such as the amount and the time period of mineral and organic fertilisers that can be applied. Farm C also produces energy and feed crops in a water protection area. Farm B produces food and energy on very fertile mineral and organic soils (fen soils). About 60 years ago, fen soils (floating grassland) were ploughed and drained and then used as farmland. This land use change has led to soil degradation, carbon loss (humus) and high mineralisation rates (see the Chap. 7). The cultivated crops are oversupplied with nitrogen and other nutrients. Without pesticides, they suffer from plant diseases,

Table 6.4 Climate, soil specifics and crop rotations

	Mean annual air temperature; Mean annual precipitation	Soil Type	Soil fertility code	Soil specific	Current crop rotations	Crop parts in % of farmland
Farm A	8° C, 720 mm	sandy loam and loam	32–82	Karst formation, water protection area, 50–60 years' sewage irrigation	1) w.rye/maize/sugar beet 2) w.wheat/maize/maize/ maize	Maize 65%; sugar beet 15%; winter rye 13%; winter wheat 7%
Farm B	9° C, 600 mm	lossial loam and organic soils	50–100	fen, used as farm land for 50 years, previously grassland	1) maize/maize/s.wheat 2) w.wheat/w.wheat/ w.wheat/sugar beet	winter wheat 53%; sugar beet 28%; maize 10%; summer wheat 6%, winter rye 3%
Farm C	9,2° C, 670 mm	sandy loam and sand	30–50	water protection area	1) maize/w.wheat/ w. triticale 2) maize/maize/maize	maize 63%; triticale 20%; winter wheat 17%

lodging and weeds; pesticide input into these soils is therefore high. Only a few crops are rotated in the crop rotation.

The most frequently cultivated crops are maize, winter wheat and sugar beet. Farms A and C respectively cultivate 65 and 63 % maize on their farmland. On farm A, maize is rotated with winter wheat; on farm B, maize is rotated with winter rye; while maize is rotated with winter triticale on farm C. On farm B, winter wheat production covers 53 % of the farmland. Winter wheat is rotated with sugar beet. On organic soil, maize is rotated with summer wheat. Owing to the high maize demand for dairy cattle and the biogas plant, farm C produces maize in rotation with triticale and winter wheat and in monoculture.

6.8.2.3 State of Ecological Challenges Regarding Current Cultivation Concepts

Table 6.5 provides an analysis of the ecological challenges on the model farms. All farmers operate their farms conventionally, which means the use of mineral fertiliser and pesticides rather than practising biological or organic farming. Owing to the very one-sided crop rotation with only a few crops (mainly maize, sugar beet and wheat), many problems can arise. If maize and sugar beet are cultivated, this means two humus-wasting crops are in a single rotation. Both of these crops start their vegetation time in April to May and are sown in wide rows, taking 4–6 weeks to build a canopy to cover the soil. During this time, maize development is specifically very affected by weeds. Weed management with herbicides or mechanical weed removal is necessary owing to young maize plants' poor competitive power against

Table 6.5 Analysis of current ecological challenges of the model farms

	Farm A	Farm B	Farm C
Crop rotation	Crop rotation with low diversity	Crop rotation with low diversity	Crop rotation with low diversity
Humus	High risk of humus-wasting crop rotation	Humus degradation on organic soils, high risk of humus-wasting crop rotation on mineral soils	High risk of humus wasting crop rotation
Diseases, pests	European corn borer (Ostrinia nubilalis) in maize	Heterodera schachtii	European corn borer (Ostrinia nubilalis) in maize
Soil compaction	Middle to high risk	High to very high risk	Low risk
Soil cultivation	Minimum tillage	Plough, conventional tillage	Plough, conventional tillage
Nitrate	High danger of nitrate leaching on karst soils	Low danger	High danger of nitrate leaching on sandy soils
Digestate recycling	Separation into solid and liquid phases, digestate back to biomass suppliers	Back to cooperation farms	Back to cooperation farms
Pesticide input	Conventional	Very high pesticide input	Conventional
Wind erosion (EFA)	Medium	Medium	Low
Water erosion (EFW)	Low susceptibility	Low to middle susceptibility	Low susceptibility
Water deficiency in summer	−63 to −5 mm	−120 to −180 mm	−130 to −84 mm
Soil water capacity	Low to medium	High	Low to medium
Ground water level	Soils with low, medium and high ground water levels	Soils with low, medium high ground water levels	Soils with low and medium ground water levels
Water protection area	All soils in area under water protection	water protection area borders soils of the farm	All soils in area under water protection and landscape protection,
Nature protection area	Few soils under fauna-flora protection		Few soils under fauna-flora protection

weeds. The uncovered soil at the outset and maize's long vegetation period (until autumn) leads to humus degradation. Furthermore, the amounts of harvest residuals that remain on the field after harvesting are very low. Humus-accumulating crops such as field grass or legumes should be followed by humus-wasting crops. On organic soils, maize and sugar beet cultivation specifically leads to strong humus degradation and enormous greenhouse gas emissions. Other problems are linked to tight rotations, including those of maize and sugar beet. Rotations with poor diversity or monoculture support crop-specific pests and diseases.

Farms A and C have problems with the European corn borer (Ostrinia nubilalis) in maize. In the larval stage, this pest hibernates in the base of maize straw, pupates in May, after which female moths deposit their eggs in clusters onto the underside of maize leaves. The borer larvae bore into the upper part of the maize plant and feed downwards inside the stalk. The older the larvae are, the further they move downwards, and the greater the damage caused. Farm B cultivates sugar beets in high concentration. The soil is infected with thc beet nematode (Heterodera Schachtii), which infects nearly all Brassicaceae species. These pests and diseases are typical effects of low-diversity crop rotations and monoculture.

Sugar beet and maize are harvested with heavy machines in late autumn (October, November) and often in unfavourable weather conditions, which promote soil compaction. The loam and organic soils of farm A and B are more at risk of soil compaction than the sandy soils of farm C.

In combination with cash crop cultivation, minimum tillage and conservation cultivation improve the soil structure and the biological life in the soil. Only farm B cultivates the soil with a plough – the other two farmers use a field cultivator and minimum tillage techniques.

The danger of nitrate leaching is very high in the karst soil of farm A and the sandy soil of farm C. Water protection areas are often allocated where these soils occur.

All three farms' biogas plant digestate is recycled and used on the fields. Since farm A's biogas plant obtains biomass from other farmers and the digestate has to be subsequently transported over a long distance to the suppliers again, a part of farm A's digestate is separated into a liquid and a solid phase. The liquid fertiliser is recycled in the nearby fields, while the solid phase is used to fertilise distant fields.

All farms use pesticides for weed, diseases and insect pest control. Farm B has an above-average input of pesticides on its organic soils. A high mineralisation rate of organic soils leads to plants that are oversupplied with nutrients. The crops are very susceptible to diseases, stem weakness leads to lodging and the high weed pressure reduces the crop yield.

The water and wind erosion susceptibility of all three farms is low. Only on a few of farm B's fields is the soil susceptible to water erosion. Water deficiency in summer is highest on farm B, but its fertile loam and organic soils have a high water storage capacity, which is counter to farm C's sandy soils with their low water storage capacity. Farm A has less water deficiency in summer due to the middle-mountain climate; however, on karst soils with low water storage capacity, early summer dryness can cause problems. The soil condition heterogeneity is reflected in the groundwater level, especially on farms A and B, where soils with low, middle and high ground water levels occur. Farms A and C cultivate their crops in water protection areas, while farm B borders on a water protection area. Furthermore, few of farm A and farm C's soils are under fauna and flora protection.

6.8.3 Implications for Sustainable Crop Cultivation Design on Model Farms

Most of the problems and challenges are due to the low crop rotation diversification. Two crops in a 2-year rotation is an undesirable situation. The minimum should be a 3-year crop rotation with three different crops to prevent crop-specific diseases and pests, humus degradation and soil compaction. The model farms' crop rotations often have a 4-year "rotation", but with only two crops (e.g., winter wheat, winter wheat, winter wheat, sugar beet; maize, maize, maize, rye). These rotations resemble a monoculture more than a crop rotation. The first aim should be to diversify the crop rotations.

Examples are given on how all the farms can optimise their crop rotations regarding their diversity, humus balance, yield stability and economical basis.

6.8.4 Examples of More Diverse Cultivation Concepts

Tables 6.6, 6.7 and 6.8 show the farms' crop rotations, humus accumulation/degradation and the contribution margins before and after the reorganisation. Farm A needs most of its agricultural land for the biogas plant. Therefore, maize, some sugar beets and rye corn were previously used as fodder for the biogas plant. During the crop rotation reorganisation, high biomass production, well-designed crop rotations to improve the crops' yield and yield stability, and achieving a balanced humus-soil content to maintain or increase the soil fertility were very important (Table 6.7). The old crop rotations wasted humus and a humus-soil balance was only possible through external purchase of manure in keeping with cross-compliance regulations. The new crop rotations are well balanced due to the field grass production (field grass after triticale biomass cultivation and ryegrass undersown in maize). The humus content is balanced by means of cereal straw incubation from the grain production and the digestate fertilisation. With the new crop rotations, the crop diversity has been increased from 4 to 7. The new crop numbers per rotation are now much higher.

Owing to the well-designed crop rotations with more favourable pre- and post-crop combinations, positive effects are anticipated on the yield and yield stability. Positive effects on the yield were quantified by using schematic classification tables to calculate a farm's contribution margin (financial revenues) before and after the reorganisation. Classification tables are normally used in organic farming systems to plan crop rotations and to estimate the pre-crop effect on subsequent crops (Kolbe 2006). Karpenstein-Machan (2010) has exceeded Kolbe's (2006) classification table with many bioenergy crops. According to Kolbe, four rankings were established: very favourable, favourable, unfavourable and very unfavourable crop combinations. Crop yields of very favourable combinations show a 10 % surplus on the yield, favourable combination a surplus of 5 %, unfavourable combinations a minus of 5 %, and very unfavourable combinations a minus of 10 % on the yield.

Table 6.6 Crop rotations, humus accumulation/degradation and the contribution margins of farm A before and after reorganization

Farm A			
Old crop rotations		**New crop rotations**	
1. w.-rye/maize/sugar beet		1. w.rape/w.triticale-fieldgras/maize/maize-untersown/w.rye corn	
2. w.wheat/maize/maize/maize		2. w.triticale-fieldgras/maize/sugar beet/summer wheat(corn)	
Cultivation area	**208** ha	**Cultivation area**	**208** ha
Maize	134 ha	W.rape	30 ha
Sugar beet	31 ha	W.triticale/fieldgras	44 ha
W.rye	27 ha	Maize	74 ha
W.wheat	14 ha	Sugar beet	14 ha
		W.rye	30 ha
		S.wheat	14 ha
Crops/farm	4	**Crops/farm**	7
Crops/rotation	2 and 3	**Crops/rotation**	5
Humus/accumulation/degradation in kg C/ha/a			
Old crop rotations		**New crop rotations**	
Crop rotation 1	-712		90
Crop rotation 2	-496		-14
Contribution margin in Euro/farm			
	Before	**After**	
Winter rape		15,495	
Triticale and fieldgrass		26,400	
Maize	54,806	33,300	
Sugar beet	36,898	18,354	
Winter rye	10868	12,000	
Summer wheat		5,364	
Winter wheat	7,953		
Total	**110,525**	**110,913**	

To quantify the crop combinations' effects on the crop yield, numerous crop rotations trails have been undertaken over the last decades (for the results, see Gliemeroth 1964; Klapp 1967; Könnecke 1967; Brouwer 1972; Bachthaler 1979; Baeumer 1990; Christen 1997, 2001). The contribution margins of the old and new crop rotations were calculated, using a farm's averaged crop yields and the exceeded classification table to adjust the crop yield to the crop rotations. The market prices of the last 5 years (2007–2011) were averaged to avoid market volatility affecting the results too much. Farm A's cultivation shows that the new, more sustainable and more diverse crop rotations are economically comparable to the older crop rotations. Through the crop rotation reorganisation, further positive effects, for instance, lower pesticide input and lower fuel energy demand during soil cultivation are anticipated due to the improved soil structure.

Table 6.7 Crop rotations, humus accumulation/degradation and the contribution margins of farm B before and after reorganization

Farm B			
Old crop rotations		**New crop rotations**	
1. Maize/maize/s.wheat		1. W.triticale-fieldgrass/maize/summer wheat	
2. W.wheat/w.wheat/w.wheat/sugar beet		2. S.oats/w.wheat/w.wheat/sugar beet	
		3. Silphie (perennial crop)	
Cultivation area	**253 ha**	**Cultivation area**	**253 ha**
W.wheat	134 ha	W.triticale/fieldgras	13.3 ha
Sugar beet	71 ha	Maize	13.3 ha
Maize	25 ha	Summer wheat	13.3 ha
Summer wheat	15 ha	W.triticale corn	50 ha
W.rye	8 ha	W.wheat	50 ha
		Silphie	13 ha
Crops/farm	4	**Crops/farm**	7
Crops/rotation	2	**Crops/rotation**	3 and 4
Humus/akkumulation/degradation in kg C/ha/a			
Old crop rotations		**New crop rotations**	
Crop rotation 1	-242	Crop rotation 1	268
Crop rotation 2	-60	Crop rotation 2	224

Contribution margin in Euro/farm	**Before**	**After**
W.wheat	78.256	78.100
Sugar beet	83.354	58.700
S.oats	0	19.950
Maize	13.325	7.767
S.wheat	8.445	8.246
TC-fieldgrass	2.400	3.990
W-rye	0	0
e.g. Silphie	0	3.900
Total	**185.780**	**180.653**

Farm B cultivates bioenergy crops on only 15 % of its farmland, because this farm operates the biogas plant in cooperation with three other farms. Its bioenergy crops can thus be mixed with food crops. Approximately 40 ha of the arable farm soils are high-yielding organic soils. For climate protection reasons, it would be better to convert these soils to wet grassland again to avoid further humus loss (see Chap. 7). However, this would imply high reductions in the farmers' income. A compromise could be the cultivation of a perennial crop such as Silphie, which needs no further soil cultivation after planting. Unlike the annual crop cultivation, this would reduce the humus degradation. In own trials, Silphie reached high biomass yields similar to the maize yield (Karpenstein-Machan, unpublished). Further investigations were necessary to evaluate the opportunities and risks for Silphie in organic soils.

Table 6.8 Crop rotations, humus accumulation/degradation and the contribution margins of Farm C before and after reorganization

Farm C			
Old crop rotations		**New crop rotations**	
1. Maize/w.barley/w.triticale		1. Maize/w.triticale-fieldgras-leg/ Maize/w.wheat/maize/w.rye corn-phacelia	
2. Maize/maize/maize			
Cultivation area	**92 ha**	**Cultivation area**	**92 ha**
Maize	58 ha	Maize	46 ha
		W.triticale-Fieldgrass-legume	15 ha
W.wheat	15 ha	W.wheat	15 ha
W.triticale	19 ha	W.rye corn	15 ha
Crops/farm	3	**Crops/farm**	6
Crops/rotation	3 and 1	**Crops/rotation**	6
Humus/accumulation/degradation in kg C/ha/a			
Old crop rotations		**New crop rotation**	
Crop rotation 1	61	Crop rotation 1	54
Crop rotation 2	-816		
Contribution margin in Euro/farm	**Before**	**After**	
Maize	23,722	22,126	
W.wheat	8,415	8,602	
W.triticale	4,465	0,000	
W.triticale-Fieldgrass-leg	0,000	4,600	
W.rye corn	0,000	3,971	
Total	36,602	39,299	

In this example (see Table 6.7), some of the organic soils were allocated for the cultivation of Silphie. Two crop rotations were designed, one with mostly energy crops and one with food crops. Biomass triticale, followed by field grass is a substitute for maize cultivation and stabilises the humus balance. The following maize is thus in a better rotation position, allowing the anticipation of higher yields. Summer wheat completes the rotation, which ensures appropriate seed time for the following winter triticale.

The food crop rotation starts with summer oats. Oats is a very good pre-crop for winter wheat, because it does not multiply "take-all" cereal diseases (Gaeumannomyces graminis var. tritici). In a cereal crop rotation, oats has the same positive effect as a crop shift. This allows a 2-year cultivation of winter wheat. Sugar beets at the end of the rotation represent a second crop shift. The late harvest of sugar beets seldom allows adequate winter crop seeds; therefore, summer oats should ideally follow sugar beets. After the reorganisation, the crop diversity has increased from 4 to 7 crops per farm, while the crop rotation diversity has increased from 2 to 3 and 4. The humus balance now tends towards humus accumulation, which is especially important in organic soils.

As the contribution margin shows, the more sustainable use of organic soils with a perennial crop sometimes leads to farmers suffering income losses. The calculated margins for Silphie in Table 6.7 are based on 15 t dry matter yield and the presently very high costs of young plants and transplantation of about 4,400 euro/ha. More cultivation experiments on a farm scale, knowledge of plant yields and long-term yield stability are necessary to verify and enhance the margins. In particular, seed quality should be improved to avoid the high nursery and transplantations costs of Silphie cultivation. If these breeding problems are solved, Silphie could be a more climate-friendly alternative to annual crops in organic soils.

Farm C needs nearly all its arable land for fodder for its dairy cattle and the biogas plant, which is operated in cooperation with others. As fodder for dairy production, maize reduces the cultivation area for maize as fodder for the biogas plant. Therefore, the farmer cultivates maize on large areas in monoculture. To fulfil cross-compliance regulations, he cultivates the winter crops wheat and triticale on a small scale. With the new 6-year crop rotation, the farmer has many options for feeding his dairy cattle and the biogas plant (see Table 6.8). Winter triticale harvested at the milky stage is very suitable as fodder for the biogas plant. After triticale, a mixture of field grass and alfalfa follows in the same year. Two harvests are possible, the first in the autumn and the second in early May. The mixture of field grass and alfalfa can be utilised either as dairy fodder or for the biogas plant. Maize is now cultivated three times in the crop rotations and is still the dominant crop, but it is now integrated into the rotation with six other crops/catch crops. Owing to maize's better position in the crop rotation, higher yields and a better soil structure are anticipated. The humus content is now in balance and the contribution margin has been increased by approximately 7 %.

6.9 Conclusion: Implementation Opportunities

New crop rotation proposals were developed with the model farms' farmers. On parts of the farmland, new crops and cultivation concepts – for example, undersown seeds, crop mixtures, perennials and herbicide-free buffer strips (Silphie, wild herbs) – were tested. As their experience increases, the farmers plan to reorganise their farms step-by-step to include more sustainable concepts. To increase the implementation opportunities on the model farms and to allow other district farmers to share in this experience, information tours have been organised to these farms (Karpenstein-Machan in Schmuck et al. 2012). Members of nature protection organisations, district politicians, administration officials, journalists and village inhabitants have taken part in these tours. The meeting of people from different groups enriches discussions, since the topics are broader than just cultivation questions, and increases understanding of the different positions. Furthermore, if farmers acquire good media coverage, this increases their motivation to pioneer and establish more sustainable cultivation concepts. Furthermore, on the district scale, energy farmers and the district landscape management can be motivated to work together to support sustainable landscaping, especially regarding the planning and implementation of new energy plants (e.g., biogas plants, woodchip-firing plants and ethanol plants). Different societal groups can thus influence the process and support the development of integrative energy crop cultivation and integrative bioenergy regions (see Chap. 11).

References

Aufhammer, W. (1999). *Mischanbau von Getreide- und anderen Körnerfruchtarten. Ein Beitrag zu Nutzung von Biodiversität im Pflanzenbau*. Stuttgart: Ulmer Verlag.

Bachthaler, G. (1979). *Fruchtfolge und Produktionstechnik*. München: BLV Verlagsgesellschaft.

Baeumer, K. (1990). Gestaltung der Fruchtfolge. In R. Diercks & R. Heitefuß (Eds.), *Integrierter Landbau* (pp. 110–135). München: BLV Verlagsgesellschaft.

Biertümpfel, A., & Conrad, M. (2013). Hinweise zur Bestandesetablierung durch Aussaat. Presentation on "Silphietag" 14. January 2013, Walsrode, Germany. Retrieved May 05, 2013, http://www.lwk-niedersachsen.de/index.cfm/portal/2/nav/74/article/21202.html

BMU/AGEE. (2010). Statistik des Bundesumweltamtes zu erneuerbaren Energien. http://www.erneuerbare-energien.de/inhalt/5468/

Brouwer, W. (1972). *Handbuch des speziellen Pflanzenbaus*. Berlin: Band 1. Paul Parey Verlag.

Christen, O. (1997). *Untersuchungen zur Anbautechnik nach unterschiedlichen Vorfruchtkombinationen*. Kiel: Habilitationsschrift an der CAU.

Christen, O. (2001). Ertrag, Ertragsstruktur und Ertragsstabilität von Weizen, Gerste und Raps in unterschiedlichen Fruchtfolgen. *Pflanzenbauwissenschaften, 5*(1), 33–39.

Conrad, M., & Biertümpfel, A. (2010). *Optimierung des Anbauverfahrens für Durchwachsende Silphie (Silphium perfoliatum L.) als Kofermentpflanze in Biogasanlagen sowie Überführung in die landwirtschafliche Praxis*. Abschlußbericht. Retrieved January 18, 2013, from http://www.tll.de/ainfo/pdf/silp0111.pdf

Federal Statistical Office. (2012). *Sustainable development in Germany* (Indicator Report 2012). Wiesbaden. Retrieved October 30, 2012, from https://www.destatis.de/EN/Publications/Specialized/EnvironmentalEconomicAccounting/Indicators2012.pdf?__blob=publicationFile

Finckh, M. R., & Karpenstein-Machan, M. (2002). Intercropping for pest management. In D. Pimentel (Ed.), *Dekker encyclopedia of pest management*. New York: Encyclopedia Entry | Print Published: 02/26/2002 | Online Published: 02/07/2002, 423–425.

FNR. (2010). *Cultivated area of energy crops*. Retrieved October 30, 2012, from http://www.nachwachsenderohstoffe.de/service/daten-und-fakten/anbau/?spalte=3

Gliemeroth, G. (1964). Nachhaltige Fruchtfolgewirkungen. *Hohenheimer Hochschultage, 64*, 11–25.

Heiermann, M., Plöchl, M., Linke, B., Schelle, H., & Herrmann, C. (2009). Biogas crops – Part 1: Specifications and suitability of field crops for anaerobic digestion. In Agricultural Engineering International: The *CIGR Journal (Eds.), Manuscript 1087, Vol. XI*. Retrieved January 18, 2013, from http://www.cigrjournal.org/index.php/Ejounral/article/viewFile/1087/1193

Herrmann, C., Heiermann, M., Idler, C., & Hertwig, F. (2009). Einfluss von Trockenmassegehalt (Abreife), Häcksellänge und Silierung auf die Biogasausbeute von Mais. *Landbauforschung Völkenrode, Sonderheft, 331*, 87–105.

Karpenstein-Machan, M. (1997). *Konzepte für den Energiepflanzenbau*. Frankfurt: DLG-Verlags-GmbH. ISBN 3-7690-0546-5.

Karpenstein-Machan, M. (2000). Nachhaltiger Energiepflanzenbau ohne Herbizide – Auswirkungen auf Erträge, Wildpflanzenentwicklung und Schadensschwellen, *VDLUFA-Schriftenreihe 53*.

Karpenstein-Machan, M. (2001). Sustainable cultivation concepts for domestic energy production from biomass. *Critical Reviews of Plant Science, Special Issue on Bioenergy, New York, 20*(1), 1 14.

Karpenstein-Machan, M. (2002). Low input energy crop rotations without herbicides. *Pflanzenbauwissenschaften, 6*(1), 36–46.

Karpenstein-Machan, M. (2004). Neue Perspektiven für den Naturschutz durch einen ökologisch ausgerichteten Energiepflanzenbau. *Zeitschrift: Naturschutz und Landschaftsplanung, 2*(36), 58–64.

Karpenstein-Machan, M. (2005). *Energiepflanzenbau für Biogasanlagenbetreiber*. Frankfurt: DLG -Verlags-GmbH. ISBN 3-7690-0651-8.

Karpenstein-Machan, M. (2009a). Bioenergie und Naturschutz im Kontext Nachhaltiger Entwicklung: Energiepflanzenbau nach ökologischen Leitlinie. In Europarc Deutschland e.V. (Hrsg.), *Bioenergie – Fluch oder Segen für nationale Naturlandschaften?* (pp. 26–31). (http://www.europarc-deutschland.de/broschueren)

Karpenstein-Machan, M. (2009b). *Umsetzung eines umweltfreundlichen Energiepflanzenbaus im Bioenergiedorf Jühnde*. Kongressband 18. Jahrestagung des Fachverbandes Biogas (Eds.): Thema: Biogas: Dezentral erzeugen, regional profitieren und international gewinnen (pp. 19–25).

Karpenstein-Machan, M. (2010). Energiepflanzenanbau für Biogasanlagen: Veränderung der Fruchtfolgen und der Bewirtschaftung von Ackerflächen in Niedersachsen. *Naturschutz und Landschaftsplanung, 42*(10), 313–320.

Karpenstein-Machan, M., & Finckh, M. R. (2002). Crop diversity for pest management. In D. Pimentel (Ed.), *Dekker encyclopedia of pest management*. New York: Encyclopedia Entry I Print Published: 02/26/2002 I Online Published: 02/07/2002.

Karpenstein-Machan, M., & Stuelpnagel, R. (2000). Biomass yield and nitrogen fixation of legumes monocropped and intercropped with rye and rotation effects on a subsequent maize crop. *Plant and Soil, 28*(1/2), 215–232.

Khoshbakht, K., & Hammer, K. (2008). Species richness in relation to the presence of crop plants in families of higher plants. *Journal of Agriculture and Rural Development in the Tropics and Subtropics, 109*(2), 181–190.

Klapp, E. (1967). *Lehrbuch des Acker- und Pflanzenbaues* (6th ed.). Berlin: Verlag Parey.

Kolbe, H. (2006). Fruchtfolgegestaltung im ökologischen und extensiven Landbau: Bewertung von Vorfruchtwirkungen. *Pflanzenbauwissenschaften, 10*(2), 82–89.

Könnecke, G. (1967). *Fruchtfolgen*. Berlin: VEB Deutscher Landwirtschaftsverlag.

LBEG. (2011). *Bodenkundliche Feuchtestufen*. http://www.lbeg.niedersachsen.de/live/live.php?navigation_id=836&article_id=793&_psmand=4

Leonhartsberger, Ch., Bauer, A., Machmüller, A., Kryvoruchko,V., Bodiroza, V., Milovanovic, D., Hrbek, R., Amon, B., & Amon, T.(2008). *Sustainable crop rotations and their potential for biogas production*. In The "Nikola Poushkarov" Institute of Soil Science and Institute of Animal Sciences, 13th RAMIRAN international conference potential for simple technology solutions in organic manure management 11–14 June 2008, Albena, Bulgaria, http://www.ramiran2008.hit.bg/index.htm

ML. (2010). Ministry of Lower Saxony of Food and Agriculture. *Biogasnutzung in Niedersachsen. Stand und Perspektiven*. http://www.erneuerbare-energien-niedersachsen.de

Pimentel, D. (2008). Renewable and solar energy technologies: Energy and environmental issues. In D. Pimentel (Ed.), *Biofuels, solar and wind as renewable energy systems: Benefits and risks* (pp. 1–17). Dordrecht: Springer.

Rode, M., & Kanning, H. (Eds.). (2010). *Natur- und raumverträglicher Ausbau energetischer Biomassepfade*. Stuttgart: Ibidem-Verlag.

Schmuck, P., Karpenstein-Machan, M., & Wüste, A. (2012). Initiating and analysing renewable energy transitions in Germany. In S. Stemke & A. Dobbelsteen (Eds.), *Sustainable energy landscapes: Designing, planning, and development* (pp. 335–343). Taylor & Francis/CPC Press. Boca Raton, Florida, US ISBN 978-1-4398-9404-0.

Vollrath, B., Kuhn, W., Werner, A., & Degenberg, M. (2011). *Was können Wildpflanzen als Biogassubstrat leisten?* Paper presented at "Fachtagung Energie aus Wildpflanzen", Berlin, April 12, 2011.

Chapter 7
Scale-Relevant Impacts of Biogas Crop Production: A Methodology to Assess Environmental Impacts and Farm Management Capacities

Wiebke Saathoff, Christina von Haaren, and Michael Rode

Abstract The cultivation of biogas crops can affect nature and landscapes in different ways. The increasing loss of permanent grassland, changes within cultivated crops, crop rotations and their spatial allocation within the landscape may have serious impacts on natural assets and commercial ecosystem services. Beneficial or impairing impacts occur at the level of interference (farm level) as well as on broader spatial and/or temporal scales. Governance problems often occur when impacts cross farm boundaries, since farmers have no interest in maintaining a service or avoiding impairments. This is due to the beneficiaries on regional and higher scales often not compensating farmers for the costs of the service at the farm level. Environmental governance should therefore deal with the discrepancies between farm activities that have transboundary relevance and administrative/property borders. Our research questions are:

(i) What kinds of transboundary impacts does biogas crop cultivation have on natural assets or ecosystem services?
(ii) How can the harmful or beneficial impacts on different spatial scales or governance levels be assessed? Where do costs and benefits occur?
(iii) Which biomass production impacts require individual and/or collective responses and which precautionary measures could be implemented to avoid possible impacts?

The purpose of this chapter is to establish an assessment methodology to identify the discrepancies between land-use-related decision competencies and the scope of the resulting impacts.

The assessment method is based on a literature analysis and is developed in three steps:

W. Saathoff (✉) • C. von Haaren • M. Rode
Institute of Environmental Planning, Leibniz University Hannover,
Herrenhäuser Str. 2, 30419 Hannover, Germany
e-mail: saathoff@umwelt.uni-hannover.de; haaren@umwelt.uni-hannover.de; rode@umwelt.uni-hannover.de

H. Ruppert et al. (eds.), *Sustainable Bioenergy Production - An Integrated Approach*, DOI 10.1007/978-94-007-6642-6_7,

1. Establishing a theoretical basis to classify the scale-related impacts of biogas crop cultivation. This theory considers the governance problems that may occur if
   - affected habitats or ecological processes cross farm boundaries;
   - the value of an affected natural asset is relevant on a broader scale (regional or even global relevance);
   - small or insignificant pressures (from the farm-level perspective) occur, as they can have a relevant impact if they occur frequently in a larger spatial context.
2. Classifying the typical pressures and impacts of biogas crops;
3. Integrating these pressures and impacts into a DPSIR framework according to their scale relevance.

This methodology provides a systematic analysis of scale-related problems of fit that occur in biogas crop cultivation. The resulting information on the required individual or collective actions supports the identification of suitable governance measures.

Keywords Biogas crop production • spatial scale • conservation • ecosystem services • biodiversity • climate protection • greenhouse gas emissions (GHG) • species protection • habitat network • on-site impact • transboundary impact • DPSIR

7.1 Introduction: State of Knowledge and Objectives

7.1.1 Impacts Through Biogas Crop Cultivation

As a consequence of different driving forces, such as the strong incentives for energy crop production, biogas crop production has expanded rapidly, accompanied by extensive land use changes. Owing to the rapid expansion of biogas crop production, the maize cultivation area grew by approximately 42 % in Germany between 1999 and 2012 (Statistisches Bundesamt 2002, 2012). This expansion of energy crop production has also increased the competition for land. The biomass production of bioenergy, food, fodder and its extensive utilisation all compete with one another and with nature conservation demands for land. In Germany, and particularly in Lower Saxony, the resulting changes include the conversion of grassland into arable land and the increased use of land that was previously set aside (Nitsch et al. 2010). These land use changes have also occurred in ecologically vulnerable areas, for instance, in areas protected by the flora and fauna habitat directive, in water protection areas, on sites vulnerable to erosion and in areas with great significance for carbon storage, such as peatlands (Nitsch et al. 2009, 2010; Buhr et al. 2010). Grasslands have increasingly been converted into cropland, particularly on sites that are relevant for CO_2 retention and species protection, such as peatlands (Nitsch et al. 2009, 2010). Further changes are caused by increasing pressure to use arable land more intensively, which is often followed by reduced crop rotation times, the introduction of new energy crops, changes in irrigation practices and an increase in plot sizes. These often have negative effects on ecosystem services, such as the impairment of habitats (definition according to Abercrombie et al. 2008) through the reduction of hedgerows and field margins, changed species composition and the deterioration of landscape amenities (Wiehe et al. 2010; Rodrıguez and Wiegand 2009).

7.1.2 Problems of Scale

The described unwanted landscape changes through biomass cultivation are partly due to scale problems. They occur if the (e.g., economic) interest on the farm level differs from that on the higher levels (e.g., regional habitat network), or if the farmer overlooks the effects on the higher scales. The terms "level" and "scale" will be used in this paper as follows: The term "scale" describes the definite spatial or temporal boundary of a quantitative entity, whereas "level" is defined as a unit of organisation (Allen 1998), which can be also spatially defined, confined by political boundaries.

Land use changes and intensification can adversely affect natural assets, such as animal species diversity and population density, if energy maize is cultivated on a large area (Rode et al. 2010; Reich et al. 2011). However, maize cropping can also result in beneficial effects if it diversifies the crop rotation, thus enriching the habitat supply for animals (Reich et al. 2011). Positive and negative effects can occur at the level of interference on the farm scale, but also on a broader spatial scale. In the latter case this occurs if, for instance, many farmers act similarly and all introduce maize resulting in large areas with monocultural maize cultivation. On a broader temporal scale (over longer time spans), such changes may contribute to gradual global warming caused by the GHG (greenhouse gas) emissions (IPCC 1996). The Brundtland Report and others have acknowledged the importance of considering temporal and spatial scales in environmental management (World Commission on Environment and Development 1991). Understanding an impact's spatial (and temporal) extension is necessary in order to identify the sources of a problem and to implement measures to prevent impacts, or to rehabilitate affected ecosystems.

7.1.3 Information and Methodology Deficits Regarding Managing Scale-Related Environmental Conflicts

Environmental impacts and their spatial dimensions caused by biogas crop cultivation are seldom foreseen or acknowledged on the spatial scale where crop cultivation decisions are made (the farm level) (see Wiehe et al. 2011). This shortcoming in forecasting is, at least partly, due to a problem of fit between the decision level for crop cultivation and the scale of the resulting impacts. According to the subsidiarity principle, it is preferable to solve environmental conflicts at the lowest possible decision tier (e.g., European Parliament 2000). Applying this principle would imply that as many impacts as possible should be prevented and reduced at the farm level. In order to enable the farmer to accept these responsibilities, he needs information about the imminent environmental impairments and compensation for the management measures he may take that are not in his economic interest. The framework conditions for such management on the farm level, or for issues that cannot be dealt with at the farm level, should be managed at higher decision tiers (EURLex 2002, Art. 174, environmental part of the EC treaty). Spatial planning is a discipline which is capable and qualified to decide on the right level of management. In Germany, spatial planning is the responsibility of forward-looking regulations and the governance of territorial functions. This includes bridging different spatial levels (counter-flow-principle) and acting according to the precautionary principle (Regional Planning Act 2009). Spatial planning has to coordinate different land use demands and deal with conflicts on different planning levels. Specifically, spatial planning, together with landscape planning, should develop, conserve and – if possible – restore soil functions, water balance, flora and fauna, climate and cultural landscapes' functions, as well as their interactions. The spatial requirements of habitat networks, climate protection (climate change mitigation) and climate change adaptation should be considered. Spatial planning should set the stage for agriculture and forestry to help conserve rural areas' natural livelihoods as well as to maintain and design nature and landscapes (e.g., ROG 2009, §2 (1, 5, 6), (Regional Planning Act 2009)). In order to follow the precautionary principle and to prevent potential spatial conflicts, the risk of such conflicts should be identified at the outset (Rode 2006). In addition, to fulfil its scale-related governance tasks, spatial planning requires competencies in managing the financial compensation of land users, who should be motivated to act against their intrinsic economic interests.

To date there has been no systematic analysis of a suitable division of tasks between the regional planning level and the farm level with respect to the scale-related problems that bioenergy production causes. The capacity of the farm level to solve problems has specifically not been systematically examined. According to the subsidiarity principle, knowledge of farm-level capacities could be the precondition to decide on the appropriateness of the decision competencies at higher governance levels. A classification of the scale effects and a methodology that can serve as a basis to identify the adverse effects or benefits of biogas crop management as well as its consequences for responses on different governance levels, are lacking. Providing farmers with knowledge of the impacts that their cultivation practices cause on different scales may improve their capacities to prevent ecological conflicts. However mere knowledge alone may not sufficiently motivate farmers to apply conservation measures. Notwithstanding, this knowledge is also an important basis for governmental institutions to supply incentives or create legal obligations that may support a farm to produce biogas crops sustainably.

7.1.4 *Objective and Outline*

In order to support regional governance institutions in their attempts to solve problems related to biogas crop production, the following questions need to be answered:

- What are biogas crop cultivation's impacts on the natural assets or ecosystem services and how can we recognise and classify transboundary impacts?
- How can the harmful or beneficial impacts on different spatial scales or governance levels be assessed?
- Which response measures are appropriate and on which institutional level should these measures be initiated or implemented?

A methodological concept is presented that helps answer these questions in concrete cases. Applying the methodology allows the spatial scale-related problems originating from biogas crop production to be assessed. The approach identifies potential options for farmers to ecologically optimise their farm management as well as the potential scale-related obstacles that may prevent them from doing so. Furthermore, the methodological concept allows an assessment of whether conservation measures can theoretically be initiated from the farm level or whether supra-local or even supra-regional scale governance initiatives are required.

Since biogas crop cultivation can affect a wide range of natural assets and ecosystem services, we will focus on species and habitat conservation (the habitat function) and the mitigation of greenhouse gas emissions (the climate regulation function) as examples. We also concentrate on the spatial scale and not on the temporal scale.

After describing the development of the methodology (Sect. 7.2), we explore the scale relevance of impacts and propose a test scheme for identifying different decision levels' responsibilities and regulation capacities (Sect. 7.3). Typical impacts of biogas crop cultivation and measures to mitigate them (Sect. 7.4) are used to integrate the test scheme (described in Sect. 7.3) into a DPSIR (**d**riving force, **p**ressure, **s**tate, **i**mpact, **r**esponse) analysis. Thereby, the scale relevance of biogas crop production's possible impacts and response options is assessed. Suggestions are made (Sect. 7.6) on how to use the test scheme and the adopted DPSIR concept to identify the right planning level for response options. Finally, the scale relevance of impacts and responses' benefits and costs is discussed (Sect. 7.7) before a conclusion is drawn about the potentials and restrictions of the methodological concept and their implications for planning and governance practice (Sect. 7.8).

7.2 Methodological Approach

A methodological framework that incorporates specific tasks and methods was developed in order to answer the questions stated above. Table 7.1 provides an overview of the tasks and methods applied to answer the research questions.

Theories about the scale relevance of environmental impacts due to agricultural land management were analysed by reviewing the relevant literature. Scale relevance, which also applies to pressures regarding biogas crop production was then classified (Sect. 7.3). Next, this classification was integrated into the DPSRI analytical framework (European Environment Agency (EEA) 2007) (see Box 7.1), where it was used to demonstrate the scale relevance of potential biogas crop production pressures and impacts. Therefore, examples of potential biogas crop production pressures and potential impacts on the habitat and climate regulation function were collected from the literature. Potential responses to these impacts as

Table 7.1 Sections of the methodological framework: questions, tasks and methods

No.	Question	Task	Method	Chapter
1.	What are biogas crop cultivation's impacts on the natural assets or ecosystem services and how can we recognise and classify transboundary impacts?	Creating a test scheme; Listing the potential impacts of and responses to biogas crop production	Literature review, relevance tree	7.3, 7.4
2.	How can the harmful or beneficial impacts on different spatial scales or governance levels be assessed?	Creating typical showcases for the scale relevance of biogas crop cultivation	Including the test scheme (Chapter 7.3) into the DPSIR analysis	7.5
3.	Which response measures are appropriate and on which institutional level should these measures be initiated or implemented?	Creating a test scheme to identify a potentially adequate decision tier to implement measures and propose an instrumental approach	Literature review, discussion, relevance tree	7.5, 7.6

Box 7.1 The DPSIR Analysis: Driving Forces, Pressures, State, Impact and Responses

The DPSIR (driving force, pressure, state, impact, response) analysis is a methodological structure to assess the impact of a specific pressure or of developments (e.g., the use of resources or land use changes), depending on the physical, chemical or biological condition of a considered site (Hák et al. 2007). Moreover, the method refers to the reason for (the driving forces of) the pressure, such as social, demographic and economic developments in societies and their influence on changing lifestyles, consumption and production patterns. In addition, measures or concepts can be listed to reduce or prevent an impact or response (Hák et al. 2007).

reported in the literature were then listed (Sect. 7.4). Examples from these lists were applied to the DPSIR analysis (Sect. 7.5).

The adopted DPSIR analysis can be used to assess the spatial scale relevance of potential driving forces and pressures of biogas crop production, the state of the affected site and the impacts on the habitat and climate regulation function. On this basis, response measures can be proposed. The DPSIR is a suitable structure for environmental impact studies and to derive practical and governance measures in concrete planning situations (Stanners et al. 2007). Integrating the scale relevance perspective into this structure is a new, still unexplored, step in the context of biogas production as well as beyond.

7.3 Criteria for the Scale Relevance of Biogas Crop Production

7.3.1 Theoretical Background: Problems of Fit

Ecological processes and interactions cross the boundaries of ecosystems and properties. Prey-predator interactions, the nutrient and water supply and other complex ecological relationships create specific vegetation patterns and biocenosis with high spatial scale sensitivity and a variety of ecological system boundaries (Veldkamp et al. 2011). In addition, the boundaries of ecological systems (e.g., cell – tissue – leaf – branch – tree – stand – forest – eco-region) (see Veldkamp et al. 2011) differ vastly from the boundaries of social systems, for instance, from governmental levels such as the local, provincial, national or intergovernmental level (see Cash et al. 2006). However, the impacts on ecological systems, which are relevant on different scales, are often not managed by the most suitable level of the societal system. For example, a habitat is managed on a local level, which has no competencies to include this habitat's function into a regional network. Such mismatches between the level of the decision-making authorities on the one hand and the spatial system levels of de facto ecological impacts, or the related pressure sources and driving forces, on the other are quite common in environmental governance (Lutze et al. 2003).

In the literature, the scale mismatch between the management institution's authority or jurisdiction and the ecological impact is commonly described as a "problem of fit" (e.g., Cash et al. 2006; Young 2002; Folke et al. 2007), or as a "cross-scale", "cross-level" (Cash et al. 2006; Gibson et al. 2000) or "transboundary" problem (Cash et al. 2006). This is especially true if the responsibility is located at a lower level than the reach of the ecological relevance. The conservation of ecological processes that transcend the boundaries of single jurisdictions, such as species migration between habitats, or the climate regulation function, is a major challenge for governance (Young 2002; Cash et al. 2006). Such discrepancies between ecological areas and processes as well as decision-making authorities' spatial scope of responsibility often result in unsustainable resource management (Folke et al. 2007). For example, the protection of a globally threatened species will always be a challenge for a regional authority where this species is still abundant. A solution could be to assign decision competences to higher administrative levels if the areas, processes, or the cumulative impacts of many single decisions (pressures) cross the borders of the own responsibility scope. Assigning decision competencies to higher governmental levels is also recommended if the affected natural asset is locally common but rare or even threatened at the higher level (Haaren et al. 2012). However, as in our example of a globally threatened species, protection would be difficult to implement from very high decision levels. Alternatively, divided competencies (e.g., legislation or incentives from higher decision tiers but implementation at a low level) could prevent problems. Not least, environmental impact management can only be successful if we know the spatial scale relevance of the pressure, state, impact and response options. Adequate information is a precondition for scale-sensitive governance. The DPSIR model can structure the modelling of future or existent ecosystem functions and services' impairments as well as the role of responses (management) (Sect. 7.2). All components of the DPSIR model also have a scale dimension. If, for example, an impact like water pollution crosses administrative boundaries because the affected ecosystem processes in a river ecosystem (state) cross these boundaries and the driving forces of the impact (economic frame conditions) are defined on yet another level, then response measures have to take these scale differences into account.

The DPSIR analysis (see Sect. 7.2) assesses the intensity of an impact according to the intensity of the pressure and the state, i.e. value and the sensitivity of the affected natural asset in relation to the considered pressure source. Not only the intensity, but also the scale of an impact is influenced by pressure and state. If we consider the scale relevance of pressure and state, we can also draw conclusion about the scale relevance of the impact and, specifically, about the required response level. This again supports targeted governance actions.

In the following, we define the relevant scale effects related to the pressure and/or state that initially determines impacts' spatial reach. In a next step, these scale effects are included in a test scheme to identify whether an impact is a transboundary or an on-site one. This information is required to identify the response level.

7.3.2 *Scale Relevance of Pressure Sources*

The pressure indicator in a DPSIR analysis describes an action's type and/or intensity, such as the use of land and other resources, as well as the release of substances and the biological and physical agents (Stanners et al. 2007). Beyond the type or intensity, the amount of responsible pressures, i.e. whether there are **single or multiple pressure sources,** also influences an impact's extent (Parker and Cocklin 1993). Individually, the undertaking of a certain farming activity (e.g., the conversion of a single grassland plot into cropland) can be without relevant negative effects for a natural asset (e.g., no complete habitat loss for a depending species, since other grasslands are nearby and migration to these is still possible). Practised by multiple individuals however (e.g., conversion of a whole grassland region), it may cause significant ecological impacts (e.g., regional extinction of species due to regional habitat loss – no habitats left to which species could migrate to) (Parker and Cocklin 1993). According to our test scheme, a transboundary impact occurs as the result of multiple pressures if multiple farmers' management jointly contributes to a compounding or additive impact that goes beyond their individual farm boundaries. We thus presume that the considered natural asset/ecosystem service is not affected by a single pressure, but that multiple pressures are required to seriously disturb the process of the service (e.g., not a single but multiple stressors releasing GHG are responsible for global warming). In the literature, the scale effects of multiple pressures have been described as "space crowding" (Roots 1988) or "structural surprises" (Noble 2010; Peterson 1987; Sonntag 1987; Hegmann et al. 1999).

7.3.3 *Scale Relevance of State*

The state indicator describes the quantitative and qualitative dimensions of the physical, biological and chemical conditions on a certain site/area (Stanners et al. 2007). The state is characterised by the values of the potentially affected ecosystem's functions and their sensitivity to influences (Schenk et al. 2007). The sensitivity describes the extent to which an affected ecosystem function responds to pressures (a positive expression would be resilience). Sensitivity becomes only relevant in case of pressure. If the ecosystem crosses farm boundaries, also pressures outside the farm may lead to on-farm changes (see Table 7.1) in case of a high sensitivity of the ecosystem and vice versa. A common example is a watercourse which will react strongly to pollution and change ecosystem functions and services in different spatial contexts.

Also the value dimension of the affected natural asset/ecosystem's is scale relevant. A transboundary, value-related impact occurs, for instance, if the impaired natural asset/ecosystem service is valuable from a political perspective, or another decision level above that of the farm level (e.g., a nationwide endangered species influenced at the farm level) (Fig. 7.1). Official directives and legislation, or technical

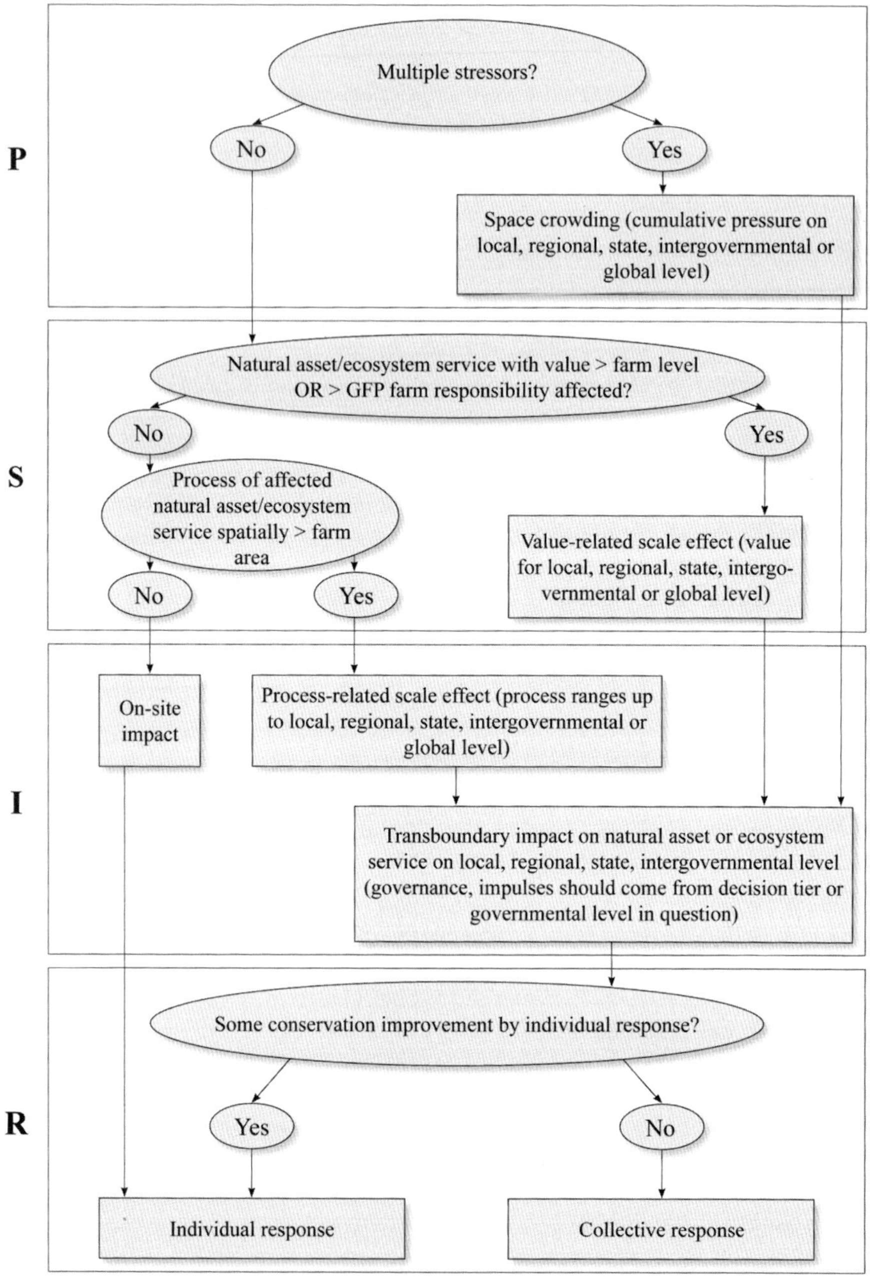

Fig. 7.1 Test scheme: Identifying the scale relevance of pressure, state, impact and response in DPSIR assessments

recommendations – such as the Kyoto Protocol (United Nations 1998), the Wild Bird Directive, (Directive 2009/147/EC on the conservation of wild birds) and the International Union for Conservation of Nature and Natural Resources (IUCN), which publishes the Red List of globally threaten species –, define the spatial value of a natural asset or ecosystem service. According to the benchmarks of species conservation regulations – such as the Directive on the Conservation of Wild Birds (2009/147/EC, Directive 2009/147/EC) – a caused impact's spatial relevance increases when a threaten species is affected. The relevance for the species' general survival is higher if it is globally threatened by extinction (e.g., according to the Red Lists (IUCN 2001)) than if it is a locally endangered population.

7.3.4 *Scale Relevance of Impacts*

The impact indicator of the DPSIR analysis describes the relevance of changes in the state of a natural asset/ecosystem service (Stanners et al. 2007). The impact's spatial extent depends on a combination of the pressure intensity, the site-specific sensitivity (Stanners et al. 2007) and the value of an affected natural asset.

Transboundary impacts can also occur if an impaired biotope or process – such as animal migration or nutrient transportation – crosses the pressure level's boundaries (e.g., a farm) (for the process-related scale effect see Fig. 7.1). We created a test scheme to check whether pressure sources from agricultural land management lead to transboundary impacts by considering all spatial scale effects, such as space crowding and value, or process-related scale effects. This scheme will answer the following questions:

1. Are multiple stressors required to cause a relevant impact on a specific natural asset/ecosystem service (for the **space crowding effect**, see Roots 1988; Parker and Cocklin 1993; Noble 2010)?
2. Does the impact affect natural assets/ecosystem services considered valuable at higher governance levels (**value-related scale effect**)?
3. Do the farm-level (on-site pressure) impacts of biological, physical or chemical processes on an ecosystem exceed farm-level boundaries (**process-related scale effect**)?

In order to answer these questions, the governmental level at which the impact may be relevant should be examined in order to identify a suitable level at which to manage and coordinate prevention or conservation measures. The answers are relevant for planning practice and other forms of governance in order to derive suitable response measures.

7.3.5 *Scale Relevance of Responses*

Land use decisions can respond to impacts by applying measures to prevent, reduce, ameliorate or compensate them, or by adapting to the changes (Stanners et al.

2007). Having identified whether a transboundary or on-site impact occurred and if more than one individual is responsible for it, the next consideration should be whether an individual effort would be sufficient to reduce/prevent this impact, or whether collective actions are required.

If just a single farmer is responsible for an impact, he or she could theoretically address the consequences of the source within his or her scope of competence. Collective efforts (a **collective approach**) are required to reduce an impact if more than one individual is responsible for this impact and if individual measure applications would not lead to improvement. Such collective approaches can be organised by the responsible group of farmers or at a higher government tier by an administration or even induced, for example through public opinion.

7.4 Assessing the Pressures, Impacts and Measures in Biogas Crop Production

At the plot level, biogas crop production's impacts do not differ significantly from those of food and fodder crop production. This is because biogas crops such as maize and cereals are also the main common food and fodder production crops (Statistisches Bundesamt 2012). However, the differences become clearer from the landscape perspective, because a biogas plant's operations may, for example, lead to a change of regional crop rotation by increasing the share of preferred substrate crops (Wiehe et al. 2010). In Germany, this is mainly maize (DBFZ 2011), which is often concentrated in monocultural cropping systems close to biogas plants (Kruska and Emmerling 2008).

The cultivation of single biogas crops such as maize often competes with other spatial demands and may impact ecosystem services such as climate regulation or the habitat function for species (Buhr et al. 2010). Table 7.2 lists the potential general impacts on the habitat and climate regulation function, the underlying pressure factors of biogas crop production and the potential response measures to prevent or reduce these impacts. The main impact of biogas crop production related to feed and fodder production is caused by its monocultural crop production close to biogas plants and its additional demand for land, which result in an intensified use of land (Wiehe et al. 2010). Consequently, the presented impacts and measures mainly refer to the reduction of intensive agriculture's negative impacts on species, habitat and climate conservation. However, the characteristic potential impacts of the biogas sector are mentioned separately in Table 7.2.

Further potential impacts can occur if food and fodder crops are replaced with biogas crop cultivation, through different cultivated crops' water consumption, through machine operations, tillage, humus depletion, pest control and fertilisation (Wiehe et al. 2010). Intensified nitrogen fertilisation may also lead to higher N_2O emissions and thus impact climate protection negatively. Intensified nitrogen input can be caused due to the cultivation of crops with higher nitrogen demands, or

Table 7.2 Potential pressures and impacts from intensive (biogas, food, etc.) crop production on the habitat and climate regulation function and response measures for impact regulation

Relevant pressures	Impact on species and habitat conservation	Impact on GHG mitigation	Potential response measures
Generally reduced share of summer corn in agricultural landscapes (this has also caused a lower share of stubble fields in the landscape) (Evans et al. 2004).	Many animal species prefer low-growing crops with low density and heterogeneous stands These habitat conditions can be provided if the share of currently rarely cultivated low growing summer crops and grain legumes is increased within crop rotation. This would also provide additional habitat structures due to their phenology, which differs from that of agricultural landscapes' dominant winter crops. A share of 10% to 30% of summer corn and grain legumes within crop rotation can increase species diversity (Fuchs & Stein-Bachinger 2008).	No relevant impact	An increasing share of cultivated summer corn and grain legumes (10% to 30%)
General decrease of stubble fields (particularly of cereals) in agricultural landscapes (Evans	Fauna also needs feeding habitats and hiding places during autumn and winter. Compared to tillage plots, stubble fields have a higher weed density and provide	No relevant impact	Maintenance of stubble fields during autumn and winter

(continued)

Table 7.2 (continued)

Relevant pressures	Impact on species and habitat conservation	Impact on GHG mitigation	Potential response measures
et al. 2004).	plant residues. Since tillage plots dominate in agricultural landscapes, stubble fields are a rare feeding source during this season for, for example, granivorous birds (e.g., Moorcroft et al. 2002). The reduction of stubble fields thus increases the lack of food supply for species in winter, which can threaten their survival in an area.		
The demand for cropland has increased with the extension of biogas crop production. Large areas of permanent (including hydromorphic) grasslands and set-aside land have been converted into cropland, also for the cultivation of biogas crops (see Nitsch et al. 2010; Rode &	The general value of a habitat for species conservation differs with its type. Cropland, for instance, is less important than grassland; intensive grassland is less important than extensive grassland (e.g., Bierhals et al. 2004). Permanent grasslands (Gardi et al. 2002), particularly extensively used pastures (Riecken et al. 2002), permanent pastures or wetland grasslands (Plantureux et al. 2005) are important habitats for many species. Maintenance of an adequate share of grasslands	Maintenance of permanent and particularly hydromorphic grasslands and set-aside areas conserves the soil's organic carbon storages (Neufeldt 2005; Höper 2008) and thus the landscape's climate regulation function.	

(continued)

Table 7.2 (continued)

Relevant pressures	Impact on species and habitat conservation	Impact on GHG mitigation	Potential response measures
Kanning 2010).	at the landscape scale secures habitat and networks for many species. Reducing grassland (especially species-rich grassland) and set-aside land due to the expansion of biogas crops diminishes a farm's overall habitat value.		
Plot enlargements and the associated expulsion of border structures such as margins (Wiehe et al. 2010; Rodrıguez & Wiegand 2009). Consequently, there is a general lack of herbal vegetation cover in agricultural landscapes.	The reduction of blossom habitat structures in agricultural landscapes reduces the gene pools of regional weed species. The clearance of field margins decreases the habitat structures of many species such as insects and their predators. These structures are particularly important in winter, because there is a general lack of overwintering herbal vegetation cover in the agricultural landscape to supply feeding and hiding habitats for a large number of animal species.	No relevant impact	
Whole crop silage within a two-cropping system is	Earlier harvest times coincide with many species' breeding seasons. Earlier harvesting	No relevant impact	Preventing the disturbance of breeding habitats

(continued)

Table 7.2 (continued)

Relevant pressures	Impact on species and habitat conservation	Impact on GHG mitigation	Potential response measures
a typical practice in biogas crop management. Harvest dates are 3 to 5 weeks earlier with regard to whole crop silage compared to food and fodder crops (Sticksel et al. 2010).	disturbs or even exterminates many individuals of a population and leads to poorer habitat conditions for arable weeds and/or animal species (Dziewiaty & Bernardy 2007; Dziewiaty & Bernardy 2010; FNR e.V. 2010)		(especially breeding birds) due to earlier harvesting by delaying harvest dates (Dziewiaty & Bernardy 2007) and/or cultivating other biogas crops[1]
Maize is the main crop for biogas crop production (Weiland 2010). In some regions, there is a tendency to monocultural cultivated maize stands close to biogas plants (Kruska & Emmerling 2008).	As maize is the main crop used in biogas plants, it displaces other crops in the crop rotation. If this leads to a contraction of the crop rotation at the farm level or landscape scale, the survival of arable weed diversity and other dependent species can be seriously affected (Marshall et al. 2003; Stevenson et al. 1997; Murphy et al. 2006). If a crop rotation is dominated by one or more crop species to a monocultural extent, the inclusion of maize can enrich a crop rotation and have positive effects for weed and	No relevant impact	Diversification of crop rotation (at least fourfold per farm, Wiehe et al. 2010)

(continued)

Table 7.2 (continued)

Relevant pressures	Impact on species and habitat conservation	Impact on GHG mitigation	Potential response measures
	other species diversity (FNR e.V. 2010). The more diverse the habitat conditions in a crop rotation in a landscape, the higher the species richness (FNR e.V. 2010).		
Expansion of the cultivation of large-growing crops such as maize, sorghum, etc.	Large-growing crops (preferred for biogas production) shade habitat network elements, such as field margins, more than low-growing crops. This can prevent xerophile species from using margins as a habitat and migration path and can thus affect the habitat network's value.	No relevant impact	Diverse crop rotation at the landscape scale, i.e. no cultivation of taller cultures on adjacent plots. Additionally, establish/and maintain sufficiently broad and sunny field margins, particularly on the south part of fields

[1] according to Dziewaty and Bernardy (2007), impacts on breeding habitats can be excluded by means of a harvest date from mid-June onward

through the conversion of a land use type with lower nitrogen demand, such as extensive grassland, into a land use type with higher nitrogen demand, such as croplands.

7.5 Integration of the Biogas Case into the DPSIR Framework

For environmentally sustainable biogas crop production, farmers need site-specific information to prove whether or not their biogas crop production causes impacts on and/or beyond their farms. Furthermore, they need to know about potential responses and whether individual implementations of various measures can successfully prevent or reduce such impacts. An adaptation of the classical DPSIR concept (see Sect. 7.2) can help decision-making authorities define whether there has been a **transboundary impact** or if an impact is restricted to the own spatial decision scope. This is relevant information in order to clarify responsibilities and check the level at which measures should be applied to prevent or reduce an impact. An analysis with the adopted DPSIR analysis can help assess:

- whether impacts occur at the farm level or whether the spatial expansion of the impacted ecosystem service has a wider reach;
- whether the reach of an impact depends on the type of pressure and its single or multiple occurrence, or
- on the site-specific sensitivity of a considered natural asset or ecosystem service and its value at different spatial levels;
- which measures can help reduce impacts;
- whether measures can be applied individually, or whether collective efforts are required to prevent or reduce an impact;
- whether the driving forces should be changed for an effective solution.

Table 7.3 shows the results of such an analysis by assessing examples of potential biogas crop production pressures on the habitat and climate regulation function.

Table 7.3 shows the dependencies between the pressure and the state of the chosen virtual site examples, which represent potential German agricultural landscapes and their spatial relevance for the climate regulation function as well as for the habitat and habitat network function.

7.5.1 Example 1: Climate Regulation Function

Substantial funding for bioenergy from renewable resources through the German Renewable Energy Source Act (EEG) has stimulated high biogas crop yields and thus increased the demand for cropland (**driving force**). Besides other reasons, such as the decrease in livestock farming, rising market prices for agricultural products and the decoupling of direct payment due to EU agricultural reform (which made land use changes possible), the biogas boom has led to the increased conversion of grasslands into cropland (Nitsch et al. 2010).

Furthermore, the grassland conversion rate in many German federal states has increased rapidly during the past few years (Behm 2008, 2011). The conversion of permanent grassland into cropland (**pressure**) has led to the decomposition of soil organic carbon and, thus, to CO_2 and – to a lesser extent – to N_2O emissions (Janssens et al. 2005; Smith et al. 2004; Soussana et al. 2004). The reduction of carbon storage affects the climate and impairs the climate regulation function of grassland areas (**impact**; Degryze et al. 2004; Del Gado et al. 2003; Lal 2003). The more grassland areas of one soil type are converted into cropland (**multiple pressure, space crowding**), the higher the GHG emissions. However, soil types

Table 7.3 DPSIR analysis: overview of examples of different potential impacts on the habitat and climate regulation function due to biogas crop production

Example	Driving force		Pressure		State[2]		Impact		Response	
	Factor	Governance level	Factor	Single or multiple (S^1) pressure (stressors)	Factor	Scale relevance (P, V^1)	Factor	Scale Relevance (transboundary; on-site)[3]	Factor	individual or collective effort required to reduce impact
1. Climate regulation function	a) EEG (market incentive) b) remuneration for cultivation	a) Federal government b) state, EU	Conversion of grassland into cropland	Single	For instance, extensive grassland on fen soil: very high sensitivity to CO_2 emissions if soil is drained or tilled.	P + V: global	Climate regulation function: increase in CO_2 (and N_2O) emissions in the atmosphere	Transboundary: global ($P + V^1$: global)	a) Conservation of permanent grassland b) avoiding tillage for grassland renovation on hydromorphic soils c) rewetting of drained peatlands	a, b, c) individual
				Multiple (e.g., S: global)	For instance, extensive grassland on fen soil: very high sensitivity to CO_2 emissions if soil is drained or tilled.		Climate regulation function: increase in CO_2 (and N_2O) emissions in the atmosphere	Transboundary: global (S, P + V: global)		

(continued)

Table 7.3 (continued)

Example	Driving force		Pressure		State[2]		Impact		Response	
	Factor	Governance level	Factor	Single or multiple (S^1) pressure (stressors)	Factor	Scale relevance (P, V^1)	Factor	Scale Relevance (transboundary; on-site)[3]	Factor	individual or collective effort required to reduce impact
2. habitat function			Intensified land use through expanded biogas crop production (Klein, Fischer & Sandkühler, 2009)	Single	For instance, diverse landscape with different crops, stubble fields, fallows, hedgerows, etc. Red Kite depends on diverse habitat structures: low sensitivity (compared to landscape with fewer but sufficient structures to provide habitat function for Red Kite (*Milvus milvus*))	P: ~15 km² (hunting ground of Red Kite); V: global	Habitat threat to Red Kite (Milvus milvus)	Transboundary: global (V: global)	a) Cultivation of summer crops (except large-growing crops such as maize, sorghum, etc.) b) conservation and establishment of landscape elements such as stubble fields, field margins, fallows	a, b) individual
				Multiple (e.g., S: regional)			Habitat function: habitat threat to Red Kite (*Milvus milvus*)	Transboundary: global (S + P regional; V: global)	a) Maintenance of structural diversity within landscape on higher scale through many single measures (cultivation of summer crops, conservation and establishment of landscape elements such as stubble fields, field margins, fallows)	a) collective

Example	Driving force		Pressure		State[2]		Impact		Response	
	Factor	Governance level	Factor	Single or multiple (S[1]) pressure (stressors)	Factor	Scale relevance (P, V[1])	Factor	Scale Relevance (transboundary; on-site)[3]	Factor	individual or collective effort required to reduce impact
3. Habitat network function			Shadowing of habitats through cultivation of high-growing biogas crops (e.g., maize, sorghum, etc.)	Single	Low: field margin (length 50 m), element of local habitat network with relevance for *Chorthippus apricarius* (target species for connectivity of field margins in open agricultural landscapes), adjacent field margin has habitat quality	P: conquerable distance ~100 m per day (Schumacher & Mathey, 1998); V: /	Habitat network function for *Chorthippus apricarius*: low threat to habitat function for population at plot level, but no threat to habitat network function for local population	On-site	a) Establishment of field margins, particularly on southern sides of plots with tall cultures b) protection of adjacent field margins and/or establishment of field margins nearby to close gaps in habitat network	a, b) individual
				Multiple (e.g., S: regional)	Very high: essential field margins for connectivity maintenance within regional habitat network with relevance for *Chorthippus apricarius*, but	P: > (differ according to species); V: intergovernmental (EU)	Habitat network function (for different xerophile target species): destruction of regional habitat network –	Transboundary: intergovern-mental (S: regional; P>;V: intergovern-mental	a) Arrangements with farmers of adjacent plots to cultivate lower crops next to tall crops; establishment of broader field margins on southern plot borders. b) protection and establishment/	a) individual; b-e) collective

(continued)

Table 7.3 (continued)

Example	Driving force		Pressure		State[2]		Impact		Response	
	Factor	Governance level	Factor	Single or multiple (S[1]) pressure (stressors)	Factor	Scale relevance (P, V[1])	Factor	Scale Relevance (transboundary ; on-site)[3]	Factor	individual or collective effort required to reduce impact
					also for other threatened xerophile species. Partly situated in FFH area (area with Europe-wide protection relevance)		threat of regional extinction of thermophile species. Within FFH area: threat to FFH habitats and species		restoration of habitat network at regional level c) … at federal estate level d) ... at national level e) ... at intergovernmental level (e.g., Europe-wide habitat network NATURA 2000)	

[1] P: process-related scale effect; V: value-related scale effect; S: Space crowding; [2] State = sensitivity and value of a considered natural asset; [3] scale relevance of global, intergovernmental, national, regional, local > above farm level; / not defined)

differ regarding their risk potential for CO_2 emissions. Grasslands with hydromorphic and, particularly, organic soils are, for instance, very sensitive to tillage, while non-hydromorphic mineral soils exhibit a much lower risk of GHG emissions due to grassland conversion (Höper 2008, 2009; Janssens et al. 2005). Therefore, a small area of converted grassland can also lead to higher emissions than those of large converted grassland areas if the smaller area exhibits a higher **risk potential** for GHG emissions due to the site conditions (**state, sensitivity**).

Climate warming is caused by multiple individuals causing GHG emissions (space crowding) on a global scale. Thereby, the impact crosses all existing administrative levels (**process-related scale effect**). According to the Kyoto Protocol, the climate regulation function of sinks and reservoirs of GHG gases is a common good of global relevance and should therefore be protected (Art. 2a, ii; United Nations 1998). Thus, the spatial value of this pressure's impact can be considered global, thus automatically crossing different decision-making levels (**value-related scale effect**).

The Kyoto Protocol proposes sustainable forms of agriculture (Art. 2a, iii; United Nations 1998). **Responses** to reduce or prevent GHG emissions due to farm management are the conservation of permanent grassland, avoiding grasslands tillage and rewetting drained peatlands. To stop global warming, the total amount of GHG should be reduced. Since it is irrelevant which source is reduced in which region of the world, each reduction will show an individual mitigation effect. Responses to mitigate GHG due to biogas crop-production-related pressures can thus also be implemented individually (see Table 7.3)

7.5.2 Example 2: Habitat Function

Expanded biogas crop production can impact the main factors that influence the landscape's habitat function for different animal and plant species (Wiehe et al. 2010). An example of the impact of extended biogas crop production on a habitat function is that of the Red Kite (*Milvus milvus*). The extended monocultural cultivation of renewable resources (**pressure**, driven by the renewable resource bonus of the EEG – **driving force**) – particularly maize for biogas and rapeseed for biofuels – is listed as a main threat to the Red Kite population in Lower Saxony (Klein et al. 2009). The Red Kite depends on diverse habitat structures such as diverse crop rotations (including summer crops) and landscape elements such as fallows, grasslands, stubble fields, etc. (Krüger and Wübbenhorst 2009). Where such feeding habitats have been displaced by maize monocultures, the Red Kite can no longer find enough food to survive (Klein et al. 2009).

Besides the **pressure** factor, the real impact on the Red Kite also depends on the **sensitivity** of the affected natural asset (see Table 7.3). The Red Kite's mobility allows it to search for food within a hunting ground of up to 15 km^2 (Bayerisches Landesamt für Umwelt 2011; Landesamt für Natur 2010). It can cover a distance between nesting and feeding sites of up to 12 km (Krüger and Wübbenhorst 2009). If this area constitutes a multi-structural landscape with sufficient feeding habitats,

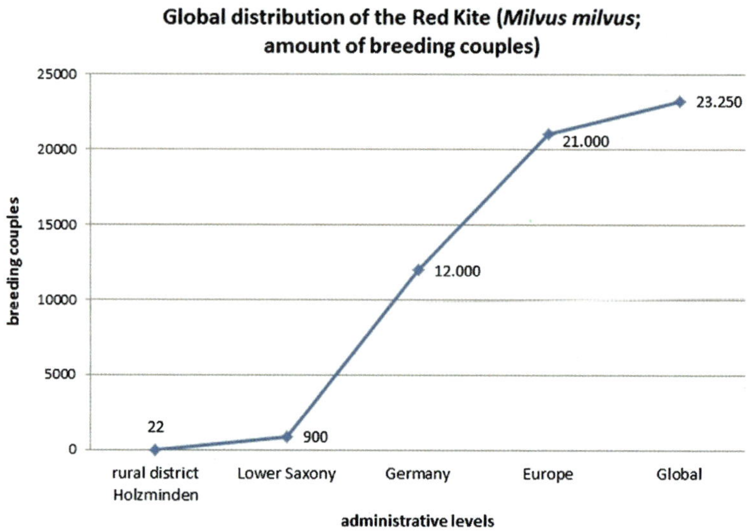

Fig. 7.2 Responsibility of different administrative levels for the Red Kite (Milvus milvus) population according to its global distribution (Data sources: Südbeck et al. 2007; Bird Life International 2011; Klein et al. 2009; Schmidt 2009)

the **sensitivity** to limited structural changes is relatively low, because many single changes are required to destroy the habitat function. Thus, growing maize on one plot will not significantly affect the Red Kite's food supply. However, this may change if a farmer has a large farm and converts larger parts of the Red Kite's hunting ground into a monocultural and monostructural cropland area, or if many farmers in the region do so (**multiple pressures – space crowding effect**). Since this reduces the food supply (smaller mammals, birds), the habitat function will probably be destroyed and the Red Kite population would be threatened. If the process affected by the pressure exceeds the own spatial decision scope – for example, if the converted farm plots previously constituted important unique feeding habitats for one or more breeding pairs of Red Kite within a broader territory – a **process-related** transboundary impact results from the structural changes.

Over 50 % of the global population of Red Kite resides in Germany (see Fig. 7.2; Bird Life International 2011; Südbeck et al. 2007). Consequently, Germany has a global responsibility to protect this bird species (Südbeck et al. 2007) and should protect it although the Red Kite is common in many German habitat regions. Since the Red Kite is listed as near-threatened on the global Red List and in Annexure I of the European Directive on the Conservation of Wild Birds (Directive 2009/147/EC), expanded monocultural biogas crop cultivation on former or potential Red Kite habitat regions in Germany (**high value**) may impact the global population (**value-related scale effect**). The value of the affected population for the maintenance of local, regional and transregional populations, or for the species as a whole, therefore defines the scale of the impact.

According to the European Directive on the Conservation of Wild Birds (Directive 2009/147/EC), the Red Kite should be protected. Specific conservation areas and measures should therefore be implemented to guarantee its survival and reproduction in its distribution areas (Directive 2009/147/EC). If the impact on the Red Kite is low and caused by single or few pressures, **responses** such as the cultivation of summer crops (other than large-growing crops such as maize, sorghum, etc.) can be applied **on individual farms**. However, if there is a broader spatial impact, it has to be reduced through a **collective response**, since the Red Kite depends on spacious structural diversity in landscapes. This would imply thc need for coordination on higher decision tiers. A single farmer's adaptation measures cannot create a connecting, diverse landscape.

7.5.3 Example 3: Habitat Network Function

Large-growing biogas crops, such as maize, sorghum, etc., shade field margins, which are important habitats and habitat network corridors for many xerophile species (Table 7.3). Shading field margins (**pressure**) can impact the habitat network function of xerophile species such as *Chorthippus apricarius* (locust species). *Chorthippus apricarius* has its main distribution in open, extensively used agrarian landscapes. It requires very high summer temperatures and ground exposed to sunlight (Grein 2005), which means its sensitivity to shading is high. This species uses field margins as habitat and as a corridor to migrate to adjacent habitats. If a formerly sunny field margin (e.g., a field of low-growing summer wheat with little shade effect adjacent to a field margin) with a *Chorthippus apricarius* population is shaded by changing the cultivation from low-growing to high-growing (energy) crops (**single pressure**), the population will probably lose this habitat. Since the species can cover a distance of approximately 100 m/day (Schumacher and Mathey 1998) and the affected field margin in the example only is only 50 m long, the population can still migrate to the adjacent field margins provided that their site conditions comply with this species' demand (low **sensitivity**). Thus, the **impact** of one shaded (shorter) field margin on the species existence will probably be low. Pursuant to the example of a single pressure on a German farm, an affected *Chorthippus apricarius* habitat would constitute an on-site impact, because the species has no particular protection status in German law, i.e. there is no **value-related scale effect**. In contrast, the impact on *Chorthippus apricarius* can be higher if many plots in one area have large-growing crops (**multiple pressures, space crowding**). The species has a very short activity radius (approx. 100 m/day, Schumacher and Mathey 1998) and shading the field margins on a broader scale will remove potential migration corridors and habitats. Thus, the affected population cannot migrate to other habitats and may become extinct there (**impact**). Since the impact of the multiple pressure within example 3 (Table 7.3) occurs partly in a flora-and-fauna habitat (FFH) area (NATURA 2000, European protection area; Council Directive 92/43/EEC of 21 May 1992) and FFH areas are affected, the result is a **transboundary impact** of Europe-wide relevance (**value**). Furthermore, the transboundary impact can also result from the pressure level exceeding the different species' activity radius (**process-related scale effect**).

Generally, the higher the number of barriers established in a habitat network, the higher the separative effect (Girvetz et al. 2007; Jaeger et al. 2007) and the smaller the chances of populations crossing over or finding new habitats and, thus, surviving (With and King 1999; Jedicke 1990). The higher the number of network corridors established, the higher the likelihood of species migrating to other habitats and maintaining a habitat network. Thus, if owners of adjacent croplands who cultivate maize and other large-growing biogas crops adjust their crop rotations and reduce their cumulated pressure, this can have a positive **response**. An additional measure to maintain a habitat network can be realised by establishing broader, extensive field margins on plots' unshaded southern sites. This measure can improve local habitat conditions, also on single fields, by providing margins large enough for a viable population (**individual effort**).

7.6 Using the DPSIR to Deduce Governance Approaches

Applying the test scheme concerning pressure and impact can help check whether the impacts of biogas crop cultivation can be solved through single-approach, initialising conservation measures on the farm level, or whether upper governmental levels should apply instruments (regulatory, financial, informative or others) to provide incentives (Fig. 7.3).

Individual farmers can prevent or reduce farm-level (on-site) impacts (see Fig. 7.1). Advice from the next administrative level on how to realise good farming practice (GFP) and cross-compliance (CC) standards, or even how to create environmental benefits from biogas crop production related to individual site conditions, can support a farmer. Single approaches can also prevent or reduce impacts if a single pressure causes a transboundary impact (**process** or **value-related scale effect**). On the one hand, measures can target the affected natural asset (spatially targeted) by, for instance, proclaiming protection zones and through agri-environmental measures to conserve a specific common good. On the other hand, they can target the individual producer (spatially untargeted) by making advice on adequate land management available or by imposing fines (e.g., if the GFP is violated). However, if an upper-level value is affected, the total impact on the natural asset or ecosystem service can probably only be detected at this upper level. Under these circumstances, governance institutions from the next level should initiate the prevention or reduction of the pressure source by, for example, organising informational support or consultation for the responsible pressure entity.

Since they are caused by multiple individuals on a broader scale (**space crowding**) (Roots 1988), many unsustainable land management practices' impacts do not become visible on a single plot or farm (Ruschkowski and Wiehe 2008; Wiehe et al. 2009; Foth et al. 2007). If a collective approach is required to solve an

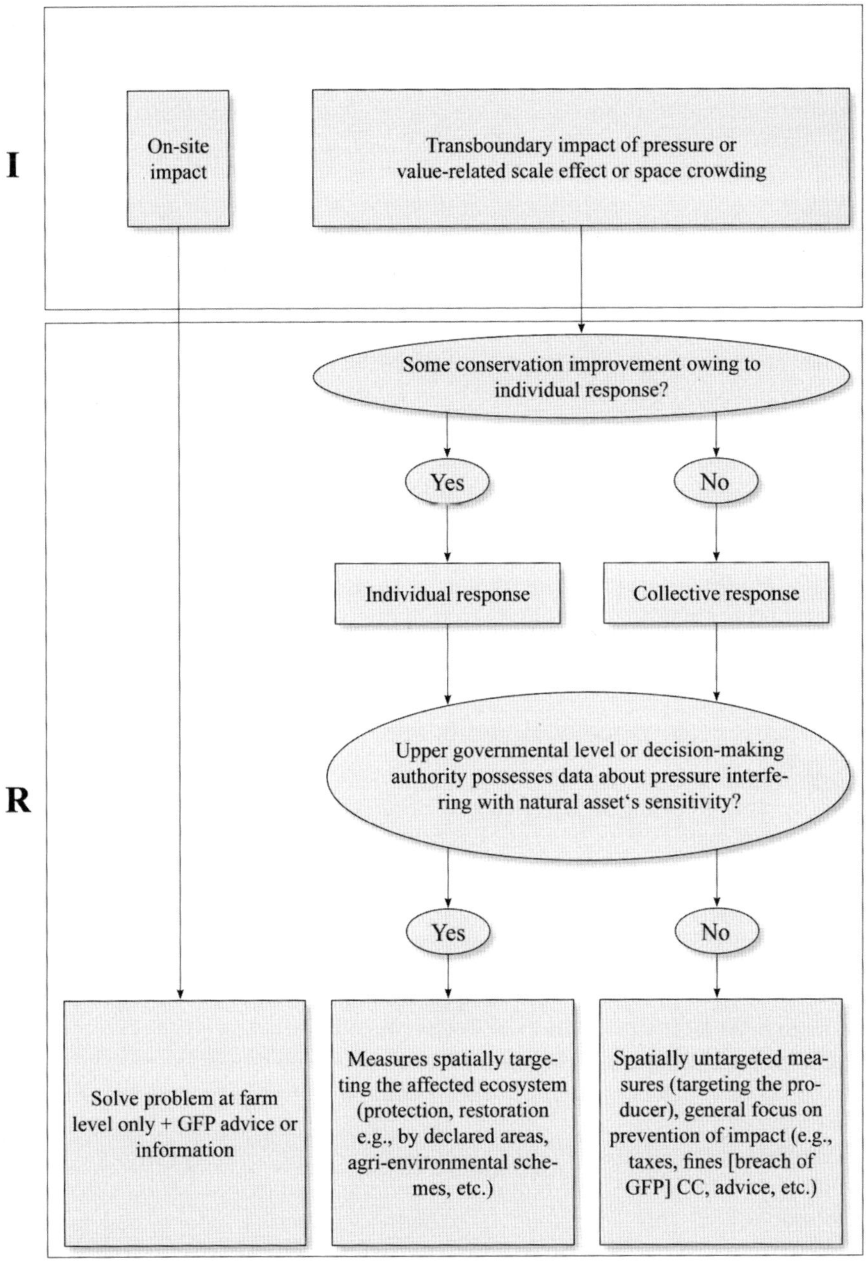

Fig. 7.3 Scale-related instrumental response approaches

impact, the majority of individuals would have to agree to improve their land management themselves. To conserve the performance and functioning of the ecosystem service for the public, a higher administrative level should supervise this by observing, managing and preventing single sources of potential cumulative (for cumulative effects assessment or CEA, see, e.g., Parker and Cocklin 1993; Dubé 2003; Cooper and Sheate 2004; Noble 2010), value-related or process-related conflicts. Authorities and planning institutions at the level in question should estimate the single and cumulated potential pressures and relate them to spatial sensitivities to assess the risk of potential impacts on the natural assets within their spatial administrative boundaries. Government coordination can help effectively design and arrange the various measures applied, if it provides broader spatial data on the environmental context and ecological demands of single sites. This is required to consider ecological interconnectivities (e.g., the network potential of different habitats). If there are data on the spatial interference of pressure and on the vulnerability of a natural asset/ecosystem service (state), governmental institutions can develop measures that target spatial site conditions (e.g., agri-environmental measures, protection areas, etc.). Spatially untargeted measures will have to be implemented if there are no spatially concrete data on how pressure and state interact. However, spatially untargeted measures, such as taxes, the GFP, etc., can also be implemented in addition to spatially targeted measures.

7.7 Scale Relevance of Benefits and Costs

Scale related problems of fit often can be expressed in economic terms. Scale related discrepancies may be cause for beneficiaries of environmental action and those who pay the cost not being identical. As farmers' decisions to apply conservation measures depend very much on the financial costs and benefits of the considered measures (Pannell et al. 2006; Mante and Gerowitt 2006) they need information about costs as well as possible benefits on farm scale. Also they should know about payment schemes for compensation if they are not the beneficiaries of environmental measures themselves. The required information about costs refers to a farmer's expenditure regarding his labour, worker wages, machine running times, fertilisers, other materials, etc., in order to apply a particular measure. Since a conservation measure's costs depend strongly on the site conditions, the cost calculations for the farmer should be site specific. The benefits on farm level may include for example to increase revenue from less productive sites by choosing a new crop which cuts cultivation costs and, for instance, reduces soil erosion. However, often the costs occur at farm level but the benefits occur on other levels and no mechanisms are in place to make beneficiaries pay the farmer for producing these benefits. Also the opposite happens: benefits happen on farm scale and costs have to be paid on higher levels. Farmers may, for example, benefit economically from permanent grassland's conversion into cropland if biogas electricity prices exceed milk prices. However, the costs of the GHG emissions released by this land use change are global due to the impacts of global warming. Mostly, governmental institutions will have to pay them to maintain the supply of ecosystem services. This may be very inefficient if the global compensation cost exceed the expenses for avoiding the impacts on farm level.

Therefore the costs and benefits of conservation measures should also be assessed in the light of their spatial distribution. The farmers need information about cost and benefits on farm scale. The government and the public need information about the amount of expenses for external costs or compensation arising for the public (on higher levels) as well as about benefits produced by farms. Such information is a precondition for taking efficient governance measures

7.8 Conclusion

Up to now scale-related problems of fit have been neglected in biogas politics. This chapter proposes methods for analysing these problems on the farm level. The DPSIR scheme has proven a suitable structure for this analysis. If pressure and impact occur on different scales this discrepancy indicates a potential problems of fit. Such a diagnosis allows for analysing or finding response measures in concrete cases as well as judging driving forces and suitable governance schemes.

The proposed assessment scheme for studying the impacts and scale relevance of biogas crop production consists of various lists of possible pressures, impacts and response options as well as the assignment of their possible or general scale relevance. The potential pressures and impacts discussed in this chapter relate mainly to biogas crop production. However, the methodology may also be applied to other agricultural land use sectors. In a concrete case, the impact and scale relevance are assessed in an integrated examination of the pressure and state (value and vulnerability). Supra-farm information about multiple pressures should also be taken into account. The proposed measures (from a general list) can be adapted to conditions of the individual farm. Adequate decision levels and governance strategies for solving problems can then be proposed from a theoretical perspective and are based on the combination of the scale relevance, the number of possible polluters, the spatial allocation and/or the limitation of the impacts. In order to comply with the subsidiarity principle and lead the way to the most efficient governance options, a concept has been developed to explore farms' and farmers' capacities to prevent or reduce their management impacts on their own.

The methodological approach to the assessment as well as the proposal of possible measures should be based on existing research on the impacts of biogas crops. In contrast, the theoretical framing in the context of the scale issue is new, as is the substantiation and adaptation of the DPSIR analysis regarding the scale-related consequences of its different components. This new classification is of great relevance in order to choose the most adequate governance strategy to solve energy plant cultivation problems. However, the assessment concept will have to be tested in future to prove its applicability. Possible difficulties could be data problems, such as missing data regarding multiple pressures. Supra-farm information on multiple pressures will have to be taken from the respective statistical data and scenarios of future development.

In addition, the theoretical approach and the proposed governance strategies will not necessarily always be the most effective way to solve problems. The strategies are based on the general assumption that regulations and decisions should always be taken on the affected (political) tier where the ecological damage and the costs of unsustainable management become clear. While there is a strong logic in this approach and other economic research results point in this direction (e.g., the theory of the tragedy of the commons) (Hardin 1968; Ostrom 1990; National research council, UN 2002), there are also good reasons for assigning as much responsibility as possible to the lowest decision level. A major argument for giving responsibility to the lowest level is that conservation measures are most successful if the individuals affected by conservation measures are involved (i.e. can participate) in the measure implementation process (Schenk et al. 2007). However, successful natural resource management cannot be managed on a single administrative level. Nested systems (see Marshall 2008; Berkes 2002; Ostrom 1990) are required, including the national and local levels and the links between them, as well as the intermediate level (Ministry of Foreign Affairs of Denmark 2007).

An intensive examination of case studies (as outlined in Sect. 7.5) is only a first step in a longer research process that sheds light on the potentials of the farm level to deal with these responsibilities. In future, case studies should lead to better hypotheses regarding the ways in which farmers can be motivated to adopt sustainable management practices and the hindrances along the way. A more extensive quantitatively oriented survey should follow in order to derive results that can be generalised and that can support governance strategies in different contexts and under different preconditions. Nonetheless, in future, it should be possible to adapt such strategies to individual farmers' capacities and willingness. A simplified and adapted version of the outlined survey may be a tool for assessing these individual capacities.

References

Abercrombie, M., Hickman, C. J., & Abercrombie, M. L. J. (2008). *A dictionary of biology*. New Brunswick: AldineTransaction.

Allen, T. (1998). The landscape "level" is dead. Persuading the family to take it off the respirator. In D. L. Peterson (Ed.), *Ecological scale. Theory and applications* (pp. 35–54). New York: Columbia University Press.

Bayerisches Landesamt für Umwelt. (2011). *Rotmilan (Milvus milvus)*. Retrieved from http://www.lfu.bayern.de/natur/sap/arteninformationen/steckbrief/zeige/90254

Behm, C. (2008). *Grünlandverluste schreiten beschleunigt voran: Schleswig-Holstein, Mecklenburg-Vorpommern und Rheinland-Pfalz haben 5 %-Grenze bereits überschritten*. Pressemitteilung Behm, C. (MdB) vom 13. November 2008. Retrieved from http://www.cornelia-behm.de/cms/default/dok/257/257613.gruenlandverluste_schreiten_beschleunigt.html

Behm, C. (2011). *Grünlandfläche nimmt in Deutschland weiter ab: Grünland-Erhaltungsverordnungen wirken, aber keine Entwarnung möglich*. Retrieved from http://www.cornelia-behm.de/cms/presse/dok/372/372285.gruenlandflaeche_nimmt_in_deutschland_we.html

Berkes, F. (2002). Cross-scale institutional linkages: Perspectives from the bottom up. In National research council (Etats-Unis) (Ed.), *The drama of the commons* (pp. 293–322). Washington (D. C.): National Academy Press.

Bierhals, E., von Drachenfels, O., & Rasper, M. (2004). Wertstufen und Regenerationsfähigkeit der Biotoptypen in Niedersachsen. *Informationsdienst Naturschutz Niedersachsen, 24*(4), 231–240.

Bird Life International. (2011). *Species: Red Kite Milvus milvus*. Retrieved from http://www.birdlife.org/datazone/speciesfactsheet.php?id=3353

Buhr, N., Kanning, H., & Rode, M. W. (2010). Raumanalyse II: Auswirkungen auf andere Raumnutzungen. In M. W. Rode & H. Kanning (Eds.), *Natur- und raumverträglicher Ausbau energetischer Biomassepfade* (pp. 91–156). Stuttgart: Ibidem-Verlag.

Cash, D. W., Adger, V., Berkes, F., Garden, P., Lebel, L., & Olsson, P. (2006). Scale and cross-scale dynamics: Governance and information in a multilevel world. *Ecology and Society, 11*(2), 8.

Cooper, L. M., & Sheate, W. R. (2004). Integrating cumulative effects assessment into UK strategic planning: Implications of the European Union SEA Directive. *Impact Assess Project Appraisal, 22*(1), 5–16. doi:10.3152/147154604781766067.

Council. (1992). Council Directive 92/43/EEC of 21 May 1992 on the conservation of natural habitats and of wild fauna and flora. *Official Journal of the European Communities, 35*(L 206), 7–50. Retrieved from http://conventions.coe.int/Treaty/en/Treaties/Html/176.htm

DBFZ – Deutsches Biomasseforschungszentrum. (2011). *Erneuter Zuwachs von Biogasausbau in Deutschland 2010. DBFZ-Umfrage zu Anlagenbestand und Flächenbeanspruchung zeigt einen deutlichen Anlagenzubau im Biogasbereich.* Retrieved from http://www.dbfz.de/web/presse/pressemitteilungen-2011/erneuter-zuwachs-von-biogasausbau-in-deutschland-2010-dbfz-umfrage-zu-anlagenbestand-und-flaechenbeanspruchung-zeigt-einen-deutlichen-anlagenzubau-im-biogasbereich.html

Degryze, S., Six, J., Paustian, K., Morris, S., Paul, E., & Merckx, R. (2004). Soil organic carbon pool changes following land-use conversions. *Global Change Biology, 10*, 1120–1132.

Del Gado, I., Six, J., Peressotti, A., & Cotrufo, F. (2003). Assessing the impact of land-use change on soil C sequestration in agricultural soils by means of organic matter fractionation and stable C isotopes. *Global Change Biology, 9*, 1204–1213.

Directive 2009/147/EC on the conservation of wild birds, European Parliament and the Council of Europe.

Dubé, M. G. (2003). Cumulative effect assessment in Canada: A regional framework for aquatic ecosystems. *Environmental Impact Assessment, 23*(6), 723–745. Retrieved from http://www.sciencedirect.com/science/article/B6V9G-494CHM6-1/2/829d87a5c5f5334bee0ecdd31b18700e

Dziewiaty, K., & Bernardy, P. (2007). *Auswirkungen zunehmender Biomassenutzung (EEG) auf die Artenvielfalt – Erarbeitung von Handlungsempfehlungen für den Schutz der Vögel der Agrarlandschaft*. Retrieved from http://www.erneuerbare-energien.de/files/pdfs/allgemein/application/pdf/endbericht_biomasse_vogelschutz.pdf

Dziewiaty, K., & Bernardy, P. (2010). Brutvögel und Energiepflanzen. In M. Reich & S. Rüter (Eds.), *Umwelt und Raum: Vol. 1. Energiepflanzenanbau und Naturschutz. Referate und Ergebnisse der gleichnamigen Fachtagung in Hannover am 30. September 2009* (1st ed., pp. 115–126). Göttingen: Cuvillier.

EURLex. (2002). *12002E174: Treaty establishing the European Community (nice consolidated version) – Part three: Community policies – Title XIX: Environment – Article 174 – Article 130r – EC Treaty (Maastricht consolidated version)*. Official Journal C 325 24/12/2002 P. 0107 – 0108. Official Journal C 340 10/11/1997 P. 0254 – Consolidated version. Official Journal C 224 31/08/1992 P. 0052 – Consolidated version. Retrieved from http://eur-lex.europa.eu/LexUriServ/LexUriServ.do?uri=CELEX:12002E174:EN:HTML

European Environment Agency (EEA). (2007). *The DPSIR framework used by the EEA*. Retrieved from http://root-devel.ew.eea.europa.eu/ia2dec/knowledge_base/Frameworks/doc101182

European Parliament. (2000). *Subsidiarity: European parliament fact sheets*. Retrieved from http://www.europarl.europa.eu/factsheets/1_2_2_en.htm

Evans, A., Vickery, J., & Shrubb, M. (2004). Importance of overwintered stubble for farmland bird recovery: A reply to Potts: Capsule without this prescription, populations of seed-eating passerines are unlikely to recover. *Bird Study, 51*(1), 94–96. doi:10.1080/00063650409461339.

Fachagentur nachwachsende Rohstoffe (FNR) e.V. (Ed.). (2010). *Standortangepasste Anbausysteme für Energiepflanzen: Ergebnisse des Verbundprojektes "Entwicklung und Vergleich von optimierten Anbausystemen für die landwirtschaftliche Produktion von Energiepflanzen unter den verschiedenen Standortbedingungen Deutschlands, EVA I"* (3., veränd. und erw). Gülzow: FNR.

Folke, C., Pritchard, L., Berkes, F., Colding, J., & Svedin, U. (2007). The problem of fit between ecosystems and institutions: Ten years later. *Ecology and Society, 12*(1), 30. online.

Foth, H., Faulstich, M., von Haaren, C., Jänicke, M., Koch, H.-J., Michaelis, P., & Ott, K. (2007). *Klimaschutz durch Biomasse: Sondergutachten ; Juli 2007*. Berlin: Schmidt. Retrieved from http://www.gbv.de/dms/zbw/540343706.pdf

Fuchs, S., & Stein-Bachinger, K. (2008). *Naturschutz im Ökolandbau: Praxishandbuch für den ökologischen Ackerbau im nordostdeutschen Raum*. Mainz: Bioland.

Gardi, C., Tomaselli, M., Parisi, V., Petraglia, A., & Santini, C. (2002). Soil quality indicators and biodiversity in northern Italian permanent grasslands. *European Journal of Soil Biology, 38*(1), 103–110. Retrieved from http://www.sciencedirect.com/science/article/B6VR7-4593YCJ-P/2/537bb13d2603999ec7a82a4a2d5518ec

Gibson, C. C., Ostrom, E., & Ahn, T. K. (2000). The concept of scale and the human dimensions of global change: A survey. *Ecological Economics, 32*(2), 217–239. Retrieved from http://www.sciencedirect.com/science/article/B6VDY-3YGDC6V-4/2/008fb8e45e8da7e478d44b807324c59e

Girvetz, E. H., Thorne, J. H., Berry, A. M., & Jaeger, J. A. G. (2007). *Integrating habitat fragmentation analysis into transportation planning using the effective mesh size landscape metric*. Retrieved from http://www.escholarship.org/uc/item/6cj9g88f

Grein, G. (2005). Rote Liste der in Niedersachsen und Bremen gefährdeten Heuschreckenarten mit Gesamtartenliste, 3. Fassung, Stand 01.05.2005. Informationsdienst Naturschutz Niedersachsen, 1/2005.

Hák, T., Moldan, B., & Dahl, A. L. (2007). *Sustainability indicators: A scientific assessment*. (International Council for Science & I. ebrary Eds.). Washington DC: Island Press.

Hardin, G. (1968). The tragedy of the commons. *Science, 162*(3859), 1243–1248. doi:10.1126/science.162.3859.1243.

Hegmann, G., Cocklin, C., Creasey, R., Dupuis, S., Kennedy, A., & Kingsley, L. (1999). *Cumulative effects assessment practioners guide: Prepared by AXYS environmental consulting and CEA working group for the Canadian Environmental Assessment Agency*. Hull/Quebec. Retrieved from http://www.coalfreealberni.ca/environmental_assessment/cumulative_effects_a_guide_for_practicioners.pdf

Höper, H. (2008). Treibhausgasfreisetzung organischer Böden. In Umweltbundesamt (Ed.), *Texte: 25/08. UBA-Workshop "Böden im Klimawandel – Was tun?!". am 22./23. Januar 2008* (pp. 105–109). Dessau: UBA.

Höper, H. (2009). Die Rolle von organischen Böden als Kohlenstoffspeicher. *NNA-Berichte, 2009* (1), 91–97.

International Union for Conservation of Nature and Natural Resources (IUCN). (2001). *IUCN red list categories and criteria: Version 3.1*. Retrieved from http://www.iucnredlist.org/documents/redlist_cats_crit_en.pdf

IPCC. (1996). *Climate change 1995: Impacts, adaptation and mitigation of climate change*. (Scientific, technical analysis, contribution of working group II to the 2nd assessment reports of the IPCC). Cambridge: Cambridge University Press.

Jaeger, J. A. G., Schwarz-von Raumer, H.-G., Essweing, H., Müller, M., & Schmidt-Lüttman, M. (2007). Time series of landscape fragmentation caused by transportation infrastructure and urban development: A case study from Baden-Württemberg, Germany. *Ecology and Society, 12*(1), 22. [online]. Retrieved from http://www.ecologyandsociety.org/vol12/iss1/art22/ES-2006-1983.pdf

Janssens, I. A., Freibauer, A., Schlamadinger, B., Ceulemans, R., Ciais, P., Dolman, A. J., & Heimann, M. (2005). The carbon budget of terrestrial ecosystems at country-scale – A European case study. *Biogeosciences, 2*(1), 15–26. Retrieved from http://www.biogeosciences.net/2/15/2005/bg-2-15-2005.pdf

Jedicke, E. (1990). *Biotopverbund: Grundlagen und Massnahmen einer neuen Naturschutzstrategie*. Stuttgart: Ulmer. Retrieved from http://www.worldcat.org/oclc/22036999

Klein, A., Fischer, M., & Sandkühler, K. (2009). Verbreitung, Bestandsentwicklung und Gefährdungssituation des Rotmilans Milvus milvus in Niedersachsen. In T. Krüger & J. Wübbenhorst (Eds.), *Informationsdienst Naturschutz Niedersachs: 3/2009. Ökologie, Gefährdung und Schutz des Rotmilans Milvus milvus in Europa. Internationales Artenschutzsymposium Rotmilan* (pp. 136–143). Niedersächsischer Landesbetrieb für Wasserwirtschaft, Küsten- und Naturschutz (NLWKN), Hannover.

Krüger, T., & Wübbenhorst, J. (2009). Vollzugshinweise zum Schutz von Brutvogelarten in Niedersachsen: Teil 1: Wertbestimmende Brutvogelarten der Vogelschutzgebiete mit höchster Priorität für Erhaltungs- und Entwicklungsmaßnahmen. In T. Krüger & J. Wübbenhorst (Eds.), *Informationsdienst Naturschutz Niedersachs 3/2009. Ökologie, Gefährdung und Schutz des Rotmilans Milvus milvus in Europa. Internationales Artenschutzsymposium Rotmilan* (pp. 201–205). Niedersächsischer Landesbetrieb für Wasserwirtschaft, Küsten- und Naturschutz (NLWKN), Hannover.

Kruska, V., & Emmerling, C. (2008). Flächennutzungswandel durch Biogaserzeugung. Regionale und lokale Erhebung in Rheinland-Pfalz. *Naturschutz und Landschaftsplanung, 40*(3), 69–72.

Lal, R. (2003). Global potential of soil carbon sequestration to mitigate the greenhouse effect. *Critical Reviews in Plant Sciences, 22*, 151–184.

Landesamt für Natur, U. u. V. N.-W. (2010). *FFH-Arten und Europäische Vogelarten: Rotmilan (Milvus milvus (L.))*. Retrieved from http://www.naturschutz-fachinformationssysteme-nrw.de/ffh-arten/de/arten/vogelarten/kurzbeschreibung/103013

Lutze, G., Schultz, A., & Kiesel, J. (2003). *Landschaftsstruktur im Kontext von naturräumlicher Vorprägung und Nutzung – ein systemanalytischer Ansatz*. Retrieved from http://www.zalf.de/home_zalf/download/lsa/agstruk/Lutze_Landschaftsstruktur_systemanalytischer%20Ansatz_2003.pdf

Mante, J., & Gerowitt, B. (2006). Characteristics of intensively used agricultural areas and their impact on biodiversity and nature conservation activities within farming practice. In B. Meyer (Ed.), *Sustainable land use in intensively used agricultural regions* (Vol. 1338. Alterra Report) (pp. 113–118), Wageningen.

Marshall, G. R. (2008). Nesting, subsidiarity, and community-based environmental governance beyond the local level. *International Journal of theCommons, 2*(1), 75–97.

Marshall, E., Brown, V., Boatman, N., Wlutman, P., Squire, G., & Ward, L. (2003). The role of weeds in supporting biological diversity within crop fields. *Weed Research, 43*, 77–89.

Ministry of Foreign Affairs of Denmark. (2007). *Community-based natural resource management*. Retrieved from http://curis.ku.dk/ws/files/33236991/Technicalnotemarts2007CommunityBasedNaturalResourceManagement.pdf

Moorcroft, D., Whittingham, M. J., Bradbury, R. B., & Wilson, J. D. (2002). The selection of stubble fields by wintering granivorous birds reflects vegetation cover and food abundance. *Journal of Applied Ecology, 39*(3), 535–547. doi:10.1046/j.1365-2664.2002.00730.x.

Murphy, S. D., Clements, D. R., Belaoussoff, S., Kevan, P. G., & Swanton, C. J. (2006). Promotion of weed species diversity and reduction of weed seedbanks with conservation tillage and crop rotation. *Weed Science, 54*(1), 69–77. Retrieved from http://www.jstor.org/stable/4539360

National research council (Etats-Unis) (Ed.). (2002). *The drama of the commons*. Washington (D. C.): National Academy Press. Retrieved from http://www.nap.edu/openbook.php?record_id=10287&page=259

Neufeldt, H. (2005). Carbon stocks and sequestration potentials of agricultural soils in the federal state of Baden-Württemberg, SW Germany. *Journal of Plant Nutrition and Soil Science, 168* (2), 202–211.

Nitsch, H., Osterburg, B., & Roggendorf, W. (2009). *Landwirtschaftliche Flächennutzung im Wandel: Folgen für Natur und Landschaft*. Retrieved from http://imperia.verbandsnetz.nabu.de/imperia/md/content/nabude/landwirtschaft/gruenland/gap-reform.pdf

Nitsch, H., Osterburg, B., Laggner, B., & Roggendorf, W. (2010). *Wer schützt das Grünland?: Analysen zur Dynamik des Dauergrünlandes und entsprechender Schutzmechanismen*. Vortrag anlässlich der 50. Jahrestagung der GEWISOLA. Retrieved from ageconsearch.umn.edu/bitstream/93940/2/A3_3.pdf

Noble, B. (2010). *Cumulative environmental effects and the tyranny of small decisions: Towards meaningful cumulative effects assessment and management*. Retrieved from http://www.unbc.ca/assets/nres/nresi_op_08_noble_2010.pdf

Ostrom, E. (1990). *Governing the commons: The evolution of institutions for collective action. The Political economy of institutions and decisions*. Cambridge: Cambridge University Press. Retrieved from http://books.google.de/books?id=4xg6oUobMz4C&printsec=frontcover&dq=governing+the+commons&hl=de&ei=FgLETYiNB8zrsgaZ7umHDw&sa=X&oi=bookresult& ct=result&resnum=1&ved=0CD0Q6AEwAA#v=onepage&q&f=false

Pannell, D. J., Marshall, G. R., Barr, N., Curtis, A., Vanclay, F., & Wilkinson, R. (2006). Understanding and promoting adoption of conservation practices by rural landholders. *Australian Journal of Experimental Agriculture, 46*(11), 1407. doi:10.1071/EA05037.

Parker, S., & Cocklin, C. (1993). The use of geographical information systems for cumulative environmental effects assessment. *Computers, Environment and Urban Systems, 17*(5), 393–407. Retrieved from http://www.sciencedirect.com/science/article/B6V9K-46639KR-C1/2/b59a6ba8eae9eae75d1117e76f2c8d14

Peterson, E. B. (1987). *Cumulative effects assessment in Canada: An agenda for action and research*. Canada: Canadian Environmental Assessment Research Council.

Plantureux, S., Peeters, A., & McCracken, D. (2005). Biodiversity in intensive grasslands: Effect of management, improvement and challenges. *Agronomy Research, 3*(2), 153–164.

Regional Planning Act (ROG). (2009). Raumordnungsgesetz vom 22. Dezember 2008 (BGBl. I S. 2986), das zuletzt durch Artikel 9 des Gesetzes vom 31. Juli 2009 (BGBl. I S. 2585) geändert worden ist.

Reich, M., Rüter, S., & Tillmann, J. E. (2011). Auswirkungen des großflächigen Anbaus von Energiepflanzen auf die Tierwelt der Agrarlandschaft: Ergebnisse des Forschungsvorhabens SUNREG III. In M. Reich & S. Rüter (Eds.), *Umwelt und Raum: Vol. 2. Auswirkungen des großflächigen Anbaus von Energiepflanzen auf die Tierwelt der Agrarlandschaft* (1st ed., pp. 5–18). Göttingen: Cuvillier.

Riecken, U., Finck, P., & Schröder, E. (2002). Significance of pasture landscapes for nature conservation and extensive agriculture. In B. Redecker (Ed.), *Pasture landscapes and nature conservation. With 167 figures* (pp. 423–429). Berlin [u.a.]: Springer.

Rode, M. K. H. (2006). Beiträge der räumlichen Planungen zur Förderung eines natur- und raumverträglichen Ausbaus des energetischen Biomassepfades. *Informationen zur Raumentwicklung, 1/2*, 103–110.

Rode, M. W., & Kanning, H. (Eds.). (2010). *Natur- und raumverträglicher Ausbau energetischer Biomassepfade*. Stuttgart: Ibidem-Verlag. Retrieved from http://www.worldcat.org/oclc/686760048

Rode, M., Buhr, N., Kanning, H., Knispel, S., Steinkraus, K., Wiehe, J., & Wolf, U. (2010). Natur- und raumverträgliche Entwicklung des Biomasseanbaus. In M. Reich & S. Rüter (Eds.), *Umwelt und Raum: Vol. 1. Energiepflanzenanbau und Naturschutz. Referate und Ergebnisse der gleichnamigen Fachtagung in Hannover am 30. September 2009* (1st ed., pp. 35–42). Göttingen: Cuvillier.

Rodrıguez, C., & Wiegand, K. (2009). Evaluating the trade-off between machinery efficiency and loss of biodiversity-friendly habitats in arable landscapes: The role of field size. *Agriculture, Ecosystems and Environment, 129*, 361–366.

Roots, E. F. (1988). *The assessment of cumulative effects: A research prospectus*. Hull: Canadian Environmental Assessment Research Council.

Schenk, A., Hunziker, M., & Kienast, F. (2007). Factors influencing the acceptance of nature conservation measures–A qualitative study in Switzerland. *Journal of Environmental Management, 83*(1), 66–79. doi:10.1016/j.jenvman.2006.01.010.

Schmidt, P. (2009). Zur Bestandessituation und Habitatnutzung des Rotmilans Milvus milvus im Landkreis Holzminden (Niedersachsen). In T. Krüger & J. Wübbenhorst (Eds.), *Informationsdienst Naturschutz Niedersachsen: 3/2009. Ökologie, Gefährdung und Schutz des Rotmilans Milvus milvus in Europa. Internationales Artenschutzsymposium Rotmilan* (pp. 151–157). Niedersächsischer Landesbetrieb für Wasserwirtschaft, Küsten- und Naturschutz (NLWKN), Hannover.

Schumacher, U., & Mathey, J. (1998). Zur Analyse der Lebensräume von Heuschrecken mit Methoden der Geoinformatik – dargestellt am Beispiel der Riesaer Elberegion. In J. Strobl (Ed.), *Angewandte geographische Informationsverarbeitung. Beiträge zum AGIT-Symposium Salzburg '98* (pp. 326–333). Heidelberg: Wichmann.

Smith, P., Ambus, P., Amézquita, M., Andrén, O., Arrouays, D., Ball, B., et al. (2004). Greenhouse gas emissions from European croplands. Retrieved from http://www.bgc-jena.mpg.de/bgc-processes/ceip/policy/files/ghg_reportss2.pdf

Sonntag, N. C. (1987). *Cumulative effects assessment: A context for further research and development*. Hull/Quebec: CEARC.

Soussana, J.-F., Salètes, S., Smith, P., Schils, R., Ogle, S., Allard, V., et al. (2004). Greenhouse gas emissions from European grasslands. Retrieved from ftp://195.37.229.5/pub/outgoing/athuille/Publications/Concerted_Papers/Grassland_GHG.pdf

Stanners, D., Bosch, P., Dom, A., Gabrielsen, P., Gee, D., & Jock, M. (2007). Frameworks for environmental assessment and indicators at the EEA. In International Council for Science & I. ebrary (Eds.), *Sustainability indicators. A scientific assessment* (pp. 127–144). Washington DC: Island Press. Retrieved from http://books.google.de/books?hl=de&lr=&id=W4o-qunretMC&oi=fnd&pg=PA127&dq=DPSIR+EEA&ots=ieyfQBCN6a&sig=kHV8ciRQ5o LtCOd75NBSyrLyHu4#v= onepage&q=DPSIR%20EEA&f=false

Statistisches Bundesamt. (2002). *Statistical yearbook 2002 for the Federal Republic of Germany. Wiesbaden*. Retrieved from http://www.digizeitschriften.de/dms/toc/?PPN=PPN635628112_2002

Statistisches Bundesamt. (2012). *Spezielle Bodennutzung und Ernte*. Retrieved from https://destatis.de/DE/ZahlenFakten/Wirtschaftsbereiche/LandForstwirtschaft/Bodennutzung/Tabellen/Ackerland-HauptfruchtgruppenFruchtarten.html

Stevenson, F. C., Légère, A., Simard, R. R., Angers, D. A., Pageau, D., & Lafond, J. (1997). Weed species diversity in spring barley varies with crop rotation and tillage, but not with nutrient source. *Weed Science, 45*(6), 798–806. Retrieved from http://www.jstor.org/stable/4045847

Sticksel, E., Salzeder, G., Eder, J., Aigner, A., Fritz, M., & Deiglmayr, K. (2010). *Zweikulturnutzungssystem (ZKNS) im Vergleich zu herkömmlichen Anbauverfahren*. Retrieved from https://biogas-forum-bayern.de/publikationen/ZKNS.im.Vergleich.zu.herkoemmlichen.Anbauverfahren.pdf

Südbeck, P., Bauer, H.-G., Boschert, M., Boye, P., & Knief, W. (2007). Rote Liste der Brutvögel Deutschlands: 4. Fassung, 30. November 2007. *Berichte zum Vogelschutz, 44*, 23–81.

United Nations (UN). (1998). *Kyoto Protocol to the United Nations Framework Convention on Climate Change*. Retrieved from http://unfccc.int/resource/docs/convkp/kpeng.pdf

Veldkamp, T., Polman, N., Reinhard, S., & Slingerland, M. (2011). From scaling to governance of the land system: Bridging ecological and economic perspectives. *Ecology and Society, 16*(1). Retrieved from www.ecologyandsociety.org/vol16/iss1/art1/

von Haaren, C., Saathoff, W., & Galler, C. (2012). Integrating climate protection and mitigation functions with other landscape functions in rural areas: A landscape planning approach. *Journal of Environmental Planning and Management, 55*(1), 59–76. doi:10.1080/09640568.2011.580558.

von Ruschkowski, E., & Wiehe, J. (2008). Balancing bioenergy production and nature conservation in Germany: Potential synergies and challenges. *Yearbook of Socioeconomics in Agriculture*, 3–20.

Weiland, P. (2010). Biogas production: Current state and perspectives. *Applied Microbiology and Biotechnology, 85*, 849–860.

Wiehe, J., von Ruschkowski, E., Rode, M., Kanning, H., & von Haaren, C. (2009). Auswirkungen des Energiepflanzenanbaus auf die Landschaft am Beispiel des Maisanbaus für die Biogasproduktion in Niedersachsen. *Naturschutz und Landschaftsplanung, 41*(4), 107–113.

Wiehe, J., Rode, M. W., & Kanning, H. (2010). Raumanalyse I: Auswirkungen auf Natur und Landschaft. In M. W. Rode & H. Kanning (Eds.), *Natur- und raumverträglicher Ausbau energetischer Biomassepfade* (pp. 21–90). Stuttgart: Ibidem-Verlag.

Wiehe, J., Rode, M., & Kanning, H. (2011). Biomasseanbau. *Naturschutz und Steuerung. Ökol Wirtschaf, 3*, 22–24.

With, K. A., & King, A. W. (1999). Extinction thresholds for species in fractal landscapes: Umbrales de extinción para especies en paisajes fraccionados. *Conservation Biology, 13*(2), 314–326. doi:10.1046/j.1523-1739.1999.013002314.x.

World Commission on Environment and Development. (1991). *Our common future*. Oxford/New York: Oxford University Press. Retrieved from http://www.worldcat.org/oclc/722107774

Young, O. R. (2002). *The institutional dimensions of environmental change: Fit, interplay, and scale*. Cambridge, MA: MIT Press. Retrieved from http://www.worldcat.org/oclc/473025182

Part IV
Economic Optimisation of Bioenergy Production

Chapter 8
Optimising Bioenergy Villages' Local Heat Supply Networks

Anke Daub, Harald Uhlemair, Volker Ruwisch, and Jutta Geldermann

Abstract Bioenergy villages' local energy facilities produce electricity and heat for their inhabitants. This electricity is fed into the public grid with the heat distributed to the households via a local hot water grid. We use a linear mathematical model to simultaneously optimise the course of the heat supply network and the selection of households to be connected to the grid. In a first step, the heat distribution system is economically optimised. In a second step, we analyse the impacts of including social criteria and of varying parameters (e.g., prices). The model is applied to a small village with 24 households.

Keywords Bioenergy village • biogas plant • heat supply network • optimisation model • sensitivity analysis

8.1 Introduction

This chapter deals with the economic optimisation of local heat supply networks for bioenergy villages. Potential heat customers could be private households, public buildings, farms, industrial buildings, hotels and recreational facilities such as swimming pools or gyms. It is assumed that an independent operating company runs the district heat supply system. The required amount of heat is purchased from a local bioenergy plant, which comprises a combined heat and power biogas plant, a central heat station burning wood chips and an additional oil-based heat generator

A. Daub (✉) • H. Uhlemair • J. Geldermann
Chair of Production and Logistics, University of Göttingen, Platz der Göttinger Sieben 3, Göttingen, Germany 37073,
e-mail: anke.daub@wiwi.uni-goettingen.de

V. Ruwisch
Interdisciplinary Centre for Sustainable Development, University of Göttingen, Goldschmidtstr. 3, Göttingen, Germany 37077,

H. Ruppert et al. (eds.), *Sustainable Bioenergy Production - An Integrated Approach*,
DOI 10.1007/978-94-007-6642-6_8,

to ensure the heat supply during very cold periods of the year. In Sect. 8.2, a linear mathematical model is presented and applied to a village with 24 potential heat customers. The model calculates which potential heat customers should be connected to the heat supply network and what the pipeline's optimal course would be. In Sect. 8.3, the results of a sensitivity analysis are shown. This analysis incorporates different scenarios regarding the villagers' willingness to be connected to the network. We also investigate how much the heat customers' initial fixed costs can be reduced by examining the profits after the optimisation. In addition, we calculate how high the price that the operating company pays for heat can rise before the network becomes unprofitable. In Sect. 8.4, we summarise the findings and describe further research steps.

8.2 The Optimisation Model

8.2.1 Components of Mathematical Optimisation Models

To set up a resource-efficient and cost-efficient heat distribution system, the planning process should be mathematically modelled. Models are a means to reduce de facto complex relationships to their essential structures in order to identify the important components, the dependencies between them and the effects that changing data have.

Mathematical planning and optimisation models consist of three basic components: the decision field (the model's variables), the planning target (the model's objective function) and the planning framework (the model's constraints) (Hillier and Lieberman 2010, p. 25 ff.). The decision variables describe the decision-maker's scope of action; by assigning a specific value to each variable, one of all the possible decisions is chosen.

On the one hand, the scope of possible actions is determined by the variables' domain (e.g., binary variables for potential heat customers – whether connected or not connected to the heat supply network –, or nonnegative real numbers for an energy plant's capacity (kW)). On the other hand, the scope of action may be restricted by constraints that should not be violated. These could be the available amount of biomass, a financial budget, or the plant's capacity. These constraints frame the set of all feasible solutions, i.e. the decision variables' region of permissible values in which all constraints are met.

An objective function has to be formulated to measure different solutions' quality. This function represents the decision-maker's preferences and consists of the performance measure (e.g., the local heat supply network's profits or the amount of emissions resulting from the biogas station's energy generation) and the direction into which these should develop (e.g., maximisation or minimisation). The optimal solution is the one with the most favourable performance measure value.

Linear optimisation models are characterised by the variables not being squared, cubed or multiplied by one another in the objective function or in the constraints, etc. Whether the region of feasible solutions is convex or non-convex depends, among others, on the variables' domain (Hu 1969). The simplex algorithm (Murty 1976), interior-point methods (Domschke and Drexl 2007) and branch-and-bound-based algorithms (Murty 1976) can be used to solve linear optimisation problems.

Before we develop an optimisation model for a specific village, we describe its structures and characteristics.

8.2.2 Selected Bioenergy Village

The village on which the following analysis is based is located in southern Lower Saxony. The villagers want to use locally produced bioenergy in future and have therefore supplied information on, for instance, their individual heat demands. The village structure and the other necessary parameters used in this analysis are real data collected in this village. Accordingly, the model described below can be used as a decision support tool to help ensure a bioenergy project's success.

This village has 24 households, each with its own heating system, which will in future receive heat from a local heat supply network. Consequently, a local hot water grid has to be installed. A local energy plant comprising a combined heat and power biogas plant will generate the required amount of heat. Electricity is fed into the national grid and heat distributed to the villagers via a local hot water grid. An additional heating system burning wood chips and an oil-based peak load boiler ensure heat supply on very cold winter days. It is assumed that there will always be sufficient energy to supply the villagers with heat. An independent operating company – such as the cooperative of farmers and villagers found in Jühnde[1] – will run the heat supply network. It will buy the heat from the bioenergy plant (at a set price of 0.03 euro per kWh_{th}) and sell it to the heat consumers (at a set price of 0.059 euro per kWh_{th}). Most of the households have signed a contract, in which they declare their willingness to be connected to the heat supply grid, with the network operating company; for various personal reasons some households have not signed this contract.

The decision situation is depicted in Fig. 8.1. The black lines show the possible course of the heat supply grid and the 24 potential heat recipients are represented by the nodes $x_1,\ldots,x_{24}$. Three nodes (x_{25}, x_{26}, x_{27}) – representing the crossroads branch points – have been introduced.

We will formulate a decision model and design a distribution system for this village for an economically optimal heat supply to the households. We do not consider the producing and selling of electricity, nor is the production system part of the planning and the decision model. Consequently, the energy biogas plant's capacity as well as its configuration and location, is considered as given.

[1] See Chap. 2 in this book.

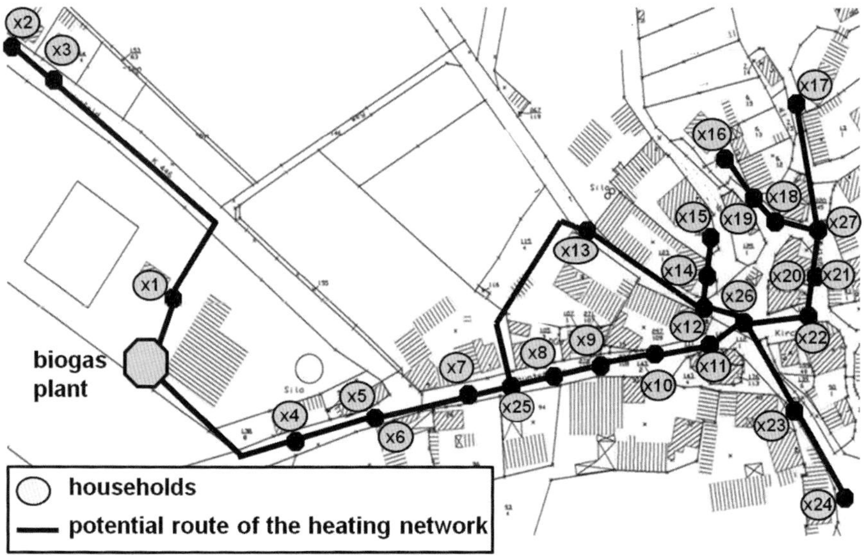

Fig. 8.1 Potential heat recipients and possible grid segments

8.2.3 *Linear Optimisation Model for a Local Heat Supply Network*

8.2.3.1 The Model's Variables

The three components of an optimisation model (variables, objective function, constraints – see Sect. 8.2.1) have to be explained and defined in terms of the bioenergy village described above.

Two decisions have to be made to construct the local heat supply grid:

- Which objects (private households, public buildings, industrial enterprises, etc.) will be linked to the grid?
- What is the grid's optimal course?

The variables can be divided into two groups: variables relating to the households[2] and those relating to the heat supply grid. The first group consists of all possible objects located within the village, irrespective of whether or not the homeowners have signed a heat supply contract. Since connecting a household to the local heat supply grid is a mere yes or no decision, the relevant variables (in this context called x_i) have to be defined as binary variables and can therefore only have the value 0 or 1. Since the village consists of 24 households, the variables $x_1, x_2, \ldots,$

[2] Here, "households" include public buildings, industrial enterprises, etc.

Box 8.1 Net Present Value (NPV)

The net present value is one of the basic key figures for investment appraisal. There are various models that support investment decisions. They can, for example, be categorised according to time (static and dynamic) and certainty (deterministic and stochastic). On the one hand, examples include the comparative cost or profit method, the comparative profitability method, or the static amortisation method. On the other hand, there are dynamic models such as the net present value method, the annuity method, the internal rate method and the dynamic amortisation method. The net present value method is used often, because it is easy to deploy and suitable to evaluate whether an investment is absolutely or relatively advantageous. More information on methods and different performance figures can be found in Götze et al. (2007).

x_{24} represent the decision whether or not a specific household will be integrated into the grid (supplemented by the variables x_{25}, x_{26}, x_{27} for the crossroads).

In the second group of variables (the grid variables), all possible courses of the grid have to be considered. The grid is divided into different network segments described by the variables y_{ij}. Each segment y_{ij} represents the link between household i and j. Contrary to the households variables, the grid variables y_{ij} are not given beforehand. Instead, the grid's different technically feasible courses have to be identified and variables have to be assigned to all the potential network segments.

8.2.3.2 Objective Function

In this case, the objective function's performance measure is the whole system's net present value (NPV, see Box 8.1). This is the value of all of an investment's present and future cash flows at the start of the planning horizon. Since the future payments cannot be compared with the current payments due to issues such as inflation, uncertainties and alternative investment possibilities, they have to be discounted by using a plausible internal discount rate. A positive net present value indicates that an investment is profitable and better than a financial investment based on a specific interest rate (Götze et al. 2007).

The local heat supply network's net present value consists of all the (usually positive) present values of the households and all the (negative) present values of the network segments.

From the operating company's perspective, the households' net present values comprise the following positive or negative payments:

- annual revenues from selling heat (product of the individual heat demand (w_i) and the difference between the selling price (p_S) and the buying price (p_B) per kWh_{th}) and an annual basic fee which the households have to pay

- one-time payments (positive payments such as the connection fee, the capital contribution, a government grant and the negative payment for the individual grid connections)
- negative annual payments for maintenance (dependent on the individual grid connections) and support.

The model for optimising the local heat supply network is based on the following assumptions:

- The planning horizon is 20 years.
- It is calculated with a 3 % internal discount rate.
- All payments are net payments.
- The problem of self-financing vs. external financing is not explicitly addressed (at least concerning the households).
- There are no heat losses when heat is conveyed through the grid.

Using the annuity present value factor to discount the (constant) payments to the start of the planning horizon (see the quotient at the end of the formula), the net present value for household i (NPV_i) can be calculated as follows (the figures in brackets below the formula show the parameters' values in this village):

$$NPV_i = (c + g + cc - h_i) + [w_i \cdot (p_S - p_B) + b - m \cdot h_i - a] \cdot \frac{(r+1)^T - 1}{r \cdot (r+1)^T}$$

with:

NPV_i:	net present value of household i	
c:	connection fee	[2,000 euro per household]
g:	government grant	[1,800 euro per household]
cc:	capital contribution	[2,500 euro per household]
b:	basic fee	[420.17 euro per year]
m:	maintenance factor	[0.02]
a:	administrative payments	[50 euro per year]
r:	internal rate of discount	[0.03]
T:	length of the planning horizon	[20 years]
p_S:	selling price	[0.059 euro per kWh]
p_B:	buying price	[0.03 euro per kWh]
h_i:	individual installation costs for connecting household i	
w_i:	heat demand of household i	

Table 8.1 lists the individual heat demands and different installation costs[3] of connecting the households to the grid.

The net present values for all the households can be calculated by using the above-mentioned formula; for example, the net present value of household 1 amounts to:

[3] Although payments are sometimes called "costs" here, "payments" is the correct term in economics theory, because only cash-effective amounts are considered.

Table 8.1 Parameters for the households

household	1	2	3	4	5	6	7	8
heat demand	**40,000**	**10,000**	**20,000**	**32,000**	**18,000**	**16,000**	**26,000**	**25,000**
installation costs (euro)	8,700	11,700	21,700	33,700	19,700	17,700	27,700	26,700
household	9	10	11	12	13	14	15	16
heat demand	**18,000**	**28,000**	**22,000**	**12,000**	**31,000**	**4,000**	**28,000**	**14,670**
installation costs (euro)	19,700	29,700	23,700	13,700	32,700	5,700	29,700	16,370
household	17	18	19	20	21	22	23	24
heat demand	**18,000**	**26,000**	**31,333**	**34,000**	**16,000**	**14,000**	**8,000**	**24,000**
installation costs (euro)	19,700	27,700	33,033	35,700	17,700	15,700	9,700	25,700

Table 8.2 Coefficients for the households

x_i	1	2	3	4	5	6	7	8
c_i	17,671	8,699	12,988	19,431	13,427	10,624	16,599	15,910
x_i	9	10	11	12	13	14	15	16
c_i	11,611	14,861	15,402	8,649	19,391	7,164	16,159	11,480
x_i	17	18	19	20	21	22	23	24
c_i	13,817	17,637	19,404	19,510	12,310	11,063	5,766	14,573

$$\begin{aligned} NPV_1 =& (2,000 + 1,800 + 2,500 - 8,700) + [40,000 \cdot (0.059 - 0.03) \\ &+ 420.17 - 0.02 \cdot 8,700 - 50] \cdot \frac{1.03^{20} - 1}{0.03 \cdot 1.03^{20}} = 17,671 \end{aligned}$$

All the net present values, which are simultaneously the coefficients c_i for the variables x_1 up to x_{24} in the objective function, are shown in Table 8.2.

To complete the objective function, i.e. to add the coefficients c_{ij} of the network segments, the corresponding net present values have to be calculated. The payments for installing the grid segments vary with the length of the single segment (in metres) and the soil type (street, grass strip, meadow, etc.), where the strip of pipeline has to be laid. Without considering the government grant of 80 euro per metre,[4] Table 8.3 lists the lengths and costs per metre for the various segments y_{ij} between the nodes i and j.

[4] This grant is addressed in the formula below.

Table 8.3 Parameters for the segments

segment	0–1	1–3	2–3	0–4	4–5	5–6	6–7
length of the segment	**35**	**120**	**15**	**63**	**25**	**0**	**32**
costs per metre (euro)	200	250	250	250	350	0	350
segment	7–25	8–25	8–9	9–10	10–11	11–26	23–26
length of the segment	**24**	**22**	**14**	**25**	**12**	**9**	**35**
costs per metre (euro)	350	350	350	350	400	400	350
segment	23–24	22–26	21–22	20–21	21–27	17–27	18–27
length of the segment	**16**	**18**	**12**	**0**	**13**	**38**	**11**
costs per metre (euro)	350	350	350	0	400	300	400
segment	18–19	16–19	12–26	12–14	14–15	12–13	13–25
length of the segment	**6**	**7**	**8**	**12**	**18**	**23**	**34**
costs per metre (euro)	400	400	400	300	250	300	200

Contrary to the payments that have to be made to connect the households, it is assumed that a credit amount, which will be paid back at a constant rate per year (interest plus redemption), is needed to install the whole grid. This annuity is the result of the net payment for the segment (the segment cost minus the government grant) multiplied by the inverse of the annuity present value factor on the basis of the credit interest rate. Furthermore, annual payments have to be considered for maintenance; these payments amount to 2 % of the payments for the main grid as a whole.

The net present values of the network segments ij (NPV_{ij}) can be calculated as follows:

$$NPV_{ij} = \left[-l_{ij} \cdot k_{ij} \cdot m - (l_{ij} \cdot k_{ij} - gn \cdot l_{ij}) \cdot \frac{f \cdot (f+1)^T}{(f+1)^T - 1} \right] \cdot \frac{(r+1)^T - 1}{r \cdot (r+1)^T}$$

with:

NPV_{ij}:	net present value of network segment ij	
gn:	government grant for the network	[80 euro per metre]
f:	interest rate for the credit	[0.05]
l_{ij}:	length of the segment between the nodes i and j	
k_{ij}:	payments (per metre) to lay the pipeline between the nodes i and j	

The net present values for all the segments can be calculated by means of this formula. For example, the net present value of the segment between node x_1 and x_3 amounts to:

Table 8.4 Coefficients for the segments

x_i–x_j	0–1	1–3	2–3	0–4	4–5	5–6	6–7
c_{ij}	7,097	33,280	4,160	17,472	10,662	0	13,647
x_i–x_j	7–25	8–25	8–9	9–10	10–11	11–26	23–26
c_{ij}	10,235	9,382	5,971	10,662	6,012	4,509	14,926
x_i–x_j	23–24	22–26	21–22	20–21	21–27	17–27	18–27
c_{ij}	6,824	7,676	5,118	0	6,513	13,372	5,511
x_i–x_j	18–19	16 19	12–26	12–14	14–15	12–13	13–25
c_{ij}	3,006	3,507	4,008	4,223	4,992	8,094	6,894

$$NPV_{13} = \left[-120 \cdot 250 \cdot 0.02 - (120 \cdot 250 - 80 \cdot 120) \cdot \frac{0.05 \cdot (1.05)^{20}}{(1.05)^{20} - 1}\right] \cdot \frac{1.03^{20} - 1}{0.03 \cdot 1.03^{20}} = 33,280$$

Table 8.4 shows all the net present values.

8.2.3.3 Optimisation Model

The optimisation problem regarding the heat supply networks can be described as a Steiner tree problem and modelled as a mixed integer program (MIP) (Uhlemair et al. 2010). In accordance with Fig. 8.1, the biogas plant (x_0) and all the potential heat customers ($x_1,\ldots, x_{24}$) are treated as nodes in the heat supply network. Three additional nodes (x_{25}, x_{26}, x_{27}) are introduced as crossroads branch points where network segments from several different directions come together. Their coefficients in the objective function are zero. As mentioned above, the heat supply grid is divided into segments y_{ij}, which link two adjoining nodes i and j. The following variables are used in the model:

$$x_i = \begin{cases} 1, \textit{if object i is connected to the grid} \\ 0, \textit{else} \end{cases}$$

$$|x_0| = \textit{number of nodes connected to the grid}$$

$$y_{ij} = \begin{cases} 1, \textit{if segment ij is installed} \\ 0, \textit{else} \end{cases}$$

Accordingly, the objective function can be described as follows:

$$\left(\sum_{i=1}^{n} c_i x_i + \sum_{i=0}^{n}\sum_{j=0}^{n} c_{ij} y_{ij}\right) \Rightarrow max \tag{8.1}$$

The constraints are:

$$x_0 + \sum_{i=1}^{n} x_i = 0 \tag{8.2}$$

$$\sum_{j=0}^{n} f_{ji} - \sum_{j=0}^{n} f_{ij} = x_i \qquad i = 0, \ldots, n \tag{8.3}$$

$$n \cdot y_{ij} \geq f_{ij} \qquad i,j = 0, \ldots, n \tag{8.4}$$

$$x_0 \leq 0 \tag{8.5}$$

$$x_i \in \{0, 1\} \qquad i = 1, \ldots, n \tag{8.6}$$

$$y_{ij} \in \{0, 1\} \qquad i,j = 0, \ldots, n \tag{8.7}$$

$$f_{ij} \geq 0 \qquad i,j = 0, \ldots, n \tag{8.8}$$

The objective function maximises the local heat supply network's overall net present value. It sums up the net present values of all objects x_i ($n = 27$ in the village) and network segments y_{ij}, which, according to the model's result, constitute the network's optimal course.

Constraint (8.3) is equivalent to the flow conservation equation of a network flow problem (Hamacher and Klamroth 2006). This constraint ensures that there is a flow to every object x_i connected to the network ($x_i = 1$). Constraint (8.4) ensures that for every flow between nodes i and j, a pipeline segment is built to transport heat from i to j. The variable f_{ij} represents this flow. It is not necessary to use the actual heat flow in kWh. Constraint (8.2) guarantees that constraint (8.3) can always be fulfilled. Constraint (8.5) requires a negative demand for heat for the production system, i.e. the bioenergy plant is the heat supplier. Initially, it is assumed that enough heat is generated in the bioenergy plant for every possible solution of the heat supply network optimisation model. Constraints (8.6), (8.7) and (8.8) are integer and non-negativity constraints.

As Steiner tree problems are NP-hard, it will be difficult to solve the problem for an increasing number of nodes and segments within a reasonable running time.[5]

In the next section, the model is applied to the village. The model can be solved by means of a branch-and-bound-based algorithm.

[5] For definitions of complexity and "NP-hard", see Eiselt et al. (1987) or Garey and Johnson (1979). Lists of the running times of different types of models can be found in Ahuja et al. (1989).

Table 8.5 Optimal values for the household variables x_i

x_i	1	2	3	4	5	6	7	8
	1	0	0	1	1	1	1	1
x_i	9	10	11	12	13	14	15	16
	1	1	1	1	1	1	1	1
x_i	17	18	19	20	21	22	23	24
	1	1	1	1	1	1	0	0

($x_i = 1$: household i is connected; $x_i = 0$: household i is not connected)

Table 8.6 Optimal values for the segment variables y_{ij}

x_i–x_j	0–1	1–3	2–3	0–4	4–5	5–6	6–7
	1	0	0	1	1	1	1
x_i–x_j	7–25	8–25	8–9	9–10	10 11	11–26	23–26
	1	1	1	0	1	1	0
x_i–x_j	23–24	22–26	21–22	20–21	21–27	17–27	18–27
	0	1	1	1	1	1	1
x_i–x_j	18–19	16–19	12–26	12–14	14–15	12–13	13–25
	1	1	1	1	1	1	1

($y_{ij} = 1$: pipeline between node x_i and node x_j is built; $y_{ij} = 0$: pipeline between node x_i and node x_j is not built)

8.2.3.4 Optimal Solution of the Model

The software programme Xpress is used to optimise the heat supply grid shown in Fig. 8.1. The results are presented in Tables 8.5 and 8.6.

The figures show that almost all the households are part of the optimised heat network. Only four households are not connected. Figure 8.2 clarifies why households 2, 3, 23 and 24 are not included in the grid. They are situated further afield; fairly long and expensive pipeline segments would therefore need to be installed to connect them to the grid. The revenues from heat sales are not high enough to compensate the costs of the required grid segments.

Dotted lines indicate households and network segments not incorporated into the grid. For instance, the link between nodes x_9 and x_{10} is not part of the network; network segment $y_{9;10}$ is unnecessary to supply both households with heat. In terms of the objective function, it is more cost effective to transport heat to household 10 via a pipeline that starts at the biogas plant und turns into the direction of household 13 at branch point 25 and into the direction of household 10 at branch point 26. If segment $y_{9;10}$ were also installed, a circle (25–13–12–26–11–10–9–8–25) would be generated in the network. On the basis of graphs theory, the optimal grid is

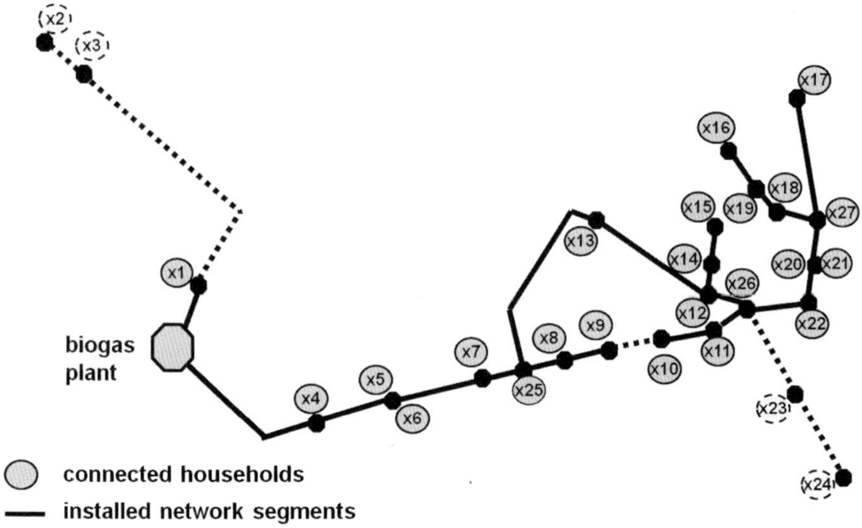

Fig. 8.2 Optimal solution

therefore a so called tree, which does not include circles (Uhlemair et al. 2010). Furthermore, it is clear that segment $y_{9;10}$ is unnecessary, which will prevent the (negative) payment for building this segment. The net present value of this solution amounts to 134,219 EUR. This is the highest possible value that the objective function can achieve.

8.3 Post-optimal Analysis

8.3.1 Overview

The solution shown in Fig. 8.2 assumes that all villagers want to be connected to the heat supply grid. It is also assumed that excluding those households whose connection to the grid would be unprofitable (from the operating company's perspective) would not be problematic.

This assumption of a "free optimisation" seems problematic, because in real life (as is the case in this village) some households do not want to receive local bioenergy. However, the calculated net present value of this "free optimum" can be used as a benchmark and the variables' values can be used as a starting point for calculating new solutions when changing the assumptions.

Making statements about the effects of changed premises based on an existing solution is called *post-optimal analysis*. It enables a decision-maker to avoid recalculating a problem completely when certain parameters or assumptions change. We consider two types of a post-optimal analysis here: suboptimal analysis and sensitivity analysis.[6]

Sensitivity analyses examine the extent to which parameters' values can vary without having structural effects on the pre-existing solution. This allows existing uncertainties concerning procurement costs or demand for heat to be considered. Statements can also be made about the critical price threshold that would put the system's profitability at risk. Furthermore, the scope for different connection fee rates can be analysed more closely. Reducing the connection fee could make the usage of locally produced bioenergy more attractive for some villagers.

In contrast, *suboptimal analysis* concentrates on the consequences of deviating from a specific pre-existing optimal solution. Referring to the planning situation specified here, suboptimal analysis seeks to estimate which net present value will apply if, for example, households without a contract for heat are excluded. Furthermore, potential heat customers whose connection to the grid is economically unviable in terms of the target function could then be connected regardless. Suboptimal analysis can show the extent to which the local heat supply grid's course and its net present value could change.

8.3.2 Suboptimal Analyses

8.3.2.1 Planning Scenario 1

In planning scenario 1, it is assumed that all households are connected to the grid, irrespective of a heat supply contract. In this case it is not considered that some households do not want to participate in the local heat supply network nor that some objects cannot be connected profitably (for instance, outlying households).

Looking at the model's formulation, a 100 % connection quota can be realised by inserting an appropriate constraint into the original model or by assigning the value 1 to all variables referring to households ($x_1,\ldots,x_{24}$).

Although the resulting solution is suboptimal compared to a free optimisation, it is the optimum with respect to the given restrictions. This solution, shown in Fig. 8.3, leads to a net present value of 117,055 EUR.

Figure 8.3 shows that the main network structure is the same as in Fig. 8.2 (same pipeline course with a break between nodes x_9 and x_{10}). Additionally, households 2, 3, 23 and 24 are connected to the grid, and the model selects the corresponding network segments.

[6] In the literature, other types are also mentioned, such as the interpretation of the optimum solution or parametric programming; see Dinkelbach (1969), Eiselt et al. (1987) or Hillier and Lieberman (2010).

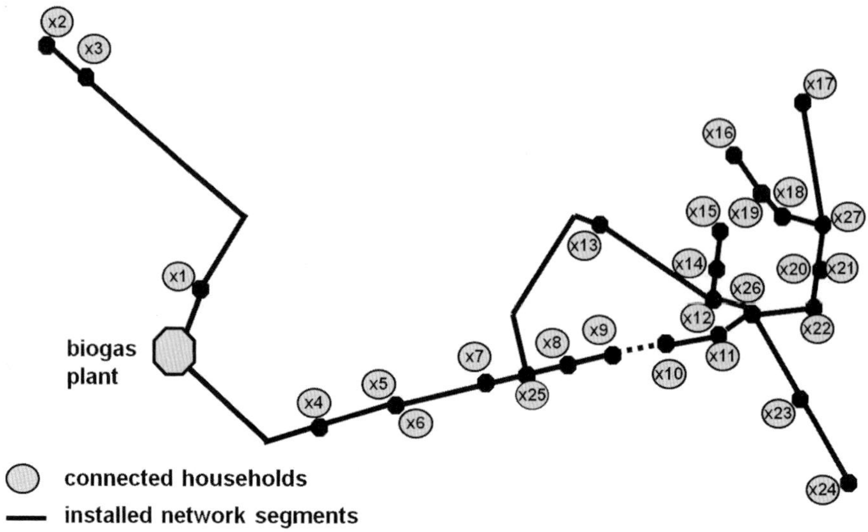

Fig. 8.3 Optimal local heating grid for planning scenario 1

8.3.2.2 Planning Scenario 2

In planning scenario 2, only those households with a signed heat supply contract are considered potential heat recipients. Households 13, 14, 18 and 19 do not want to use local bioenergy and therefore did not sign a heat supply contract. Consequently, these households are excluded and a value of 0 (not connected to the grid) is assigned to their binary variables (x_{13}, x_{14}, x_{18} and x_{19}). They are therefore no longer part of the set of (changeable) variables, and the optimisation process concentrates on the remaining households. All the other households have signed a heat supply contract and are therefore treated as potential heat customers. However, potential heat customers are not automatically connected to the heat supply grid. The optimisation model selects profitable heat customers and excludes unprofitable households. Whether a certain household can be profitably integrated into the network is mainly a question of heat demand and the costs of installing the necessary pipeline segments (see the net present values for the households and network segments in Sect. 8.2.3.2).

Figure 8.4 shows a situation in which only a route from the biogas plant straight through the village to household 17 (supplemented by the branches to object 1 and object 15) will be realised. In contrast to the previous planning scenarios, the link between nodes x_9 and x_{10} will be built, but the route section x_{25}–x_{13}–x_{12} seems too expensive and will be omitted. In advance, it had not been certain whether the branch to node x_{15} would be part of the grid, because household 14 had not signed a contract. However, object 15's heat demand is apparently high enough to make the pipeline from node x_{12} to node x_{15} profitable, although household 14 is not provided with heat.

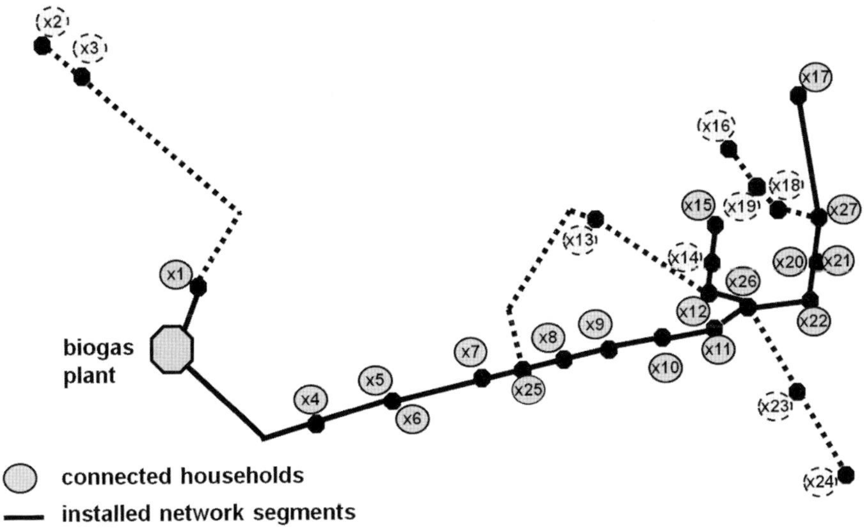

Fig. 8.4 Optimal local heating grid for planning scenario 2

The local heat supply grid in Fig. 8.4 leads to a net present value of 81,561 EUR. A comparison of this result with planning scenario 1's net present value is particularly interesting in this case. If all households are connected to the heat supply grid (planning scenario 1), including those that cannot be profitably integrated into the heat supply network, this results in a remarkably higher net present value than when only those with a heat supply contract (planning scenario 2) are considered. This leads to the conclusion that it would be extremely worthwhile convincing indecisive homeowners, or even those households that still prefer not to use local bioenergy, to become part of the group of local heat consumers.

8.3.2.3 Planning Scenario 3

Finally, it is considered that – following the idea of a bioenergy village – all those who have signed a heat supply contract will be offered the opportunity to receive bioenergy from the local heat supply grid. Whether or not this is economically viable in specific cases (from the operating company's perspective) is not taken into consideration. A value of 0 is assigned to the binary variables of households 13, 14, 18 and 19, because they do not want to be connected to the heat supply grid. Since all the other households have signed a heat supply contract, a value of 1 is assigned to their variables. Figure 8.5 shows the optimised heat supply grid with a net present value of 57,785 euro.

The grid's course is similar to that in planning scenario 2; the two networks differ only in the connection of the (unprofitable) households 2, 3, 23 and 24 in planning scenario 3.

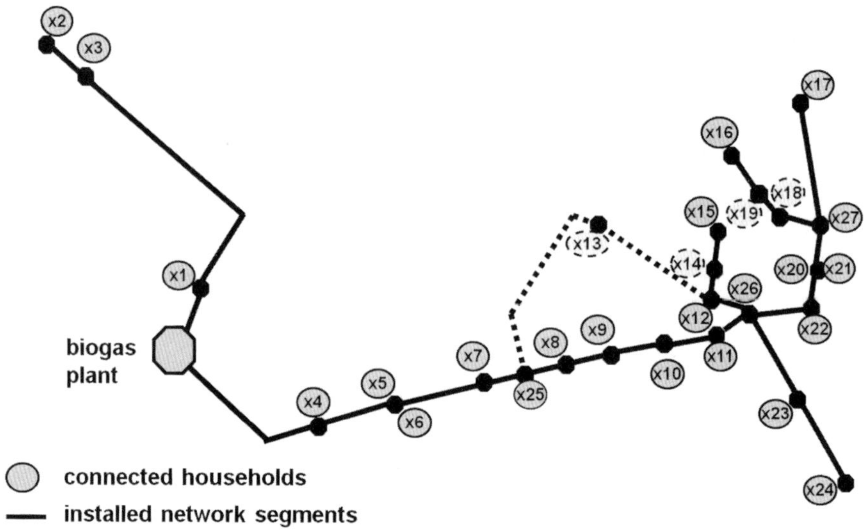

Fig. 8.5 Optimal local heating grid for planning scenario 3

Planning solution 3, rather than planning solutions 1 and 2, is expected to be characterised by the lowest net present value, because the values of the variables for all the households are determined prior to the optimisation and only the network's course is left to be optimised. However, it is interesting to compare this planning solution's net present value with that of planning solution 1 (Sect. 8.3.2.1). Planning solution 1 does not reflect reality, since several objects (nodes x_{13}, x_{14}, x_{18} and x_{19}) are part of the heat supply grid although their owners have not signed a contract. Nevertheless, the calculated solution can be used as a benchmark for further analysis. In fact, the difference between the two net present values indicates the financial scope of increasing the households' willingness to be integrated into the local heat supply network (e.g., by reducing the connection fee).

8.3.3 *Sensitivity Analyses*

8.3.3.1 Reducing the Connection Fee

On comparing planning solution 1, in which all the households form part of the network, with planning solution 3, in which a heat supply contract is a precondition for integration into the grid, the respective net present values reveal a large difference (117,055 EUR − 57,785 EUR = 59,270 EUR). Hence, it would be desirable with respect to the idea of the bioenergy village and for economic reasons to convince the remaining villagers to consider signing a contract.

If they are not convinced, efforts should be made to point out the positive impacts of a local heat supply grid and the concept of a bioenergy village.[7] If they have economic reasons for not wanting to be connected to the grid, it is important to have financial incentives (e.g., reducing fees) to convince these villagers to sign a contract.

We now analyse how far the connection fee (for all the villagers)[8] could be cut without overly reducing the operating company's profit. The difference between the net present values in planning scenarios 1 and 3 can be used as a maximum financial margin, because the lower one (57,785 euro, planning scenario 3) is the best that can be achieved if the unwilling owners do not wish to be connected to the grid.

Following this argument, a financial scope of (59,270 EUR/24 =) 2,470 EUR per household can be used as an incentive to use locally produced bioenergy. This implies that the connection fee for each household in the village can be decreased by 2,000 EUR. Without a connection fee, planning scenario 1's local heat supply network would lead to a net present value of 69,055 EUR.[9]

8.3.3.2 Variation in the Buying Price

It is assumed that the necessary amount of heat is available at 0.03 euro per kWh. Nevertheless, there may be changes to the price the operating company pays for the heat. Although rising prices are taken into account by using the internal discount rate (which can contain a certain risk surcharge, among others),[10] their impact on profitability should be analysed separately.

The starting point for the following sensitivity analysis is planning solution 3, in which only those households with a signed contract are part of the heat supply network. Without going into detail – many factors influence the buying price –, it is crucial to identify the critical price above which the system would no longer be profitable.

When profitability is considered, the net present value is again the key figure: If this value becomes negative, the system will be unprofitable. Therefore, there is a financial scope of up to 57,785 EUR (planning solution 3's net present value, based on a buying price of 0.03 EUR per kWh) that can compensate for potential price increases.

The net present value can be divided into those components that vary with the buying price (summand 1) and those components that do not depend on the buying price (summand 2):

---

[7] Chap. 10 deals with bioenergy villages' acceptance.

[8] For reasons of fairness, the other villagers cannot be excluded from the fee reduction.

[9] This analysis does not consider the question of liquidity. As noted, it is assumed that the capital needed to connect the households to the grid is completely self financed. Decreasing the connection fee may therefore require some external financing.

[10] The various functions of the internal discount rate are described by Götze et al. (2007).

$$\textit{net present value (NPV)} = \textit{price-dependent components} + \textit{fixed components}$$

The second summand consists of all the payments for building the grid ($\Sigma_i\Sigma_j$ NPV_{ij}), the one-time payments for connecting the objects to the grid ($c + g + cc + \Sigma_i h_i$), and the constant annual payments such as the basic fee and the payments for maintenance and support (discounted to the beginning of the planning horizon). The first summand includes the heat demand, the difference between the selling price and the buying price for heat, and the annuity present value factor for discounting the payments (last quotient in the formula below). In detail, it looks as follows:

$$\textit{summand}\,1 = \sum_{i \in I^*} w_i \cdot (p_S - p_B) \cdot \frac{(r+1)^T - 1}{r \cdot (r+1)^T}$$

with: I^*: set of households with a heat supply contract

When looking at that critical price $p_{B,crit}$, which leads to a net present value of 0, the analysis can concentrate on summand 1, because rising buying prices only affect this summand. Consequently, when answering the question of how high the buying price can go without leading to a negative net present value (planning scenario 3), the critical buying price at which the difference between the current value of summand 1 (using a buying price of 0.03 EUR per kWh) and the value of summand 1 using $p_{B,crit.}$ equals the net present value of 57,785 EUR needs to be calculated.[11]

The equation below describes these considerations.

$$NPV = \sum_{i \in I^*} w_i \cdot (p_S - p_B) \cdot \frac{(r+1)^T - 1}{r \cdot (r+1)^T} - \sum_{i \in I^*} w_i \cdot (p_S - p_{B,crit.}) \cdot \frac{(r+1)^T - 1}{r \cdot (r+1)^T}$$

Filling in planning scenario 3's data, the following equation provides:

$$57{,}785 = 423{,}670 \cdot (0.059 - 0.03) \cdot \frac{1.03^{20} - 1}{0.03 \cdot 1.03^{20}} - 423{,}670 \cdot (0.059 - p_{B,crit}) \cdot \frac{1.03^{20} - 1}{0.03 \cdot 1.03^{20}}$$

When solving the equation for $p_{B,crit.}$, one can see that the critical buying price is 0.0392 EUR per kWh. Thus, based on the initial price, an increase of up to 30 % can be dealt with without descending into unprofitability.

These findings can be used to evaluate the risk of the heat supply network becoming unprofitable if the price that the operating company pays for heat varies. Clearly, there are other uncertainties (especially with regard to the long planning

[11] At the same time, this buying price leads to a net present value of 0 if both summands are taken into account.

horizon of 20 years) that may lead to negative impacts in terms of the heat supply network's profitability. Technical problems in the energy station and in the heat supply network as well as biomass availability problems regarding the energy facility are examples of the insecurities that need to be analysed in future research.

8.4 Conclusion

We have developed an optimisation model for a local heat supply network. In terms of the objective function, the best grid course has been identified and the profitable heat recipients have been selected. In addition, optimal solutions were calculated for different scenarios regarding people's willingness to use bioenergy conveyed by a local heat supply network. Further, the consequences of changing parameters (e.g., the price of heat and the connection fee) have been analysed and the break-even point at which the investment would lose its profitability has been calculated. We will carry out further sensitivity analyses regarding governmental grants and the internal discount rate used in the calculations of the net present values. It may be reasonable to increase the internal discount rate so that the calculation of the system's profitability follows the principle of caution.

In future research, the distribution system will be enhanced by the production system. So far, it has been assumed that enough heat will be available. The production facility was not considered. The next research steps will be to develop a model that simultaneously optimises the distribution system (as described in this chapter) and the bioenergy facilities' capacities and configuration. The combined heat and power biogas plant will be supplemented by further heating stations and a peak load boiler to ensure heat supply over the coldest days of the year. When integrating the production system into the optimisation model, important factors such as sustainable energy crop cultivation, the impact on biodiversity when using biomass as a renewable energy source, general biomass availability and the special logistics issues associated with biomass usage should also be considered.

References

Ahuja, R. K., Magnanti, T. L., & Orlin, J. B. (1989). Network flows. In G. L. Nemhauser et al. (Eds.), *Optimization. Handbooks in operations research and management science* (Vol. 1, pp. 211–369). Amsterdam: North Holland.

Dinkelbach, W. (1969). *Sensitivitätsanalysen und Parametrische Optimierung*. Berlin: Springer.

Domschke, W., & Drexl, A. (2007). *Einführung in Operations Research*. Berlin: Springer.

Eiselt, H. A., Pederzoli, G., & Sandblom, C.-L. (1987). *Continuous optimization models*. Berlin: de Gruyter.

Garey, M. R., & Johnson, D. S. (1979). *Computers and intractability: A guide to the theory of NP-completeness*. New York: Freeman.

Götze, U., Northcott, D., & Schuster, P. (2007). *Investment appraisal*. Berlin: Springer.

Hamacher, H. W., & Klamroth, K. (2006). *Lineare Optimierung und Netzwerkoptimierung*. Wiesbaden: Vieweg.

Hillier, F. S., & Lieberman, G. J. (2010). *Introduction to operations research*. New York: McGraw-Hill.

Hu, T. C. (1969). *Integer programming and network flows*. Reading: Addison-Wesley.

Murty, K. G. (1976). *Linear and combinatorial programming*. New York: Wiley.

Uhlemair, H., Körner, M.-C., & Geldermann, J. (2010). Optimization of local heat networks for bioenergy villages. In The International Association of Science and Technology for Development (IASTED) (Eds.), *Proceedings of the Third IASTED African Conference* (pp.188–191), Botswana

Part V
Bridging Bioenergy Production and Society

Chapter 9
Growth of Biogas Production in German Agriculture: An Analysis of Farmers' Investment Behaviour

Karol Granoszewski, Christian Reise, Achim Spiller, and Oliver Musshoff

Abstract There has been a dramatic increase in renewable energy production from biogas in German agriculture. However, farmers have shown varied investment behaviour regarding biogas plants. In particular, policymakers and local authorities need a better understanding of decision-making structures at the farm level in order to estimate the future investment potential of biogas production. Socio-economic patterns, such as the perceived conflict potential, have been found to determine the future production potential. The determinants of investment behaviour were identified in two complementary sub-studies based on a survey of 160 German farmers. The first sub-study focused on land use competition as a negative impact of the increased biogas production in agriculture. Using a multinomial logit regression, we thereafter explore this impact on farmers' willingness to invest. The second sub-study confronts farmers with a hypothetical investment option in order to investigate their decision-making behaviour. Our findings indicate that risk aversion and bounded rationality explain the different decision-making outcomes. Furthermore, we find some evidence that the extrinsic factor, namely the economic benefits that the hypothetical investment's funding policy provides, overshadows the intrinsic factors such as ecological awareness. Knowledge of farmers' decision-making structures is helpful when revising the current funding policy as well as for the development of models forecasting the future potential of biogas production in agriculture.

Keywords Biogas • bounded rationality • Investment behaviour • land use competition

---

K. Granoszewski (✉) • C. Reise • A. Spiller • O. Musshoff
Department of Agricultural Economics and Rural Development, University of Göttingen, Platz der Göttinger Sieben 5, 37075 Göttingen, Germany
e-mail: karol.granoszewski@agr.uni-goettingen.de; christian.reise@agr.uni-goettingen.de; a.spiller@agr.uni-goettingen.de; Oliver.Musshoff@agr.uni-goettingen.de

H. Ruppert et al. (eds.), *Sustainable Bioenergy Production - An Integrated Approach*, DOI 10.1007/978-94-007-6642-6_9,

9.1 Introduction

In view of the world's climate change, many states have developed adaptation and response strategies to reduce greenhouse gas emissions and to reduce energy from fossil resources (IPCC 2007). In this context, the expansion of the production of energy from renewable resources (renewables) has become a key factor (IRENA 2010). In Germany, the policy framework for renewables has been gradually improved over the last few years. The policy focusses on the Renewable Energy Source Act (REA), which has been internationally used as an example and has been adopted in many other European countries. The recently amended REA promotes the generation of electricity from renewable sources.

Entrepreneurs have increasingly invested in renewables, such as wind energy and anaerobic digestion (biogas production) (Reiche and Bechberger 2006). A steady increase in investments in renewables has been observed since 2000 – especially in the agricultural sector (Plieninger et al. 2006a). The REA scheme reduces potential investors' risk of revenue loss. However, farm managers respond very differently to these incentives and they are heterogeneous in their exploitation of this 'new' investment potential. Some are very quick to invest in such plants, while others are more cautious.

This chapter's main objective is to identify the differences in investment behaviour. We aim to analyse farmers' investment behaviour with regard to renewable energy, particularly biogas production, in order to understand the decision-making process at the farm level in this context.

If policymakers do not have sufficiently detailed knowledge of farm managers' decision-making structures, they run the risk of misestimating market developments. Over the last few years, such misinterpretations have led to alternating boom and bust periods in the field of renewables (Granoszewski et al. 2011). Energy and environmental policymakers thus need to estimate the investment volatility to assess biogas production's further expansion potential. Local stakeholders preparing biogas projects require knowledge of specific agricultural decision-making structures to accurately estimate farmers' response and investment behaviour. This is particularly important because the German policy has set ambitious goals to increase bioenergy production in the form of biogas in the coming years. By 2020, bioenergy's share of the primary energy consumption will have increased from 4.9 % in 2007 to 11.0 % (BMU 2009). Biogas production plays a crucial role in this context (BMU 2009). The regulations for access to gas supply networks (Gasnetzzugangsverordnung) aim to substitute natural fossil fuels with biogas. According to § 31, biogas production should be expanded to six billion m^3 by 2020 to support this substitution (JURIS 2011). By the end of 2010, approximately 40,000 m^3 had been substituted (DENA 2011).

The biogas processing energy sector is strongly linked to agriculture. While some farms produce biomass and convert this into biogas on their own, others sell raw material as a substrate directly to biogas energy processors. Therefore, farmers' willingness to engage in this form of clean energy will influence the future of biogas production.

A number of studies have examined farmers' general decision-making and investment behaviour (see Kool 1994; Willock et al. 1999). Furthermore, several authors have analysed the economics of biogas plants (see Heissenhuber and Berenz 2006; Keymer 2009) and the production of agricultural raw materials for these plants (see Karpenstein-Machan 2005). To the best of our knowledge, no quantitative study has been undertaken on farmers' decision-making behaviour in the context of investments in biogas plants. Thus, no models are known that can explain the characteristics of bioenergy investments' different implementations.

In recent years, biogas production has had a strong impact on local agriculture. Since the beginning of the last decade, the progressive construction of biogas plants has been followed by an increased demand for biomass. However, the supply is restricted by limited arable farmland. Consequently, energy and food-producing farmers are in competition for farmland. This land use competition causes conflicts, which are likely to increase in future, given the current trajectory of biogas development (Granoszewski et al. 2011).

The conflict potential between farmers due to biogas production may have an influence on their engagement in this sector. Thus, land use competition and its conflict potential might threaten the future supply of biomass (energy crop cultivation) as well as the expansion of biogas production in the long run. Section 9.4 investigates the impact of land use competition on farmers' investment decision behaviour. Here, a theoretical model was developed and validated in a survey of German farmers. The following questions are discussed:

1. What do farmers perceive as the negative effects of biogas production and to what degree is the land use competition between them related to biogas?
2. To what extent does the land use competition influence farmers' decision-making behaviour regarding investments in biogas production compared to other determinants?

Farm-specific benefit and cost effects associated with investing in such plants might cause the observed differences in investment behaviour regarding biogas plants. For example, the production of biogas requires the cultivation of energy crops – mostly maize. This could have various economic effects and should be seen in the context of the existing cultivation of maize and the changes in crop rotation. However, farmers may simply make suboptimal decisions due to incomplete information and their limited cognitive abilities to process information, a phenomenon Simon (1956) refers to as 'bounded rationality' (see Gigerenzer and Selten 2001; Kahneman 2003; Selten 1990; Simon 1959). Frör (2008) links the concept of bounded rationality to environmental valuation. According to this concept, decision-makers may come to different conclusions even if they have the same entrepreneurial objectives and face identical business conditions. Bounded rationality does not refer to a deviation from the profit maximisation goal, but rather an inconsistency in decision-making behaviour. To estimate the consequences of the afore-mentioned economic incentives promoting bioenergy production, it is important to understand farmers' decision-making behaviour, including bounded rationality's impact. The decision whether or not to invest in a biogas plant may be explained by farmers' different risk attitudes. Furthermore, changing funding conditions may determine a decision in favour of engagement in biogas production. Farmers' decision-making behaviour will therefore be analysed in terms of the phenomena: bounded rationality, individual risk aversion and investment subsidies. To this end, farm managers were interviewed and confronted with a hypothetical investment in a specific biogas plant. The substrate required for the biogas plant would be cultivated on land currently used to produce wheat. The following questions were discussed:

1. Which conversion threshold – measured as the (critical) price for wheat, the competing crop – is necessary to motivate the respondent to invest in the biogas plant and change the existing production programme?
2. What are the driving factors (e.g., the individual risk attitude, the valuation of the sustainability effects, etc.) that influence this conversion threshold? To what extent can farmers' observed decision-making behaviour be attributed to bounded rationality?
3. Could an investment subsidy realistically promote the expansion of bioenergy? How do farmers value this subsidy in terms of their investment decision?

This paper consists of two sub-studies based on data gathered in the same survey.[1] Section 9.2 presents the diffusion process of bioenergy in agriculture as well as the state of the art. The survey is described in Sect. 9.3 and an overview given of the data. The first sub-study focussing on land use competition is described in Sect. 9.4. Farmers' investment behaviour is investigated with a descriptive approach from a socio-economic point of view in order to estimate the further expansion of biogas production. In Section 9.5, the second sub-study analyses risk aversion, bounded rationality and investment subsidies in terms of confronting farmers with a hypothetical investment. The paper concludes with a description of the implications and conclusions of both sub-studies' findings in Sect. 9.6.

9.2 Diffusion and State of the Art of Biogas Production in German Agriculture

The agricultural sector in Germany is still of considerable economic importance. Agricultural and forestry enterprises generate 54.2 billion EUR annually – about the same as the clothing, textile and paper industries together (59.8 billion EUR). If the production values of the agribusiness and food industries are included, the overall turnover rises to 215.9 billion EUR, which indicates the low added value of primary agricultural production (DBV 2010). This is reflected in the high income disparity

[1] The sub-studies were conducted as part of the interdisciplinary research project 'Sustainable Use of Bioenergy: Bridging Climate Protection, Nature Conservation and Society' at the University of Göttingen. The project was financed by the Lower Saxony State Ministry of Science and Culture.

between the primary agricultural production and the other sectors (Plieninger et al. 2006b). The majority of farms are individually owned family businesses, with the profit flowing directly into the family household. Protecting farm viability is therefore crucial for rural development (Henningsen et al. 2005). According to the investment and economic barometer of the German Farmers Association (Deutscher Bauernverband e.V.), farmers felt 'miserable' in 2010 due to the bad economic situation on most farms. They did, however, have great expectations regarding renewable energy (DBV 2010). Hence, alternative strategies, such as diversification through the production of energy from renewables, have spread to more farms in the interim.

Over the last decade, political measures have improved the economic conditions for energy production from renewables. The compensation rates for produced energy have gradually increased since the introduction of the Electricity Feed Act (Stromeinspeise-Gesetz) in 1991 and the much amended REA. The current REA determines the purchase price of produced energy for a period of 20 years at a comparatively high and fixed price (§ 21 EEG). Consequently, energy producers can calculate their revenue very precisely, which is a strong incentive for investors (DBFZ 2010).

Farmers' first engagement in the production of renewable energy was the installation of wind turbines in the 1990s. After the millennium, the REA's funding scheme changed in favour of other renewables, such as energy from biomass and photovoltaic systems. In principle, all investors can assess the REA subsidies. However, farms have certain structural advantages regarding bioenergy production, such as land ownership, appropriate machinery and storage facilities, access to credit, etc. Farms also have an adequate infrastructure, such as barn roofs, for the installation of photovoltaic systems. Moreover, they already have access to a direct supply of biomass as raw material for biogas generation. Farmers' interest in bioenergy production has therefore increased significantly compared to their interest in other sectors (Heissenhuber and Berenz 2006). The active diffusion of these innovations since 1991 reflects farmers' high level of investment activity (Mautz 2007).

The changed REA and farms' structural advantages led to a greater diffusion of two forms of renewables in German agriculture: energy from biomass in the form of anaerobic digestion (biogas production) and energy production from solar radiation (photovoltaics) (Mautz 2007). By the end of 2009, 4,960 biogas plants were operating in Germany, almost all of which were located on farms (DBFZ 2010). The photovoltaic market is more heterogeneous; the agricultural share of the total produced photovoltaic energy is only 19 % (BS and EUPD 2009). Given the greater number of qualified enterprises from other sectors, however, the agricultural market share is remarkably high.

The diffusion process of renewables is extremely rapid in Germany's agriculture. For instance, the biogas expansion and German farmers' significant investment activities are closely linked to the changes in the REA (Ehlers 2008; Mendonca 2007). The improved feed-in tariffs in 2000, 2004 and 2009 were followed by an increase in the construction of agricultural biogas plants (see Fig. 9.1).

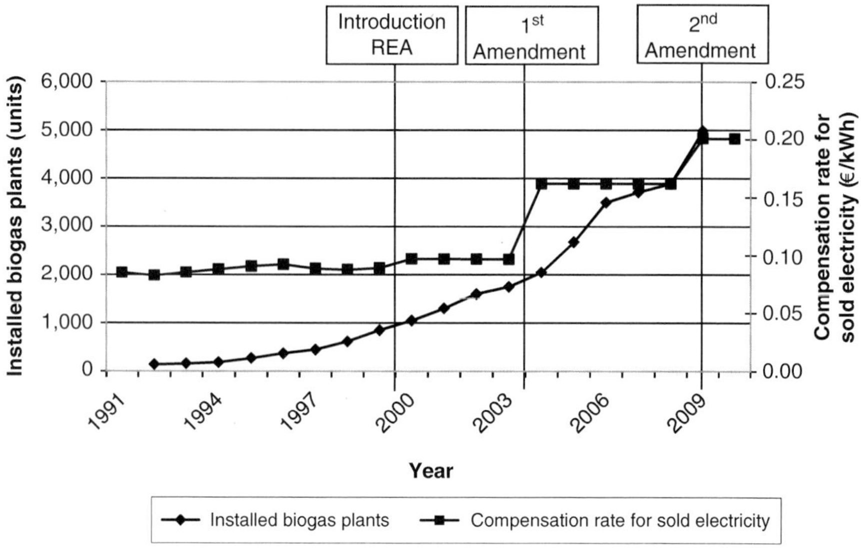

Fig. 9.1 Diffusion of renewable energies – The case of biogas production (Source: authors' elaboration (data from BMU 2000, 2004, 2008, Eurostat 2010, FvB 2010))

The close connection between the compensation rates and the installed plants indicates that market-based aspects are not only influencing the biogas market, but that it is also largely shaped by the policy (Jacobsson and Lauber 2006; Reiche and Bechberger 2006). Therefore, it is important to have a more detailed understanding of farmers' investment responses to financial incentives, which was the objective of the following empirical study.

9.3 Survey Description

Ample research has been conducted on farm management behaviour in general. However, to the best of our knowledge, there have been no studies on renewables. In order to provide a better understanding of farmers' investment behaviour in this specific context, we conducted a first explorative empirical study. The study consists of a two-step analysis. In the first sub-study, the analysis focusses on the framework of land use competition. The second sub-study focusses on risk aversion, bounded rationality and investment subsidies as determinants of farmers' investment behaviour. These two sub-studies are based on the same survey.

In total, 160 farm managers in Germany were personally interviewed with a standardised questionnaire between August and September 2009. The questionnaire had been developed on the basis of intensive discussions with experts and pre-tests. Hypotheses were operationalised through statements that the respondents rated on a five-point Likert scale. An experiment from behavioural economics was also undertaken.

The questionnaire consisted of three parts:

- Part 1: Decision making regarding bioenergy and biogas. The factors influencing investments in renewables, attitudes to renewables and the perceived level of land use competition.
- Part 2: Farmers' investment behaviour as based on a hypothetical investment experiment.
- Part 3: The socio-demographic and structural characteristics of the farm.

A representative survey was not feasible due to financial and time restrictions. However, the targeted selection of three subgroups was undertaken:

- Farmers who had invested in biogas production (biogas investors).
- Farmers who had invested in another form of renewables (other renewables (RE) investors).
- Farmers who had made a negative investment decision and had not (yet) invested in renewables (non-investors).

The group of 'other RE investors' consisted of farmers who did not produce biogas, but had chosen to produce other forms of renewables. These were mostly photovoltaic technology and, more rarely, wind power. The survey region was North West Germany as it has a large variety of agriculture production branches. Livestock farmers as well as cash crop producers were interviewed. Animal husbandry has strong interactions with biogas production. In regions with intensive livestock farming, a large amount of manure is used as substrate in the biogas production process. Accordingly, a large number of biogas plants have been constructed in these regions, which has led to debates about the positive and negative effects of biogas production in these regions (see Sect. 9.4.1.1). To further explore this, we surveyed both intensive livestock farmers and small-scale livestock farmers, such as cash crop producers (production of livestock and cash crop together). North West Germany was therefore a particularly suitable region.

Since no data were available to identify these target groups in advance, the selection of the respondents was carried out according to a 'pyramid scheme' (snowball). The student interviewers who conducted the survey were required to first make contact with farmers and thereafter extend the number of respondents in each group through their personal efforts. The interviewers were trained at a launch event during which an interview guideline was used to ensure uniform interview conditions. The farmers needed approximately 45 min to respond to the questionnaire.

The collected data were corrected for outliers by using box plots and the single linkage method for visualisation and identification. One anomalous response was detected and excluded from the survey. Overall, 159 responses were available for further analysis.

The socio-demographic and farm structural characteristics of the dataset do not reflect the German average exactly. The amount of farmland area per farm is on

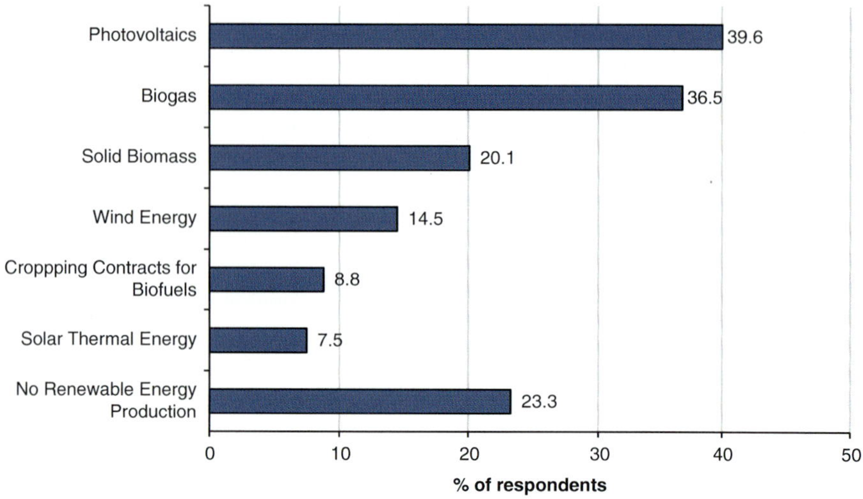

Fig. 9.2 Forms of renewable energy produced (N = 159, multiple responses)

average 173 ha, with a relatively high standard deviation of 238 ha. Thus, the farmland per surveyed farm is considerably higher than the national German average of 48.5 ha (2007) and the mean of 54.6 ha (2007) of Lower Saxony, where the survey was conducted (BMELV 2009a). While149 farms were managed as a main occupation (full-time), only ten were managed part-time, which differs from the national average. According to the national statistics, there are an equal number of part-time and full-time farmers (BMELV 2009a). Five farmers farmed organically. On average, three workers were employed on each farm.

In terms of the farms' production portfolio, about 9 % of all the farms were forage growers, 28 % cash crop producers, 38 % livestock producers, 23 % produced mixed crops and 2 % produced other types of crops. The perceived average soil quality on the farms was 44.0 points (on a scale of 0–100, with 100 equalling the best quality). This is similar to the Lower Saxony average of 42.5 points (NLS 2001).

On average, the respondents were 45 years old and well educated. Approximately 4 % had no agricultural education, whereas 20 % of the respondents held a university degree. Only five were female. Seven farms expected to sell their business in the near future. In total, 42 farms' future was unclear. The remaining managers had only recently taken over.

Most of the farmers were positive (fairly positive) about bioenergy production in agriculture. Only 24 % had a negative (fairly negative) opinion. They considered photovoltaics and biogas, followed by solar thermal energy, the best production opportunities for their farm in terms of renewables. The production of biofuels and the use of geothermal energy were considered the least favourable.

These estimations are in line with the actual production of renewables on the farms (see Fig. 9.2). The majority of managers had invested in biogas and/or photovoltaic systems, followed by solid biomass. In this case, solid biomass refers

Table 9.1 Distribution of respondents in terms of their investment decision regarding renewable energies

Total sample	Non-investors	Biogas investors	Other RE investors
n = 159	n = 37	n = 58	n = 64
100 %	23.3 %	36.5 %	40.3 %

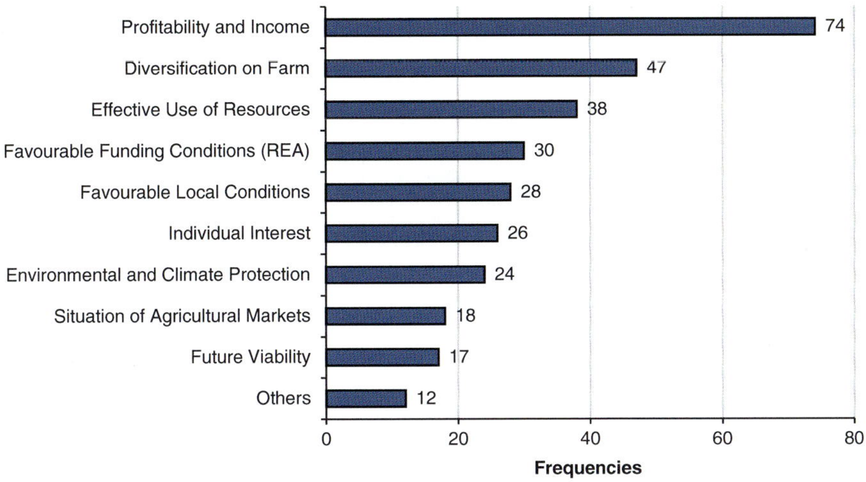

Fig. 9.3 Farmers' reasons for engaging in renewable energies

to small-scale heat production in low-power combustion systems to cover household demands. Only about 23 % of the 159 farmers did not produce any form of renewables.

As some farms produced more than one type of renewable, it is useful to group the farmers a priori into sub-samples according to their target groups. The sample was therefore divided into three types of farmers on the basis of their investment decision (see Table 9.1).

We investigated the background of the farmers' investment decisions. An open question was asked to determine the respondents' reasons for being for or against investing in renewables. The farmers provided a variety of reasons for their investment decisions.

An investment is mainly linked to economic motives and the possibility of diversifying the farm as well as the possibility to exploit existing resources, such as farmland and labour capacity, more effectively (see Fig. 9.3). While the farmers' individual interests and ecological motives were less important, they were not insignificant.

The situation changed for those farmers who did not invest. The farmers' individual attitudes and interests were the key determinants of their restraint and decision not to invest (see Fig. 9.4). The most mentioned restraints were the high capital requirements and adverse regional production conditions. Potential investors

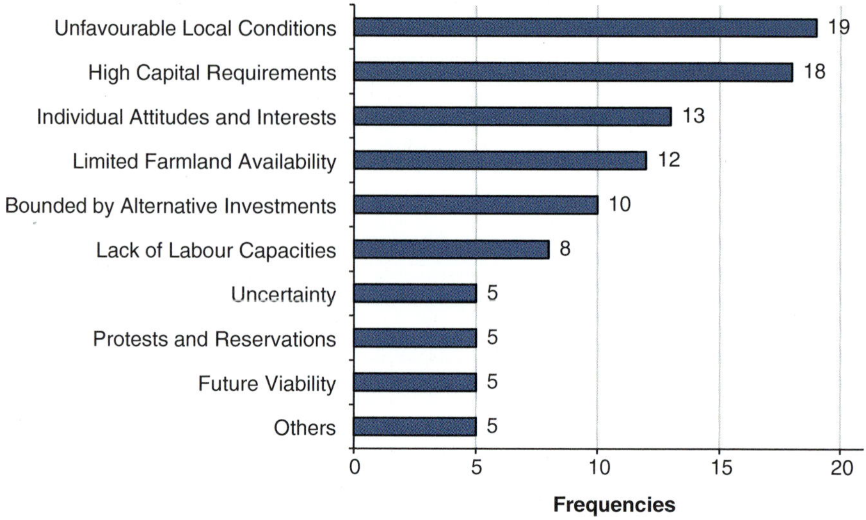

Fig. 9.4 Farmers' reasons for not engaging in renewable energies

considered the limited farmland and adverse local conditions as the main barriers to biomass production and bioenergy utilisation.

Around 23 % of the non-investing farmers definitely intended to invest in renewables in the future. In total, 25 % of the non-investors had negative or very negative attitudes to these investments. The majority of the non-investors had made long-term investments on their farm and were bound by these. Only 11 of the non-investing farmers had not made larger investments during the previous 5 years.

9.4 The Role of Land Use Competition

The theoretical background of land use competition and investment behaviour was introduced at the beginning of this first sub-study (Sect. 9.4.1). Section 9.4.2 describes the study design and methodology. In Section 9.4.3, the findings related to land use competition and other factors determining the farmers' investment behaviour are presented. The sub-study concludes with a discussion of the results and the study limitations.

9.4.1 Theoretical Background

9.4.1.1 Land Use Competition at the Farm Level

Wibberley (1959) investigated the competition for rural land due to urban growth and produced one of the first studies suggesting allocation strategies to reduce the level of competition. However, in recent years, the discussion has changed from urban growth to biofuel and bioenergy as competitive drivers. The recent increased production of biofuels is a result of globalisation, improved trade opportunities and an early action against climate change and the depletion of scarce fossil fuel resources. Energy generation from biomass has a positive impact on climate change projections (Berndes et al. 2003). However, certain of its negative impacts on nature, economics and social networks are becoming ever more apparent (Domac et al. 2005; Dornburg et al. 2010).

Besides benefits such as improving investment conditions, the strong diffusion of biogas plants on farms also has ecological and social impacts. These externalities may be the feedback effects of agricultural production and beyond, which may influence the further expansion of energy production from biomass (Dehnhardt and Petschow 2004; Mautz 2007).

Externalities or external effects can be described as the side effects of actions that unrelated third parties experience as a consequence of an economic activity (Buchanan and Stubblebine 1962). If this concept is applied to biogas production, the externalities are disorders due to the increasing competition between the different stakeholders. Competitive conditions intensify due to the differences in the interests in and limitations of available resources, such as environmental goods or production factors. These conditions may have negative social, ecological or economic consequences (DBFZ 2009; Mautz 2007; SRU 2007). Within the bioenergy pathways, discussions on its negative effects centre on biogas production (DBFZ 2009). Table 9.2 provides a classification of the negative external effects of agricultural biogas production.

Intra-sectoral effects occur in the form of increasing competition between farmers, which is the focus of the first sub-study (DBFZ 2010). The negative intra-sectoral effects are closely related to biomass production for digestion (WBA 2007). Food and biogas-producing farmers compete due to biogas production's relatively high and secure revenues, which the REA guarantees. Biogas-producing farmers are therefore better able to pay for production factors (e.g., farmland) than their food-producing colleagues, which has led to an increase in land lease rates (Berenz et al. 2008; Heissenhuber et al. 2008). The excessive subsidisation of biogas production, compared to that of food production, has led to politically induced competitive distortions between the two. We observed farms restructuring their production portfolio in favour of biogas. In some German regions, such as the northwest, biogas production has increased to such a level that it has replaced cash crops and animal husbandry as the dominant product on certain farms, some of which are becoming just energy producers.

The competition for agricultural land causes conflicts (Granoszewski et al. 2011). The emerging disputes could influence farmers' acceptance of and willingness to invest in bioenergy production. These externalities could therefore threaten the future expansion of biogas production and the biomass supply (energy crop cultivation) in the long run. At the same time, competition has emerged between biogas producers, as the large number of biogas plant installations restricts the producers' access to low-cost biomass in the local area.

Table 9.2 Systematisation of biogas production's negative externalities

Dimension	Competitive situation	Background	Feedback effect
Intra-sectoral	[a]*Food producing farms vs. biogas producing farms*	*Scarcity of farmland, competitive distortions caused by biogas production subsidies*	*Farmers' acceptance of bioenergy decreases, resources and relationship conflicts*
	Food producing farm vs. biogas producing farm	Local bounded supply of raw materials endangered, higher costs of raw materials	Revenue from biogas production decreases
Inter-sectoral	Local residents	Worsening quality of life (Nimby effect), personal reservations	Protests, citizens' action initiatives delay or prevent biogas plant construction
	Ecological environment	Negative ecological effects and concerns about biogas production (nature conservation vs. climate protection)	Ecological standards increase, less potential support
	Food-processing industry	Limited local supply of raw materials, local bounded processing of raw materials	Transaction costs of procuring raw materials increase, price transmission to consumers is difficult
	Existing fossil energy power supply companies	Farmers as new competitors in the energy market	Difficult market access for biogas producers

Source: authors' elaboration (based on DBFZ 2009; Mautz 2007; SRU 2007)
[a]Focus of empirical study

There are many interactions between the agricultural sector and other sectors; these connections are affected when biogas production expands. The biomass for anaerobic digestion is usually produced by energy crop cultivation, particularly maize. Changing crop rotations to increase maize production has a strong influence on biodiversity, which could experience negative impacts (Karpenstein-Machan 2005, see also Chaps. 6 and 7), especially in regions where the maize cultivation area is extended. Consequently, nature conservation NGOs are becoming more critical of biogas production (WBA 2007). Another frequently discussed biogas issue is the local residents' reservations regarding such installations. These protests have similarities to barn construction on agricultural land, which residents believe affects their quality of life i.e.through smell nuisance negatively (Mackenzie and Krogman 2005; Mann and Kögl 2003).

9.4.1.2 Farmers' Decision-Making and Investment Behaviour

According to investment theory, the main task of investments through expansion and diversification strategies is to strengthen the competitiveness and viability of enterprises in markets (Jorgenson 1968). Entrepreneurs make investment decisions based on a cognitive decision-making process, which involves the decision-maker's serious consideration of the investment's specific characteristics (East 1993). However, in order to understand an investment decision, it is necessary to analyse decision-making behaviour.

Many disciplines, such as psychology, sociology and economics, examine the decision-making process. The analysis is centred on a decision as an outcome of a specific human cognitive function, which occurs according to specific rules and is intended to make a choice between various alternatives (East 1993). Decision-makers have to be self-motivated to find a solution for decision-making problems themselves.

Approaches to explain entrepreneurs' decision behaviour offer complementary normative (prescriptive) and descriptive decision theories. The prescriptive decision theory aims to understand decision behaviour on the basis of formalised rules and procedures under the assumption of rationally correct (optimal) decisions. On the other hand, the descriptive approach aims to reflect realistic decision making by considering it in a broader context (Bell et al. 1988).

The economic decision theory presupposes the rationality of the decision-maker and decision making is therefore a largely rational analysis (Bell et al. 1988). However, studies by experimental economists have questioned the assumption that humans are rational and show that we are unable to solve decision-making problems with a totally rational approach due to our limited cognitive abilities (Simon 1979). Behavioural decision research addresses this finding. Edwards (1954), Simon (1959) and Kahneman and Tversky (1979) found that managerial decisions do not strictly follow the rational goal of economic profit maximisation, but can also be influenced by other psychological elements, such as intrinsic (e.g., satisfaction or risk-taking) or social factors (e.g., desired behaviour) as extrinsic objectives. If the decision-making situation is complex, the actual decision behaviour differs greatly from the expected formal normative behaviour (Simon 1959). This may explain why decision behaviour is not only influenced by an investment's economic benefits, but also by other factors. An investment in renewables is a multidimensional issue; which complicates the decision process more.

This contribution is not aimed at an economic evaluation of an investment decision on the basis of quantities that can easily be calculated, such as (opportunity) costs and government-guaranteed payments (subsidies), but rather on the basis of behavioural and other influential elements that should be taken into account in a realistic investigation. This part of the study takes a descriptive approach. In the empirical study, the influence of intrinsic and extrinsic variables on the outcome, namely the investment decision, is tested.

Behavioural studies of decision making on farms mostly analyse farmers' motivations, goals and attitudes. However, their decision behaviour is regarded as the result of a combination of motivational and external factors as well as the farm's structure (Burton 2004). Within the external dimension, Solano et al. (2003) focus on social network structures. The strong social impact that other people have on farmers' work or family environments is characteristic of agricultural behaviour and differs significantly from that of other sectors (Solano et al. 2003).

Compared to other sectors, farms and their primary production are very dependent on and influenced by natural resources and the environment. Hence, farmers face a variety of complex decision situations, which allow for only a limited formalisation (Nuthall 1999, 2010).

> As a farmer must deal with most aspects of biology, economics, the weather, organisations, people and so on, they face very complex decision situations with only a modicum of support in an immediate office sense (Nuthall 1999: 17).

Willock et al. (1999) developed a basic model to explain farmers' behaviour. The study explores the relationship between farmers' personality and their behaviour. The authors found that personal factors, such as personal character traits, influence farmers' behaviour indirectly by means of attitudes and objectives. Consequently, four different kinds of behaviour were observed: production-oriented business behaviour, environmentally oriented behaviour, stressed behaviour and business development behaviour. These behaviours confirm that there are differentiated behavioural and decision-making structures. Willock et al. (1999) and Burton (2004) pointed out that external, physical or situational effects as well as intrinsic drivers strongly influence decision situations, such as specific investments – the outcome behaviour. The additional factors do not influence farmers' attitudes, but have a direct impact on their decisions (Willock et al. 1999; Burton 2004)

9.4.2 Data and Methods

9.4.2.1 Research Design

Based on these theoretical assumptions and taking the dynamics of the diffusion of renewables in agriculture into consideration (see Sect. 9.1), we created a basic model to explain farmers' individual decision behaviour regarding investments in renewables (see Fig. 9.5).

Hypotheses (H_1 to H_7) were posited to operationalise and measure each construct in the empirical analysis. The model describes the potential factors influencing decision-making behaviour. There are three core elements: individual, farm-internal and farm-external factors.

Individual Factors

In her qualitative approach, Trojecka (2007) found that ecological awareness is a motivational factor that influences farmers' behaviour. Farmers' desire for

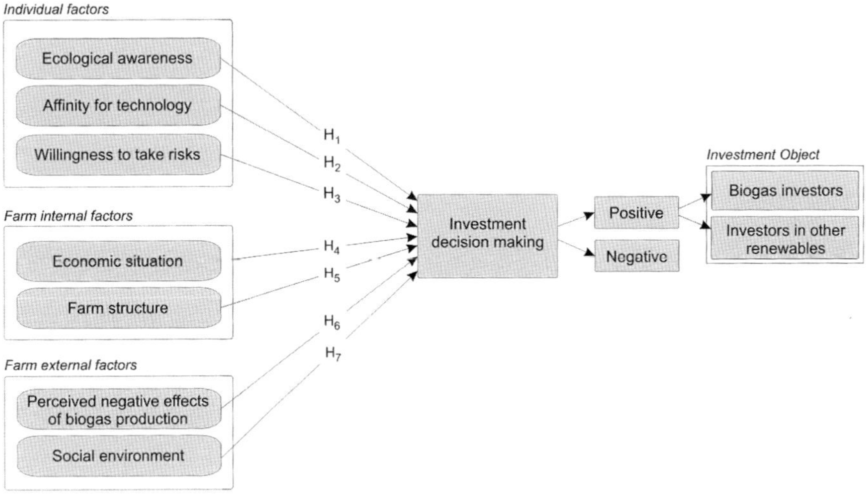

Fig. 9.5 Empirical model explaining farmers' decision behaviour regarding investments in renewable energies

CO_2-neutral, ecologically friendly regenerative energy production could be a driver of bioenergy investments. Lynne and Rola (1988) refer to a sense of responsibility for the environment, which has a strong influence on farmers due to its central importance in their mode of production. Therefore, we assume that a high level of ecological awareness influences the probability that farmers will invest in renewables positively.

H_1: *A high level of environmental awareness has a positive impact on the probability of investing in renewables.*

It is well known that farmers with a high technical interest adopt new production techniques (Austin et al. 1998). In order to select an appropriate renewables technology, farmers should also have extensive knowledge and understanding of construction and the operation of such facilities. We thus expect that a farmer with a high affinity for technology will be more likely to make an investment than one who is not interested in technology.

H_2: *A high affinity for technology has a positive impact on the probability of investing in renewables.*

Sauer and Zilbermann (2009) state that some decision-makers are not willing to take risks in order to advance their business, which results in a delay in the uptake of innovations. Therefore, we presume that risk aversion affects the decision-making process.

H_3: *A high willingness to take risks has a positive impact on the probability of investing in renewables.*

Farm-Internal Factors

The implementation of new production procedures or branches of industry in an existing firm is restricted by their structural conditions. Schramm (1977) indicates that entrepreneurs pay more attention to their factor endowments than other determinants when making business decisions. La Due et al. (1991) emphasise financial conditions' relevance for farm expansion and, hence, for the agribusiness sector.

H_4: *A high level of satisfaction with the economic situation has a positive impact on the probability of investing in renewables.*

Langert (2007) uses the example of energy crops cultivation to confirm that structural conditions are a major factor when making farm production decisions. Therefore, the production structure is assumed to have a strong influence on investment considerations, especially in biogas production.

H_{5a}: *High quality farmland has a positive impact on the probability of investing in renewables.*

H_{5b}: *A large amount of farmland has a positive impact on the probability of investing in renewables.*

H_{5c}: *High labour capacity has a positive impact on the probability of investing in renewables.*

Farm-External Factors

Biogas production has complex impacts on local agriculture, as described in Sect. 9.4.1.1. The competition between food and (exclusively or partly) energy-producing farmers has increased. However, farmers are strongly linked and dependent on one other (Fehr and Schmidt 1999). In Granovetter's (1985) *Theory of Embeddedness*, economic actions are examined in the social context. He found evidence of pro-social behaviour, which means that entrepreneurs' final decisions are not exclusively driven by monetary considerations, but can be influenced by relational aspects. We assume that strongly perceived externalities, such as increasing land use competition and politically induced competitive distortions, have a negative impact on farmers' individual investment decisions, since they have a high potential for social conflicts. Farmers who are aware of these phenomena are unwilling to enter these complex social conflicts. Consequently, they reject the opportunity to investment in biogas production, or favour another type of renewable energy.

H_6: *Strongly perceived negative external effects of biogas production have a particularly negative impact on the probability of investing in biogas production.*

Farmers are key actors in the rural communication network and are in close contact with many non-agricultural groups (Retter et al. 2002).

H_{7a}: *In general, the social environment has a strong impact on the decision-making process.*

In their *Theory of Reasoned Action*, Fishbein and Ajzen (1975) describe the social environment's strong impact on behaviour. Human behaviour is affected by people's perception of how others would view them if they acted in a certain way, especially if the behaviour is socially visible. The opinion of family, friends and local residents of a specific situation affects farmers' behaviour significantly (Solano et al. 2003).

H_{7b}: *The social environment's negative opinion of the object of investment (biogas production) has a negative influence on the probability of investing in biogas production.*

In the empirical model, attitudes are represented by individual factors. The farm's internal dimension reflects its given structural and economic situation. The farm's external dimension includes all the outside determinants that affect the decision. These include the social environment and competition between biogas and food production as described in the literature. The outcome of the model is the investment decision for (positive) or against (negative) an investment in renewables. Additionally, the type of renewables is considered in the model. The various types of renewable technologies have different impacts on agriculture. Owing to the strong interactions between biogas production and its effects on agriculture, this technology is considered separately. Consequently, biogas (biogas production) and other renewables sources (other renewables), such as the investment object, are integrated into the model.

9.4.2.2 Methodology

The empirical validation of the model is based on the previously mentioned survey of German farmers (see Sect. 9.3). To the best our knowledge, no other quantitative studies investigate farmers' decision behaviour regarding investments in renewables from a behavioural approach. Moreover, biogas production's negative externalities, which affect farmers' individual investment behaviour, require further research.

Based on the model, the three respondent groups (non-investors, biogas investors and other RE investors) were interviewed on the topics of land use competition and their investment behaviour.

The starting point is the real investment decision that the farmers take from an ex-post perspective. The analysis focusses on the predictors' influence on the decision-making outcome. The complex decision-making process itself is not investigated in this sub-study. After removing one outliner from the data, 159 cases (58 biogas investors, 64 other RE investors and 37 non-investors) were used for further analysis (see Sect. 9.3). A multivariate analysis of variance was used to perform a multiple comparison between the three identified groups of farmers in terms of their perception of biogas production's effects in their region. The level of significance was adjusted according to the Bonferroni correction. This post-hoc test identifies in which sub-samples the differences are statistically proven.

The data dimension was reduced by an explorative factor analysis to further validate the empirical model. A multinomial logistical regression revealed the effect that the direction and strength of the independent variables has on investment behaviour.

9.4.3 Empirical Results

9.4.3.1 Intensity of Land Use Competition

Firstly, farmers' perception of the negative effects of the biogas production, as described in Sect. 9.4.1.1, should be mentioned. The majority of the interviewed farmers are highly aware of biogas production in their region. An average of four biogas plants is located within 10 km of 86 % of all the respondents' farms.

Significant differences in farmers' perceptions of biogas production's effects on the local agriculture were identified between the subsamples (see Table 9.3). The increasing land rental rates and land scarcity particularly challenged farmers who had neither invested in the production of biogas (non-investors), nor in a different form of renewables (other RE investors). Thus, the non-investors perceive an increasing level of competition with their biogas producing colleagues. The farmers did not, however, believe that biogas production in the region could lead to an increase in regional feed prices.

In a further item, all the respondents were asked how they would feel if a biogas plant were built close to their farm. The very highly differentiated responses suggest that the different regional conditions and individual situations on farms play a large role in farmers' conflict perspective.

Except for the labour capacities, no significant differences were observed between the structural and socio-demographic characteristics of the three groups. With a mean of 4.13 (full-time) employees, the biogas investors have more employees than the other RE investors, who operate with 2.37 employees. The significant difference (F-value: 3.88, $p < 0.05$) can be explained by the high demand for labour for the cultivation of energy crops as raw materials for biogas production.

Table 9.3 Farmers' perceptions of the impacts of biogas production on agriculture

	Total sample		**Non-investors (A)**		**Biogas investors (B)**		**Other-RE investors (C)**		
	N=159		n=37		n=58		n=64		F-Statistics
	M	SD	M	SD	M	SD	M	SD	
In my area[a]									
...biogas plants force up land lease rates.	0.70	1.18	1.31	0.79	-0.12	1.07	1.10	1.04	30.77 AB***; BC***
...biogas plants lead to problems complying with nutrient limits.	-0.39	1.11	-0.17	1.06	-0.86	0,92	-0.10	1.16	8.97 AB**; BC***
How do you assess the potential impacts of biogas plants located near your farm as mentioned below?[b]									
Land scarcity.	0.27	1.33	0.84	1.34	-0.17	1,18	0.36	1.35	6.23 AB**; BC**
Increasing land lease rates.	0.37	1.31	0.78	1.26	-0.21	1.26	0.69	1.20	9.29 AB**; BC**
Increasing feedstuff rates.	-0.53	1.14	-0.24	1.24	-0.78	0.10	-0.44	1.18	2,34
Problems with manure utilisation.	-1.01	1.10	-0.67	1.24	-1.41	0.73	-0.80	1.22	6.22 AB**; BC*
Increasing competition between farmers.	0.25	1.18	0.68	1.14	-0.23	1.02	0.47	1.20	8.12 AB***; BC**
The energy crop cultivation in my area will lead to...[a]									
...an increased level of competition with livestock farmers.	0.36	1.37	0.69	1.40	-0.07	1.25	0.56	1.38	4.81 AB*; BC*
...an increased level of competition with cash crop farmers.	0.54	1.18	0.67	1.15	0.14	1.22	0.81	1.08	5.48 AB+; BC**
...an increased level of competition with nature conservation.	-0.37	1.06	-0.11	0.89	-1.00	0.78	0.05	1.12	19.83 AB***; BC***
Biogas plants...[a]									
...are bothering nonbiogas producers in my area.	0.21	1.26	0.89	0.95	-0.54	1.14	0.50	1.16	21.61 AB***, BC***
...are crucially important for my area.	0.06	1.06	-0.29	1.10	0.58	0.89	-0.22	1.02	12.50 AB***; BC***

Annotations: M = Mean; SD = Standard deviation; [a]Rated on a scale '-2 = fully disagree' to '2 = strongly agree'; [b]Rated on a scale '-2 = very low' to '2 = very high'; $***p \leq 0,001$; $**p \leq 0.01$; $*p \leq 0.05$; $^{+}p \leq 0.10$; [AB] Significance between groups A and B; [BC] Significance between groups B and C

9.4.3.2 Identifying the Potential Determinants of Investment Behaviour

An exploratory factor analysis was carried out to reduce the data dimension and aggregate variables. Following the theoretical model (see Fig. 9.5), all recorded and appropriate variables that explain the differentiated decision behaviours were included in the analysis. The final factor analysis comprised 18 variables from five factors (see Table 9.4). With the exception of the fifth factor 'affinity for technology', all the constructs have satisfactory indicator values for validity and reliability (Cronbach's alpha > 0.6; MSA > 0.6) (Field 2009).

The factor 'perceived negative external effects of biogas production' reflects biogas production's side effects on agriculture in the course of the diffusion of renewables through biogas production. The dimension 'economic situation' involves farmers' economic self-assessment. The assessment of farms' viability is included in this factor. The construct 'social influence' reflects the influence of entrepreneurs' social environment on decision making. Entrepreneurs include business people as well as external contacts from the area surrounding the farm, such as the local residents. The family of the entrepreneur is not included in this factor. We aimed to have two dimensions in this construct: Firstly, we determined how people from farmers' social environments influence them when making investment decisions (Hypothesis H_{7a}). We questioned the farmers on the extent to which other people influence their decision making. Secondly, we consider the phenomena of driving or inhibiting impacts on investment behaviour (Hypothesis H_{7b}). However, only the factor reflects the first dimension. Therefore, the rest of the analysis only focusses on Hypothesis H_{7a}. The aggregation 'environmental awareness' reflects farmers' attitudes to the environment and the entrepreneurs' ecological orientation. The factor 'affinity for technology' represents individuals' attitudes to new technologies. Individuals' willingness to become early adopters of innovations is integrated into this factor.

The factor analysis does not represent all of the explanatory model's constructs. Therefore, based on suitable logical considerations, individual items were also further analysed. The following variables represent the outstanding constructs:

- Willingness to take risks: 'The improvement of existing production branches on the farm is less important than investing in unknown areas.'[2]
- Farm structure: the soil quality rated in points, the cultivated farmland area in ha and the labour capacity in full-time employees (family employees included)

9.4.3.3 The Impact of Land Use Competition and Other Determinants on Investment Behaviour

The multinomial logistic regression reveals independent variables' direction as well as their impact on investment behaviour. A positive investment decision is determined by various factors. Farmers' actual investment decision is used as a

[2] This variable was recoded.

Table 9.4 Results of the explorative factor analysis

Factors or items	M	SD	R
Factor 1: Perceived negative effects of biogas production[a] explains 23.8% of variance, Cronbach's α: 0.87			
Biogas plants increase land lease prices in my area.	0.68	1.18	0.81
The cultivation of energy crops increased competition with livestock producers in my area.	0.36	1.37	0.80
Biogas plants are bothering nonbiogas producing farmers.	0.20	1.26	0.79
There are already far too many biogas plants in my area.	-0.25	1.15	0.72
Biogas plants cause problems with the compliance of nutrient input limits in my area.	-0.40	1.10	0.70
The energy crops cultivation leads to an increased competition with the cash crop cultivation in my area.	0.54	1.18	0.68
The energy crops cultivation leads to an increased competition with the material usage of renewable resources in my area.	0.04	1.07	0.67
The energy crops cultivation leads to an increased competition with nature conservation.	-0.37	1.06	0.63
Factor 2: Economic situation[b] explains 10.8% of variance, Cronbach's α: 0.68			
Our farm income enables greater investments.[4]	0.57	1.17	0.87
I am satisfied with the current overall situation on my farm.	0.67	1.03	0.77
My farm will still exist in 20 years' time.	0.89	0.99	0.66
Factor 3: Social influence[c] explains 9.7% of variance, Cronbach's α: 0.61			
How important are your employees' opinions for your investment decision.	-0.03	1.15	0.78
How important are your shareholders' opinions for your investment decision.	0.61	1.49	0.75
How important is are local residents' opinions for your investment decision.	-0.64	1.04	0.73
Factor 4: Ecological awareness[d] explains 8.8% of variance, Cronbach's α: 0.67			
I personally make sure my production is sustainable and respects nature and the environment.	1.49	0.71	0.85
As a farmer, I have a special responsibility towards the environment.	1.45	0.71	0.85
Factor 5: Affinity towards technology[b] explains 7.4% of variance, Cronbach's α: 0.40			
I am very interested in new technologies.	0.81	0.78	0.75
I am the first to invest in new technologies.	-0.55	0.78	0.73

Annotations: n = 159; MSA = 0.713; R^2 = 60.56%; M = Mean; SD = Standard deviation; R = Factor loading; [a]Scale from -2 'totally disagree' to +2 'totally agree'; [b]Scale from -2 'does not apply at all' to +2 'completely applies'; [c]Scale from -2 'not important at all' to +2 'very important'; [d]This variable was recoded.

dependent variable in the logit model. More specifically, the variable indicates whether the farmer has invested in renewables production, specified as biogas production (coded as 1, subsample biogas investors), any other form of renewables (2, other RE investors), or has not (yet) invested in any form of renewables (3, noninvestors). As covariates, the identified factors and the additional items were used as explanatory variables. Despite its limited reliability, the factor 'affinity for technology' was integrated into the regression model due to its theoretical relevance for behaviour (Austin et al. 1998).

Seventeen respondents were excluded from the regression analysis due to missing values in the socio-demographic and farm structure variables, resulting in a sample of 142 respondents.

The regression model fulfils the required quality criteria concerning reliability and validity (see Table 9.5). With a total variance (Nagelkerkes R^2) of 53.0 %, the model's overall explanatory power is good. In total, 45.7 % of non-investors, 66.1 % of other RE investors, and even 79.2 % of biogas investors are correctly predicted. Overall, 66.5 % of the cases in the model are classified correctly and are well above the proportional (34.7 %) and maximal random probability (41.6 %) (Field 2009).

Table 9.5 shows how the odds ratio (equivalent to exponential B) of belonging to a certain group changes if the value of the dependent variable increases by one unit.

A comparison of the statistical significance of regression coefficients between the farmers who have invested in renewables (biogas and other renewables) and those who have not shown that three factors influence group affiliation. Satisfaction with the economic situation and a positive assessment of farm viability are strong predictors of the two groups of investors in renewables. Compared to the non-investors (reference group), farmers' likelihood of investing in renewables (i.e. to be assigned to one of these two groups) increases 2.8 and 2.5 times, respectively, if the economic satisfaction increases by one unit (H_4 rejected). In respect of the two groups of investors, separate consideration have to be undertaken regarding how other variables can influence the group to which a farmer belongs. The probability of an investment in biogas production is 2.8 more likely if the farmer has a high technological interest (H_2 confirmed). The negative coefficient of soil quality, but especially of the 'perceived negative effects of biogas production', means a reduction in the probability of belonging to the group of biogas investors. An increase in soil quality decreases the probability of being an investor. This might be due to the high opportunity costs of good farmland (H_{5a} rejected). An increase in the perception of the negative impacts of biogas production by one unit results in farmers being 6.3 times less likely to be willing to invest in biogas production.

This decreasing investment effect is caused by the focused political and financial promotion of biogas, which leads to higher competition between farmers (increasing land rental rates) and shows the diffusion of renewables' relevance to individual investment behaviour (H_6 confirmed). Interestingly, this effect is only slightly significant for the other forms of renewables.

The impact of farmers' social environment on decision making is not significant (H_{7a} confirmed). However, the opinions of people in the farmers' networks (not tested in our model) may have an influence on their investment behaviour (see Hypothesis H_{7b}).

The differences between the forms of technology/investments and between biogas and other RE investors regarding decision making, were considered in a second step of the logit regression, in which the other RE investors were chosen as the new reference group. In Table 9.5, this second step estimation is called the

Table 9.5 Influencing factors on farmers' decision behaviour regarding investments in renewable energies

		Investment decision				**Investment object**	
		Biogas investors vs. non-investors[a]		**Other RE investors vs. non-investors**[a]		**Biogas investors vs. other RE investors**[a]	
		B	**exp(B)**	**B**	**exp(B)**	**B**	**exp(B)**
Individual	**Ecological awareness**	0.20	1.23	0.41	1.51	-0.21	0.81
	Affinity for technology	1.01**	2.76	0.46[+]	1.58	0.56[+]	1.75
	Willingness to take risks	0.37	1.45	0.18	0.84	0.55*	1.73
Farm internal	**Economic situation**	1.03***	2.81	0.93***	2.54	0.10	1.10
	Farm structure: soil quality	-0.38*	0.96	-0.01	0.99	-0.03*	0.97
	Farm structure: cultivated farm land area	-0.00	1.00	0.00	1.00	-0.00	1.00
	Farm structure: labour capacity	0.19	1.21	-0.07	0.94	0.25[+]	1.29
Farm external	**Perceived negative effects of biogas production**	-1.85***	0.16	-0.56[+]	0.58	-1.30***	0.27
	Social influence	0.41	1.50	-0.07	0.93	0.48[+]	1.61
	Absolute term	1.64		1.11		0.54	

Annotations: n=142 (biogas investors=48; other renewable energy (RE) investors=59; noninvestors=35); ***p≤ 0.001, **p≤ 0.01, *p≤ 0.05; [+] nonsignificant trend; [a] reference group; Chi^2=88.67 (p<0.001); Cox&Snell-R^2=0.46; Nagelkerkes-R^2=0.53

'investment object'. The biogas and other RE investors differ from each other in terms of the factor 'perceived negative effects of biogas production' on a significant level (odds ratio 0.27, $p < 0.001$). Farmers who perceive such negative effects are 3.7 times (1/0.27) less likely to invest in biogas production, which reflects a high awareness of land use competition. These farmers are engaged in less criticised forms of renewables, as wind and photovoltaic energy, which do not lead to resource conflicts between farmers such.

Furthermore, a better level of soil quality (odds ratio 0.97, $p < 0.05$) has a slight positive effect on the probability of investment in other renewables. Unlike the factor 'willingness to take risks', farmers who are less risk-averse are 2.7 times more likely to belong to the group of biogas producers (H_3 is thus confirmed).

A comparison between the farmers' 'ecological awareness', 'labour capacity' and 'cultivated farm land area' reveals no significant differences (H_1, H_{5b} and H_{5c} are thus rejected).

9.4.4 Discussion

The findings of the multi-group comparison reveal that biogas extension has a strong impact of on agriculture. There are large differences between biogas-producing farmers and food producers. However, the perceptions of farmers who produce other forms of renewables also differ from those of biogas producers. The most crucial points that the farmers mentioned are the farmland scarcity and the resulting increase in land lease rates. The interviewed farmers were very concerned about biogas expansion. Biogas production thus contributes to an increase in the already high level of competition between German farmers. The threat potential for food-producing farms is obviously high, revealing a very high conflict potential.

The results show that, on the basis of the selected model, a large number of different factors influence farmers' decision behaviour with regard to investments in renewables. Thus, the results of Willock et al.'s (1999) and Burton's (2004) studies on agricultural decision making are basically confirmed.

A decisive contribution to the adoption of renewables is the self-assessment of the economic situation and the farm's viability. Our hypothesis that economically successful farmers are more willing to invest in renewables than their less successful colleagues was confirmed. This may be due to solvent farms having a good basis for capital-intensive investments, such as renewables. Therefore, farmers' financial perception of their entrepreneurial behaviour (La Due et al. 1991) is proven relevant in the case of innovations, such as renewables.

Lynne and Rola's (1988) as well as Trojecka's (2007) studies indicate ecological motivations' strong influence on farmers' behaviour. According to our model, ecological awareness does not influence decision behaviour regarding renewables. Considering the recent discussions on climate change, these are quite unexpected findings. They may be explained by German farmers' general scepticism regarding environmental issues (Pongratz 1992). In recent years, all farms have been faced with higher environmental requirements. Non-organic farms are very sceptical of ecological innovations (Pongratz 1992). However, findings from the multi-group comparison on the effects of biogas production indicate that the investor groups' attitudes to nature conservation change when the environmental effects of energy crop cultivation are taken into consideration. This brings us to the conclusion that farmers are ecologically aware; however, this is not relevant for their investment decision. Farmers' ecological awareness should rather be understood in the specific context of energy crop cultivation. Farmers are quite aware of the negative ecological effects of biogas production, such as the decreased level of biodiversity caused by monocultures. This ecological effect and their impact on farmers' investment behaviour may become more important in the course of expanding biogas production.

Biogas producers are willing to take higher risks than investors in other renewables (e.g., photovoltaics). Therefore, attitudes towards risks are of crucial importance when choosing the technology type. Investments in biogas production are associated with significantly higher technical uncertainties than those in other forms of renewables.

Contrary to Retter et al.'s (2002) and Solano et al.'s (2003) findings, the social environment did not influence business decisions in our model. This may be due to the social dimension being latently integrated into the first factor (the perceived negative effects of biogas production).

Farmers' investment behaviour is not only affected by intrinsic motives and the farm-internal factor 'economic situation', but primarily by biogas production's perceived negative external factors. The perceived negative external factors' strong influence on decision making confirms the current controversial debate on the usefulness of biogas production on farms (WBA 2007). Farmers perceive the effects of biogas production's expansion as increasing the competitive pressure with other local farmers. This has a negative impact on the general decision to invest in biogas production. Hence, these farmers are less willing to invest in this technology. Moreover, we observe that these externalities have an impact on technology selection. Farmers perceiving a high level of land use competition are more likely to invest in other renewables than in biogas production.

The considerable investment-inhibiting effect of land use competition can be explained from a resource and a social point of view. Firstly, in the resource-based attempt to explain the increased demand for farmland for biomass production (energy crop cultivation) and the limited land supply, the higher competition leads to increasing productive land's cost. Thus, in many regions with intensive agriculture, prices have increased on the land lease market (Bahrs and Held 2007; Heissenhuber et al. 2008). The progressive land use competition is problematic, because farms generally have a high proportion of leased farmland. Hence, they are very sensitive to changing land lease prices.

Secondly, energy production has a higher added value than food production, which results in biogas-producing farmers exhibiting a higher willingness to pay for land leases than their food-producing colleagues (Bahrs and Held 2007). Agricultural energy producers are a force to be reckoned with in the land market. Therefore, the competitiveness of existing agriculture production branches on farms (food production) is at risk. Food producers are critical of the short and medium-term effects; they therefore decide not to invest in biogas production, which leads to the existing problems in the bioenergy market.

Furthermore, social effects can explain the investment-inhibiting effect of land use competition. The intra-agricultural resource conflict is problematic because less competitive farmers are not willing to sell their farms. Most farms are owned and operated by a family, which ensures family's income. These findings result in a higher 'willingness to survive' than that found in firms in other sectors (Inhetveen and Schmitt 2010). The farm business is continued even if their primary economic circumstances no longer permit this. Furthermore, many farmers are very emotional and feel that tradition links them to their farms (Roessingh and Schoonderwoerd

2005). From an economical point of view, this irrational behaviour is also observed in other farm production branches. For instance, despite the price decline in German dairy production over the last few years and the subsequent decline in dairy farms' profitability, dairy farming has not been excessively abandoned (BMELV 2009b). This could be explained by farmers' 'willingness to survive', which highlights the earnestness of competition between farmers. The availability of farmland due to farm sales has therefore not increased and, consequently, the pressure on the land lease market is still high. Given this increased competition, there is a higher potential for conflict in the long term (Mautz 2007). This type of competition has an explosive nature, especially if resource conflicts change into relationship conflicts, which are more complex (Feindt et al. 2004). Under these conditions, farmers choose other forms of renewables with a lower conflict potential (e.g., photovoltaics). Thus, our results confirm Granovetter's (1985) evidence of socially driven behaviour.

In addition, many farms have close, mutually beneficial ties with other local farms. An example of such positive network externalities is the joint purchase of machinery. If some of these farms restructure from food to energy production, the new operating structure could result in a loss of cooperation partners in the network. Consequently, the transition costs will increase.

This sub-study explores the determinants of farmers' investment behaviour in renewables and how they contribute to a better understanding of decision making at the farm level. However, when interpreting the findings, some limitations should be taken into account. Firstly, the study is limited by its regional focus on North West Germany and the small sample size. In Germany, the diffusion of biogas production has to date been rapid and far-reaching. In many other countries, in which biogas production is only in its infancy, the level of competition between the farmers differs. These limitations should be taken into account when transferring the results to other regions.

We pointed out that external factors, such as localisation and the level of competition between farmers, predict their investment behaviour better than intrinsic individual factors. Therefore, each farm's specific situation should be considered. The findings should be complemented by an analysis of the effects of increased biogas production at the individual and local level. The necessity for this is evident in the high standard deviations in the multi-group comparison through the multivariate analysis of the variance and the in-factor analysis. This confirms that the farmers have very different perceptions. However, we point out the high relevance of socio-economic patterns (the level of competition) for biogas production in agriculture. The findings show biogas production's potential for further expansion in agriculture and the crucial effects thereof.

Furthermore, our explorative study's findings indicate possible approaches to further research:

- The lack of influence that 'environmental awareness' has on the investment decision-making process is surprising in view of the climate change debate and should be validated in a large sample study.

- To date, the decision-influencing factors have been considered separately. However, there are indications that, even among the factors themselves (e.g., between the factors 'social influence' and 'perceived negative effects of biogas production', see Sect. 9.4.3.3), there is path dependency. It may therefore be useful to examine these effects in an extended structural equation model.
- The large number of identified determinants and the high standard deviations indicate heterogeneity among the farmers. Clustering these agricultural entrepreneurs according to the identified behavioural determinants may prove these differences.
- Knowledge of unintended negative effects' impact on entrepreneurial decision-making behaviour should be the basis of a further analysis of the policy implications in terms of a policy analysis by means of reflexive evaluations. According to Le Grand (1991), a stable policy relies on such knowledge.

9.5 The Role of Risk Aversion, Bounded Rationality and Investment Subsidies

This section presents the second part of the sub-study. The chapter proceeds as follows: Firstly, the theoretical background is described (Sect. 9.5.1). Thereafter, the design of the second part of the survey and the methodology are described in detail. Section 9.5.3 presents the results of the hypothetical investment's investment threshold and its potentially influencing factors. We determine the explanatory power of the influencing factors and illustrate the investment subsidy's effect on the investment behaviour.

9.5.1 Theoretical Background

The term investment describes a long-term (lasting for more than one production period) monetary investment for economic purposes. Real investments, such as an investment in a biogas plant or in buildings and machinery, lead to changes in the company's equipment with regard to producer durables, which are also called durable means of production. Each investment is characterised by the cash flow, which comprises cash resources that, in relation to an investment, flow out of or into the company. Usually, an investment starts with a major pay-out, namely the purchase price. Pay-outs can also arise at a later stage in the form of repairs, operating supplies, insurances or the maintenance of a biogas plant's input substrate. In contrast to these pay-outs, one or more deposits arise from the investment in the form of products' sales revenue or services produced during the investment's useful lifetime. The investment object's sales revenue (residual value), which may accrue at the end of a useful lifetime, is also a deposit.

Several aspects are of particular importance when evaluating an investment's (investment analysis's) economic advantageousness: Besides the immediate pay-out in the form of the purchase price, investments generate return flows during their multi-annual useful lifetime. The latter often comprises 10, 20 or even 30 years. Hence, payments that accrue at different times should not be compared nominally (according to their numerical value). In order to economically evaluate payments, they need to first be financially and mathematically comparable (Sect. 9.5.1.1). If the investments are planned for the distant future, uncertainty will be particularly high. While the prices and revenues expected in the near future can be estimated quite accurately, the risks increase with an expanding planning horizon. Consequently, investors' individual risk attitudes are of considerable importance for the evaluation of an investment (Sect. 9.5.1.2). Additional aspects might have an impact on the investment decisions. Section 9.5.1.3 presents the examples of bounded rationality as well as the effects of the soil and the environment.

9.5.1.1 Basics of Investment Analysis

Financial/Mathematical Fundamentals

In a broader sense, the calculation of interest, which is also denoted as financial mathematics, is an indispensable prerequisite for the profitability analysis of investments. Assuming you invest 10,000 EUR in a bank over a period of $N = 5$ years with an interest rate of $i = 5\ \%$ per year, how much money will you have after the 5-year period? To achieve an exact result, it should be taken into account that the interests obtained over these years also yield interest. This is done by 'correctly' adding the unaccrued interest. After a year, one will already have accumulated 10,500 EUR (10,000 + 10,000 5 % = 10,000 1.05). In the second year, this sum of 10,500 EUR can be invested at 5 % interest. At the end of year two, of the amount will have grown to 11,025 EUR (10,000 1.05^2). After 3 years, capital to the value of 11,576 EUR will have been obtained. At the end of 5 years, the capital will have grown to 12,763 EUR (10,000 1.05^5) as a result of this compound interest effect.

The factor with which the initial amount C_0 is multiplied by a given number of years N and at a given interest rate i, is called the compounding factor ($AF_{i,N}$). The compounding formula describes, in general terms, how to calculate the future value C_N. The latter equals the amount of money C_0 available at time zero after N years, including the compound interest:

$$C_N = C_0 \cdot (1+i)^N = C_0 \cdot AF_{i;N} \tag{9.1}$$

The present amount of money C_0 is also referred to as the present value. The future value of a present payment is higher the farther the future point in time and the higher the interest rate.

Hence, the answer to the reverse question of how much capital one has to invest today in order to accumulate 12,763 EUR at 5 % interest within 5 years is already known – one has to invest 10,000 EUR. However, if one wants to calculate the present value of an equal current amount of money C_0 starting with a future amount

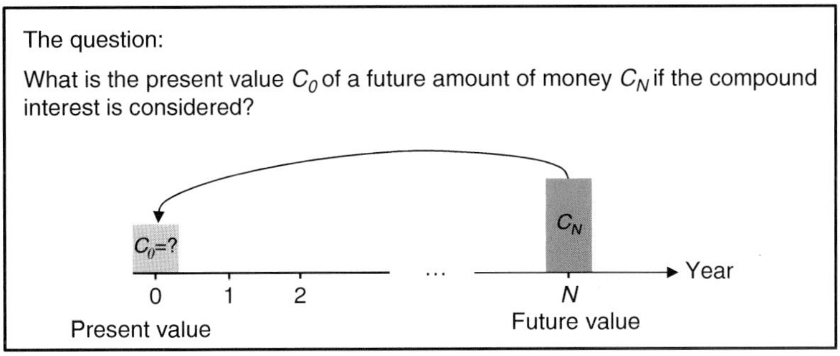

Fig. 9.6 Discounting of a future payment

C_N, the compounding formula needs to be converted. This reverse of compounding is referred to as discounting.

The factor with which the future amount of capital is multiplied when discounting it at a given number of years N and at a given interest i is referred to as the discount factor ($DF_{i,N}$). This factor is the reciprocal of the compounding factor. The discounting formula describes, in general terms, how to calculate the present value C_0 of a future amount of money C_N under consideration of the compound interest (see Fig. 9.6):

$$C_0 = C_N \cdot (1+i)^{-N} = C_N \cdot \frac{1}{AF_{i;N}} = C_N \cdot DF_{i;N} \tag{9.2}$$

The higher the interest rate and the longer the observed period of time, the lower the discounting factors. In other words, in the present, the more future payments are and the higher the interest rate is, the less they are worth.

In order to compare the alternatives with payments that accrue at different times, they have to be applied to *one* point in time. The time of investment does not matter with regard to the decision support: Either all payments are applied to the future by means of compounding or they are applied to the present by means of discounting. When calculating the net present value, all surpluses of future deposits are applied to the present.

Net Present Value

The net present value (NPV) is the present value of all deposit surpluses triggered by the investment that is the sum of the *discounted* deposit surpluses (see Brealey et al. 2008):

$$\begin{aligned} NPV &= (e_0 - a_0) \cdot (1+i)^{-0} + (e_1 - a_1) \cdot (1+i)^{-1} + \cdots + (e_N - a_N) \cdot q(1+i)^{-N} \\ &= \sum_{t=0}^{N} (e_t - a_t).(1+i)^{-t} \end{aligned} \tag{9.3}$$

$(e_t\text{-}a_t)$ is the deposit surplus at the respective time t and $(1 + i)^{-t}$ is the discounting factor for the respective year; thereby, i denotes the interest rate. Since each investment represents a series of payments, which starts with a payout or a negative deposit surplus ($a_0 > 0$ and $e_0 = 0$), the situation can be summarised with the following formula:

$$NPV = \underbrace{-a_0}_{investment\ costs} + \underbrace{\sum_{t=1}^{N} (e_t - a_t) \cdot (1+i)^{-t}}_{present\ value\ of\ the\ future\ investment\ returns} \tag{9.4}$$

In the calculation of the net present value, the initial value is compared to the present value of the future investment returns. This present value is calculated on the basis of the adequate target rate. In the case of equity financing, the investments to evaluate are therefore compared implicitly with the alternative of 'not implementing the investment and not investing with a bank'. The net present value of the returns from this alternative does not have to be determined separately. It is zero because the interest rate of the investment is equivalent to the adequate target rate.

The net present value denotes today's value of the whole operational investment project. It indicates, in the form of an absolute value, how much more the investment will earn than used capital costs are. From a profitability point of view, an investment is worth implementing, if its net present value is greater than zero, as it will generate more profit than costs. Otherwise, the investment should be discarded. If the net present value is only marginally greater than zero, an investment should only be made if the entrepreneur solely pursues a profit maximisation goal. If additional corporate goals as well as any non-monetary efforts related to the investment need to be considered, entrepreneurs are likely to demand a net present value considerably greater than zero before they invest.

From a technical point of view, it is worth mentioning that, for investments with homogenous future deposit surpluses ($e_1 - a_1 = e_2 - a_2 = \cdots = e_N - a_N = e - a$), the net present value formula Eq. 9.3 can be simplified as follows:

$$NPV = -a_0 + (e - a) \cdot CF_{i;N} \tag{9.5}$$

The net present value of an investment is significantly influenced by the following factors:

- The deposit surpluses. The higher the deposit surpluses in the specific years during an investment's useful lifetime, the higher the net present value of the investment ceteris paribus.
- The level of the adequate target rate. The lower the costs of the deployed capital, the higher the net present value of an investment ceteris paribus.
- The temporal structure of the accruing payments. The earlier deposits accrue and the later pay-outs accrue, the more profitable the investment ceteris paribus.

9.5.1.2 Individual Risk Attitude

It has often been pointed out that entrepreneurial decisions in general and investment decisions in particular are dependent on the decision-maker's risk attitude (see Bard and Berry 2000; Harwood et al. 1999). If the word 'risk' is used in the context of economic activity, different social groups associate different matters with it. Non-entrepreneurs primarily think of the dangers posed to society, which result, for example, from the use of genetic engineering, fossil fuels or pesticides (social risk perspective). For entrepreneurs, however, the word 'risk' has a totally different primary definition (entrepreneurial perspective). For the latter, changes in the institutional and legal framework conditions as well as developments in the markets represent uncertain environmental conditions and, therefore, sources of risk that make economic activities' success less certain. Here, risk denotes that nobody knows how much money he or she will earn or lose in the future.

Although these two perspectives could not differ more at a first glance, they have something in common: In both cases, risk describes the probability distribution of a target-relevant quantity. The social perspective is about the probability that society as whole will have to bear economic activities' future adverse effects. From an entrepreneur's perspective, risk is, conversely, all about the probability distribution of the entrepreneurial success that results from the prevailing framework conditions as well as from the input and output prices. Since the present section deals with entrepreneurial decisions, we focus on the entrepreneurial perspective despite the obvious relevance of the social risk perspective.

When examining the net present value in the previous section, we did not take into account that we live in an uncertain world in which many influencing factors are stochastic variables. Entrepreneurial decisions are always made under uncertainty. This means that prices, revenues, etc. are generally stochastic variables. The latter can adopt different future values, referred to as variations or states of the environment. If the factors relevant to success are uncertain, the overall success of an activity that an entrepreneur has selected also becomes a stochastic variable whose variations cannot be definitely predicted. As in many cases, the term 'risk' is here equated with the term 'uncertainty', although one would actually have to refer to it as uncertainty in both a narrower and broader sense in order to contrast the specific meaning (= quantifiable risk) with the superordinate meaning (= uncertainty in total).

With regard to the two objectives 'income' and 'certainty', the decision-maker's subjective risk attitude is of particular importance. If two alternative courses of action have the same risk but expect different amounts of income, all entrepreneurs would choose the alternative with the higher income expectation. If, however, the alternative with the higher income expectation also bears a higher entrepreneurial risk, decision making is not so simple. The risk-averse decision-maker experiences conflicting objectives in terms of striving for income and striving for certainty. Despite the lower expected income, this decision-maker may prefer a less risky alternative if the lower risk constitutes more benefits than the loss the decision-maker can make through his or her expected reduction in income.

The meaning of risk attitude and especially risk aversion can be effectively explained by a lottery example: If a person participates in a coin tossing game to receive payment for a stake, there is an equal chance that the coin will either land on 'heads' or 'tails'. The participant will earn nothing for 'heads', while he or she will receive 1,000 EUR for 'tails'. Thus, the expectation value of the game's payoff is 500 EUR. How much would you be prepared to pay to participate in this game?

If you are prepared to pay more than 500 EUR to participate, you are very prepared to take risks. The expected winnings of the game are negative but the risk provides an additional benefit. If, given a participation fee of 500 euro, you are indifferent regarding participation, you are risk neutral. In this case, in your opinion, the amount that you are willing to pay to participate is equivalent to the expectation value of the uncertain payoff of the game. However, if you are only prepared to pay less than 500 EUR to participate, you are risk-averse. If you are willing to pay a maximum of, for example, 400 EUR, you are only prepared to participate in the game because you can expect an average winnings of 100 EUR per round played. In other words, the participation fee of 400 EUR that you definitely have without participating in the game is of the same value to you as the expectation value of 500 EUR that can be won in the game. In this context, one often refers to a premium that the risk-averse decision-maker (risk premium) demands for the risk; in this case, the premium is 100 EUR. The risk premium is generally an amount of money that a risk-averse decision-maker demands for taking a risk.

In reality, it is assumed that entrepreneurs are risk-averse although to varying degrees. This assumption explains why entrepreneurs voluntarily take out insurance in a risky environment even though, due to administration costs as well as the insurance's profit margin, it on average generates more costs than income over the years. Hence, the insurance reduces the expected amount of income rather than increasing it. Action alternatives, ceteris paribus, become increasingly less beneficial for risk-averse decision-makers, the riskier they are. Conversely, action alternatives, ceteris paribus, are more beneficial for risk-taking decision-makers, the riskier they are. This corresponds with the assumption that the decision-maker is prepared to invest an amount of money to experience risk (negative risk premium). People who like to gamble at casinos are examples of such decision-makers. Anybody who takes 100 EUR along to a casino knows that the expectation value of the money he or she will take home is significantly less than 100 EUR as, statistically, casino gamblers lose more than they win. Consequently, gambling at a casino is not a sound business strategy for earning money. However, some people obviously derive benefits, for example, the amusement value, from the risk. For these people, the benefits are worth the investment. The risk would not matter at all for risk-neutral decision-makers. Their benefit only depends on the expectation value of the income. In this case, the risk premium is zero. In order to take risk into account, a premium can be used to calculate the net present value (see Sect. 9.4). Alternatively, it is possible to increase or reduce the risk-free interest rate.

9.5.1.3 Further Aspects

Simon (1956, 1957) introduced the 'bounded rationality' concept, i.e. the notion that the availability of information and the individual's cognitive abilities limit the rationality of individual decision making. From a standard rational choice perspective, which assumes that individuals maximise their utility, bounded rational decision making is considered a behavioural 'anomaly'. Since Simon (1956) first published his pioneering work, many economists have adopted the concept of bounded rationality, i.e. the notion that, in many situations, optimisation is beyond humans' cognitive abilities (see Cyert and March 1963 and Sauermann and Selten 1962 as examples of early adopters).

Daniel Kahneman, a psychologist, and Vernon L. Smith, who established laboratory experiments as a tool in economic analysis, propagated the idea that behavioural 'anomalies' are ubiquitous, and thus a relevant field of research in economics, most successfully. Kahneman and Smith jointly won the Nobel Prize in Economics in 2002. Nonetheless, nearly 35 years after Simon's seminal work, Selten (1990: 649) urged economists to 'put their effort into the further development of the bounded rationality approach to microeconomics'.

Gigerenzer (2000) emphasises that, in decision making, people use heuristics derived from an adaptive learning process. He furthermore argues that a relevant feature of bounded rationality is that many people are unable to interpret relative figures, such as percentages, correctly due to their 'figure blindness' (Gigerenzer 2002). This is relevant for financing decisions, because banks use the effective rate of interest (a relative figure) to enable comparisons between different loans. Farmers who do not recognise the monetary differences between alternative loan offers when the information is presented as a difference in the effective rate of interest, may therefore reveal bounded rationality.

Moreover, effects concerning the soil and environment (see Muradian et al. 2010; Willms et al. 2009) may be relevant for the evaluation of an investment in a bioenergy plant.

9.5.2 Data and Methods

9.5.2.1 Research Design

In order to adequately predict the effects of political changes on investment conditions, it is essential to understand farmers' decision-making behaviour. The observation of farmers' decisions is of little use in this context. On a farm, investment decisions related to a capital-intensive object (such as a biogas plant) are relatively rare. Moreover, basic conditions differ between farms (e.g., the financial resources), making it difficult to draw comparisons (Gardebroek and Oude Lansink 2008).

We therefore confronted farmers with a hypothetical situation: They had to decide on the implementation of a hypothetical investment. We used this type of standardised experiment to make the surrounding conditions and activities manageable (Just and Wu 2009; Starmer 1999). While this method reduces the external validity, it has the advantage of increasing the internal validity (Roe and Just 2009). Furthermore, we gave all respondents the same information to focus our analysis on the conversion threshold and on certain driving factors as well as to examine limited cognitive abilities to process information as a facet of bounded rationality. The economic experiment was designed to allow us to calculate a normative benchmark to which the empirical data could be compared.

The research is based on the survey described in Sect. 9.3. The hypothetical decision situation was as follows: All the respondents were asked to imagine managing a 200 ha crop farm. They were also asked to assume that they had 600,000 EUR in capital. Two investment alternatives were available. The money could be placed in a bank for 20 years and earn 5 % interest per year, or could be invested in a recently built biogas plant next to their farms. The farmers had to choose between these two alternatives.

The biogas plant was described as having the capacity to generate 150 kW of installed electrical power. The expected operating life of the plant was said to be 20 years and there would be no residual value. Furthermore, the farmers had to assume that the investment was tax neutral and that the expected annual cash inflow resulting from the generation of power and heat would be 200,000 EUR. The expected cash outflow for labour costs, maintenance, electricity and insurance would amount to 100,000 EUR annually, excluding the cost of the biogas plant's input substrate. Thus, in each of the 20 years of use, the biogas plant would provide an expected net cash flow of 100,000 EUR before the cost of the substrate was subtracted. The farmers were asked to assume that silo maize was the only substrate that could be used. The operation of the plant would require an input of 30,000 decitonne (one tenth of a metric ton) of maize each year. The maize would be produced on land currently used to cultivate wheat. We used wheat as the competing product because it is very popular in Germany. To make the critical wheat price – the price on which the investment had to be decided – independent of a site-specific yield level, the farmers were asked to assume that there was a fixed relationship between the maize and wheat yields: The production of 6 decitonnes of maize replaces 1 decitonne of wheat. Therefore, the wheat price is independent of the site-specific yield level because the wheat production on the farm would be reduced by 5,000 decitonnes per year. The variable costs of the maize production were assumed to equal the variable costs of the wheat production. The resulting digestate was an 'item in transit' because the value of the fertiliser was presumed to correspond exactly to the costs of its application.

While we are aware of the described situation's hypothetical character, care was taken to achieve plausible conditions. Table 9.6 summarises the cash flows associated with the investment. This presentation was not shown to the surveyed farmers.

After the biogas plant was described to the farmers, they were asked to indicate their investment threshold. This 'trigger price' was defined as the average (critical) wheat price (in euro per decitonne) over the operating life of the plant that would convince the farmers to invest in the above-mentioned biogas plant and change the existing production programme. Consequently, the wheat price had to decrease

Table 9.6 Cash flow structure of the considered biogas plant

Period	0	1	2	...	20
Cash inflow for:					
The generation of power and heat		200	200		200
Cash outflow for:					
Investment costs	600				
Others (maintenance, power, etc.)		100	100		100
Opportunity costs (lost revenues from selling the 5,000 decitonnes of wheat)		?	?		?

Annotations: n = 142; in thousands of euro; the variable costs of the production of wheat and maize are assumed to be equal and are therefore not included

until the biogas plant was selected as the investment alternative. The trigger price shows the subjective value that each decision-makers expect from the investment and from which they estimate their individual utility.

Further questions examined the trigger price more closely and provided information about factors that were expected to affect the price (see Sections 9.5.1.2 and 9.5.1.3). These 'influencing factors' take the farmers' individual preferences into consideration. The following questions were asked:

1. Imagine that instead of investing in the biogas plant, you put the 600,000 EUR in the bank, earning an annual interest of 5 %. Please estimate the annuity that you could withdraw over a 20-year period under the assumption that the total amount will be consumed at the end of this period.
2. What is the maximum annual insurance premium that you would have to pay to have a guaranteed yearly incremental cash flow of 100,000 EUR from the biogas plant?
3. At what wheat price would you start investing if the cultivation of maize for the biogas plant had no impact on the soil fertility?
4. At what wheat price would you start investing if the biogas plant had no environmental effects?
5. At what wheat price would you invest if the government supported your investment in the biogas plant with an investment subsidy of 100,000 EUR?

The first question was asked to determine the cost of capital; the second, the risk premium; the third, the soil fertility premium; the fourth, the environmental premium and the fifth, the effect of the investment subsidy on the critical wheat price. Various pre-tests showed that using an arrow with predetermined intervals and asking the farmers to mark the appropriate place with a cross helped with the queries about the wheat prices in questions 3–5. All aids were allowed when answering the questions and there was no time restriction.

9.5.2.2 Determining a Normative Benchmark

Can calculations based on rational choice models adequately explain farmers' investment behaviour? To answer this question, we first determined normative benchmarks. Based on the assumptions that decision-makers act perfectly rationally and that profit maximisation is the only entrepreneurial objective, a biogas plant's net present value (*NPV*) and homogenous investment returns per period were calculated as follows (see Eq. 9.6):

$$NPV = -a_0 + (e - a) \cdot CF_{i;N}, \quad \text{with}\, a = q^W \cdot p^W + a^{\circ S} \,\text{and}\, CF_{i;N} = \frac{(1+i)^N - 1}{(1+i)^N \cdot i} \tag{9.6}$$

Above, a_0 denotes the investment costs, e is the cash inflow and a is the cash outflow. The cash outflow is composed of the substrate costs and other cash outflows $a^{\circ S}$. The substrate costs correspond to the displaced wheat yield q^W multiplied by the wheat price p^W. The risk-free interest rate is denoted by i; the expected useful life of the biogas plant is described by N and the capitalisation factor is $CF_{i;N}$.

The wheat price triggering the investment is the price of the crop replaced by the substrate cultivation. This price can be calculated by setting the *NPV* equal to zero and solving Equation Eq. 9.6 to obtain the wheat price p^W:

$$p^W = \frac{e - a^{\circ S} - a_0/CF_{i;N}}{q^W} \tag{9.7}$$

Therefore, the biogas plant considered in this study requires a reduction in wheat production of 5,000 decitonnes and investment costs of 600,000 EUR, while it generate an expected investment returns of 100,000 EUR p.a. Therefore, the corresponding wheat price is:

$$p^W = \frac{200,000 - 100,000 - 600,000/12.46}{5,000} = \frac{51,854}{5,000} = 10.37 \tag{9.8}$$

The opportunity costs of the land, which amount to 51,854 EUR p.a., pertain to the lost 5,000 decitonnes of wheat. With an assumed wheat price of 10.37 EUR per decitonne, the biogas plant investment results in a net present value of zero. In other words, the annual cash inflow of 20 EUR per decitonne of wheat (= (200,000 EUR – 100,000 EUR)/5,000 decitonnes) allows the cost of the substrate to be 10.37 EUR per decitonne. The remaining 9.63 EUR per decitonne are needed to cover the investment costs.

In addition to the costs of capital and the annual investment returns from the investment in a biogas plant, which are all incorporated into Eq. 9.7, there are other cost components that may influence decision making. The cost component 'risk premium' (*RP*) is influenced by the subjective perception of the risk resulting from the investment as well as each farmer's individual risk attitude (see Sect. 9.5.1.2). Some farmers may frame their subjective perception of investment risk in terms of the supply of the substrate (the variability of the silage maize yield), or the amount of energy output (the technical default risk of the biogas plant). However, other farmers may view risk in terms of the risk reduction that the diversification effects of the new branch of farm business create. The larger the mentioned value of the risk premium, the bigger the reduction in capital expenditure will be.

The cost component 'soil fertility premium' (*SP*) is determined by sustainability aspects that are relevant from a production factor point of view. For instance, the decision-maker may fear negative effects due to the introduced (expanded) cultivation of maize because the productivity of his land may decrease in the long term due to the negative effects on crop production. The cost component 'environmental premium' (*EP*) is relevant when decision-makers have non-economic aims that are affected by their investment in a biogas plant. For example, the appreciation of climate-friendly energy production at the biogas plant may, from the farmer's point of view, be reflected in a negative environmental premium. In contrast, if negative environmental effects, such as the ploughing up of grassland, are expected, a positive environment premium is indicated. This leads to the following extension of Eq. 9.9:

$$p^W = \frac{e - a^{oS} - a_0/CF_{i;N} - RP - SP - EP}{q^W} \tag{9.9}$$

Equation 9.9 clarifies that the investment reluctance increases when the critical wheat price decreases; the higher the risk, the higher the soil and environmental premiums. In addition to these premiums, there are other factors that may increase the investment reluctance, for example, the expectation of inflation or of a possible farm succession. Such factors were not considered in this investigation. Figure 9.7 provides an overview of the three normative benchmarks.

Although it is not taken into account in Eq. 9.9, it is evident that an investment subsidy would reduce the investment costs, (normatively) resulting in a higher trigger price and a higher willingness to invest. Technically, the existing investment costs a_0 are replaced by $a'_0 = a_0 - z$, where z denotes the investment subsidy.

9.5.3 Empirical Results and Discussion

Section 9.5.3.1 presents the results of the trigger price and the influencing factors. In Section 9.5.3.2, the explanatory power of the influencing factors is examined using a regression model. Section 9.5.3.3 shows the effect of an investment subsidy on investment behaviour. In a first step, all the data were converted into euro per decitonne of wheat.

9.5.3.1 Survey Results of the Trigger Price and Influencing Factors

The survey results show that farmers have various conversion thresholds for potential investments in bioenergy production. In our example, the conversion thresholds range from 5 to 30 EUR per decitonne of wheat (see Fig. 9.8), although

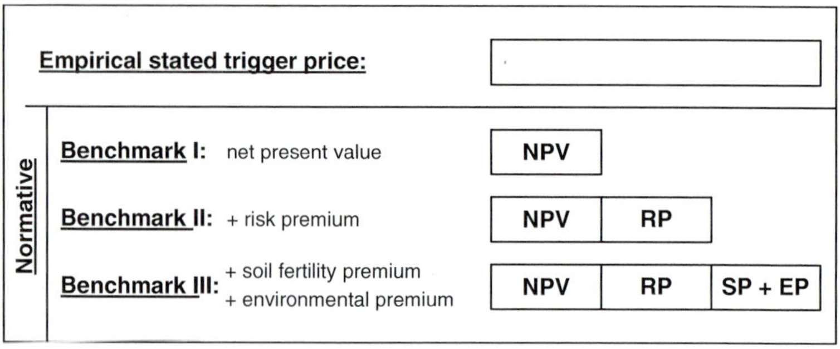

Fig. 9.7 Three normative benchmarks to analyse the trigger price

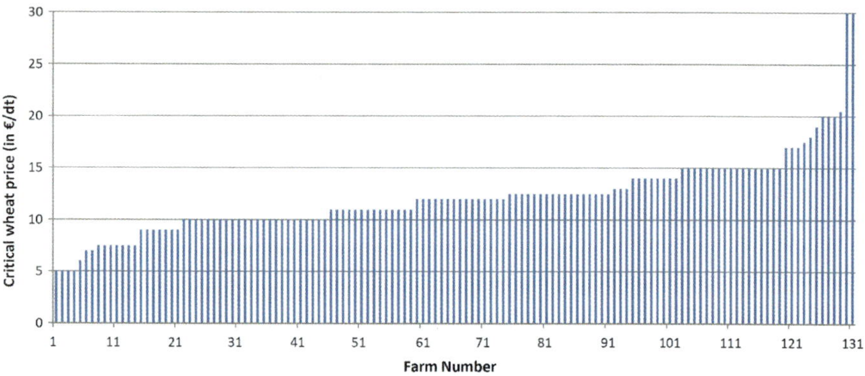

Fig. 9.8 Critical wheat price

the farmers were confronted with the same investment scenario and the site-specific conditions had no effect on the critical wheat price. This indicates that the farmers' behaviour regarding potential investments in a biogas plant was very heterogeneous. Figure 9.9 presents a chart of the historical wheat price from 2005 to 2010 as a comparison standard. At a historical price of 5 EUR per decitonne, the investment was not favourable because the market price was always higher. In contrast, the farmers who quoted a trigger price of 30 EUR per decitonne should have invested because the market price was always lower. Overall, the range of empirical results seems plausible.

Table 9.7 shows the mean, standard deviation, minimum and maximum values of the trigger price and the influencing factors.

The average critical wheat price stated by the farmers is 12.14 EUR per decitonne (standard deviation: 3.78 EUR per decitonne). If the farmers are myopic profit maximisers, they invest too early because the normative critical wheat price is

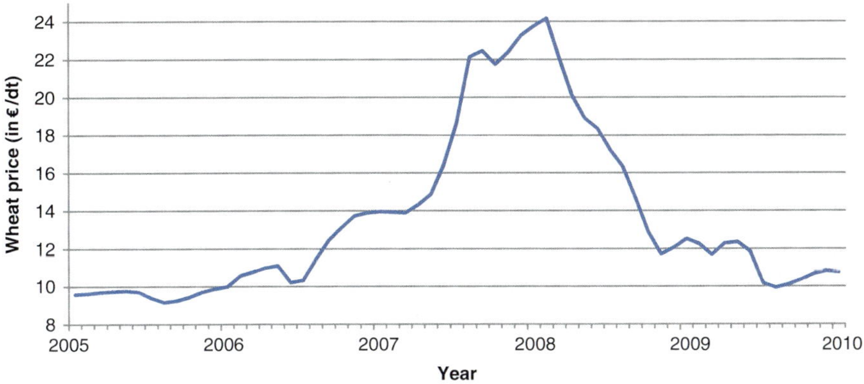

Fig. 9.9 Historical price chart for wheat from 2005 to 2010 (Data based on GJAE different volumes)

Table 9.7 Survey results of the trigger price and influencing factors

		Influencing factors				
	Trigger price (wheat)	Cost of capital	Risk premium	Soil fertility premium	Environmental premium	Others
Mean	12.14	11.86	1.12	0.19	−0.04	−0.99
			13.13			
Standard deviation	3.78	2.64	1.05	1.00	1.20	4.68
Minimum	5.00	5.00	0.05	−5.00	−7.00	−11.00
Maximum	30.00	19.70	5.00	5.00	5.00	16.20

Annotations: n = 135; in euro per decitonne; Assuming a pure profit-maximising decision-maker, the normative critical wheat price is 10.37 EUR per decitonne. Of the cash inflow of 20 EUR per decitonne, 9.63 EUR per decitonne remain to cover the cost of capital

10.37 EUR per decitonne (see Eq. 9.8). This significant difference (p-value < 0.001; two-sided t-test) of 1.77 EUR per decitonne (= 12.14 EUR per decitonne – 10.37 EUR per decitonne) implies that the investment in the hypothetical biogas plant has a net present value (see Eq. 9.6) of −110,235 EUR. This highly negative figure is caused by the plant's 20-year operating life and the yearly requirement of 5,000 decitonnes of wheat equivalent. This seems to question the assumption that farmers are profit maximisers. It is important, however, to note that these results only prove that farmers' decision-making behaviour is not in accordance with Eq. 9.6. One cannot conclude that they act under bounded rationality regarding investment decisions in the context of biogas plants. Farmers may expect equally large benefits from the total investment than those arising from the positive effects that these benefits have on their farms' risk profile. In the same manner, the 'soil fertility premium' and 'environmental premium' could be relevant.

Hence, the effects of the risk, the environment and the soil fertility need to be analysed in addition to the potential existence of bounded rationality. The total cost of the investment was 9.63 EUR per decitonne. The farmers estimated the cost of capital to be an average of 11.86 EUR per decitonne (standard deviation: 2.64 EUR per decitonne), thus misjudging the value by 2.23 EUR per decitonne. This difference between the empirically measured and the normatively determined cost of capital deviates significantly from zero (p-value < 0.001; two-sided t-test). It should be pointed out that 2.23 EUR per decitonne results in a net present value of −138.954 EUR. Taking the correctly estimated cost of capital into account, the capital value of a myopic profit maximiser would, ceteris paribus, total 28,719 EUR (= −110,235 EUR + 138,954 EUR). The underestimation of the cost of capital could, for example, be due to the decision-makers' lack of skills to adequately take the interest and compound interest effects into account. Gigerenzer (2002) emphasises that many people have difficulty with correctly calculating relative values, such as percentage values, due to 'number blindness'. Musshoff et al. (2009) show that farmers underestimate the interest and compound interest effects. When considered in isolation, this underestimation leads to an overinvestment in bioenergy.

The average of the risk premium is 1.12 EUR per decitonne (standard deviation: 1.05 EUR per decitonne) and is positive. This indicates a reduction in the willingness to invest compared to a simple profit orientation. Farmers seem to be risk-averse and expect an increase in the overall corporate risk resulting from an investment in a biogas plant. Therefore, they invest later. Soil fertility effects were relevant for about 11 % of the interviewed farmers and they revealed an average soil fertility premium of 0.19 EUR per decitonne (standard deviation: 1.00 EUR per decitonne). Environmental effects, which were important to approximately 12 % of the farmers, result in a mean environmental premium of −0.04 EUR per decitonne (standard deviation: 1.20 EUR per decitonne). In this regard, the negative premium shows a slightly positive perception of the investment. Just over 10 % of the farmers specified a soil and environmental premium. This may have been due to their belief that they were already following procedure in these areas and therefore had little room to improve, even without investing in a biogas plant.

If the trigger price specified by the farmers is taken as a basis, an overall effect can be inferred. The influencing factors total 13.13 EUR per decitonne, which is 0.99 EUR per decitonne higher than the trigger price of 12.14 EUR per decitonne. This deviation differs significantly from zero (p-value $= 0.016$; two-sided t-test). Therefore, Eq. 9.4 does not appropriately describe the interviewed farmers' investment behaviour. Furthermore, the difference of −0.99 EUR per decitonne could have three possible causes:

1. Bounded rationality in the context of determining the cost of capital, which has already been confirmed.
2. Bounded rationality regarding the aggregation of the separate influencing factors.
3. Additional influencing factors that were not explicitly addressed in the interviews (e.g., effects that include ethical or image considerations).

Table 9.8 Aggregation of the influencing factors based on the empirical and normative cost of capital

	Trigger price (wheat)	Influencing factors				
		Cost of capital	Risk premium	Soil fertility premium	Environmental premium	Others
Mean with the empirical cost of capital	12.14	11.86	1.12	0.19 13.13	−0.04	−0.99
Mean with the normative cost of capital	12.14	9.63 10.90	1.12	0.19	−0.04	1.24

Annotations: n = 135; in euro per decitonne

The afore-mentioned difference of −0.99 EUR per decitonne is overcompensated by 1.24 EUR per decitonne when the underestimated cost of capital (2.23 EUR per decitonne) is taken into account. The residual value of 1.24 EUR per decitonne (see Table 9.8) is probably divided between the two other causes mentioned above.

With regard to the interpretation of the results, it is important to bear in mind that decision-makers in the real world – where real money is involved – have stronger incentives to make optimal decisions. Therefore, real-life decisions are often based on detailed assessments that, if necessary, are made with the help of consultants. In a hypothetical decision situation, such incentives cannot be given. The influence of bounded rationality may therefore have been overestimated. However, the literature emphasises that, in principle, effects found in non-incentive scheme experiments remain, even if the incentives are increased. With regard to bounded rationality, Schoemaker (1982: 553 f.) concludes: '[There is] no evidence that suboptimal laboratory behaviour improves when committing subjects financially to their decisions' (see also Frey and Eichenberger 1989).

To investigate how a change in incentives affects the extent of bounded rationality in a decision situation, we integrated the afore-mentioned questions into a written university examination on investments. Fifty-nine 'prospective (farm) managers' – agriculture and economics students in their fourth to sixth term – participated, thereby ensuring the educational requirements necessary for capitalisation. However, they had little practical experience with running a business. Incomplete responses were not included in the analysis. The cost of capital was evaluated accurately by 25 students. The remaining 34 students were unable to answer the question correctly. The students calculated the average cost of capital as 10.08 EUR per decitonne (standard deviation: 6.05 EUR per decitonne). Compared to the normative benchmark (9.63 EUR per decitonne), they underestimated the costs of the capital by about 0.45 EUR per decitonne. In contrast, the farmers' absolute deviation from the normative benchmark was 2.23 EUR per decitonne. Therefore, the level of bounded rationality regarding the cost of capital decreases the larger the incentive is, but is not totally eliminated.

Table 9.9 Results of the linear regression with the trigger price as dependent variable

	Cost of capital	Risk premium	Soil fertility premium	Environmental premium
Regression coefficients	0.909	0.923	−0.322	−0.083
Standardised regression coefficients	0.869	0.112	−0.026	−0.008
t-value	22.041	2.838	−0.889	−0.272
(p-value)	(0.000)	(0.005)	(0.375)	(0.784)
R^2 (adjusted R^2)	0.895 (0.892)			
F-value (p-value)	279.722 (0.000)			

Annotation: n = 135

9.5.3.2 Explanatory Power of the Potential Influencing Factors

The hypothetical decision-making situation might have been too abstract for the farmers, or they might not have understood the decision-making situation. Therefore, we analyse whether there are significant correlations between the trigger price for the described biogas plant and the potential influencing factors. Based on a multiple linear regression analysis, we examine the extent to which the influencing factors (independent variable) affect the trigger price (dependent variable):

$$y_i = \sum_{j=1}^{j} a_j \cdot x_{ij} + \chi_i, \quad \text{with } i = 1, 2, \ldots, I \tag{9.10}$$

y_i denotes the dependent variable of the i-th observation and a_j is the corresponding regression coefficient for the j-th independent variable. x_{ij}. χ_i is the error term of the regression. The regression coefficients a_j are estimated using the least squares method. Table 9.9 shows the results of the regression.

The coefficient of determination R^2 is 0.895 (adjusted $R^2 = 0.892$); it therefore presents a high explanatory potential. The four regressors can explain about 89 % of the variation in the trigger price. Furthermore, the F-test indicates a highly significant correlation. Thus, the global quality criterions demonstrate that the regression model is not misspecified. In addition, they indicate that the farmers' answers were not based on pure guesswork and that the farmers understood the complex hypothetical decision-making situation sufficiently.

In addition to the influencing factors' overall explanatory potential, each influencing factor's input is of particular interest. The significance of each regression parameter is analysed using a t-test. The cost of capital and the risk premium significantly influence the trigger price at a probability of error of less than 1 %. The impacts of the soil fertility premium and the environmental premium are, however, not significant. The standardised regression coefficients show that the cost of capital, followed by the risk premium, offers the highest explanatory potential for investment behaviour concerning bioenergy plants.

Table 9.10 Consequences of the investment subsidy

	Trigger price *before* subsidy	Trigger price *after* subsidy		Changes in propensity to invest		Empirically unanticipated part of subsidy
	Empirical	Normative	Empirical	Normative	Empirical	
Mean	12.14	13.74	13.03	1.60	0.89	0.71
Standard deviation	3.78	–	3.73	–	2.17	2.17
Minimum	5.00	–	2.00	–	−7.50	−4.90
Maximum	30.00	–	29.00	–	6.50	9.10

Annotations: n = 135; in euro per decitonne; The investment subsidy is 100,000 EUR

A separate regression analysis was conducted with bounded rationality in the context of the cost of capital as a determinant (see Sect. 9.5.3.1). However, as independent variables, personal characteristics, such as age or the level of education, showed no significance.

9.5.3.3 Impacts of Investment Subsidies

Specific bioenergy policies are designed to increase the production of energy from renewable resources. How can farmers' willingness to invest in bioenergy plants be increased? With the net present value in mind (see Eq. 9.9), there are three main possibilities. In addition to reducing the risk of the investment returns (e), which the REA guarantees, a low-interest credit (i) can lead to a cutback in capital expenditure. A further alternative to create investment incentives is to lower the investment costs (a_0) by means of a subsidy (z).

As an example, we investigate the third possibility. We therefore asked the farmers to state their trigger price under the assumption of an investment subsidy of 100,000 EUR for the biogas plant. If the subsidy is taken into consideration, the reduced investment costs amount to 500,000 EUR. From a macro-economic point of view, a support programme of investments may cause a misallocation of resources (Brümmer and Loy 2000); however, this aspect is not considered here.

Under the assumption of perfect rationality with profit maximisation as the only entrepreneurial objective and if, according to Eq. 9.7, the application costs for the subsidy are subtracted; the reduced investment costs increase the trigger price by 1.60 EUR per decitonne. In other words, as expected, the investment will then be made earlier, ceteris paribus. The subsidy will have a positive impact because the average critical wheat price of 12.14 EUR per decitonne that the farmers mentioned, will increase to more than 13.74 EUR per decitonne if all else is equal.

Table 9.10 presents the means, standard deviations and ranges of the trigger price before and after the implementation of the investment subsidy.

With the inclusion of the investment subsidy, the range of the empirical trigger price changes slightly from 2 to 29 EUR per decitonne compared to 5 to 30 EUR per decitonne without the subsidy. Therefore, the minimum and maximum trigger prices did not shift in the expected direction. Owing to the investment subsidy, the average critical wheat price increased from 12.14 EUR per decitonne to 13.03 EUR per decitonne (standard deviation: 3.73 EUR per decitonne). Therefore, as expected, the subsidy generally increased the farmers' willingness to invest. In particular, 83 of the 135 farmers mentioned a higher trigger price, 30 were not affected by the subsidy and 22 mentioned a lower trigger price. However, the net present value of the average price would be −65,692 EUR if the wheat price was 13.03 EUR per decitonne (without the subsidy: −110,235 EUR). Thus, the net present value would increase by about 45,000 EUR if the investment subsidy was 100,000 EUR.

Although the normatively expected change in the trigger price is 1.60 EUR per decitonne, only 0.89 EUR per decitonne of that amount is reflected in the increase in the average willingness to invest. The remaining 0.71 EUR per decitonne did not have the expected effect. Hence, about 45 % of the investment subsidy is not reflected in the increase in the willingness to invest. Based on a comparison of the means, the anticipated and the unanticipated parts of the subsidy differ significantly from zero (p-value < 0.001; two-sided t-test). We could exaggerate and say that, in the considered investment decision, the investment subsidy would have to be approximately 200,000 EUR to affect an incentive of 100,000 EUR.

The results can be explained in two ways: Firstly, some farmers may feel the investment would require too much effort due to the bureaucracy and the work (including the loss of farm labour hours due to desk work) involved in the investment subsidy. Consequently, even if a subsidy were offered, their willingness to invest would not increase to the extent that the normative prognosis model predicts, and could even decrease. Secondly, the bounded rationality of the decision-makers may mean that they do not understand the actual value of an investment subsidy.

9.6 Implications and Conclusions

The decision-making behaviour at the farm level is crucially important in the context of bioenergy expansion. Both sub-studies help us understand how farmers make their investment decisions. Thus, our research findings contribute to the literature on farmers' entrepreneurial behaviour in general. In the context of renewables, knowledge of the different determinants of farmer decision outcomes is relevant to estimate their future engagement in innovations such as bioenergy. Understanding individual farmers' decision making improves our ability to determine the future development of bioenergy production and market potentials. Therefore, decision-making variables should be incorporated into the design of forecasting models for energy and environmental policies.

The first sub-study confirms negative externalities' occurrence in the context of expanded bioenergy production. The intensive discourse among experts from research and practice on the intra and inter-sectoral consequences resulting from the political support of biogas production indicates that the government is unable to predict far-reaching effects, such as the unintended land use competition. Such side effects illustrate the limits of political control and politically driven markets' high vulnerability (Wolf 1987).

The externalities are linked to the REA scheme. In many European countries, a similar fixed price system was adopted. In these countries – especially those with a strong agricultural background, such as France or Spain – land use competition may only occur later on. Our findings should be considered to optimise the funding policy in order to indicate or prevent externalities.

However, in its biomass allocation roadmap, the policy has established objectives for the further expansion of biogas production (BMU and BMELV 2009) (see Chap. 1). Our findings imply that the further diffusion of biogas production has been overestimated and is not realisable under present conditions. The policymakers would be well advised to restructure the German REA to reduce the competitive distortions among farmers. In this context, a slight reduction in the compensation for energy from biogas may decrease biogas production's economic profitability. However, the structure of the compensation rates also needs to be amended. The current additional compensation for using biomass from farmland (energy crops) is driving competition. Reducing this benefit will promote conflict reduction.

Over the last decade, the policy has focused exclusively on economic investment incentives as an extrinsic factor to motivate farmers to invest. Our findings show that farmers have little scope for creating intrinsic motivations, such as ecological awareness. This is somewhat problematic, because non-economic dimensions, such as ecology, then become less important. In fact, a large number of biogas plants' operations have been energetically inefficient for many years (Pöschl et al. 2010). Biogas plants were, and some still are, largely unaware of waste heat utilisation. Policy and the agricultural advisory services should therefore provide better ecological and energetically guidance and not only focus on economic incentives.

In the second sub-study, farm managers were confronted with a hypothetical decision situation regarding an investment in a biogas plant. The survey results confirm that farmers have various conversion thresholds (trigger prices) for potential investments in bioenergy. This explains why they respond very differently to economic conditions. Moreover, farmers who have actually invested in a biogas plant also invested earlier on in the experiment than the others.

Furthermore, the farmers' answers were compared to three types of normative benchmarks with different components to isolate the driving factors that influence the conversion threshold. The first benchmark only contained the cost of capital. A risk component was added to the second benchmark. The third benchmark was extended to include soil fertility and environmental premiums. These potential components of the trigger price showed different levels of influence. The investigated effects with regard to soil sustainability, altered substrate cultivation and non-economic objectives, did not sufficiently justify the trigger price. In contrast, the individual assessment of the risk and the cost of capital had a high impact on the trigger price. Bounded rationality was another essential influencing factor regarding the evaluation of the cost of capital, resulting in the capital costs often being underestimated.

The impact of an investment subsidy was analysed to determine investment incentives' potential. On average, the farmers only perceived 55 % of the amount of the total investment subsidy. Investment subsidies are already viewed critically in agricultural policy and will be called into question further if future studies confirm this effect.

The results show that bounded rationality has an effect on real decision-makers' behaviour and that they do not follow normative forecast models. Therefore, rational choice is not always a suitable explanation for economic decision making (see Faucheux and Froger 1995). This indicates, firstly, that additional profits could be earned from those arising from attractive (unattractive) but unrealised (realised) investments. Secondly, a decision-making aid, such as training (capacity building), could be helpful to counteract wrong decisions. It should be noted that support for decision making is not about influencing or changing farmers' preferences, but aims to increase their utility by reducing their bounded rationality and enabling them to make decisions that are better aligned with their individual preferences.

The behavioural economic aspects of bounded rationality require more in-depth research. This involves, for instance, collecting more detailed information on the farmers' socio-economic background, such as their education, age, income and family background. It also implies investigating the decision-making process to discover the algorithms, heuristics and calculi used to make decisions.

On a more general level, bioenergy should be considered a sustainable energy supply system. Although bioenergy production plays an important role, it requires room for innovation (Madlener and Stagl 2005: 162) and public subsidies that are effective, efficient and transparent.

References

Austin, E. J., Willock, J., Deary, I. J., Gibson, G. J., Dent, J. B., Edwards-Jones, G., & Morgan, O. (1998). Empirical models of farmer behaviour using psychological, social, and economic variables – Part I: Linear modeling. *Agricultural Systems, 58*(2), 203–224.

Bahrs, E., & Held, H. (2007). *Steigende Nachfrage auf den Energie- und Agrarrohstoffmärkten – Konsequenzen für die niedersächsische Landwirtschaft, die Bodenmärkte und die Agrarpolitik*. Retrieved March 12, 2011, from http://www.ml.niedersachsen.de/download/3269

Bard, S. K., & Berry, P. J. (2000). Developing a scale for assessing risk attitudes of agricultural decision maker. *International Food and Agribusiness Management Review, 3*(1), 9–25.

Bell, D. E., Raiffa, H., & Tversky, A. (1988). *Descriptive, normative, and perspective interactions in decision making*. Cambridge: Cambridge University Press.

Berenz, S., Hoffmann, H., & Pahl, H. (2008). Konkurrenzbeziehungen zwischen der Biogaserzeugung und der tierischen Produktion. In T. Glebe, A. Heissenhuber, L. Kirner, S. Pöchtrager, & K. Salhofer (Eds.), *Agrar- und Ernährungswirtschaft im Umbruch – Schriften der Gewisola 43* (pp. 497–516). Münster-Hiltrup: Landwirtschaftsverlag.

Berndes, G., Hoogwijk, M., & van den Broek, R. (2003). The contribution of biomass in the future global energy supply: A review of 17 studies. *Biomass and Bioenergy, 25*(1), 1–28.

BMU (Federal Ministry for the Environment, Nature Conservation and Nuclear Safety). (2000). *Erneuerbare-Energien-Gesetz vom 29.03.2000*. Berlin: BMU.

BMU (Federal Ministry for the Environment, Nature Conservation and Nuclear Safety). (2004). *Gesetz zur Neuregelung des Rechts der Erneuerbaren Energien im Strombereich vom 21.07.2004*. Berlin: BMU.

BMU (Federal Ministry for the Environment, Nature Conservation and Nuclear Safety). (2008). *Erneuerbare-Energien-Gesetz vom 25.10.2008*. Berlin: BMU.

BMU (Federal Ministry for the Environment, Nature Conservation and Nuclear Safety). (2009). *Nationaler Biomasseaktionsplan für Deutschland*. Berlin: BMU.

BMU (Federal Ministry for the Environment, Nature Conservation and Nuclear Safety) & BMELV (Federal Ministry of Food, Agriculture and Consumer Protection). (2009). *Nationaler Biomasseaktionsplan für Deutschland. Beitrag der Biomasse für eine nachhaltige Energieversorgung*. Berlin: BMU.

Brealey, R. A., Myers, S. C., & Allen, F. (2008). *Principles of corporate finance* (9th ed.). New York: McGraw-Hill.

Brümmer, B., & Loy, J.-P. (2000). The technical efficiency impact of farm credit programmes: A case study of northern Germany. *Journal of Agricultural Economics, 51*(3), 405–418.

BMELV (Federal Ministry of Food, Agriculture and Consumer Protection). (2009a). *Ausgewählte Daten und Fakten der Agrarwirtschaft 2009*. Retrieved March 22, 2011, from http://berichte.bmelv-statistik.de/DFB-0010000-2009.pdf

BMELV (Federal Ministry of Food, Agriculture and Consumer Protection). (2009b). *Ergebnisse der Viehbestandserhebung Nov. 2009 Rinder – Weiter stagnierender Rinderbestand*. Retrieved March 22, 2011, from http://berichte.bmelv-statistik.de/WBB-3800001-2010.pdf

BS (German Solar Industry Association) & EUPD (EUPD Research). (2009). *Solarstandort Deutschland 2009 – Pressekonferenz Intersolar München 2009*. Retrieved March 15, 2011, from http://www.solarwirtschaft.de/fileadmin/content_files/Intersolar_BSW_EuPD.pdf

Buchanan, J. M., & Stubblebine, W. C. (1962). Externality. *Economia, 29*(116), 371–384.

Burton, R. J. F. (2004). Reconceptualising the 'behavioural approach' in agricultural studies: A sociopsychological perspective. *Journal of Rural Studies, 20*(3), 359–371.

Cyert, R. M., & March, J. G. (1963). *A behavioral theory of the firm*. Englewood Cliffs: Prentice Hall.

DBFZ (German Biomass Research Center). (2009). *Identifizierung strategischer Hemmnisse und Entwicklung von Lösungsansätzen zur Reduzierung der Nutzungskonkurrenzen beim weiteren Ausbau der energetischen Biomassenutzung*. Leipzig: Zwischenbericht.

DBFZ (German Biomass Research Center). (2010). *Monitoring zur Wirkung des Erneuerbare-Energien-Gesetz (EEG) auf die Entwicklung der Stromerzeugung aus Biomasse*. Leipzig: Zwischenbericht.

DBV (German Farmers' Association). (2010). *Situationsbericht 2010 – Trends und Fakten zur Landwirtschaft*. Berlin: DBV.

Dehnhardt, A., & Petschow, U. (2004). Nobody is perfect! Erneuerbare Energien, externe Effekte und ökonomische Bewertung. *Ökologisches Wirtschaften, 2004*(5), 24–25.

DENA (German Energy Agency). (2011). *Biogaspartner – Marktentwicklung in Deutschland*, Retrieved March 15, 2011, from http://www.biogaspartner.de/index.php?id=11871&L=1\

Domac, J., Richards, K., & Risovic, S. (2005). Socio-economic drivers in implementing bioenergy projects. *Biomass and Bioenergy, 28*(2), 97–106.

Dornburg, V., van Vuuren, D., van de Ven, G., Hans Langeveld, H., Marieke Meeusen, M., Banse, M., van Oorschot, M., Ros, J., van den Born, G. J., Aiking, H., Londo, M., Mozaffarian, H., Verweij, P., Lyseng, E., & Faaij, A. (2010). Bioenergy revisited: Key factors in global potentials of bioenergy. *Energy and Environmental Science, 3*(3), 258–267.

East, R. (1993). Investment decisions and the theory of planned behaviour. *Journal of Economic Psychology, 14*(2), 337–375.

Edwards, W. (1954). The theory of decision making. *Psychological Bulletin, 51*(4), 380–417.

Ehlers, M.-H. (2008). Farmers' reasons for engaging in bioenergy utilisation and their institutional context: A case study in Germany. In T. Glauben, J. H. Hanf, M. Kopsidis, A. Pieniadz, & K. Reinsberg (Eds.), *Agri-food business: Global challenges – Innovative solutions*. Studies on the Agricultural and Food Sector in Central and Eastern Europe 46 (pp. 106–117). Halle (Saale).

EUROSTAT. (2010). *Electricity prices by type of user*. Dataset Code TSIER040. Retrieved March 15, 2011, from http://epp.eurostat.ec.europa.eu/portal/page/portal/product_details/dataset?p_product_code = TSIER040

Faucheux, S., & Froger, G. (1995). Decision-making under environmental uncertainty. *Ecological Economics, 15*(1), 29–42.

Fehr, E., & Schmidt, K. M. (1999). A theory of fairness, competition, and cooperation. *The Quarterly Journal of Economics, 114*(3), 817–868.

Feindt, P. H., Canenbley, C., Gottschick, M., Müller, C., & Roedenbeck, I. (2004). *Konflikte des Agrarsektors – Eine Landkarte. Empirische Ergebnisse einer konflikttheoretischen Fundierung der Nachhaltigkeitsforschung*. BIOGUM Forschungsbericht Nr. 12. Retrieved March 28, 2011, from University of Hamburg. Research Centre for Biotechnology, Society and the Environment Web site: http://www.uni-hamburg.de/onTEAM/grafik/1107511876/biogum_fb_2004_13.pdf.

Field, A. P. (2009). *Discovering statistics using SPSS*. London: Sage Publications.

Fishbein, M., & Ajzen, I. (1975). *Belief, attitude, intention, and behavior: An introduction to theory and research*. Reading: Addison-Wesley.

Frey, B., & Eichenberger, R. (1989). Zur Bedeutung entscheidungstheoretischer Anomalien für die Ökonomik. *Jahrbücher für Nationalökonomie und Statistik, 206*(2), 81–101.

Frör, O. (2008). Bounded rationality in contingent valuation: Empirical evidence using cognitive psychology. *Ecological Economics, 68*(1–2), 570–581.

FVB (German Biogas Association). (2010). *Branchenkennzahlen*. Retrieved March 15, 2011, from http://www.biogas.org/edcom/webfvb.nsf/id/DE_Branchenzahlen

Gardebroek, C., & Oude Lansink, A. G. J. M. (2008, January–February). *Dynamic microeconometric approaches to analysing agricultural policy*. Paper presented at the 107th Seminar of the European Association of Agricultural Economists, Sevilla.

Gigerenzer, G. (2000). *Adaptive thinking: Rationality in the real world*. Oxford: Oxford University Press.

Gigerenzer, G. (2002). *Calculated risks: How to know when numbers deceive you*. New York: Simon & Schuster.

Gigerenzer, G., & Selten, R. (2001). *Bounded rationality: The adaptive toolbox*. Cambridge, MA: MIT Press.

GJAE (*German Journal of Agricultural Economics*) (2005–2010). Wirtschaftszahlen, Volumes (2005). 54(8), 369–371. (2006). 55(8), 377–379. (2007). 56(8), 269–272. (2008). 57(8), 397–400. (2009). 58(8), 377–379. (2010). 59(4), 283–286.

Granoszewski, K., Spiller, A., Reise, C., & Mußhoff, O. (2011, January). *The diffusion of renewable energies in German Agriculture – Investment behaviour in a politically induced market*. Paper presented at the 10th annual international conference of marketing trends of the International Marketing Trends Association, Paris).

Granovetter, M. (1985). Economic action and social structure: The problem of embeddedness. *The American Journal of Sociology, 91*(3), 481–510.

Harwood, J., Heifner, R., Coble, K., Perry, J., & Somwaru, A. (1999). *Managing risk in farming: Concepts, research, and analysis*. Market and trade economics division and resource economics division, Economic Research Service, U.S. Department of Agriculture, Agricultural Economic Report No. 774.

Heissenhuber, A., & Berenz, S. (2006). Energieproduktion in landwirtschaftlichen Unternehmen. In I. Darnhofer, H. K. Wytrzens, & C. Walla (Eds.), *Alternative Strategien in der Landwirtschaft* (pp. 135–144). Wien: Facultas.

Heissenhuber, A., Demmler, M., & Rauh, S. (2008). Auswirkungen der Konkurrenz zwischen Nahrungsmittel und Bioenergieproduktion auf Landwirtschaft, Gesellschaft und Umwelt. *Technikfolgenabschätzung – Theorie und Praxis, 17*(2), 23–31.

Henningsen, A., Henning, C. H. C. A., Struve, C., & Müller-Scheessel, J. (2005). Economic impact of the mid-term review on agricultural production, farm income and farm survival: A quantitative analysis for local sub-regions of Schleswig-Holstein in Germany. In F. Arfini (Ed.), *Modelling agricultural policies: State of the art and new challenges* (pp. 757–772). Parma: Monte Universita.

Inhetveen, H., & Schmitt, M. (2010). Prekarisierung auf Dauer? Die Überlebenskultur bäuerlicher Familienbetriebe. In A. D. Bührmann & H. J. Pongratz (Eds.), *Prekäres Unternehmertum – Unsicherheiten von selbstständiger Erwerbstätigkeit und Unternehmensgründung* (pp. 111–136). Wiesbaden: VS Verlag.

IPCC (Intergovernmental Panel on Climate Change). (2007). *Climate change 2007: Impacts, adaptation and vulnerability*. Cambridge: Cambridge University Press.

IRENA (Preparatory Commission for International Renewable Energy Agency). (2010). *Our Mission*. Retrieved March 10, 2011, from http://www.irena.org/ourMission/index.aspx?mnu = mis

Jacobsson, S., & Lauber, V. (2006). The politics and policy of energy system transformation – Explaining the German diffusion of renewable energy technology. *Energy Policy, 34*(3), 256–276.

Jorgenson, D. W. (1968). The theory of investment behavior. In R. Ferber (Ed.), *Determinants of investment behavior* (pp. 129–188). New York: UMI.

JURIS (German Legal Information Service). (2011). *Verordnung über den Zugang zu Gasversorgungsnetzen (Gasnetzzugangsverordnung – GasNZV)*. Retrieved March 10, 2011, from http://www.gesetze-im-internet.de/bundesrecht/gasnzv_2010/gesamt.pdf

Just, D. R., & Wu, S. Y. (2009). Experimental economics and the economics of contracts. *American Journal of Agricultural Economics, 91*(5), 1382–1388.

Kahneman, D. (2003). Maps of bounded rationality: Psychology for behavioral economics. *American Economic Review, 93*(5), 1449–1475.

Kahneman, D., & Tversky, A. (1979). Prospect theory: An analysis of decision under risk. *Econometrica, 47*(2), 263–291.

Karpenstein-Machan, M. (2005). *Energiepflanzenbau für Biogasanlagenbetreiber*. Frankfurt am Main: DLG-Verlag.

Keymer, U. (2009). *Rentabilität der Biogaserzeugung – Erläuterungen, Übersichten und Musterkalkulationen zum Erneuerbaren-Energien-Gesetz (EEG)*. Schriftenreihe des Hauptverbandes der landwirtschaftlichen Buchstellen und Sachverständigen, *183*. Sankt Augustin: HLBS Verlag.

Kool, M. (1994). *Buying behavior of farmers*. Wageningen: Wageningen Academic Publishers.

La Due, E. L., Miller, L. H., & Kwiatkowski, J. H. (1991). Investment behavior and farm business expansion. *Review of Agricultural Economics, 13*(1), 73–84.

Langert, M. (2007). *Der Anbau nachwachsender Rohstoffe in der Landwirtschaft Sachsen-Anhalts und Thüringens – Eine innovations- und diffusionstheoretische Untersuchung*. Dissertation. University of Halle-Wittenberg, Halle.

Le Grand, J. (1991). The theory of government failure. *British Journal of Political Science, 21*(4), 423–442.

Lynne, G. D., & Rola, L. R. (1988). Improving attitude-behavior prediction models with economic variables: Farmer actions toward soil conservation. *The Journal of Social Psychology, 128*(1), 19–28.

Mackenzie, J., & Krogman, N. (2005). Public involvement processes, conflict, and challenges for rural residents near intensive hog farms. *Local Environment, 10*(5), 513–524.

Madlener, R., & Stagl, S. (2005). Sustainability-guided promotion of renewable electricity generation. *Ecological Economics, 53*(2), 147–167.

Mann, S., & Kögl, H. (2003). On the acceptance of animal production in rural communities. *Land Use Policy, 20*(3), 243–252.

Mautz, R. (2007). The expansion of renewable energies in Germany between niche dynamics and system integration – Opportunities and restraints. *Science, Technology & Innovation Studies, 3* (2), 113–131.

Mendonca, M. (2007). *Feed-in tariffs: Accelerating the deployment of renewable energy*. London: Earthscan.

Muradian, R., Corbera, E., Pascual, U., Kosoy, N., & May, P. H. (2010). Reconciling theory and practice: An alternative conceptual framework for understanding payments for environmental services. *Ecological Economics, 69*(6), 1202–1208.

Musshoff, O., Hirschauer, N., & Wassmuss, H. (2009, August). *The role of bounded rationality in farm financing decisions - First empirical evidence*. Paper presented at the 27th international conference of the International Association of Agricultural Economists, Beijing.

NLS (State Office for Statistics of Lower Saxony). (2001). *Die Ermittlung der amtlichen Hektarerträge für Feldfrüchte in Niedersachsen und Ernteergebnisse 2001*. Retrieved March 15, 2011, from http://www.nls.niedersachsen.de/Tabellen/Landwirtschaft/bee_text/e_stat.htm

Nuthall, P. L. (1999). *The psychology of decision making in farm management*. Retrieved March 15, 2011, from Lincoln University, Farm Management Group, Web site: URL: http://researcharchive.lincoln.ac.nz/dspace/bitstream/10182/51/1/fhmg99-03.pdf

Nuthall, P. L. (2010). *Farm business management: The human factor*. Oxfordshire: CABI.

Plieninger, T., Bens, O., & Hüttl, R. F. (2006a). Perspectives of bioenergy for agriculture and rural areas. *Outlook on Agriculture, 35*(2), 123–127.

Plieninger, T., Bens, O., & Hüttl, R. F. (2006b). Landwirtschaft und Entwicklung ländlicher Räume. *Aus Politik und Zeitgeschichte, 37*, 23–30.

Pongratz, H. (1992). *Die Bauern und der ökologische Diskurs*. München: Profil Verlag.

Pöschl, M., Shane, S., & Owende, P. (2010). Evaluation of energy efficiency of various biogas production and utilization pathways. *Applied Energy, 87*(11), 3305–3321.

Reiche, D., & Bechberger, M. (2006). Policy differences in the promotion of renewable energies in the EU member states. *Energy Policy, 32*(9), 843–849.

Retter, C., Stahr, K., & Boland, H. (2002). Zur Rolle von Landwirten in dörflichen Kommunikationsnetzwerken. *Berichte über Landwirtschaft, 80*(3), 446–467.

Roe, B. E., & Just, D. R. (2009). Internal and external validity in economics research: Tradeoffs between experiments, field experiments, natural experiments, and field data. *American Journal of Agricultural Economics, 91*(5), 1266–1271.

Roessingh, C., & Schoonderwoerd, A. (2005). Traditional farmers or modern businessmen? Religious differentiation and entrepreneurship in a Mennonite Community in Belize. *Journal of Developmental Entrepreneurship, 10*(1), 65–77.

Sauer, J., & D. Zilbermann. (2009, September). *Innovation behaviour at farm level – Selection and identification*. Paper presented at the 49th annual meeting of the German Association of Agricultural Economics and Sociology, GEWISOLA, Kiel.

Sauermann, H., & Selten, R. (1962). Anspruchsanpassungstheorie der Unternehmung. *Zeitschrift für die gesamte Staatswissenschaft, 118*, 577–597.

Schoemaker, P. J. (1982). The expected utility model: Its variants, purposes, evidence and limitations. *Journal of Economic Literature, 20*(2), 529–563.

Schramm, R. (1977). The influence of relative prices, production conditions and adjustment costs on investment behaviour. *The Review of Economic Studies, 37*(3), 361–376.

Selten, R. (1990). Bounded rationality. *Journal of Institutional and Theoretical Economics, 146*(4), 649–658.

Simon, H. A. (1979). Rational decision making in business organizations. *The American Economic Review, 69*(4), 493–513.

Simon, H. A. (1959). Theories of decision-making in economics and behavioral science. *The American Economic Review, 49*(3), 253–283.

Simon, H. A. (1957). *Models of man*. New York: Wiley.

Simon, H. A. (1956). Rational choice and the structure of environments. *Psychological Review, 63*(2), 129–138.

Solano, C., Leon, H., Perez, E., & Herrero, M. (2003). The role of personal information sources on the decision-making process of Costa Rican dairy farmers. *Agricultural Systems, 76*(1), 3–18.

SRU (German Advisory Council on the Environment). (2007). *Klimaschutz durch Biomasse – Sondergutachten*. Berlin: Erich Schmidt Verlag.

Starmer, C. (1999). Experimental economics: Hard science or wasteful tinkering? *The Economic Journal, 109*(453), F5–F15.

Trojecka, A. E. (2007). *Landwirte als Energiewirte? Bedingungen einer ökologischen Modernisierung der Landwirtschaft am Beispiel der Biogaserzeugung*. Dissertation, University of Osnabrück, Osnabrück.

WBA (German Scientific Advisory Board on Agricultural Policy). (2007). *Nutzung von Biomasse zur Energiegewinnung – Empfehlungen an die Politik*. Berlin: BMELV.

Wibberley, G. P. (1959). *Agriculture and urban growth, a study of the competition for rural land*. London: Michael Joseph Ltd.

Willms, M., Deumlich, D., Hufnagel, J., Reinicke, F., Wagner, B., & v. Buttlar, C. (2009). Anbauverfahren für Energiepflanzen – Auswirkungen auf Boden und Umwelt. In Fachagentur Nachwachsende Rohstoffe e.V. (Eds.), *Biogas in der Landwirtschaft – Stand und Perspektiven*, Gülzower Fachgespräche, 32 (pp. 148–162). Gülzow.

Willock, J., Deary, I. J., McGregor, M. M., Sutherland, A., Edwards-Jones, G., Morgan, O., Dent, B., Grieve, R., Gibson, G., & Austin, E. (1999). Farmers' attitudes, objectives, behaviors, and personality traits: The Edinburgh study of decision making on farms. *Journal of Vocational Behavior, 54*(1), 5–36.

Wolf, C., Jr. (1987). Market and non-market failures: Comparison and assessment. *Journal of Public Policy, 7*(1), 43–70.

Chapter 10
Social Acceptance of Bioenergy Use and the Success Factors of Communal Bioenergy Projects

André Wüste and Peter Schmuck

Abstract This chapter analyses the social acceptance for bioenergy resources and bioenergy utilisation based on the following three studies: (1) A quantitative study on 678 rural Germans and Austrians attitudes towards bioenergy, based on a standardised questionnaire; (2) a study in 13 villages, analysing data on 2,200 inhabitants readiness to support a bioenergy project; (3) a qualitative interview study analysing the success factors as well as impediments to establish decentralised, communal bioenergy projects. Interviews were conducted with the initiators or participants in 25 bioenergy villages in Germany. This chapter focuses on changes in the individual and social well-being during the planning of a bioenergy village. Through the three studies, we seek to gain insights into Germany's very dynamic development of bioenergy production facilities, not all of which meet sustainability criteria: A growing number of people in Germany's rural areas are directly or indirectly affected by the increasing development of bioenergy utilisation. In many cases, only the economic aspects of bioenergy plants are considered prior to their being built; local population and other stakeholders are not involved. Increasing fears, caused by the local population's lack of information, often lead to conflicts, resistance and declining acceptance of bioenergy projects. The studies in this chapter seek to open potential avenues in order to have local population's support for sustainable bioenergy projects.

Keywords Acceptance • Bioenergy village • Success factors

A. Wüste (✉) • P. Schmuck
Interdisciplinary Centre for Sustainable Development, University of Göttingen, Goldschmidtstraße 1, 37083 Göttingen, Germany
e-mail: Andre.Wueste@geo.uni-goettingen.de; peterschmuck@gmx.de

H. Ruppert et al. (eds.), *Sustainable Bioenergy Production - An Integrated Approach*, DOI 10.1007/978-94-007-6642-6_10, © Springer Science+Business Media Dordrecht 2013

10.1 Introduction

This chapter focuses on social scientific studies as an activity that helps solve problems during the transformation of our conventional energy system on the basis of renewable energy sources. Our study bases on the Göttingen approach of sustainability science (see Sect. 2.3). In the selected problem field of bioenergy use – one option in a future global renewable energy scenario – traditional research results should be used as a general basis for solving problems. However, few such results are available. The development of bioenergy production installations in Germany has been a swift and dynamic process. This has prevented researchers from using the usual sequence, in which, prototypically, several years of basic research result in a substantial knowledge pool that serves as the basis for practical applications. Given that sustainability science has to solve global problems under severe time pressure, swift changes have forced scientists to undertake research and apply the result in parallel.

The situation becomes more concrete if one knows that, 10 years ago, Germany had only a few biogas plants, but since the introduction of the 2000 Renewable Energy Sources Act, the number has increased by several hundred per year to more than 7,000 today. Therefore, public acceptance of these plants may change substantially from year to year: There are different advantages, but also negative side effects, such as maize monocultures, which may critically influence public opinion. If one intends to support sustainable bioenergy projects precise data are required on the public acceptance of different production and consumption alternatives. These data include qualitative information about the different argumentation pattern types (for and against) within Germany's rural population. Such scientific knowledge allows a researcher, when planning with practitioners, to anticipate and adequately respond to biased arguments against bioenergy projects or to incorrect information in the media.

Our studies also analyse the success factors of already completed sustainable bioenergy projects in Germany. Applying these success factors to on-going projects has proven a very powerful mechanism for transferring sustainable models to our projects and to the regional and national levels. In 2000, we analysed the social success factors of pioneering communal renewable energy projects and then successfully applied these principles (Eigner and Schmuck 2002), when establishing five bioenergy villages in the Göttingen district between 2002 and 2010. In 2008 we documented our experiences in Jühnde as well as those in various following projects. The resulting publication (Ruppert et al. 2008) was distributed nationwide and supported the developing of more than a hundred of other bioenergy villages in Germany.

In this on-going project, we follow the principles of our Göttingen approach of sustainability science at the regional level, widening our focus on solutions for sustainable renewable energy models from villages to rural districts, each of which include several communities and villages (see Chap. 11).

10.2 Acceptance of and Social Barriers to the Development of Bioenergy Usage

Since the first amendment to the Renewable Energy Sources Act, there has been a boom in biogas production in Germany. Today, there are more than 7,000 biogas plants in Germany (FNR 2012). However, this rapid growth has partly led to the rural population's declining acceptance of these plants and to dissent regarding their suitability in agriculture.

Thus, in addition to bioenergy plants' technical, financial, administrative, organisational and infrastructural challenges, the perceptions of and acceptance by the affected population represent a massive obstacle to their implementation (Roesch and Kaltschmitt 1999, p. 348) that needs to be overcome to facilitate the shift to renewable energies.

There are many definitions of acceptance. According to Endruweit and Trommsdorff (2002), acceptance is an attribute of an innovation's introduction in order to achieve positive responses from the concerned people. Dethloff (2004) considers acceptance the positive adoption of an idea, a status, a product or service, thus defining willingness. Acceptance is therefore not merely toleration and tolerance (attitude level), but also comprises readiness to act (behavioural level) as a criterion. The opposite of acceptance is rejection or non-acceptance, and if the rejection is linked to defensive actions, this leads to active resistance or reactions (Dethloff 2004, p. 18). At the very least, acceptance is a tolerant attitude or even a consensus-oriented process (Jenssen 2010, p. 197).

A nationwide survey by the German Forsa Institute ascertained that more than 95 % of Germans approve of the increased development of renewable energies (AEE 2011). On the other hand, hundreds of citizens' action groups have been formed against bioenergy projects in Germany. For example, the construction of a biogas plant in the northern Hessian village of Wommen was prevented by a citizens' action group. There are even citizens' action groups against bioenergy in our three selected districts. In the city of Burgdorf, in the Hannover region, a citizens' action group was formed to oppose a large biogas plant (1.5 MW) (see Chap. 11).

The causes of social protest against bioenergy projects are multifaceted. Fears that local residents' current quality of life – especially due to unwanted odours from the bioenergy plant – could be affected play a major role in this regard. Further concerns include rising costs, loss in value of immovables and of other tangibles (Mautz et al. 2008, p.107), traffic nuisance owing to biomass transport, monocultures' effect on the landscape and fear of accidents. These fears are potential causes for the well-known NIMBY (not in my back yard) conflicts. NIMBY has meant that while rural inhabitants considered bioenergy technology very important and useful in principle, they nonetheless often oppose bioenergy plants in their surroundings.

The increase in conflicts over renewable energy and especially bioenergy may partly be derived from the shift from small plants to large-scale industrial biogas production. Pooling individual bioenergy plants in bioenergy parks, which occurred in Penkun (north of Berlin), for example, requires extremely area-intensive and transport-intensive logistics. Increasing resistance may therefore be expected from the inhabitants of such park's surroundings. In addition, mainly non-agricultural investors, such as power supply companies, implement such large-scale projects. Furthermore, as a rule, local citizens do not participate in industrial-scale biogas production. The profits from local raw materials use therefore do not remain in the region, but mainly benefit investors who are often not from the region (Mautz et al. 2008).

To prevent the NIMBY phenomenon, local residents need to be involved in decision-making and implementation processes (Aretz et al. 2009, p. 49). This allows them to openly discuss their fears, which could then perhaps be overcome. Zoellner et al. (2008) showed that there are significant correlations between the fairness perceived by the implementation processes and the acceptance. Furthermore, it is crucial for the implementation process to be transparent, since citizens tend to be oppose a bioenergy project if they are not involved in the planning and decision-making processes (see Zoellner et al. 2008, p. 4140).

To date, there are very few scientific results concerning bioenergy acceptance – only a few relating to wind energy. Egert and Jedicke (2001) investigated wind energy acceptance in relation to the landscape of a northern Hessian region. A team of environmental psychologists from the University of Magdeburg analysed renewable energy acceptance in four different regions, focussing on photovoltaic, wind and biomass energy (Zoellner et al. 2008). Griesen (2010) investigated biogas plant acceptance factors by surveying two German regions. He identified the following key acceptance factors: (1) the ethical appraisal, (2) the distance between the biogas plant and the local residents' homes and (3) the residents' perceptions of bioenergy.

In our research project, we undertook a bioenergy acceptance study in Germany's rural areas that will provide findings on the acceptance of current bioenergy production and consumption options and bioenergy usages perceived and expected opportunities and risks. However, our research project has already contributed findings about the social criteria for multicriteria decision analysis (MCDA) (see Chap. 12). We next outline a few of the first study results.

10.3 Bioenergy Acceptance in Germany: A Nationwide Acceptance Survey

This acceptance study, which took place between the summer of 2010 and February 2011, focuses on the different bioenergy production alternatives (e.g., small biogas plants, major industrial bioenergy plants and biofuel plants) as well as the different biomass resources (e.g., wood, straw, liquid manure and energy crops).

10.3.1 Methods

10.3.1.1 Description of the Sample and the Investigated Regions

Six respondent sub-samples from residents in rural areas of Germany were surveyed: The main sample (n = 377) comprised people living in villages without bioenergy production. This sample gives an overview about general acceptance or concerns regarding bioenergy. Furthermore, several smaller samples were collected from residents in areas surrounding the following special pathways of bioenergy production and use:

- local communal biogas projects (n = 66)
- major industrial bioenergy plants fuelled with energy crops (n = 98)
- major industrial biofuel plants (n = 55)
- organic farming in combination with biogas production (n = 30)
- short-rotation plantations (n = 52).

10.3.1.2 Design of the Questionnaire

A partly standardised questionnaire was created for the survey, with ten groups of questions. Question complex one contains 15 items related to different biomass resources for producing energy (e.g., bio waste, straw, energy crops, tree-cut, liquid manure) with three response categories ("I am in favour because . . .", "I am only in favour if . . ." and "I reject this because . . ."). Each respondent also had the opportunity to write a short statement. Question complex two contains nine items concerning different bioenergy consumption opportunities (e.g., communal biogas plants with or without a heating concept, large industrial biogas plants, biofuel plants, wood heating plants). The response categories in question complex two are the same as those in complex one. In question complex three, there are open-ended questions on the potential opportunities and risks of using bioenergy. Question complexes four and five contain closed questions relating to the expected consequences of a bioenergy plant's construction in the main sample and in the smaller samples to identify the perceived consequences in areas with specific bioenergy production lines. Five-point Likert scales were used evaluate statements regarding these questions. Question complexes six and seven are semi-open questions on experiences with bioenergy in the respondents' local surroundings and their attitudes towards a possible bioenergy plant in their villages. Question complex eight contains nine items concerning other energy generation opportunities, such as petroleum, coal, solar energy and wind energy; the response categories are the same as those in complexes one and two. Question complex nine relates to the respondents' actual and planned energy supply. In the question, we requested complex, demographic data.

10.3.2 First Results

The questionnaire analysis is still on-going; we therefore focus on descriptive results.

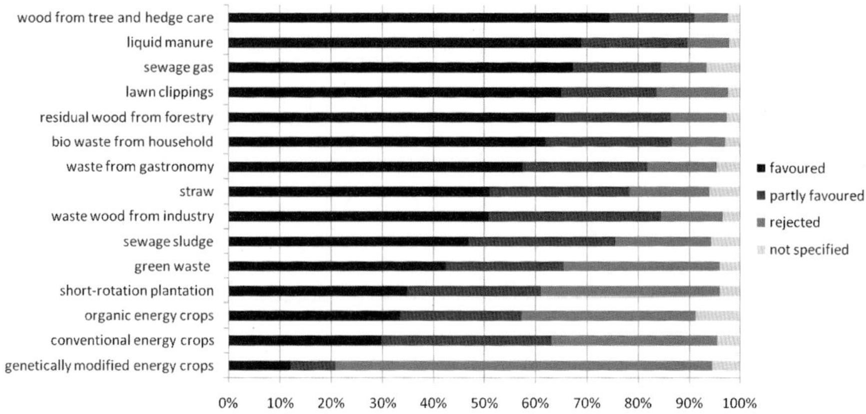

Fig. 10.1 Percentages of subjects' preference for potential bioenergy resources in the bioenergy acceptance survey

10.3.2.1 Acceptance of Different Biomass Resources

An impressive result is the participants' preference for biomass resources that consist of waste materials. Wood from trees and hedges to generate bioenergy is favoured by 75 % of the respondents , liquid manure by 69 %, sewage gas by 67 %, lawn clippings by 65 % and residual wood from forestry by 64 % (see Fig. 10.1). The respondents mention *waste materials' usefulness* and *waste reduction* as a main reason for this high endorsement of waste materials. Further positive outcomes associated with waste material use for bioenergy are its *environmental* and *financial benefits*.

The general endorsement and the rejection of the use of short-rotation plantations and organic energy crops are in balance. The respondents mention *competition between food production and energy crop production* as a main reason for their rejection of energy crops and short-rotation plantations. A further reason is the *risk of monocultures*. Positive arguments for energy crops use are that energy crops are *renewable sources* and that they can help *conserve fossil fuels*. The highest rejection figure (at 74 %) is for genetically modified energy crops, reflecting the Germany population's assessment of genetic engineering having too many unforeseeable risks.

10.3.2.2 Acceptance of Current Bioenergy Consumption Options

The respondents favour smaller, communal plants regarding their acceptance of current bioenergy consumption opportunities. Small biogas plants with a heating concept and heating plants with residual wood are greatly favoured at 72 and 69 % (see Fig. 10.2). The respondents explain that these are *useful* and *decentralised* bioenergy consumption opportunities with *benefits for the environment*.

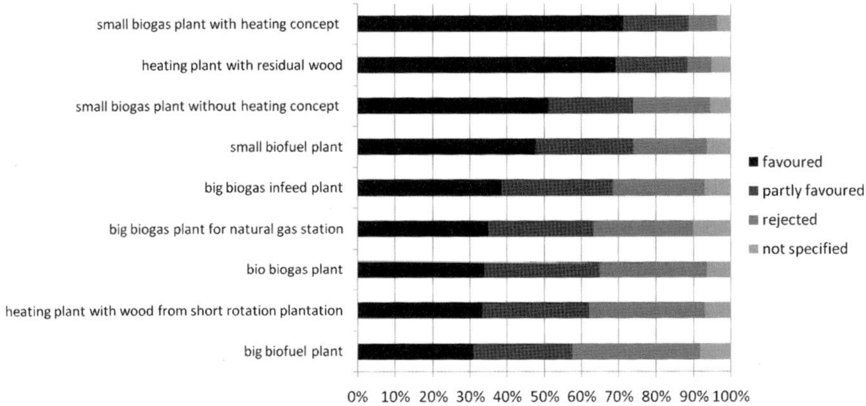

Fig. 10.2 Percentages of subjects' preference for potential bioenergy production lines in the bioenergy acceptance survey

A relatively small majority (50 %) supports small biogas plants without a heating concept, while almost 20 % do not, owing to the *lack of a heating component.* About 35 % approve large industrial bioenergy plants. The respondents refer to concerns relating to *competition between food production and energy crop production and impacts on quality of life* (e.g., traffic nuisance, impact on the landscape), but mention *independence from fossil fuels* and *environmental benefits* as positive aspects.

10.3.2.3 Acceptance of Other Energy Sources

The survey results show a general support for renewable energies. The highest approval is for solar heating (more than 80 % of respondents) and photovoltaic energy (approximately 74 %), followed by geothermal energy (more than 67 %) and hydropower (65 %). The respondents especially mention *benefits for the environment and the global climate* as the main reason for their positive appraisal of renewable energies. A relatively small majority of respondents (51 %) are supportive of wind power plants.

The fossil fuel and nuclear resources are less accepted. Especially nuclear energy receives a high rejection rate (63 %), while that of coal energy is almost 50 % (see Fig. 10.3). The main reasons for the high rejection of fossil and nuclear fuels are *negative environmental effects* and the *finite nature* of these resources. The *unclear situation concerning the disposal of nuclear waste* is another aspect that the respondents mention.

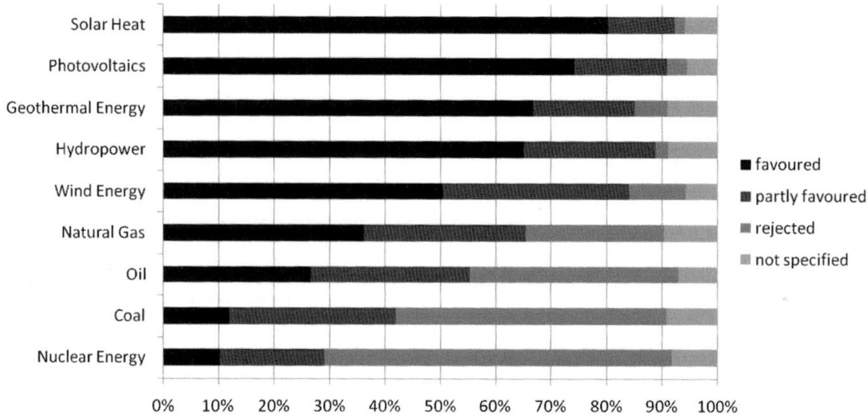

Fig. 10.3 Percentages of subjects' preference for fossil, nuclear and renewable energy resources in the bioenergy acceptance survey

10.4 Acceptance of Bioenergy Villages in Göttingen District

After the successful implementation of the bioenergy village Jühnde, the district of Göttingen initiated a follow-up project in 2006 to establish more bioenergy villages in the district. As part of a village selection process, 34 villages were interested in biomass-based power and heat supply. Representatives from the bioenergy village's project team and from the district administration organised meetings and information sessions in these villages. Through selection criteria such as "agricultural and forestry potential", "actors' high motivation" or a "compact village structure", 13 potentially suitable villages were selected: Barlissen, Ellershausen, Erbsen, Gelliehausen, Hemeln, Krebeck, Landolfshausen, Lödingsen, Reiffenhausen, Renshausen, Sattenhausen, Scheden and Wollbrandshausen. In these villages, further village meetings were organised and working groups, which the university team moderated, were initiated. The working groups comprised interested and active villagers, who analysed the local biomass potential and potential plant locations, informed and mobilised other villagers. A survey with the following research questions was conducted in the single households to get information about their readiness to participate on the bioenergy project:

- What opportunities, expectations, risks or fears regarding the implementation of a bioenergy village do the villagers express? Which main motives lead to the approval or rejection of the bioenergy village concept?
- How willing are the residents to participate in the planning phase?
- How do the residents assess the bioenergy village's feasibility as a shared task of the village community?

10.4.1 Methods

A two-page household survey was done in 2006 in 13 candidate bioenergy villages in Göttingen (n = 2,061). The survey consisted of quantitative and qualitative questions addressing:

- The villagers' willingness to connect their households to the planned local heat supply system
- The villagers' assessment of the notion of a bioenergy village (with explanatory statements)
- The villagers' willingness to participate in working groups
- The village community's assessment of the chances of implementing a bioenergy village project successfully.

This study focuses primarily on the open questions, which were analysed by means of a content analysis. This is one of the classical approaches to analyse text material (Flick 2004). The central element of a content analysis is creating a category system. The category system can be developed through (1) a deductive approach with the categories being developed before the analysis, or (2) an inductive approach with the categories generated on the basis of the text material without reference to pre-formulated theory concepts (Mayring 2008). A deductive theory-driven approach is combined with an inductive material-based approach to develop the category system. After defining the categories, the text material is compared with the category system by noting the occurrence of the categories in the text. Based on the text material, open-coding – which seeks to summarise data and phenomena by dividing them into units of meaning (Flick 2004, p. 259) – is chosen. Single statements are defined as the coding units. Quantitative working steps let us to arrange the categories according to the frequency of their occurrence in the material.

Two examples illustrate the category system's development: In keeping with the three-pillar model of sustainability, a theory-driven and deductive approach is used in respect of the question regarding the motives for agreeing to a bioenergy village concept. Hence, the categories *economic*, *ecological* and *social* motives are conceived, which prove useful to assign the data. The reasons for the rejection of the bioenergy village concept do not correspond with these categories; here, categories are developed inductively from the text material (e.g., *ethical doubts* or *limited quality of life*). After repeated reading, the categories are confirmed and arranged according to the frequency of their occurrence.

10.4.2 Results

10.4.2.1 Question 1: Willingness to Connect to the Local Heating Network

The survey reveals 52 % of the respondents' general readiness to connect to the communal heating system. The highest connection readiness was in village H (67 %), and the lowest in village J (only 31 %) (Fig. 10.4).

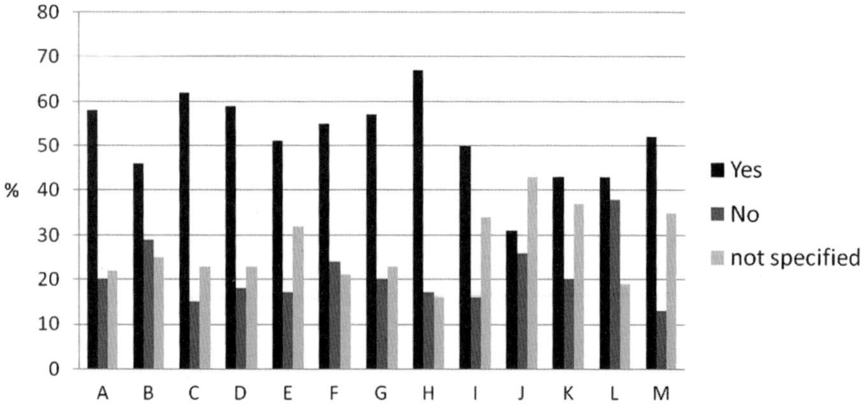

Fig. 10.4 Percentages of respondents intending and not intending to take part in the bioenergy village project in 13 villages (A–M) of the Goettingen district

In 9 of the 13 villages, more than 50 % of the villagers expressed a desire to join the project. In these villages, the process of social and technical support for the later steps, such as the feasibility study, continued (IZNE 2007).

10.4.2.2 Question 2: Assessment of the Notion of a Bioenergy Village with Explanatory Statements

74 % of the respondents in the 13 villages expressed a positive opinion of the bioenergy village concept, 11 % were undecided and only about 1 % (or 34 of the surveyed households) rejected the project (Fig. 10.5).

The reasons for a positive assessment of the bioenergy village concept, namely ecological motives, economic motives and social motives, are used as possible categories and are later confirmed through a data review. This division is inspired by the three-pillar model of sustainability (Ott 2009). Other categories are *increased comfort* and a residual category formulated during the content analysis process. The categories regarding a negative assessment of the bioenergy village concept are *economic motives*, *limitation of living standards*, *lack of experience*, *ethical concerns* and a residual category.

The quantitative analysis of the categories results in the following findings: Economic motives (e.g., *savings in heating costs* or *energy independence*) are the category mentioned most often (58 %) regarding a positive assessment of a combined power and heat supply. Ecological reasons (e.g., *reduction of greenhouse effect and climate change*) follow as they are mentioned by 31 % of the respondents. Social reasons (e.g., *stabilisation of the village community*) and increased comfort (*heating oil no longer needed*) are given less often.

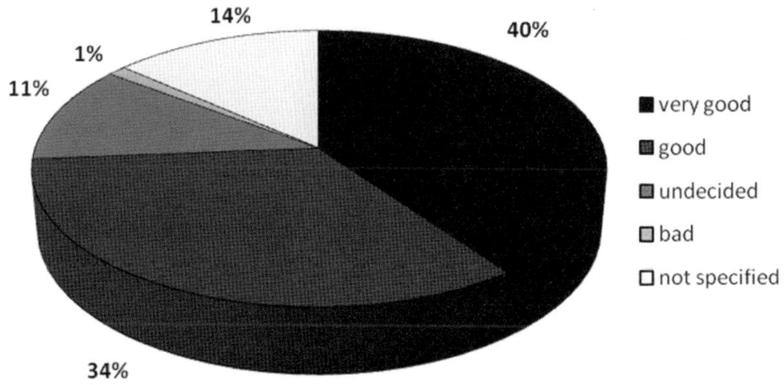

Fig. 10.5 Percentage of the respondents' opinion of the bioenergy village plan in 13 villages of the Goettingen district

Furthermore, the rejection of a bioenergy village is mainly justified for economic reasons (43 %). The perceived limitation of living standards (e.g. *odour from manure or traffic nuisance*) is mentioned by 23 % of the respondents. Ethical concerns (e.g. *the burning of food*) and a lack of experience (e.g. *the technology is not fully developed*) are not often mentioned. Other reasons include the *perceived dependency on farmers* (Fig. 10.6).

10.4.2.3 Question 3: Willingness to Participate in Working Groups

In total, 474 persons (25 %) agreed to participate in planning working groups – an average of 36 persons per village (variation: 16–55). The highest willingness to participate in planning was 44 % – in village K. The lowest interest in active participation was found in village L (17 %) (Fig. 10.7).

The reasons for a lack of willingness to participate in planning are assigned to the following categories: high age, health concerns, lack of time, information deficit, distance between home and work, and other reasons. The main reason for most villagers' lack of willingness to participate in the working groups is a lack of time, followed by high age. Medical concerns, lack of information and the distance between home and work are also mentioned (an equal share of 6 % each) (Fig. 10.8).

10.4.2.4 Question 4: Assessment of the Village Community

The question about the village community's ability to establish a bioenergy village as a shared project drew an affirmative response from most villagers (on average, 85 %) in the 13 villages. Even in village A, the least optimistic village in this regard,

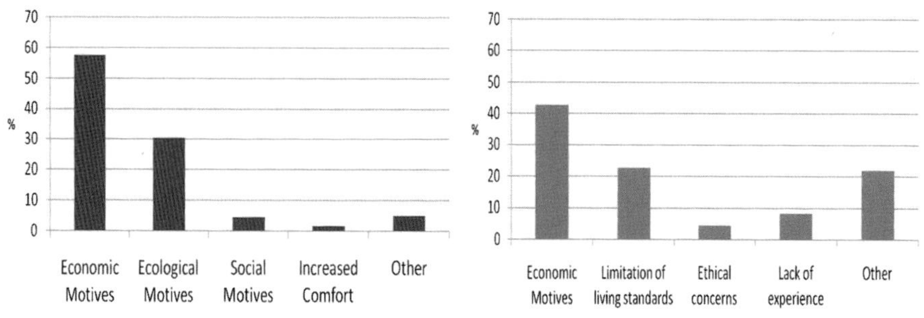

Fig. 10.6 Percentages of respondents' motives for supporting (on the *left*) and rejecting (on the *right*) the bioenergy village plan in 13 villages of the Goettingen district

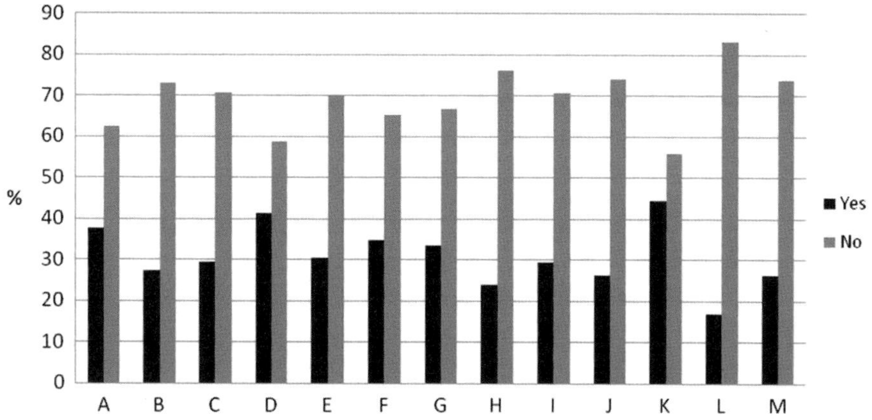

Fig. 10.7 Percentages of respondents intending and not intending to participate in working groups in the bioenergy village project in 13 villages (A–M) of the Goettingen district

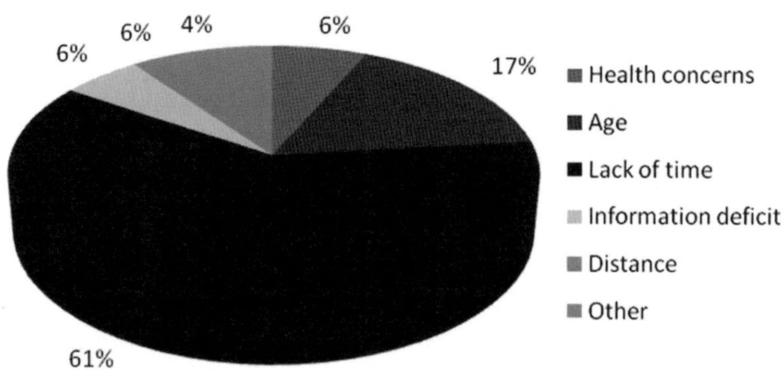

Fig. 10.8 Percentage of respondents' reasons for their lack of willingness to participate in the working groups in the bioenergy village project in 13 villages of the Göttingen district

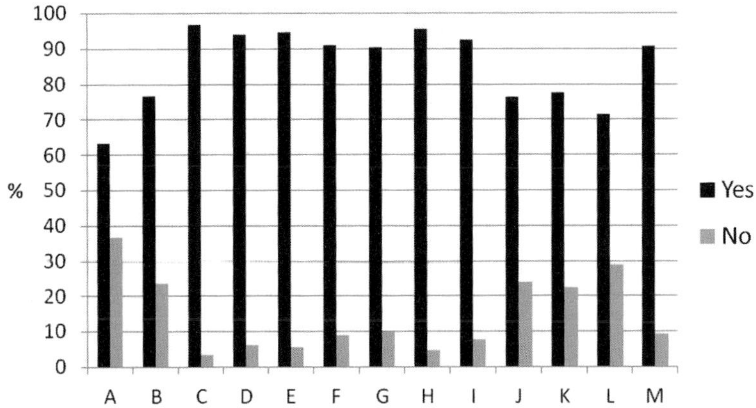

Fig. 10.9 Percentages of respondents in 13 villages (A–M) of the Goettingen district who are optimistic and not optimistic that their village will successfully complete the bioenergy village project

66 % of the analysed households were of the opinion that the village community could implement a bioenergy village project (Fig. 10.9).

The following are our theoretical considerations when categorising the respondents' reasons: The collective self-efficacy construct is a potential motive for faith in the shared project's success. According to Bandura's (1992, 1997) the social cognitive theory, cognitive processes, motivational processes, emotional processes, and behavioural processes are controlled by self-efficacy expectations. Self-efficacy is described as the extent of the subjective expectation of being able to perform a required behaviour to achieve a desired result. Individual expectations and collective expectations differ. High collective self-efficacy is based on the assumption that the group has trust in the team's capacities and on an optimistic perception of the accomplishment of future stress-producing events (Eigner-Thiel and Schmuck 2010). Schwarzer and Schmitz (1999) describe collective self-efficacy as the subjective certainty that new or difficult requirements can be managed by means of the group's shared competences. The appropriate category can be confirmed in the responses. During the analysis, the category *positive experiences* is formed and anchored in personal perceptions of previous community activities in the village.

The positive statements about the village community were primarily explained (46 %) by the perceived collective self-efficacy. Positive experiences with community projects were also mentioned, for instance, "building a community house", "renovating a swimming pool" as well as "village festivals" (Fig. 10.10).

The following categories were formulated for negative attitudes about the shared project's success: "disinterest" in a community development, the (negative) "personal experience" resulting in a lack of faith in the project, and a residual category. Remarkably, in the category "other" reasons (see Fig. 10.11), structural aspects of the village, such as the small "size of the village", the "too large buildings" and the "new buildings" were specifically mentioned. This suggests that some respondents

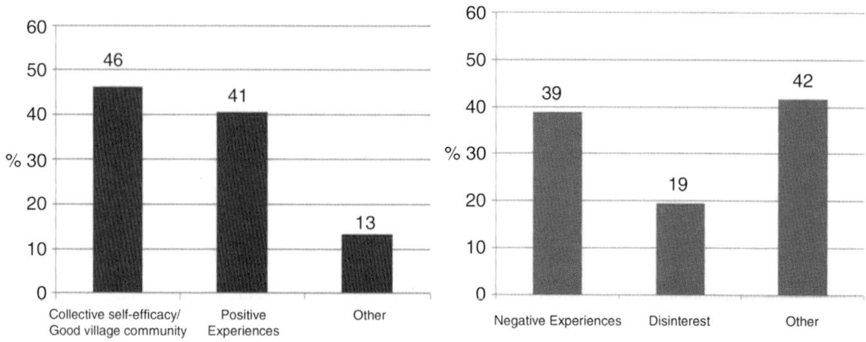

Fig. 10.10 Percentages of respondents' reasons in 13 villages of the Göttingen district for optimism (on the *left*) or lack of optimism (on the *right*) regarding a successful creation of a bioenergy village

did not understand the question, since the question sought to determine the village community's perception of the feasibility of a bioenergy village as a village community project.

10.4.3 Discussion

An average of 70 % of the villagers' answers point to strong interest in the bioenergy village concept. This high percentage may reflect the university members' successful public relations activities as well as the positive influence of the well-known bioenergy village Jühnde, which is very close to the other villages. Even the lowest rate of 51 % in village J was impressive.

10.4.3.1 Motives for the Assessment of the Bioenergy Village Concept

Economic and ecological reasons dominate regarding the positive statements about the bioenergy village concept, while social motives play a minor role. According to Stern et al. (1993), these motives can also be interpreted as egocentric (self-centred), biocentric (nature-related) or anthropocentric (related to the community of all humans). This means that the economic reasons mainly reflect egocentric motives, because heating cost savings are expected once the bioenergy village has been realised. The expected increase in comfort by having an own heating system and the lower maintenance costs may also be assigned to egocentric motives. Biocentric motives play an important role as they were approximately one-third of all the reasons mentioned. Anthropocentric motives have little significance. This could be explained as a ceiling effect: In these villages, a strong sense of community predated the start of the on-going project (see above). Another aspect that may

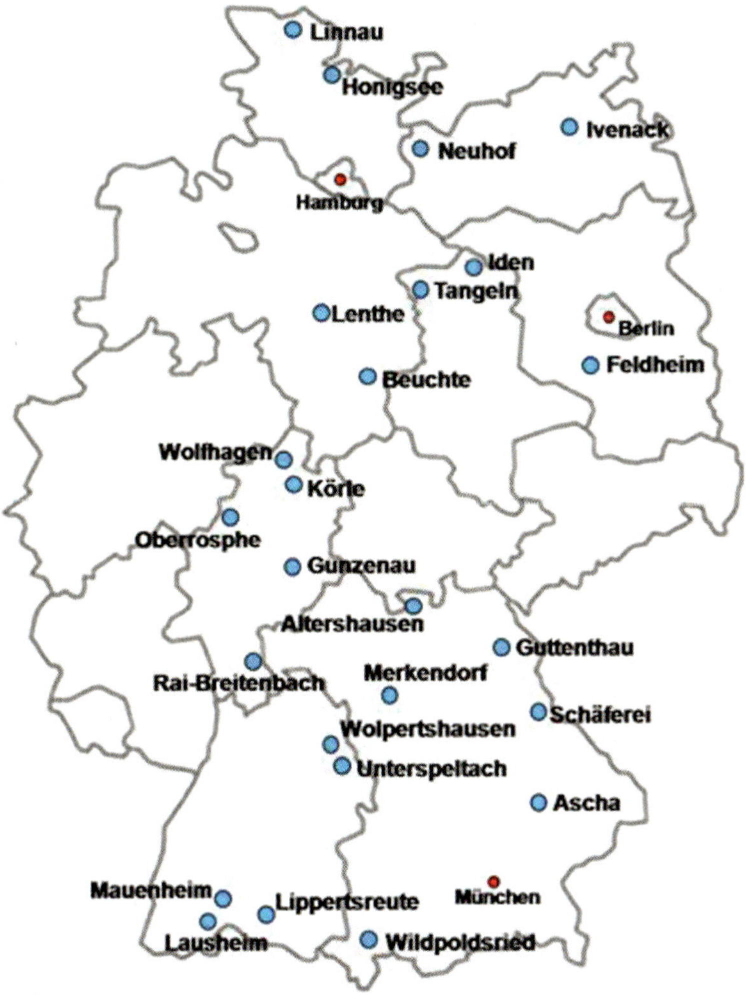

Fig. 10.11 Location of the 25 German communal renewable energy projects whose initiators were interviewed

exert a positive influence on the assessment of the bioenergy village concept is transfer effects and positive experiences with the bioenergy village Jühnde. However, these are rarely mentioned in the survey (under the category "other"). In the case of the negative statements, the economic reasons and the expected limiting of living standards can be assigned to egocentric motives. The "ethical concerns" that lead to the rejection of a bioenergy village can be interpreted as an anthropocentric motive, because "the burning of food" competes with the consumption of food and is thus detrimental to humanity. The category "lack of experience" reflects suspicion of the "new technology" and the associated fear of supply uncertainties.

However, such fears can be countered by targeted public relations activities, for example, by visiting best practice projects. In some cases, the new dependence on the farms providing the material led to the rejection of the bioenergy village concept.

The following recommendations can be derived from these findings for future projects: In the first information sessions with villagers, it seems to be important to stress the broad range of positive motivations for such projects. In addition to financial aspects, the benefits to the community and the ecological advantages should be addressed in detail to strengthen the biocentric and anthropocentric motives. A motivational mix rather than a single motivation seems to increase the likelihood. According to the findings, the rural population has different motivations for participation. On the other hand, critical arguments should be addressed early and systematically. If possible, such arguments should be refuted in public discussions or invalidated by visiting best practice models to show that certain concerns are unfounded.

10.4.3.2 Social Feasibility

In all the villages, the evaluation shows slight willingness of the inhabitants to join working groups to organise the conversion process. Since the required activities are mainly unsalaried, limited available time is a barrier conditioned by work and family responsibilities. Furthermore, the offered jobs are mostly outside the villages, so that many residents are dependent on a daily or even weekend commute between home and work. Our recommendation is that, with regard to project implementation, the main actors in the operating company should discuss the possibility of a shift from voluntary activities to financed activities. The statements concerning the positive assessment of the village community were equally due to collective self-efficacy and positively perceived experiences. For future projects, this means that the likelihood of establishing a bioenergy village should be based on a systematic survey of the village community's opinions. If there are positive experiences and attitudes, the likelihood of implementation are probably higher. However, this is a beneficial, but not sufficient, condition for the project's success; our study shows villages, in which optimism was present, but the project has not as yet been implemented.

In short, in these 13 villages, there was a very high social acceptance of the bioenergy village concept. Nevertheless, the low willingness to connect the own house with a heating network shows that there is a discrepancy between the general support and the de facto implementation of a bioenergy village. Despite the relatively equal initial conditions in the 13 villages, only four villages have to date successfully converted into a bioenergy village.

10.5 Success Factors for Communal Bioenergy Projects

Following the principle of a community-related energy supply, many villages and communities take control of their energy production. There are approximately 140 bioenergy villages in Germany, with many more in progress. For example, the state of Baden-Wurttemberg is funding the development of 100 bioenergy villages until 2020 and the state of Mecklenburg-Western Pomerania is planning 500 bioenergy villages until 2020. On the other hand, many project plans have not been realised. Therefore, analysing the conditions for success in such projects may be a valuable instrument to increase the likelihood of a systematic transfer of the idea of self-sustainable renewable energy communities.

The following section deals with the different paths to implement communal bioenergy projects and its success factors. These are based on a qualitative interview study. This study mainly sought to elaborate on motives for engagement, motivation, organisation and implementation strategies, the factors supporting and hindering a bioenergy project's implementation and the consequences of an established project. Therefore, interviews were held with initiators in 25 bioenergy villages or communal renewable energy projects (see Fig. 10.12)

10.5.1 *Methods*

Witzel's problem-centred interview method, an established qualitative method, was used to collect data. The interview allows the interviewee to speak as freely as possible, thus allowing an open discussion. However, it is centred on the interviewer introducing a specific problem. The interviewer prepares certain aspects of this

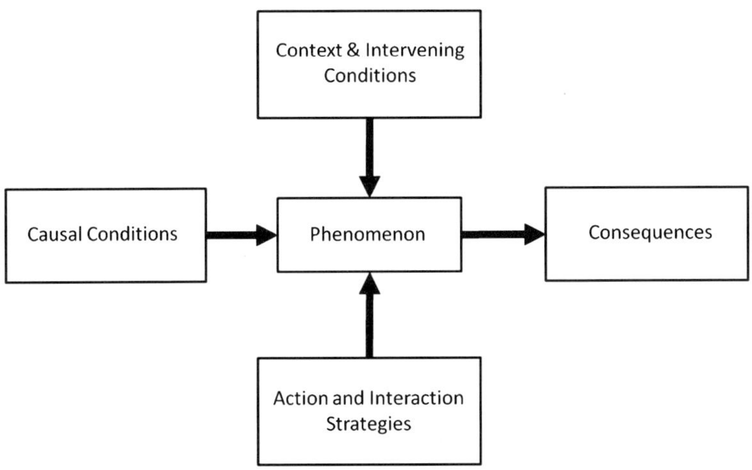

Fig. 10.12 Paradigm model (see Strauss and Corbin 1996)

problem beforehand, having compiled these as an interview guide. The principle of openness is important for the interview procedure. The interviewee can respond without predetermined response alternatives (Mayring 2002, p. 67).

In 20 bioenergy villages and five "integrative" bioenergy projects that combine bioenergy use with other renewable energies, we contacted interviewees who were substantially involved in projects' initiation, development and implementation. We then transcribed the recorded interviews.

10.5.1.1 Analysis of the Interviews

The interview transcripts were analysed using the grounded theory method (Strauss and Corbin 1996). Grounded theory is not a method, but a style of research and a strategy to discover a theory on the basis of empirical, mostly qualitative, data (Legewie 2005, p. 12). Its central element is the encoding process. Encoding means assigning one or multiple codes (keywords, items) to a text passage. During the analysis, the codes are not only derived from the data, but are also linked together and combined into superior categories (Legewie 2005, p. 16). It is useful to classify the categories in a coding scheme in order to determine their relationship. Strauss and Corbin (1996) proposed a particularly common coding paradigm. In addition, the analysis has a central phenomenon to which the other categories are related. The causal conditions are events that help develop the phenomenon. Furthermore, the phenomenon is embedded in a context with intervening conditions. Action and interaction strategies refer to the actions and reactions that occur as the result of the phenomenon and, finally, these actions and reactions' outcomes are the results (Strauss and Corbin 1996).

10.5.2 Results

The illustration of the results relates to Strauss and Corbin's (1996) paradigm model.

The following five main categories were formed as causal conditions: "local conditions", "impulses", "individual motives", "other participants' motives" and "tackling problems with dynamism". Verve concerning the context and intervening conditions, the main categories were "impeding factors", "internal barriers", "support factors", "cooperation" and "synergy effects". In the field of the action and interaction strategies, the following main categories were developed: "looking for information", "information strategies", "communication strategies", "project implementation strategies" and "organisation". The consequences were reflected in the subcategories "project effects", "personal effects" and "new perspectives and aims" (Fig. 10.13).

We next describe the different main categories on the basis of the proposed subcategories.

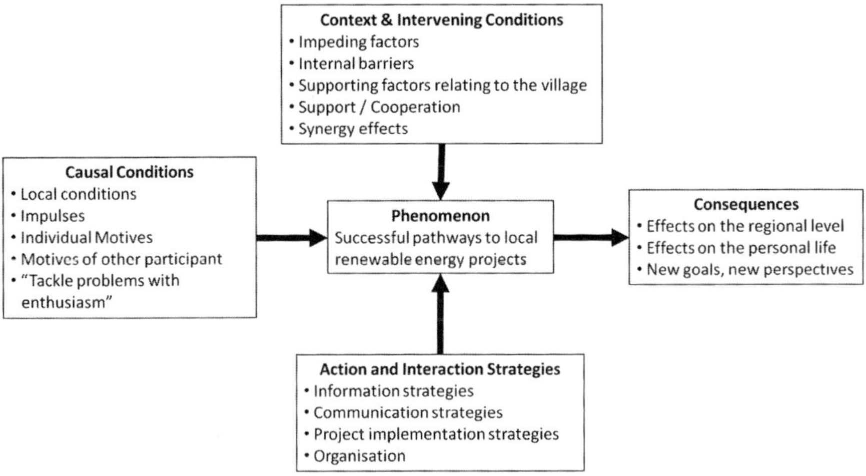

Fig. 10.13 Paradigm model of the results of the interviews with the initiators of renewable energy projects

10.5.2.1 Phenomenon

The phenomenon is referred to as *successful pathways to local renewable energy projects*, because the interview study focuses on determining the different success factors as well as transferring them to and applying them in our own projects (see Chap. 11). Some of the investigated villages combined bioenergy use with other renewable energies; the phenomenon therefore does not only focus on bioenergy projects.

10.5.2.2 Causal Conditions

We subsequently describe the conditions responsible for project initiation.

Local conditions: This main category describes important requirements that support the implementation of a bioenergy village project. The local inhabitants' peaceful coexistence is also a relevant condition to successfully establish the project. Many interviewees mentioned the availability of agricultural area and biomass for energy production as a fundamental requirement for the project. Another requirement is the interviewed initiator and other local persons acting as a driving force.

Impulses: This category describes the various initial sparks that lead to the project's initiation. Transfer effects from other, already established, bioenergy villages in Germany or Austria, such as the bioenergy village Jühnde and the energy self-sufficient district Guessing, were considered crucial. These projects' positive effects influenced the interviewee and other inhabitants, who conveyed these ideas to their own village. In some villages, the impetus to realise a bioenergy village came from the villagers. In almost all 25, the interviewees mentioned the active search for alternatives to fossil or nuclear energy based fuels. It became clear that there were different initiator motives and participant motives.

Individual motives and other participants' motives: During the analysis of the interviews, a spectrum of different motives was identified. The main reasons for engagement in a bioenergy village project were ecological motives (e.g., carbon dioxide reduction), social motives (e.g., to strengthen community life), economic motives and self-sufficiency (e.g., agricultural added value).

Tackling problems with dynamism: This category describes the constant efforts and endurance required for a sustainable and local energy supply to improve living conditions in the village and eventually culminating in the transformation of society.

10.5.2.3 Context and Intervening Conditions

We next present the general conditions that influence the development of a bioenergy village project.

Impeding factors: This main category describes the influencing factors that negatively affect the project development process. On the one hand, there were price fluctuations in the global market (crop price increase or oil price decrease), which influenced the bioenergy village project negatively. Another negative factor was the uncertainty concerning the project financing, especially the acquisition of financial support. These aspects therefore have an impact on the project economy. Some interviewees mentioned their uncertainty concerning the economic viability of the project and contradictory economic interests as impeding influencing factors. The initiators also mentioned the lack of support by policymakers and administrative bodies as a negative influencing factor.

Internal barriers: This main category relates to impeding factors concerning the village and the local conditions generally. The inhabitants of all 25 villages had doubts about the project. The initiators in particular have to grapple with questions concerning costs and energy supply security. Certain villagers' envy of others was another problem. In some villages, doubt and envy led to negative propaganda about the bioenergy village project. Disinformation (rumours) was mentioned as an impeding factor.

Supporting factors relating to the village: In nearly all of the 25 villages, the initiators mentioned open-minded inhabitants as an essential supporting factor in the village. In some villages, discharge pipe or road construction works were planned, so that the installation of the local heating grid could be undertaken in combination with the roadworks. These synergy effects had a positive impact on the project acceptance and project economy.

Support/Cooperation: Constructive cooperation with supporters at different levels was a key factor. Especially support from the local council and the mayor was considered necessary for the successful implementation of a bioenergy village project. Assistance from outside the village was also important if, for instance, the district administration and the permit authorities supported the project. Some of the interviewed initiators perceived different organisations' support as helpful (e.g., association of cooperatives, German Biogas Association (Fachverband Biogas)). Some interviewees appreciated constructive cooperation with planning offices and funding bodies.

10.5.2.4 Action and Interaction Strategies

We subsequently describe the envisaged strategies that contribute to project success.

Information strategies: In all the examined villages, the initiators planned information sessions and village meetings to inform the inhabitants about the project and mobilise them to participate in the planning process. Best practice visits were organised to already established, communal renewable energy projects. Face-to-face conversations were very important, especially to convince sceptics and opponents. In some villages, a significant contribution was conversations with inhabitants and word-of-mouth recommendations by them.

Communication strategies: This main category describes ways to discuss and communicate with the inhabitants about the bioenergy village project. Most of the 25 interviewed initiators emphasised the principle of transparency, especially relating to project finances and project economy. That means that all problems and points of criticism were discussed openly with the local inhabitants. In some villages, an independent moderator was included in the communication process.

Project implementation strategies: Most interviewees considered the involvement of the villagers in the planning and implementation process to be important. As a result, one or more working groups were established in the villages. The villagers' competencies were not only included in the planning process, but also in the construction work of the heat supply system. In a later phase, it was necessary to obtain professional support, such as planning offices, for a feasibility study. Some initiators recommended a cross-party approach. It is very important that the project is not exploited for the interest of only one sub-group of villagers.

Organisation: This main category contains important organisational steps relating to the bioenergy village project. This includes the choice of the type of company, the acquisition of subsidies, the organisation of biomass and cooperation with financial institutions.

10.5.2.5 Consequences

In the following we describe the individual-level consequences and effects.

Effects on the project: Nearly all the interviewed initiators reported positive ecological effects as a result of the project, especially carbon dioxide reduction. Furthermore, the projects added much value in the region, because energy expenditure remain in the region instead of it being paid to energy companies outside the region. Nearly all the interviewees noted an improved communal life, a feeling of togetherness in the village and that new inhabitants had been integrated into their communal life. Only in two cases did the initiators not notice any effect on their communal life. Furthermore, in some villages, the inhabitants identified with the project and appreciated the work of the initiator and the main actors.

Another positive effect was more nationwide publicity. Many interviewees reported numerous visitors to the villages. In addition, many bioenergy village projects won awards, for instance, from environmental organisations or from federal state governments.

Personal effects: This category focuses on the personal effects that the interviewed initiators experienced during and after the implementation process. Many were proud of the achieved result and reported stronger feelings of well-being. Some interviewees noticed improvements in their social skills. Furthermore, the initiators gained professional knowledge in the field of renewable energies. Some of them are now highly sought as experts in the development of a communal renewable energy supply. The interviewees also reported a higher quality of life owing to the more secure local or regional energy supply.

New perspectives and goals: This category describes further developments in these 25 bioenergy villages. Newly established goals include the expansion of the local heating grid, the implementation of other renewable energies, developing the region into a renewable energy region and the construction of renewable energy charging stations for electric cars.

10.5.3 Discussion: Success Factors and Recommendations for Future Projects

Similar to the experiences of the bioenergy village Jühnde, this interview study confirmed the findings of Eigner-Thiel and Schmuck (Eigner and Schmuck 2002; Eigner-Thiel 2005; see Chap. 2). We subsequently focus on the relevant success factors derived from the interview study results.

10.5.3.1 Individual Motives

The motives for initiator engagement in the investigated villages are multifaceted. The interviewees primarily mentioned ecological aspects, mostly in combination with other motives, such as the regional added value, independence from fossil fuel resources, and intergenerational justice. The diversity of the personal motives for engaging in communal renewable bioenergy projects confirms Dörner's (1999) findings that ecological actions require a mix of motives. Self-centred motives, such as the desire for self-realisation, also play an important role.

10.5.3.2 Create Awareness

There are different reasons for individual inhabitants participating in a communal renewable energy project. A holistic message is therefore important to communicate that various objectives, for example, a balance between environmental protection and regional added value, can be achieved with these projects.

10.5.3.3 Financing Aspects

All the projects could only be implemented on the basis of the funding opportunities that the federal government's Renewable Energy Source Act offered. Further financial support from different funding programmes was also necessary for investment.

10.5.3.4 Support from Politicians and the Administration

Political support is a very important factor for the successful implementation of a communal renewable energy project. Support from the local council and the mayor is especially necessary. High-level political support (e.g., the administrative district, federal state government and federal government) assists with the implementation process. In some cases, the projects were considered lighthouse projects, which made it easier to obtain subsidies and permissions.

Given that initiators of and participants in such communal bioenergy projects work on an honorary basis, a stronger knowledge-based, logistic and financial support from competent authorities is helpful.

10.5.3.5 Democratic Structures

A democratic organisational structure, such as a registered cooperative society, is recommended for the operating company. Some bioenergy villages show that energy supply on the basis of renewable energies may help democratise society's energy supply.

10.5.3.6 Transparent Communication Policy

A major challenge for the initiators was finding appropriate information and communication strategies. Village meetings are suitable for conveying the initial information. New information and results arising from the process can also be made public in village meetings. One-on-one conversations are very helpful, especially when dealing with sceptics. Transparency should be applied during the information and communication process. Furthermore, visiting model plants and already established communal renewable energy projects is a very successful way of obtaining information and convincing people to participate in a project.

10.5.3.7 Involvement of Inhabitants

The early involvement of inhabitants in the planning and organisation process increases the likelihood of success. Inhabitants can contribute their different competences and knowledge, and the work can be divided over many people. In some villages, working groups similar to the Jühnde model (see Chap. 2) were founded.

10.5.3.8 Personal Contribution

In most villages, the inhabitants' personal contribution to the project proved a success factor. Not only it is an opportunity to save costs (e.g., the excavation work), but it can also strengthen the village community.

10.6 General Discussion and Outlook

The Göttingen approach of sustainability science includes the close interconnection between social scientific research results and their applications in transdisciplinary projects. We could directly apply some of these study results in the planning workshops in our model regions of Wolfenbüttel, Goslar and the Hannover district (see Chap. 11). For example, the results (success factors) of the interview study in the 25 communal renewable energy projects were presented in a planning workshop in the district Wolfenbüttel. Based on these experiences and the experiences of the bioenergy villages of the Göttingen district, the workshop participants started an initiative to convince the district government of Wolfenbüttel to provide financial and political support for the development of bioenergy villages in their districts. Convinced of the impact that visiting model plants has, our team invited district politicians to a best practice tour of the bioenergy villages Barlissen, Krebeck and Wollbrandshausen in the Göttingen district. As a result, the initiative received funding to start a bioenergy village support process in the Wolfenbüttel district.

The nation-wide acceptance survey results are useful to predict the acceptance of different bioenergy consumption options. Consequently, they are useful when policymakers seeking to help develop bioenergy use in their region need to make strategic decisions. The broad concerns regarding energy plants or wood farms call for careful reflection of which bioenergy resources to prioritise, given that rural populations accept biowaste resources more easily. Further, the limited arable area available for both food and bioenergy production calls for a more complex consideration of the interplay between the different renewable energy production lines than that on which our project has focused to date. The follow-up phase of the on-going project will therefore emphasise the combination of different renewable energy lines. For instance, in the Goslar district, the combination of wind and water power, bioenergy from degraded soils and the storing of wind electricity in underground pump power stations is thought to create a stable and locally-based energy supply, in order to successfully progress to a post-fossil and post-nuclear age.

References

AEE (German Renewable Energies Agency). (2011). *Erneuerbare-Energien-Projekte in Kommunen* (5th ed., 47 pp). Retrieved December 14, 2012, from http://www.kommunal-erneuerbar.de/fileadmin/content/PDF/AEE_KommunalErneuerbar_Aufl05_web.pdf

Aretz, A., Hauber, J., Kress, M., Ruppert-Winkel, C., Schlager, P., Schmieder, K., Stablo, J., & Trommler, M. (2009). Regionale Selbstversorgung mit erneuerbaren Energien. *Ökologisches Wirtschaften, 24*(4), 47–50.

Bandura, A. (1992). Exercise of personal agency through the self-efficacy mechanism. In R. Schwarzer (Ed.), *Self-efficacy: Thought control of action* (pp. 3–38). Washington, DC: Taylor & Francis Inc.. ISBN 10: 1560322691.

Bandura, A. (1997). *Self-efficacy: The exercise of control* (604 pp). New York: Worth Publishers, ISBN-10: 0716728508.

Dethloff, C. (2004). *Akzeptanz und Nicht-Akzeptanz von technischen Produktinnovationen*. Dissertation (340 pp). Lengerich: Pabst Science Publishers.

Dörner, D. (1999). *Bauplan für eine Seele* (831 pp). Reinbeck: Rowohlt-Verlag, ISBN-10: 3498012886.

Egert, M., & Jedicke, E. (2001). Akzeptanz von Windenergieanlagen. *Naturschutz und Landschaftsplanung, 33*, 373–381.

Eigner, S., & Schmuck, P. (2002). Motivating collective action: Converting to sustainable energy sources in a German community. In P. Schmuck & W. Schultz (Eds.), *Psychology of sustainable development* (pp. 241–257). Boston: Kluwer Academic.

Eigner-Thiel, S. (2005). *Kollektives Engagement für die Nutzung erneuerbarer Energieträger – Motive, Mobilisierung und Auswirkungen am Beispiel des Aktionsforschungsprojekts "Das Bioenergiedorf"*. Studien zur Umweltpsychologie (310 pp). Hamburg: Verlag Dr. Kovac, ISBN-10: 3830020465.

Eigner-Thiel, S., & Schmuck, P. (2010). Gemeinschaftliches Engagement für das Bioenergiedorf Jühnde – Ergebnisse einer Längsschnittstudie zu psychologischen Auswirkungen auf die Dorfbevölkerung. *Zeitschrift für Umweltpsychologie, 14*(2), 98–120.

Endruweit, G., & Trommsdorff, G. (2002). *Wörterbuch der Soziologie* (2nd ed., 754 pp). Stuttgart: UTB, ISBN-10: 3825222322.

Flick, U. (2004). *Qualitative Sozialforschung – Eine Einführung*. Reinbek: Rowohlt Taschenbuch Verlag GmbH.

FNR (Fachagentur Nachwachsende Rohstoffe). (2012). Entwicklung Biogasanlagen. Retrieved December 14, 2012, from http://mediathek.fnr.de/grafiken/daten-und-fakten/bioenergie/biogas/entwicklung-biogasanlagen.html

Griesen, M. (2010). Akzeptanz von Biogasanlagen. Dissertation, University of Bonn. Bonner Studien zur Wirtschaftssoziologie 34 (252 pp). Aachen: Shaker Verlag, ISBN: 978-3-8322-9616-2.

IZNE. (Interdisciplinary Centre for Sustainable Development) (2007). *Entwicklung weiterer Bioenergiedörfer im Landkreis Göttingen*. Unpublished report of Projektgruppe Bioenergiedorf, IZNE, Uni Göttingen.

Jenssen, T. (2010). *Einsatz der Bioenergie in Abhängigkeit von der Raum- und Siedlungsstruktur* (396 pp). Wiesbaden: Vieweg + Teubner, ISBN-10: 3834808199.

Legewie, H. (2005). *Qualitative Forschung und Ansatz der Grounded Theory* (23 pp). Retrieved December 14, 2012, from http://www.ztg.tu-berlin.de/download/legewie/Dokumente/Vorlesung_11.pdf

Mautz, R., Byzio, A., & Rosenbaum, W. (2008). *Auf dem Weg zur Energiewende* (175 pp). Göttingen: Universitätsverlag Göttingen, ISBN: 978-3-938616-98-7.

Mayring, P. (2002). *Einführung in die qualitative Sozialforschung* (5th ed., 170 pp). Weinheim, Basel: Beltz Verlag, ISBN-10: 3407252528.

Mayring, P. (2008). Neuere Entwicklungen in der qualitativen Forschung und der Qualitativen Inhaltsanalyse. In: P. Mayring & M. Gläser-Zikuda (Eds.), *Die Praxis der Qualitativen Inhaltsanalyse* (2nd ed., pp. 7–20). Weinheim und Basel: Beltz Verlag, UTB, 2. Aufl., ISBN 978-3-407-25502-0.

Ott, K. (2009). Leitlinien einer starken Nachhaltigkeit. Ein Vorschlag zur Einbettung eines Drei-Säulen-Modells. *Gaia, 18*, 25–28.

Roesch, C., & Kaltschmitt, M. (1999). Energy from biomass – do non-technical barriers prevent an increased use? *Biomass and Bioenergy, 16*(5), 347–356.

Ruppert, H., Eigner-Thiel, S., Girschner, W., Karpenstein-Machan, M., Roland, F., Ruwisch, V., Sauer, B., & Schmuck, P. (2008). *Wege zum Bioenergiedorf – Leitfaden für eine eigenständige Strom- und Wärmeversorgung auf Basis von Biomasse im ländlichen Raum* (120 pp), with DVD, ISBN 978-3-9803927-3-0. Retrieved December 14, 2012, from http://www.fnr-server.de/ftp/pdf/literatur/pdf_318-leitfaden_bioenergiedorf_2010_web_neu.pdf

Schwarzer, R., & Schmitz, G. S. (1999). Kollektive Selbstwirksamkeitserwartung von Lehrern: eine Längsschnittstudie in zehn Bundesländern. *Zeitschrift für Sozialpsychologie, 30*(4), 262–274.

Stern, P. C., Dietz, T., & Kaloff, L. (1993). Value orientations, gender and environmental concern. *Environment and Behavior, 25*(3), 322–348.

Strauss, A., & Corbin, J. (1996). *Grounded theory: Grundlagen qualitativer Sozialforschung* (240 pp). Weinheim: Beltz, ISBN : 978-3-621-27265-0.

Zoellner, J., Schweizer-Ries, P., & Wemheuer, C. (2008). Public acceptance of renewable energies: Results from case studies in Germany. *Energy Policy, 36*, 4136–4141.

Chapter 11
Applying the Sustainability Science Principles of the Göttingen Approach to Initiate Renewable Energy Solutions in Three German Districts

Peter Schmuck, Marianne Karpenstein-Machan, and André Wüste

Abstract This chapter reports on an interdisciplinary and transdisciplinary action research project that applies sustainability science principles and supports the conversion of the energy supply from fossil and nuclear fuels to biomass and other renewable energies in three districts in Lower Saxony, Germany. The project began in 2009 and is still continuing. The first steps were: (1) A partner district selection to identify districts highly likely to realise the intended changes. A suitability criteria list was compiled and three districts were selected. (2) In these districts, a detailed analysis was performed of the de facto state of biomass use for energy production, with special focus on existing personal networks, bioenergy potentials, related conflicts, and actual plans. (3) Planning workshops were arranged with local politicians, regional administration staff for agriculture and environment and other stakeholders, such as farmers and nature conservation activists, who articulated their regional conversion goals, developed concrete projects and discussed ways to realise these plans. The setting was consensus-oriented and moderated by the team of scientists. They also supported this energy conversion process and performed parallel research.

Keywords Action research • Sustainability science • Bioenergy regions

11.1 Background

This chapter focuses on the application aspects of sustainability science in our project. In line with Chapter 2, we continue with the process we worked on between 2000 and 2008 in the bioenergy villages, but now focus on the rural district level.

P. Schmuck (✉) • M. Karpenstein-Machan • A. Wüste
Interdisciplinary Centre for Sustainable Development, University of Göttingen, Goldschmidtstr. 1, 37077 Göttingen, Germany
e-mail: peterschmuck@gmx.de; mkarpen@gwdg.de; awueste@gwdg.de

DOI 10.1007/978-94-007-6642-6_11, © Springer Science+Business Media Dordrecht 2013

The conversion of the energy supply from fossil and nuclear fuels to bioenergy and other renewable energies is practically and scientifically supported (for the theoretical background, see Schmuck and Schultz 2002 as well as Sheldon et al. 2000). The following paragraphs follow the Göttingen approach's sequence of activities (see Chap. 2, Fig. 2.1).

The first problem-solving steps are described in Chap. 2, with energy as the problem field selected from the global problem map. During the past few years, no scientific evidence has indicated a decrease in this problem's significance; on the contrary, climate scientists have begun to formulate "tipping points" in our earth climate system that will accelerate climate change effects nonlinearly if we do not search for and implement post-fossil fuel alternatives (Lenton et al. 2008). Examples are the instability of the Greenland ice sheet with an accelerated ice-melting and its interactions with the atmospheric and oceanic circulation or the increase of irregularities of the Indian Monsoon patterns. Our project's goal is to replace finite fossil fuels with renewable energies and focuses on biomass for regions and districts in Germany.

The search for financial and political support for our project plan was far more straightforward than the first step in 1998, when the notion of a bioenergy village was still a very innovative and seemingly risky endeavour. In 2007, the Ministry of Science and Culture of the federal state of Lower Saxony asked the team to continue our bioenergy villages success story (Ruppert et al. 2008; Chap. 2 in this book) at the district level in Lower Saxony. An independent commission evaluated and accepted the submitted project proposal; the project could therefore begin in 2009.

The rest of the chapter describes the next steps in our approach and our action research activities (Whyte 1991; Mills 2000; Brydon-Miller et al. 2003; Kasemir et al. 2003). We describe the search for practice partners through a competition between districts, the competition results, the pilot projects in the three selected districts in Lower Saxony and the transfer of knowledge between the districts. Finally, we provide an evaluation of the planning procedure.

11.2 Search for Practice Partners: A Competition Between Districts

The problem of finding practice partners for the project was solved as follows: Based on our experience at the village level, we launched a competition between the 38 districts in Lower Saxony. The three winner districts would receive our support to establish a sustainable bioenergy supply. Rather than directly asking the districts' political heads for support, our method had the advantage that the cooperation activity and responsibility would be balanced between the practitioners and the scientists, thus ensuring that districts with the best-motivated politicians applied.

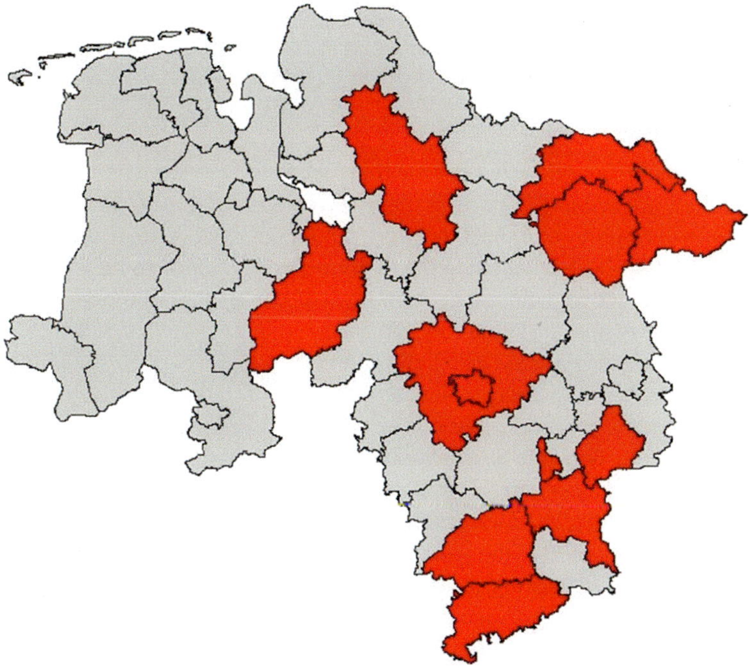

Fig. 11.1 Ten districts within the federal state of southern Lower Saxony applying for cooperation in the bioenergy project

We contacted the political leaders of the 38 districts, offering our scientific support for and expertise to the three winning districts once they had described their bioenergy use's status quo, had indicated their political will to change to renewable energies and had described the strengths and weaknesses of the district's approach.

Twelve districts applied for cooperation. We held meetings in ten districts that had completed their applications (Northeim, Wolfenbüttel, Uelzen, Goslar, Göttingen, Lüchow-Dannenberg, Hannover, Diepholz, Lüneburg and Rotenburg/Wümme, see Fig. 11.1). The leading representatives of agriculture and forestry, local politics, administration, nature protection, electric power companies, bioenergy networks, biogas plant planners and operators, and the heads of relevant research projects were present at these meetings.

At these meetings, we asked for details of actual bioenergy development in the districts to identify the de facto preconditions that best fit our research demands and goals. This would promote sustainable bioenergy production and distribution. On the whole, the information provided would allow us to systematically compare the interested districts and select the three partner districts with the best realisation prospects.

This fit was evaluated by means of a systematic list of five primary and two secondary suitability criteria that the scientists had developed. The criteria served as a basis to compare the districts. These criteria and indicators were:

Primary criteria:

- *1. The political will to foster renewable energies, including bioenergy, and to protect the climate.* The indicators of this political will were (a) relevant political resolutions, which the highest political level of the district supported or initiated, (b) relevant steering instruments developed to measure environmental data (the landscape framework plan) and (c) the administration's active support of sustainability projects.
- *2. The motivation of the districts' primary actors to cooperate with a team of scientists.* The indicators included the variety of the stakeholder groups at the first meeting, the interest in our research project topics and the readiness to discuss and cooperate as expressed verbally during the initial meeting.
- *3. The contact persons' cooperation readiness and reliability as well as the interpersonal fit.* The indicators included the organisational quality of the first meeting as well as the completeness and timeliness of the application documents. Our empathy with the contact persons indicated the interpersonal fit.
- *4. The agriculture representatives' cooperation readiness and reliability.* The indicators were their willingness to complete data sheets on the agricultural practices in the own company and the willingness of single farmers from the district to cooperate with our team to establish a "model bioenergy farm".
- *5. The social cohesion among the district's different stakeholder groups.* The indicators included the presence of relevant actors at the first meeting, their engagement during this meeting and a constructive discussion climate during this meeting.

Secondary criteria:

- *6. The representativeness.* The three selected districts had to have a broad diversity of landscape forms, agricultural structures and bioenergy development levels to ensure later on the transferability of the scientific experiences to other districts.
- *7. The fit with our scientific profile.* The three selected districts had to give us the opportunity to fully apply our team's different competencies concerning bioenergy villages, degraded soils, nature protection topics and the combustion of straw and wood.

The following tables show the districts' ranking regarding the five primary criteria. Where subjective evaluations were required, the three authors, who were present at all ten meetings, undertook the ratings individually and independently of one another. Our individual ratings differed in a few fields of the evaluation matrix (10 villages * 5 criteria), we subsequently discussed these. In most cases, we could then agree on one rating score. In the remaining three cases (see Table 11.1), we

Table 11.1 Political will to foster renewable energies including bioenergy, and to perform climate protection activities

District	Resolutions to support renewable energies/ bioenergy	Support by administration	District head supports the process	Responsibility for environment, landscape framework plan	Sum
A	–	0	–	0	−2
B	+	++	++	++	7
C	–	–	–	++	−1
D	+	++	+	–	3
E	+	++	++	0/+	5.5
F	++	++	++	+/++	7.5
G	++	++	0	++	6
H	0	0	0	++	2
I	++	++	0	+/++	5.5
K	0	–	–	++	0

Criteria: – not fulfilled, 0 partially fulfilled, + fulfilled, ++ very well fulfilled, empty space not applicable or irrelevant

provided two ratings, which were averaged to form the sum score. The criteria and evaluation results were presented and discussed at several plenary meetings of the team of scientists. This contributed to the team reaching a final consensus on which districts were best suited for cooperation.

Five districts (sum score > 5) showed a strong political commitment compared to the other five districts with lower scores (Table 11.1).

In Table 11.2 we find a very clear bifurcation: In three districts, the motivation to cooperate is clearly higher than in the others.

The sum scores of Table 11.3 again show a fairly high variability, indicating differing willingness to cooperate in the ten districts.

Table 11.4 shows that four districts had to be excluded from cooperation with our project, because cooperation with agriculture representatives was a necessary precondition for the collaboration.

The indicators shown in Table 11.5 – the presence of a variety of relevant actors at the first meeting and their engagement during the meeting by providing constructive discussion – again showed significant differences between the districts.

Table 11.6 shows that three districts have (1) positive sum scores across all criteria, and (2) the highest scores in the sum of all the criteria. These three districts thus form the top three in the ranking. Given that all districts that had not fulfilled, or only partially fulfilled, one or more criteria had to be disregarded, this leads to seven districts (sum scores marked in grey) being eliminated. Therefore, the secondary criteria, which had been developed in case more than three districts fulfilled all the main criteria, were not applied.

The districts selected for cooperation were Wolfenbüttel, Goslar, and Hannover.

Table 11.2 Motivation of the districts' main actors to cooperate with a team of scientists

District	Representatives from agriculture	Representatives from forestry	Mayors	Environmentalists	Research project heads	Regional planner	Sum
A	0		+			+	2
B	++		++		++	+	7
C	–	–	0	0		–	–3
D	++	++	+		+	+	7
E	–	0	++	0			1
F	0		0		0	0	0
G	+		+	+	++	++	7
H	0		+	+			2
I	–	+		++			2
K	–	0		+			0

Criteria: – not fulfilled, 0 partially fulfilled, + fulfilled, ++ very well fulfilled, empty space not applicable or irrelevant

Table 11.3 Direct contact persons' cooperation readiness and reliability as well as interpersonal fit

District	Application documents on time	Application documents complete	Organisation of the first meeting	Empathy, positive feeling	Sum
A	+	+	+	0	3
B	+	0	++	++	5
C	–	+	–	–	–2
D	+	+	++	++	6
E	–	+	++	–	1
F	+	++	++	++	7
G	–	0	++	++	3
H	+	0	+	0	2
I	+	+	+	+	4
K	+	+	+	–	2

Criteria: – not fulfilled, 0 partially fulfilled, + fulfilled, ++ very well fulfilled, empty space not applicable or irrelevant

Table 11.4 Agriculture representatives' cooperation readiness and reliability

District	Data sheets completed and returned	Declarations of readiness to cooperate with our team to establish a "model bioenergy farm"	Sum
A	++	++	4
B	++	++	4
C	–	–	–2
D	++	+	3
E	+	+	2
F	0	+	1
G	0	+	1
H	–	–	–2
I	–	–	–2
K	–	–	–2

Criteria: – not fulfilled, 0 partially fulfilled, + fulfilled, ++ very well fulfilled, empty space not applicable or irrelevant

Table 11.5 Social cohesion of different stakeholder groups in the district

District	Variety of the stakeholder groups at the first meeting	Constructiveness of the discussion	Sum
A	++	0	2
B	+	++	3
C	++	–	1
D	+	++	3
E	+	–	0
F	+	++	3
G	+	++	3
H	+	+	2
I	0	++	2
K	0	–	–1

Criteria: – not fulfilled, 0 partially fulfilled, + fulfilled, ++ very well fulfilled, empty space not applicable or irrelevant

Table 11.6 Summary of Tables 11.1, 11.2, 11.3, 11.4, and 11.5: Sum scores of the five main fit criteria for the ten districts

District	Political will	Main actors' motivation	Fit with contact persons	Farmers' cooperation readiness	Social cohesion	Sum	Rank
A	-2	2	3	4	2	9	8
B	7	7	5	4	3	**26**	**1**
C	-1	-3	-2	-2	1	-7	10
D	3	7	6	3	3	**22**	**2**
E	5,5	1	1	2	0	9,5	7
F	7,5	0	7	1	3	18,5	4
G	6	7	3	1	3	**20**	**3**
H	2	2	2	-2	2	6	6
I	5,5	2	4	-2	2	11,5	5
K	0	0	2	-2	-1	-1	9

Grey: criteria with sums of 0 or lower, indicating that the criterion is not fulfilled or, at best, partially fulfilled. Bold numbers: Sum scores and ranks of the top three districts

11.3 Pilot Projects in Three Districts – Different Lighthouse Projects

Our next step was a detailed analysis of the de facto bioenergy use's status quo in each of the districts. Competent main actors were interviewed with regard to the bioenergy projects, their project plans and conflicts with existing plans. The following sections describe the first steps of our work and the on-going cooperation activities in the three districts (for further results see Schmuck et al. 2012).

11.3.1 District Goslar

District Goslar is located in eastern Lower Saxony and includes the western part of the Harz Mountains and its northern forcland. Approximately 145,000 people live on an area of 965 km^2. Today, approximately 35 % of this district area is used for agriculture, while 60 % is covered in forests. District Goslar has both fertile soils and stony soils with low fertility. Typical crops are wheat, sugar beets, and brewing barley.

District Goslar has six biogas plants in operation, with another planned. Most of these plants are not very efficient because the heat produced in the biogas combustion is only partially used or not used at all. In District Goslar we were confronted with a substantial conflict, which we had to solve before we could start. A company from a neighbouring district planned a waste combustion plant in a village in the district. There a citizens action committee was formed to protest against the plans, which contributed to the plan being abandoned. However, during the starting phase of our project, a consultant from the neighbouring district invited bioenergy actors from the district, to which the protesting village belongs, to found a bioenergy network. Interestingly, at the meeting, the relevant company, which also planned another energy plant in the same area, was the official sponsor of the meals. Our team could not collaborate with this initiative because many actors questioned the consultant's independence and credibility. We therefore started our activities independently of these activities.

In the first planning workshop, the main goal was to formulate the goals and visions for the district's future bioenergy production. Our team members provided presentations on the bioenergy acceptance data in the rural population and on the ecologically and socially sound promotion of bioenergy projects. Afterwards, many expectations, hopes and visions were formulated regarding bioenergy's role in 2020. These included:

- 100 % renewable energy without any CO_2 emissions, power stations combining wind power, water power and bioenergy sources and the installation of energy storage facilities (i.e. dam reservoirs as pump storage stations).
- Using polluted soils to produce energy plants.
- Cooperatives running power plants as well as electric mobility solutions with associated filling stations.
- Compilation of bioenergy resources' potential to develop other plants.
- Initiation of ten bioenergy villages.
- Using 50 % of the available straw to heat the schools in the district.

These visions were made more concrete in a later step by formulating the following goals:

- Local power plants based on renewable energy and organised by communities produce energy for the district.
- Electricity produced in the district is mainly merchandised in the district.
- Biogas pipelines collect biogas from small biogas plants; the gas is then used directly, or converted to fossil gas quality, fed into the fossil gas pipeline and used at another place in the district.
- Installing hot water pipelines for efficient distribution of heat energy in districts with a high demand for heat.
- Making the external costs of fossil and nuclear energy transparent to improve renewable energy's profile.

Local actors then presented two bioenergy projects to provide potential starting points for the planned processes:

The farmer of our model farm formulated the development goals of his biogas plant: increased acceptance of his plant within his village and full use of the heat produced in the plant. The group provided possible solutions for these intentions, including a satellite heat and power station connected to the existing plant via a biogas pipeline at a point in the village where heat is needed. Such a solution would fulfil both the farmer's and the villagers goals.

The second project was presented by a farmer with a biogas plant and by the mayor of the village that intends to use the electricity generated by the biogas plant and wind power generators near the village. The idea is to have a stable electricity supply that is independent of the four big power companies that deliver most of Germany's electricity from fossil and nuclear sources. The group developed possible helpful activities, including the search for partners for local power plants, the search for best practice projects, and the search for information on communities' takeover of public power supply networks.

The final step was to plan our next workshop. The group agreed to start by elaborating the necessary steps for the second project, which would focus on further renewable energy development in village I. In two workshops (I and II), our team delivered information on communities' takeover of public power supply networks. At the second workshop, an expert who organised all the electricity and heat supply for a village delivered an impromptu presentation.

Planning workshop II started with a presentation of a best practice project, the first village (Feldheim, near Berlin) in Germany with an own electricity supply network fed completely with local renewable energy (in addition to a hot water pipeline delivering heat from local bioenergy).

Inspired by this presentation, possible steps for implementing goals in the village and on the district levels were written on moderation cards and pinned to a board, discussed and then evaluated by means of priority scores. Every group member received five points that could be placed on these activity cards, which contained the actions most likely to be realised within a short time. Finally, the participants were invited to take responsibility for specific actions. The results are shown in Tables 11.7 and 11.8.

Table 11.7 Results of the brainstorming regarding future activities in the village I., district Goslar

Activity	Priority scores	Who is Responsible
Analysis of actual state		
Analyse the demand of electricity in the village	8	Group of activists in the village I.
Cooperation with pupils of the local school in collecting corresponding data	4	Group of activists in the village I.
Calculating available amount of electricity of the biogas plant and the wind power generators	4	Group of activists in the village I.
Activities directed at implementation		
Feasibility study for a hot water pipeline fed by a wood chip combustion plant	7	Mr. W. from local power company and Group of activists in the village I.
Management of an accounting grid for the wind power used directly in the village	4	Mr. W. from local power company
Foundation of a civil society i.e. cooperative to finance feasibility studies	3	Group of activists in the village I.

Table 11.8 Results of the brainstorming regarding future activities in the district Goslar

Activity	Priority scores	Who is Responsible
Analysis of actual state		
Producing of maps of the district showing existing biogas plants and further agricultural potential for biogas, existing fossile gas pipelines and hot water grids	2	University group, Mr. W. from local power company
Analyses of amounts of agricultural waste products and biowaste in industry and restaurants	2	Company "Fritz-Planung"
Publicity work		
Public information actions regarding renewable energy via mass media, internet, supported by chambers of crafts, guilds, engineering companies	9	Ms. G., association "X with energy"
Activities directed at implementation		
Surface solar and wind power stations on areas not usable for agriculture (degraded soils), local citizens contributing to financing	5	
Founding of a community power company (several neighbored cities and villages around city L.) with regional partners	4	Major of city L., Mr. S.
Enforced usage of wood and water power to gain energy, discussions with responsible administration	3	

In the spring of 2011, the activist group in the village independently organised meetings of the villagers and working groups. This indicates that we had successfully initiated the transformation process.

11.3.2 District Wolfenbüttel

District Wolfenbüttel is situated north of the Harz Mountains in eastern Lower Saxony. Approximately 125,000 people inhabit an area of 722 km^2. Today, approximately 68 % of the district is used for agriculture. District Wolfenbüttel has very fertile soils. Typical cultivated crops are wheat, sugar beets and brewing barley as well as vegetables. Approximately 18 % of the district area comprises forests.

District Wolfenbüttel has six biogas plants in operation, with another two planned. Similar to District Goslar most of these plants are not very efficient, because the heat from the combined heat and power stations is only partially used or not used at all. However, in the District Wolfenbüttel exists one bioenergy lighthouse project: the bioenergy village B. It is supplied with heat by a central heating plant burning wood chips. A biogas plant is under construction and will also soon deliver heat to households. The electric power will be fed into the public grid. Currently, approximately 60 households are supplied with heat and hot water. Another specific characteristic of District Wolfenbüttel is the contaminated farmland soil in the floodplains of the rivers O and I. In this context, energy plant cultivation for bioenergy can be an alternative to food production in these areas (see Chap. 14). In this district, the planned projects of intensive animal husbandry in order to use the waste material for bioenergy production was another conflict hotspot. This led to citizens' initiatives against these projects.

In planning workshop I in District Wolfenbüttel, our team of scientists provided input presentations with regard to energy crop cultivation, the usage of contaminated soils for the energy production, bioenergy's social acceptance in the rural population and the creation of integrative and sustainable bioenergy districts.

The main goal of the participants in this workshop was to formulate future goals and visions concerning renewable energies and bioenergy production for District Wolfenbüttel. They were invited to draw colour pictures on paper to visualise their visions or hopes for District Wolfenbüttel in 2020. The following visions and hopes were formulated (see also Fig. 11.2):

- The integration of bioenergy resources (wood, energy plants) with other renewable energies (wind, water, solar and geothermal energy).
- Energy storage in disused underground mines.
- The shared organisation of renewable energy projects.

In a next step, two small working groups elaborated these visions and formulated more concrete goals. One group worked on the topics "acceptance of bioenergy, decentralized bioenergy projects and the integration of other renewable energy sources" with the following results:

Fig. 11.2 Drawings by representatives of district Wolfenbüttel, showing their hopes and visions regarding energy supply with renewable energy by 2020

- The development of bioenergy villages in District Wolfenbüttel: A village competition to initiate bioenergy villages combined with an information session about bioenergy villages.
- The use of a biogas plant's unused heat for a heating network.
- Strengthening tourism by improving the aesthetic of the landscape by improving the diversity within energy plant cultivations.
- Taking the characteristic landscape into account when planning energy bioenergy facilities.
- Using roadside vegetation for bioenergy production.

The other working group worked on the topic "using contaminated soils in the floodplains for the production of energy plants". Its results were:

- Using contaminated soil instead of non-contaminated soil areas for energy plant production.
- Investigating the heat, which may be produced from plants in the contaminated areas (is wood gasification possible?).
- Establishing short-rotation plantations on contaminated soils as a flood attenuation and to prevent water contamination by eroded soil material.
- Establishing cooperation models.

In the final step, the themes for the next planning workshops were defined. The participants agreed to investigate the possibilities to use contaminated soils for bioenergy production. Between these workshops, our team collected relevant information on the contaminated soils in District Wolfenbüttel. A specialist in our team created maps of the contaminated floodplains, water protection areas and flood protection areas (see Chap. 14).

Planning workshop II started with two natural scientists in our team offering presentations on bioenergy production on contaminated soils, energy plant cultivation and flood protection. Two working groups were then formed. One working group discussed the opportunities for energy plant cultivation in relation to flood protection, water pollution control, nature protection, and compensating measures. Three scenarios and working steps were discussed for renewable energy usage on contaminated soils:

- Short-rotation plantation on contaminated soils. It is important to identify the flood areas, because short-rotation trees do not provide flood protection.
- Biogas production. In this context, it is necessary to check whether the digestate on these soils can be fertilized (for details, see Chap. 14).
- Photovoltaic power installations, including extensive grassland farming under these installations.

The participants realised that there is a lack of scientific analysis to allow one to safely argue for or against any of these three scenarios. A legal bioenergy usage guideline on contaminated soils was proposed as a mid-term goal.

The second working group discussed the potential heat demand in the communities with contaminated soil areas. The following points were discussed:

- Do enough residents want to join a local heat supply system?
- How can the heat be distributed?
- What are optimal locations for a biogas plant or a bioenergy village?
- Which operating structures are possible?
- More information sessions for farmers cultivating energy plants on contaminated agricultural areas.
- All communities in these areas are already supplied with natural gas. This could be a point of conflict.

The aim of the next workshop was to create detailed maps indicating water protection areas, flood protection areas, nature protection areas, and gas pipelines. The theme of the next planning workshop was establishing bioenergy villages in District Wolfenbüttel. Our team prepared input presentations and invited an expert from a bioenergy village in the Göttingen district.

Planning workshop III started with three input information presentations. The first presentation, which our team provided, focussed on the experiences from implementing five bioenergy villages in Göttingen, emphasizing the necessity of political support for the initiation of such a process. The second presentation dealt with the different pathways to successfully establish renewable energy projects and bioenergy villages in Germany. A representative of the bioenergy villages Krebeck and Wollbrandshausen delivered the third presentation. He focussed on the economic opportunities that implementing a bioenergy village would bring to the village and district. Based on his personal experience, he also explained the social and ecological advantages for the village communities.

Maps that our team prepared were then presented; these showed the eight biogas plants in the district, the water protection areas and natural gas grids. Based on these maps, our scientific team suggested that approximately 30 villages and communities could be suitable bioenergy villages. This recommendation was based on the following criteria:

- There is no main natural gas supply; competition with a fairly environmentally friendly energy supply can thus be avoided.
- The distance to the next bioenergy plant is more than 3 km; competition for farmland can thus also be minimized.
- There are ample farmland and grassland in the surrounding areas.

During a planning meeting, the participants discussed supporting villages in District Wolfenbüttel with the district government's financial resources. In this regard, our university team guaranteed its support for a public meeting on the opportunities to establish bioenergy villages in District Wolfenbüttel. Furthermore, the university team organised a best practice tour to two bioenergy villages in Göttingen for District Wolfenbüttel's political and administrative staff; it took place in May 2011.

During the spring of 2011, the district government decided to provide substantial financial support for a bioenergy village competition in District Wolfenbüttel.

11.3.3 District Hannover

District Hannover has 1.1 million inhabitants. It is situated in the vicinity of the city Hannover in southern Lower Saxony. Agriculture plays a significant role in District Hannover. It has approximately 1,800 farms with an agricultural area of 116,000 ha. The farms comprise an average area of 66 ha. In contrast to Districts Goslar and Wolfenbüttel, District Hannover's soils are less fertile. The main cultivated crops are winter wheat, sugar beets, winter rape, winter barley, winter rye and maize. Of the 1,800 existing farms, approximately 500 farms engage in animal husbandry (300 cattle-breeding farms and 200 pig farms). A climate protection agency promotes bioenergy and other renewable energies in District Hannover. This agency's quest is a CO_2 reduction of 40 % by 2020 by increasing energy efficiency, through energy savings and the enforced development of renewable resources such as wind, biomass, solar and geothermal energy. Fifteen communities in District Hannover have already signed a climate protection programme. A position paper for biomass use has been drawn up; it points out that the district's biomass potential should be used for sustainable energy production, with simultaneous consideration of the landscape's ecological balance. By 2010, 14 biogas plants had been implemented in the district, with 17 further plants planned. Once all the projects are realised, there will be an electrical capacity of 16 MW. Besides manure, which can be digested, approximately 127,000 t of biomass (dry matter) are necessary to run the digestion plants. crops from arable land have to be cultivated to feed the biogas plants. Nearly 10 % of the arable land has to be used for this purpose. The following describes exemplary projects in the district.

The City of Uetze: The authorities are very ambitious and innovative in promoting and implementing renewable energy in their community. They plan to establish their own public energy services. Two biogas plants have been already implemented. Both energy plants have very good heat energy usage concepts. Communal properties, private houses, and industrial buildings are connected to a hot water pipeline carrying heat energy from the combined heat and power station situated close to the biogas plant. In one project, farmers financed the biogas plant and hot water pipeline were privately; in the other, a farmer financed the biogas plant and the community the hot water pipeline. The community supports sustainable energy crop cultivation with diverse crops. Besides biomass energy, a wind park, which an external investor financed, is being implemented in the community and a further park, which the Uetze community will finance, is planned for the near future.

The Community of Lenthe: Energy production from biomass started in this community with a biogas plant that a farmer financed and a hot water pipeline that the community financed. The farmer delivers the electricity to the public grid and supplies the heat energy via a hot water pipeline to Lenthe's households. The plant and the outside facilities are very well designed and are integrated into the landscape. The farmer has upgraded the biogas plant to produce more bioenergy and to supply more households with heat energy.

The city of Burgdorf: In this community, four farmers plan to invest in a large biogas plant with an electricity capacity of 1.5 MW. The intention is to build a biogas treatment facility that separates the methane from other gases. This nearly pure methane can then be fed into the natural gas grid, which is operated by an international energy supplier. However, a citizens action committee was formed in the city to oppose the farmers' plan. The committee is opposed to too much maize cultivation, as well as the bad odours, high transportation frequency and the loss of life quality in their residential area that could result from the planned plant. The committee members call the plant the "gas factory" to highlight its industrial nature and indicate their rejection of such a plant in the rural landscape. The city administration, which is also the licensing authority, plans to arrange a public hearing with the biogas plant applicants, the action committee and other people from the community. After this hearing, the authority has to decide about the licensing application. If the application is formally correct and adheres to all regulations, the local authorities have to agree to the plant being built, even if some citizens oppose it. A compromise might be the best way forward. One possibility would for the farmers to transparently communicate their energy crop cultivation plans and to operate the plant accordingly. If they can convince people that they do not only cultivate maize as an energy crop, can avoid bad odours through high technical standards and have transportation that does not impact negatively on the residential areas, they could build the facility with the locals' approval. If not, the farmers and their families will lose their standing in the community, even if they do obtain the license to build the facility.

Fig. 11.3 Planning workshop with farmers and administration staff in district Hannover

In the first planning workshop in district Hannover (see Fig. 11.3), the main focus was on changes to and problems in the energy crop cultivation. The scientific team members provided presentations on relevant topics such as (1) data on the energy crop cultivation status quo in district Hannover, (2) the biomass potential in district Hannover, (3) sustainable energy crop rotations and possibilities to diversify the cultivation with intercropping, mixed cropping and undersown crops and (4) information on the nature protection aspects of energy crop cultivation.

After the presentations and discussion, two working groups were established to work on various issues.

Subgroup 1 worked on the topic 'Energy crop cultivation/crop rotation – the chances and risks of a diverse crop rotation design with different energy crops and its implementation chances in practice'. The farmers in this group pointed out that, in practice, there is already sufficient crop rotation with energy crops. On sandy soils, this rotation comprises winter rye as the energy crop. On more fertile soils, triticale yields are better than that of rye. Sorghum is cultivated as a second crop after winter rye. The following three crop rotations belong the a typical crop rotation cycles:

- winter rye – (1a) fieldgrass – (2) maize
- winter rye – (1a) sorghum – (2) maize with undersown fieldgrass – (3) maize
- winter rye – (2) maize – (3) maize (organic fertiliser for the humus balance).

The farmers note that, in maize cultivation, far fewer pesticide treatments are necessary than in food crops such as winter wheat, rape and sugar beet.

The following problems were outlined: The farmers complain that people demonise maize cultivation and always associate humus degradation and soil erosion with it. Maize plants' height is also a problem for some citizens, because the plants block peoples' view of the landscape. Farmers argue that maize concentration in the district is very low compared to that of other districts and regions. They note that even if they try to involve their neighbours and inform them about their projects, people are often simply opposed to bioenergy. Biogas farmers also point out that farmers without a biogas plant are envious and fear the higher rentals that energy crop cultivation can trigger.

Members of nature protection organisations feel under-informed about farmers' crop rotation and the crops that farmers want to cultivate for their biogas plant. More communication between the different actors is therefore called for. In our workshop's working group, farmers and nature protection organisation members met for the first time and gave their impressions of and opinions on bioenergy projects. They agreed to regularly talk *with* one another rather than *about* each other.

The following ideas were proposed during discussion to solve problems:

- More public relations work, such as reports on positive bioenergy examples in the media, and suggestions for alternatives to maize cultivation.
- The building of hot water pipelines in the neighbourhood of biogas plants to supply houses with inexpensive heat energy.
- Farmers should organise their crop rotation cycles systematically, thus avoiding a high maize concentration in the district.

Subgroup 2 worked on the topic 'Acceptance of bioenergy projects: How to optimise nature protection and licensing regulations'. Here, the following problems and barriers were mentioned:

- The other actors' (administration, farmers, citizens, members of nature protection organisations) perspectives are often not well known and therefore not considered.
- Administration authorities often do not know the details of bioenergy projects; some have never even visited a bioenergy plant.
- People notice visual disturbances triggered by bioenergy projects.

The following ideas were formulated as possible solutions:

- Establishing networks between farmers and the administration to improve communication and sensitivity.
- Introducing informal networking: a regulars' table, "fireplace meetings" and conferences.
- Establishing a coordination position in the administration to arrange communication between the groups/actors.
- Establishing steering and project groups that the district government manages and finances to enable the transfer of positive experiences of and acceptance by bioenergy projects in other districts.
- Inviting people to information days.
- Providing administration members with opportunities to visit biogas plants.

- The subsidies for maize cultivation via the Renewable Energy Sources Act (EEG) should be reduced, but increased for other crops.
- The local and regional government should set a maximum percentage limit for maize cultivation in the district.
- The local and regional administration should compensate activities such as flower strips between energy crops financially.
- The media should be provided with information on sustainable and accepted projects (e.g., a participative planning process that includes a biogas plant's neighbours in the plant location, biomass transportation, heat supply offers to the plant's neighbours, etc.)
- The communication of advantages in press, for example, showing sustainable, well-designed and aesthetically pleasing crop rotation, thereby reducing people's prejudices against energy crops; communicating the district's the low maize concentration.

Topics for the next planning workshop were determined in the workshop: an in depth analysis of the status quo of the current biogas plants in district Hannover, drawing up a list with plants' most important characteristics (the electricity capacity, input materials and crops, the radius of the crops cultivated for the plant, and the heat energy utilisation) to show the differences between plants. This will allow one to recommend a specific plant to administration visitors, politicians and citizens.

Another goal is to discuss the biogas agency's position paper on bioenergy. Are this paper's theses and demands a basis for sustainable bioenergy projects? Could the paper form the basis of better agreements on district Hannover's bioenergy scenarios?

11.4 Knowledge and Experience Transfers Between Districts

Whenever applicable, we created opportunities for the actors in our model districts to gain direct contact with experienced persons from other German villages that have successfully implemented sustainable renewable energy projects, including bioenergy projects. An important advantage of such direct contacts is that experienced people – often farmers or local politicians – speak the same "language" as the participant actors.

One way to enable such contacts is to invite these persons to workshops. For example, we invited the Feldheim project's coordinator. In Feldheim, 100 % of the heat and electricity demand is produced through a combination of different renewable energy plants. This achievement inspired our workshop attendants to start similar lighthouse project in their district. In many of our workshops, we organised similar presentations, which strengthen the audience's trust that the intended projects could be realised. Best practice trips are another efficient way to enable the direct transfer of experience. We organised one such trip for farmers, politicians and administration staff from districts Goslar and Wolfenbüttel. They were invited to three bioenergy villages in our home district of Göttingen, where the mayors of the bioenergy villages Barlissen, Krebeck and Wollbrandshausen guided the visitor groups through the plants and villages, conveying historical and other project details. The people from our model districts found these talks very convincing. Some of the participants told us that the decision by district Y's government to start a bioenergy village support fund was partly the result of this. However, such anecdotal evidence should be proven by means of social scientific analyses, on which we focus next.

11.5 Research: Evaluation of Support Activities

The research activities of the Göttingen approach are designed to take place during (1) the start of a practical sustainability project in order to ensure that the first decisions have the best scientific basis possible; during (2) the conversion process, they continuously check the progress towards the goals and to look for unexpected side-effects; and finally, (3) after the conversion has been completed. At the end of the on-going 3-year project phase, during the social science activities, structured interviews are planned with active stakeholders in the three model districts participating in the workshops. These interviews will mainly focus on the question of how efficiently the principles and parts of the procedures we applied in the workshops as well as other activities in the districts contributed to the participatory activities. We will also survey the importance of personal contacts, the preference for email rather than paper communication, the balance between strategy planning and detailed planning during the workshops, the impact that the presentations offered by the scientists, practitioners and external experts from other bioenergy projects had, the adequacy of the workshop group size and diversity and the efficiency of the plenum meetings compared to small group activities in workshops. The results of these interviews will be the subject of a later publication.

11.6 Conclusion

In this project, beside assuming their traditional scientific role as objective analysers, the authors also act as the initiators of the conversion processes to sustainable development, basing their activities on the application of their scientific research results. This dual role within sustainability research seems to be effective for scientists to contribute to the challenges emerging from the indispensable transformation to renewable energy systems.

The first results of our work show that scientists from different disciplines can cooperate with different actors and groups in the chosen districts to promote these villages and districts' conversion to renewable energy sources. Our work demonstrates that action research, a tool with a long tradition in social science (Lewin 1946), can also be embedded in the broader framework of sustainability science, thus contributing to sustainable and accepted bioenergy projects – one of the key aspects of sustainable development.

References

Brydon-Miller, M., Greenwood, D., & Maguire, P. (2003). Why action research? *Action Research, 1*, 9–28.

Kasemir, B., Jäger, J., Jaeger, C., & Gardner, M. (2003). *Public participation in sustainability science: A handbook*. Cambridge: Cambridge University Press.

Lenton, T. M., Held, H., Kriegler, E., Hall, J. W., Lucht, W., Rahmstorf, S., & Schellnhuber, H. J. (2008). Tipping elements in the Earth's climate system. *Proceedings of the National Academy of Sciences (PNAS), 105*, 1786–1793.

Lewin, K. (1946). Action research and minority problems. *Journal of Social Issues, 2*, 34–46.

Mills, G. (2000). *Action research: A guide for the teacher researcher*. Upper Saddle River: Prentice Hall.

Ruppert, H., Eigner-Thiel, S., Girschner, W., Karpenstein-Machan, M., Roland, F., Ruwisch, V., Sauer, B., & Schmuck, P. (2008). *Wege zum Bioenergiedorf. Leitfaden*. Gülzow: Fachagentur für Nachwachsende Rohstoffe.

Schmuck, P., & Schultz, W. (2002). Sustainable development as a challenge for psychology. In P. Schmuck & W. Schultz (Eds.), *Psychology of sustainable development* (pp. 3–19). Boston: Kluwer Academic.

Schmuck, P., Wueste, A., & Karpenstein-Machan, M. (2012). Initiating and analyzing renewable energy transitions in Germany. In S. Stremke & A. Dobbelsteen (Eds.), *Sustainable energy landscapes: Designing, planning, and development* (pp. 335–354). Boca Raton: Taylor & Francis.

Sheldon, K., Schmuck, P., & Kasser, T. (2000). Is value-free science possible? *American Psychologist, 55*, 1152–1153.

Whyte, W. (Ed.). (1991). *Participatory action research*. Thousand Oaks: Sage Publications.

Chapter 12
Assessment of Different Bioenergy Concepts in Terms of Sustainable Development

Swantje Eigner-Thiel, Meike Schmehl, Jens Ibendorf, and Jutta Geldermann

Abstract This chapter focuses on the assessment of different concepts' sustainability regarding the energetic use of biomass in rural areas. The aim is to provide decision support, while taking environmental, economic, social, and technical perspectives into consideration. Possible (technical and organisational) concepts include biogas plants operated by electric service providers, a single biogas plant owned by a farmer, or bioenergy villages owned by a village cooperative. We describe the development of suitable ecologic, economic, social and technical criteria to assess the sustainability of different concepts and the adaption of existing indicator systems to the special requirements of sustainable biomass use for energy. The results of this sustainability assessment illustrate the different biomass concepts' advantages and disadvantages, which are compared by means of multi-criteria decision analysis methods. This decision support tool faciliates the decision process for mayors, district administrators, farmers and investors, who have to choose the most sustainable concept for a certain area. Furthermore, the sustainability assessment of bioenergy concepts has specific requirements with regard to their visualisation if such an assessment is to support the decisions of interested stakeholders in communities.

S. Eigner-Thiel (✉)
Interdisciplinary Center for Sustainable Development, University of Göttingen, Goldschmidtstr. 1, Göttingen D-37077, Germany

HAWK – University of Applied Sciences and Arts Göttingen, Büsgenweg 1a, D-37077 Göttingen, Germany
e-mail: eigner-thiel@hawk-hhg.de

M. Schmehl • J. Geldermann
Department of Production and Logistics, University of Göttingen, Platz der Göttinger Sieben, 3, Göttingen D-37073, Germany
e-mail: Meike.Schmehl@wiwi.uni-goettingen.de; geldermann@wiwi.uni-goettingen.de

J. Ibendorf
Interdisciplinary Center for Sustainable Development, University of Göttingen, Goldschmidtstr. 1, Göttingen D-37077, Germany
e-mail: jibendo@gwdg.de

H. Ruppert et al. (eds.), *Sustainable Bioenergy Production - An Integrated Approach*, DOI 10.1007/978-94-007-6642-6_12, © Springer Science+Business Media Dordrecht 2013

Keywords Bioenergy concepts • Multi-criteria decision analysis (MCDA) • Social criteria • Sustainable development • Visualisation

12.1 Introduction

The use of biomass to produce energy is gaining increasing attention from policy-makers, energy supply companies and the public (Edenhofer et al. 2011; BMU 2009; Leitl 2007). There are several reasons for this: Owing to bioenergy's potentially lower carbon dioxide emissions, it is expected to contribute less to climate change than fossil energy resource use. Furthermore, using biomass to produce energy could preserve fossil energy reserves. In addition, bioenergy is the only renewable energy source that addresses all three energy sectors: the supply of electricity, heating/cooling and fuels for transportation (Vis et al. 2008). Fourthly, it can support rural development by giving farmers an alternative source of income besides food production, and – in the case of bioenergy villages – by involving villagers in direct democracy (i.e. via their participation in the decision processes in their village or district) and giving them a satisfying sense of community (Eigner-Thiel 2005). Finally, by using local biomass to produce energy, the domestic energy supply should be stabilised, thus reducing dependency on other – potentially political unstable – countries for the import of energy resources (oil, uranium, natural gas, etc.) (IEA 2004; Van Loo and Koppejan 2008).

Nevertheless, discussions on the sustainable development of biomass use to produce energy, do not only mention the positive effects. There are also concerns that the use of monocultures will increase due to a higher demand for energy crops, which could result in massive land use changes to accommodate high-productive crops such as maize.

Currently, the use of maize to produce energy has resulted in heated discussions. In addition, this could increase transport activities in rural areas, which would worsen the air pollution and lead to unwanted disturbances. Another critical point is energy plants' direct emissions, such as particulate matter and sulphur dioxides, which could be hazardous to human health. The designation of areas for energy crop production is also highly controversial. The ethical aspects of converting food production, nature conservation or grassland areas for the production of energy crops will lead to criticism, as will the environmental effects of direct and indirect land use changes (e.g., more carbon dioxide emissions due to the ploughing of grassland and a reduction in the biodiversity) (Jessel 2008; Fritsche et al. 2009).

In the meantime, several concepts for biomass use for energy, such as individually or collaboratively organised biogas plants or large-scale plants, have been realised or planned in Germany. These types of concepts are the main focus of the analysis in this chapter. However, economic, ecological and social aspects should be considered when following sustainable development principles (for our definition of sustainable development, see Sect. 12.3). Therefore, the decision process concerning the type of bioenergy plant and its dimensions has become increasingly complicated. Multi-criteria decision models may need to be applied to arrive at optimal agreements (Buchholz et al. 2009; Oberschmidt et al. 2010). The crucial management of considerable amounts of diverse data is linked to the decision model. The coordination of these data and their processing to arrive at different visualised results constitute a challenge because the data originate from different scientific fields (biology, physics, chemistry, geology, meteorology, psychology, social and economic sciences) that are not only extensive, but also time and space dependent (see Rautenstrauch 1999; Page and Rautenstrauch 2001). Thus, decision support methods should collect data from heterogeneous sources and condense them into different formats.

Many bioenergy supply concepts have been developed at a local scale in Germany over the past years. One of these is the idea of an energy self-sufficient village, which a group of scientists at the Interdisciplinary Centre for Sustainable Development (IZNE) at Göttingen University developed in 1998 (see Projektgruppe Bioenergie 2010, Chap. 2 in this book and in Box 12.1).

Box 12.1 Bioenergy Village Jühnde, Lower Saxony, Germany

In Germany, the bioenergy village concept has been around since 2000. The main aim of a village self-sufficient in energy is to have such a village produce at least as much electric energy as the residents and local industry need. The heat production should cover at least two-thirds of the village's demand. Another requirement is that the heat customers and the farmers providing the biomass should actively help plan the conversion of the village energy supply. With this idea in mind, the relevant scientists chose a suitable village in the Göttingen district as a pilot project from 17 other appropriate and interested villages. Thus, in October 2001, Jühnde was chosen as the model village. Jühnde has 780 inhabitants, nine farmers, an agricultural area of 1,300 ha and a forest area of 800 ha (Ruppert et al. 2008). Its advantage was that it was of a suitable size, which was of economic importance as the village was large enough to build a bioenergy plant that would be profitable. It was also socially important, since it was still small enough to ensure that everyone in the community could be kept informed. It was also a suitable area for biomass production as the village farmers were willing to use their land for biomass production. Moreover, it had a strong village community with many active associations, which spread the idea throughout the village and motivated enough households to participate in the project (Eigner-Thiel 2005).

After diverse planning stages (informing the inhabitants, closing contracts with the farmers and energy consumers, obtaining building and operating licences, etc.), construction began in 2001 and was completed in 2004. The technical concept's central component is the biogas plant in which microorganisms turn liquid manure and other wet biomass into biogas by

(continued)

Box 12.1 (continued)
means of wet fermentation. In the combined heat and power plant (CHP), the biogas is turned into electricity and heat. Electricity is fed into the public grid, and the heat is used to warm water, which the district's heating network pipes to the connected households. To cover the high demand for heat in winter, the plant is supplemented by a woodchip heating plant and an oil heating plant as contingency reserves.

Heat distribution in Jühnde began in September 2005. Today, 72 % of the households receive about 2,800 MWh of heat per year from the biogas plant. The remainder (approximately 1,500 MWh of heat per year) is supplied by the woodchip heating plant. The production of electricity is about 4,000 MWh per year, which the local energy supply company purchases. The Renewable Energy Sources Act (EEG) regulates the price of electricity.

Using biomass as a substitute for oil has had various impacts on the individuals, society, economy and ecology in Jühnde. The ecological benefit can be quantified as a 70 % per person reduction in the carbon dioxide emissions. After a financial deficit in 2005 when the plants were installed and started up, the operating company recorded a positive annual surplus. Since then, the heat customers (households) have saved approximately €800 per year. Psychological research has shown that those who were actively engaged in the planning process experienced the village community more profoundly as well as increased individual learning. Different methods of public relations, participatory planning and planning workshops were also realised and documented (Eigner-Thiel 2005, 2010). On the whole, Jühnde's inhabitants are very satisfied with the heat supply and the bioenergy village concept (Eigner-Thiel and Schmuck 2010; Ruppert et al. 2008; Ahl et al. 2007).

This example of a bioenergy village clarifies that planning a bioenergy village requires data from different sources. First, technical data on the bioenergy plant's output are required. In addition, each household's heat requirement has to be identified. Geographical data are also needed to identify suitable crops for the areas and to calculate the potential yields. Furthermore, economic data are required to calculate the investment and operating costs. For the realisation of a bioenergy village, social factors, such as people's motivation to engage in the planning process and the quality of the social networks are crucial, because as many people as possible should be involved and kept informed during all the planning stages. Without people's willingness, such a project cannot be implemented.

However, during the decision process for the optimal bioenergy concept, ecological, economic and social objectives were sometimes at loggerhead: One of the social aims was, for instance, to connect as many houses as possible to the bioenergy plant. From an economic perspective, this was not, however, always the best solution; for example, certain houses were too isolated from the others and too far away from the biogas plant (Eigner-Thiel and Geldermann 2009). These conflicts are indicative of all the other biomass paths: Each concept has its advantages and disadvantages. This is typical of complex decision situations when common sense becomes overburdened if a large number of criteria has to be considered. Here, multi-criteria decision analysis (MCDA) methods offer a way to structure the decision problem; therefore, this chapter will also show how this problem can be solved. Section 12.2 describes how MCDA works.

Why is MCDA needed for the choice between different biomass alternatives and why should these be assessed in terms of multiple sustainability dimensions?

Potential initiators of bioenergy projects draw on the limited experiences with existing bioenergy concepts. Diverse life-cycle assessment (LCA) studies were compiled after a methodological analysis of environmentally relevant material and energy flows for the use of biomass use for energy. However, natural scientists have criticised the assessment of the impacts, as the interrelationships are usually too complex to be modelled along linear impact factors. In addition, these studies do not show the impact on the affected local stakeholders, as economic and social perspectives as well as local aspects are usually underrepresented (see Hofstetter 1998; Kempener et al. 2009).

General sustainability criteria can initially be used to comprehensively assess the different bioenergy concepts pertaining to the economic, ecological and social aspects. However, their actual application may lead to very different and even conflicting results. Chapter 40 of Agenda 21 (BMU 1992) states that the usual indicators such as a country's gross national product or the unemployment rate do not sufficiently describe the status of sustainability development. Therefore, beyond the existing economic characteristics, further indicators should be developed to represent the three dimensions of sustainable development (economic growth, ecological protection and social equality) (see Box 12.3) as accurately as possible. There are, however, many indicator systems for assessing sustainable development (see, e.g., Breitschuh et al. 2008; Gamba 2008; Hoffmann 2007; Rösch et al. 2009; WBGU 2008).

Nevertheless, there are significant difficulties with defining such indicator systems, specifically if bioenergy's sustainable use needs to be assessed, as the possible indicators might not be precise, specific or comprehensive enough to reflect the local and regional developments' sustainability (see Heiland et al. 2003; Fleury 2005). After preliminary theoretical considerations of the definition and the formulation of sustainability criteria, actual significant and quantifiable criteria should be chosen for the specific area of biomass use for energy. Currently, there is no general system for specifying the indicators, due to the specificity and the complexity of the issues. Thus, besides an orientation towards the principles of sustainability (see Agenda 21, BMU 1992), it is crucial to describe the indicator system requirements transparently (e.g. see Reul 2002; Werheit 1996). Section 12.4 provides a description of our actual development of a criteria system.

First, sustainable development's increasing requirements – owing to the increasing awareness of the climate change impacts and the need for a reliable future energy supply system – mandate consulting interdisciplinary expert groups, who

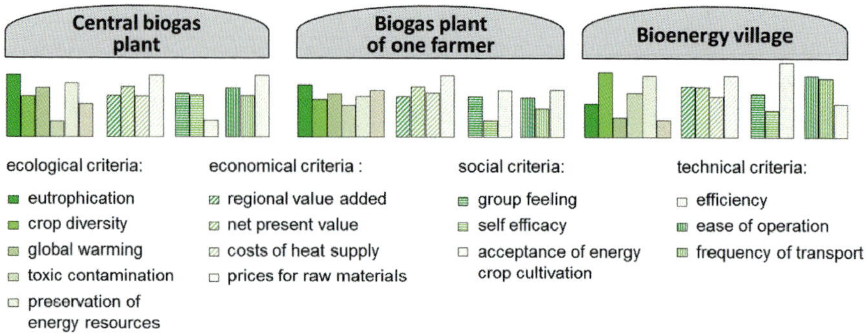

Fig. 12.1 Schematic presentation of conflicting targets of different bioenergy concepts

will take the numerous, and partly conflicting, objectives of and criteria for bioenergy assessment into consideration. General objectives, such as sustainability, economic viability and technical feasibility, have to be broken down into operational criteria that can be measured and are decision-relevant, i.e. that will allow one to distinguish between alternatives. In most decision situations, no dominating alternative meets all of the objective sustainable development criteria to allow it to be unanimously chosen from all the other alternatives. In fact, most of the alternatives have their strengths and weaknesses, but these are measured in different units; some kind of trade-off is therefore required. Figure 12.1 outlines a typical decision situation with regard to a central biogas plant, a farmer's biogas plant and a bioenergy village as examples of bioenergy concepts and the focus of this chapter; the bars' different heights show the various attributes' values within each concept's catalogue of criteria. The alternatives are described in some detail in Sect. 12.5.

Many aspects must be considered for a comprehensive assessment of the different biomass paths. Their varying priorities also need to be weighted, since not all of them are equally relevant.

First, absolute judgement span and immediate memory span impose severe limitations on the amount of information we are able to receive, process and remember (Miller 1956). Many aspects are considered during the process of balancing and condensing information,, which can quickly lead to a situation in which common sense no longer suffices (Dörner 2003; Vester 2003). The larger the number of people involved in a decision process in complicated situations, the more support is needed to objectively and efficiently arrive at decisions. Decision models with several objectives often describe reality better than models with only one objective. This has led to the development of numerous new approaches to multi-criteria decision support over the past 30 years (Figueira et al. 2005; Hwang and Yoon 1981; Yue and Li 1998; Munda 1995; Oberschmidt et al. 2009). In the theory of decision support and multi-criteria analysis, weighting (see Sect. 12.6) is one of the most disputed steps due to its relatively subjective character. However, decision trees and objective hierarchies can be used to operationalise ecological, economic, social and technical criteria and represent them in terms of certain attributes (e.g., their global warming potential).

12.2 Decision Support for Sustainable Biomass Use for Energy with Multi-criteria Decision Analysis (MCDA)

The MCDA framework is applied in research projects in this case to compare various concepts of bioenergy villages and alternative bioenergy supply solutions. This approach seeks to establish a decision support tool that increases the transparency of the decision process and lessens decision problems. However, with a minimum of technical and energetic effort, this tool should help those deciders who have to determine the most sustainable biomass concept for a certain area, as well as mayors, district administrators, farmers and bioenergy investors. We therefore outline the MCDA structure and describe the criteria development process. In addition, a comprehensive list of criteria is introduced and considered in Sect. 12.4.

The complex group decision process for sustainable biomass use for energy in a specific rural area can be simulated in an MCDA model.

The aim of MCDA is the ex ante assessment of a few individual options by explicitly considering a decision-maker's subjective preferences with regard to decision support and planning (monitoring and control vs. planning and choice) (Belton and Stewart 2002).

MCDA has been widely applied in an environmental context but hardly in bioenergy contexts. In these contexts, Mustajoki et al. (2003) describe the usage of this method in lake regulation policy, Malczewski (1999) establishes a link between spatial approaches (GIS) and MCDA, while Buchholz et al. (2009) provide a comprehensive overview of MCDA's application in the context of bioenergy.

The MCDA process can be divided into six steps, which might be somewhat iterative and interdependent due to the growing insight into the underlying decision problem:

1. Define and specify the overall objective in some detail in the criterion hierarchy.
2. Compile alternatives that can meet the defined objective.
3. Model and process information – investigate and calculate the values of the attributes (lowest-level criteria) (see Fig. 12.2 below) for the alternatives.
4. Assign a relevant weight, i.e. depending on each attribute's relevance, assign weights to certain values.
5. Calculate the results with operations research methods (MCDA algorithms).
6. Make the results visible with graphs and charts to assess the alternatives, then choose one.

Accordingly, the formulation of the overall decision objective is the starting point of decision support. In most cases, the overall objective is very general ("sustainable biomass use for energy") and needs to be broken down into operational attributes. A criterion hierarchy (see Fig. 12.2) shows the top-down approach. It starts with the overall objective (sustainable biomass use for energy) and expands this by adding more detailed targets, which should cover all the ecological, economic, social and technical aspects adequately without creating redundancy. Below the targets, there are attributes; these can operationalise the objective on an ordinal or cardinal scale (Belton and Stewart 2002).

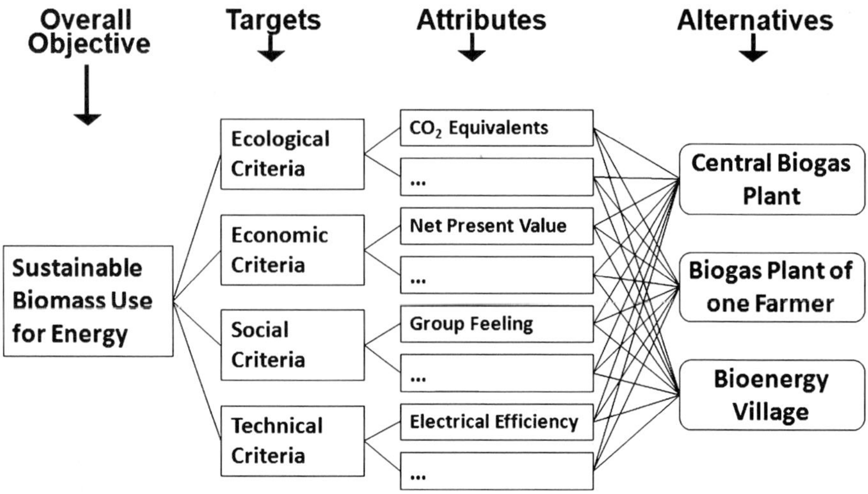

Fig. 12.2 Criterion hierarchy of sustainable biomass use for energy

In our case, the different concepts for biomass use for energy will be assessed according to the primary objective "sustainable development".

The MCDA process and the compilation of the criteria hierarchy are iterative actions; therefore, the criteria development procedure presented in this chapter is only a preliminary one. Nevertheless, the information development procedure is already applicable and we therefore demonstrate this procedure here.

12.3 Definition of Sustainable Development

A suitable definition is required to assess the different bioenergy alternatives' effects on sustainable development. There are many different definitions of sustainable development. The most popular is the one by the Brundtland Commission (United Nations 1987), the World Commission on Environment and Development (WCED): "Sustainable development is development that meets the needs of the present without compromising the ability of future generations to meet their own needs" (p. 43).

The definitions that UNCED (1992) and the Helsinki Conference (1993) used highlighted three issues as the cornerstones of sustainability. According to these definitions, for development to be sustainable, it must be economically profitable, biologically proper and socially acceptable.

To assess a sustainable development issue, the concept must be broken down into single indicators or criteria. Some definitions only refer to ecological aspects (e.g., the indicator system SCOPE (1995)). The most comprehensive indicator system, which also considers social and economic aspects, is the UN Commission's sustainable development (CSD) indicator system. With its approximate 130 indicators, it is also a good foundation for the comparison of bioenergy concepts – but nothing more, as not all of its indicators are useful for our aim. For example, one of the indicators is that child labour should be avoided, which is obviously not applicable in Germany's biomass sector.

The EU set a minimum sustainable standard for biofuels with the Renewable Energy Directive (European Parliament 2009). This focuses on ecological aspects, seeks to reduce greenhouse gas emissions and tries to ensure that the biodiversity will be maintained. Social criteria are only mentioned in relation to the production of biofuels in developing countries and not with regard to biomass use in industrial countries like Germany. The criteria mentioned reflect the world's different agricultural situations, which are not totally applicable to a comparison of the different biomass concepts in Germany. However, the Directive has already been converted into legislation in different European countries, including Germany (BioSt-NachV 2009). The relevant criteria therefore need to be defined in greater detail.

In the scientific literature, a distinction is made between strong and weak sustainability: Weak sustainability means that a single dimension's value can be substituted by another (e.g., high economic values can substitute low ecological ones), whereas strong sustainability means that no substitution is possible between dimensions (Ott 2003; Wuppertal-Institut für Klima, Umwelt, Energie 1997; Daly 1999).

In our approach, we refer to the Brundtland Commission's (1987) definition of sustainability and to the sustainability concept comprising the three aspects economic growth, environmental protection and social equality (UNCED 1992; see Box 12.2) because they form a good basis for breaking down the concept into different categories. The technical dimension was added to the ecological, social and economic dimensions to enhance criteria development and assessment transparency, because the technical conversion of biomass is one of the distinct criteria for assessing different bioenergy concepts. In addition, the strong sustainability definition was chosen for the comparison of bioenergy concepts due to its above-mentioned benefits.

We break the sustainable development concept down into different criteria, which are addressed in the following section. We fill first apply the criteria to the assessment of three concrete alternatives described in some detail in Sect. 12.5.

Box 12.2 The Brundtland Commission (1987)

The increasing deterioration of the human environment and natural resources led the former UN Secretary General to appoint Gro Harlem Brundtland as chairman of The Brundtland Commission (formerly the World Commission on Environment and Development (WCED)) in 1983. The purpose of the The Brundtland Commission was to rally countries to pursue sustainable development together. Gro Harlem Brundtland, a former Prime Minister of Norway, was chosen to head the Commission due to her strong background

(continued)

Box 12.2 (continued)
in the sciences and public health. After releasing the Brundtland Report in October 1987, the Brundtland Commission was officially dissolved in December 1987. The organisation Centre for Our Common Future was founded in April 1988 to replace the Commission.

The Centre for Our Common Future seeks to create a united international community with shared sustainability goals by identifying global sustainability problems, raising awareness about them, and suggesting the implementation of solutions. Its report – *Our Common Future* – strongly influenced the Earth Summit in Rio de Janeiro, Brazil in 1992 and the third UN Conference on Environment and Development in Johannesburg, South Africa in 2002. It is also credited with creating the most prevalent definition of sustainability: "Sustainable development is development that meets the needs of the present without compromising the ability of future generations to meet their own needs" (p. 43). The three main pillars of sustainable development are economic growth, environmental protection and social equality.

12.4 Compiling a Criterion List

Although our introduction refers to indicator systems (e.g., see Breitschuh et al. 2008; Gamba 2008; Hoffmann 2007; Rösch et al. 2009), the present approach uses the term 'criteria', because the aim is not to compare the same aspect over time (which is usually the aim of an indicator system), but to compare different kinds of biomass paths at a single point in time, which is a fairly static approach.

From the different disciplines' perspectives, the bioenergy concept's evaluation criteria were collected from experiences gained with local bioenergy projects (mainly the bioenergy village Jühnde project and other village projects in the Göttingen district in lower Saxony; Projektgruppe Bioenergiedörfer 2010; Ruppert et al. 2008), from the literature and from discussions with experts (project-internal and project-external experts).

Moderated discussions were organised to collect criteria from different expert groups:

- ecology experts: geographers, earth scientists, agronomists, soil scientists, forestry scientists and plant scientists;
- economy experts: business administration, agricultural economists and industrial engineers;
- social aspects experts: psychologists and sociologists;
- technical experts: practitioners and scientists with bioenergy experience.

The greatest challenge of collecting criteria was to make everybody in the discussion groups understand that a criterion is only decision-relevant if its parameter value differ from comparative alternatives. The comparison of various bioenergy concepts would focus on special preconditions for agricultural land, because each region has special underlying basic geographical and socio-cultural conditions. These can include:

- temperature
- precipitation
- soil type
- social traditions
- price of the land

Since these preconditions cannot be changed in one location and cannot distinguish between the different concepts, these criteria are not decision-relevant. However, these preconditions influence the criteria specifications.

The implemented workshops to collect and weight the criteria (see Sect. 12.6) allowed for an iterative definition of the relevant decision criteria. If one of the scientific disciplines found a more significant criterion, or an easier-to-measure criterion, the criteria hierarchy was adjusted accordingly. The moderation of the groups guaranteed effective structuring, systematic management as well as the decision-making process's transparency. This would ensure that all the experts possessed the same level of information. Nevertheless, during the interdisciplinary discussions, difficulties were experienced with understanding what certain people really meant, even *within* one discipline. After many discussion forums, the relevant data and information were presented as a hierarchy of criteria. This hierarchy's structure and organization form the basis of a systematic and quantitative assessment – a decision table.

The following criterion list was drawn up to assess the sustainability of bioenergy alternatives (see Tables 12.1, 12.2, 12.3, and 12.4). As it is a work in progress, this list has a preliminary status. Decision support is an iterative process, and changes in the criteria hierarchy are thus to be expected when the decision table, with all the criteria scores, is completed. In the following, the ecological, economic, social and technical criteria are introduced and explained.

12.4.1 Ecological Criteria

Nature conservation refers to the protection of the ecosphere from negative impacts by human activities, including the use of biomass for energy. These have effects on the quality of the environmental media, i.e. the air, soil and water, as well as on the non-renewable resources and biodiversity. To quantify these impacts on the environment, methodological impact assessment approaches can be used as part of the life-cycle assessment (Guinee et al. 2002; Geldermann et al. 1999). Suitable impact categories and their characterisations are found in, for example, Schmitz and Paulini (1999) or SETAC (1996). The methodology of life-cycle assessment involves considering the entire product life-cycle. The impact assessment of a bioenergy concept should also include resource extraction, the agricultural production of biomass and its conversion into energy. Several studies have therefore

Table 12.1 Ecological criteria related to biomass concepts

Ecological targets 	Sub-targets	Attributes	Unit	Manifestation of the attribute for sustainable development (Min./Max.)
(1) Air and climate	(1.1) Climate change	Global warming potential	kg CO_2 equivalents	Min.
	(1.2) Toxic contamination	Mass of respirable particulate matter	kg PM_{10}	Min.
		Mass of benzo(a) pyren	kg benzo(a) pyren	Min.
		Mass of inorganic reference substance	kg of inorganic reference substance	Min.
	(1.3) Acidification	Acidification potential	kg SO_2 equivalents	Min.
(2) Water	(2.1) Aquatic eutrophication	Mass of applied fertiliser – nitrogen	kg fertiliser – nitrogen	Min.
		Mass of applied fertiliser – phosphor	kg fertiliser – phosphor	
	(2.2) Toxic contamination	Mass of applied pesticides	kg pesticides	Min.
(3) Soil	(3.1) Erosion	Cultivation method	Points on ordinal scale	Min.
		Land cover level	%	Max.
	(3.2) Terrestrial eutrophication	Mass of applied fertiliser – nitrogen	kg fertiliser – nitrogen; kg fertiliser – phosphor	Min.
		Mass of applied fertiliser – phosphor		Min.
	(3.3) Soil contamination	Accumulation of heavy metal reference substance	kg mobilised reference substance	Min.
		Mobilisation of heavy metal reference substance	kg mobilised reference substance	Max.

(continued)

Table 12.1 (continued)

Ecological targets 	Sub-targets	Attributes	Unit	Manifestation of the attribute for sustainable development (Min./Max.)
(4) Preservation of resources	(4.1) Energy	Scarcity of energy ressources	kg crude oil-equivalents resource	Min.
		Cumulative energy demand	MJ	Min.
	(4.2) Minerals	Demand for phosphate	kg phosphate	Min.
	(4.3) Land area consumption	Demand for space	m^2	Min.
	(4.4) Water consumption	Demand for water	m^3 water	Min.
(5) Protecting biodiversity		Quantity of cultivated crops	Quantity	Max.
		Mass of applied pesticides	kg pesticides	Min.
		Nitrogen fertiliser type	Points on a ordinal scale	Min.

Table 12.2 Economic criteria related to biomass concepts

Economic targets 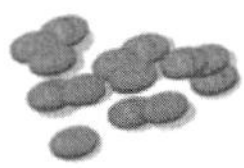	Attributes	Unit	Manifestation of the attribute for sustainable development (Min./Max.)
(1) Operating company's perspective	Net present value	€	Max.
	Supply contract duration	Years (points)	Max.
(2) Employee's perspective	Profit sharing	Yes/no (points)	Max.
	Possibility of additional fee	Yes/no (points)	Max.
(3) Heat clients' perspective	Annual heat supply costs	€ per year	Min.
	Minimum deposit	€	Min.
	Connection and conversion fee	€	Min.
(4) Farmer's perspective	Input in pricing	Points	Max.
	Operational flexibility	Per year one point	Max.
(5) Regional perspective	Regional value added (investment)	% of the investment sum	Max.
	Regional value added (current)	€	Max.
	Tax revenue	€	Max.

Table 12.3 Social criteria related to biomass concepts

Social targets

	Attributes	Unit	Manifestation of the attribute for sustainable development (Min./Max.)
(1) Acceptance	Cultivation concept scenery (aesthetics)	Points	Max.
	Scenery of the technical plants (aesthetics)	Points	Max.
	Smell	Points	Min.
	Noise (factory)	Points	Min.
	Noise (transport)	Points	Min.
(2) Participation	Planning	Points	Max.
	Information	Points	Max.
	Decisions concerning finances	Points	Max.
(3) Psychological effects	Feeling of independence from electricity supplier	Points	Max.
	Feeling of independence from fossil energy	Points	Max.
	Solidarity	Points	Max.
	Self-efficacy	Points	Max.
	Pride, fun, meaning	Points	Max.
	Image of the village	Points	Max.
	Assessment of accidents	Points	Min.
(4) Employment	Additional workplaces	Number	Max.
	Possibility to work part-time	Points	Max.

Table 12.4 Technical criteria related to biomass concepts

Technical targets

	Attributes	Unit	Manifestation of the attribute for sustainable development (Min./Max.)
(1) Plant efficiency	Thermal efficiency factor	%	Max.
	Electrical efficiency factor	%	Max.
	Use of heat in summer	Yes/no (points)	Max.
	Modularity	Points	Max.
(2) Transport	Frequency	Points	Min.
	Point in time	Points	Min.
(3) Administrative effort	Duration of licence	Days	Min.

analysed the environmental effects of bioenergy product chains (Roedl 2010; Fritsche et al. 2009; Kimming et al. 2011; Schmehl et al. 2012).

The list of ecological criteria below were presented by Weber-Blaschke et al. (2002). These criteria describe recent environmental indicator systems and identify various structures according to their environmental media, problem areas, sectors, spatial dimensions and socio-economic indicators.

In this study, the first level of the ecological criteria's hierarchy is structured according to the environmental media and resources. The subordinate criteria level lists the associated problem areas (see Table 12.1), which are quantified by attributes at the bottom level. Firstly, the decision-relevant criteria are briefly explained. Table 12.1 summarises the set of ecological criteria with their units of measurement.

(1) Air and climate

The emission of specific substances into the air contributes to climate change, has toxic effects on humans, animals and plants, as well as acidifying effects on terrestrial and aquatic ecosystems. The impact categories ozone depletion and photochemical ozone creation are not yet considered in the context of this study. The main origins of these environmental issues are chlorofluorocarbons and hydrocarbons, whose emission is estimated as not relevant for the comparison of different bioenergy product chains.

(1.1) Climate change

At present, climate change is one of the most discussed environmental issues. Although potential individual impacts on humans and ecosystems have been analysed in different studies (Hughes 2003; Thuiller et al. 2011), the entire extent of the future effects have not yet been estimated.

Global warming potential is one commonly accepted indicator with which to quantify emitted greenhouse gases' contribution to climate change. Recent indicator values for greenhouse gases with a time horizon of 100 years are listed in the IPCC (2007), which considers the emissions of all greenhouse gases from the bioenergy chain's life-cycle. The target is to minimise the global warming potential.

(1.2) Toxic contamination of the air

The burning of solid biomass – which is, for example, part of the bioenergy village concept – causes emissions with potentially toxic effects (Ferge et al. 2005); therefore, the conversion process is used to assess the degree of air contamination. Since our study cannot carry out a detailed exposition analysis, the assessment is restricted to the following three attributes (see also Chap. 14):

(a) Particulate matter

Particulate matter has a dusty and gaseous nature and is thus a potential risk for living organisms' health. An increased concentration of particulate matter can lead to respiratory and cardiovascular diseases in humans and to a high mortality rate. The sources of particulate matter are: industrial processes, road transport and the burning of biomass. PM_{10} (particles with an aerodynamic diameter smaller than 10 μm), which characterises respirable particulate matter, has commonly been used as an indicator of particulate matter in the air (Hewitt and Jackson 2003; World Health Organization 2006). Therefore, PM_{10} has been chosen as a suitable attribute of particulate matter within the bioenergy process chain.

(b) Organic pollutions

The working group of the German Research Center for Environmental Health analyses the hazardous organic substances emitted in the biomass burning process and has identified Benzo(a)ypyren as a reference parameter that might be cancer-causing (Lenz 2010) (see also Chap. 13). This reference substance is also taken as the attribute of organic pollutions in the MCDA.

(c) Inorganic pollutions

The inorganic pollutions due to the biomass burning process are analysed by colleagues in our research project (see Chap. 13). The group of inorganic pollutants includes heavy metals such as cadmium, arsenic and lead, which might also be carcinogenic (Lenz 2010). A reference substance has not yet been determined. A suitable attribute for assessing inorganic pollutions will therefore be added in the course of the project.

(1.3) Acidification

Several air pollutions such as sulphur dioxide, hydrogen sulphide, ammonia and other sulphur and nitrogen compounds react in an oxygen environment with acid and contribute, among others, to damage forests and lakes (Nixon et al. 2000). Although the environmental impacts refer to water and land, acidification criteria are not listed for these environmental media but for air as the origin of emission. Acidification is a commonly used impact category in life-cycle assessment. In line with de Haes (1996) acidification potential has therefore been chosen as an attribute.

(2) Water

Nixon et al. (2000) emphasise four significant water quality issues: eutrophication, persistent organic pollution of rivers, acidification (see above) as well as nitrate and pesticide contamination of groundwater. In this study, these issues are combined in two sub-categories of aquatic eutrophication and water contamination with persistent toxic substances. Since bioenergy chains do not seem to have significant effects on the organic pollution of rivers, this aspect is omitted.

(2.1) Aquatic eutrophication

Aquatic eutrophication is caused by nutrients that lead to increased algae growth. As a result of bioenergy chains, important nutrients, such as mineral fertiliser and manure, enter ecosystems by means of run-off in the agricultural process stage.

Within the outline of this study, it should be sufficient to consider minimising the mass of nitrogen and phosphor due to the application of mineral fertiliser and manure for biomass production as the criterion goal.

(2.2) Toxic contamination of water

In the context of bioenergy, there is a risk of toxic contamination of the groundwater through pesticide application during the energy crop cultivation phase. The higher the mass of pesticides applied for a special energy crop per area, the higher the risk of groundwater contamination and the less sustainable the bioenergy concept.

(3) Soil

Erosion is the main threat that the environmental medium soil – which includes soil life – faces; however, nutrient enrichment and heavy metal pollution have also been identified as threats (Bouwman et al. 2002; Rodrígues et al. 2008; Rusco et al. 2008). A further problem is agricultural soil compaction due to the use of heavy agricultural machines in cultivation. However, compaction is not expected to have an effect on different bioenergy concepts and it is consequently omitted.

(3.1) Erosion

Erosion leads to agricultural soil losing its functionality (Pimentel 2006). Besides site-characteristic factors, such as the soil texture, the precipitation regime and slope, agriculture management also plays a significant role (Kort et al. 1998; Lobb et al. 1999). The cultivation method and covering the land with crops are considered important soil stabilization factors.

(a) Cultivation method

There are several cultivation methods to prepare agricultural soil for sowing. All these methods increase the risk of erosion. In this study, three methods are defined: (1) direct saw, (2) grubbing, and (3) ploughing. Direct saw is considered the best and ploughing the worst regarding minimising the risk of erosion.

(b) Land cover level

Crops stabilise the soil and reduce the risk of erosion. The higher the land cover level throughout the year, the better the protection against erosion and the better the sustainability assessment.

(3.2) Terrestrial eutrophication

The deposition of aerial nitrogen compounds leads to increased vegetation growth accompanied by a decrease in biodiversity and the vegetation's increasing sensitivity to disease, drought, frost and herbivore increases (Gallego Schmid 2009). The same approach is used to calculate terrestrial eutrophication and aquatic eutrophication. The target is to minimise nitrogen and phosphor feeding by means of fertilisers.

(3.3) Soil contamination

Hazardous heavy metals affect soil, especially agricultural soil, which is the basis of life. In the bioenergy chain, the most relevant contamination source is fertiliser application (mineral fertiliser, digestate and manure) during the energy crop production phase. The fewer the heavy metals introduced into the soil through fertiliser, the better the sustainability assessment. In the context of this project, two aspects are relevant: the accumulation and mobilisation of heavy metals.

(a) Accumulation of heavy metals

The introduction of heavy metals through fertiliser is a criterion that represents heavy metal accumulation in soil. The fewer the heavy metals applied to agricultural soil, the less the risk of heavy metal accumulation.

(b) Mobilisation of heavy metals

Since potentially contaminated sites can also be considered for energy crops, heavy metal mobilisation is another attribute (see Chap. 14). Unlike amelioration activities, the removal of pollutants is not used in this project's assessment approach (see Chap. 14). As within the biogas chain, a minimum of heavy metals transferred to an energy crop ultimately leads to a low heavy metal content in the digester. The mobilised quantity of a specific heavy metal is quantified for each alternative bioenergy concept. The fewer the heavy metals mobilised, the better the bioenergy concept.

(4) Preservation of resources

The need for mineral and energy resources is an essential part of all industrial processes and should 'also be considered in bioenergy concepts. As these resources are finite and the principle of equal opportunities for future generations should be respected, resource consumption evaluation cannot be omitted from sustainability analysis. Given the particularly strong association of bioenergy with agricultural energy crop cultivation, further – renewable – resource types should also be covered; the resources land area and water also belong to this aspect.

(4.1) Energy

Energy resources can be classified as non-renewable and renewable. Renewable sources of energy are solar energy, hydropower, wind power, geothermal energy and biomass. Non-renewable resources can be divided into fossil energy (oil, gas and coal) and nuclear power.

(a) Scarcity of energy resources

The consumption of non-renewable fossil energy resources, such as oil, gas and coal, leads to a potential scarcity. Given the static range and specific calorific value, scarcity can be quantified with respect to crude oil as the tonnes of crude oil resource equivalent (Gromke and Detzel 2006; Monier and Labouze 2001; Schmitz 1995). The higher the crude oil resource equivalents' value, the higher the extraction effect.

(b) Cumulative energy demand

The cumulative energy demand (CED) is possibly an important characteristic value with which to evalute energy criteria. The CED is defined as the entire primary energy demand in Joule that can be allocated to an economic good's life-cycle (VDI 1997); in this case, the specific bioenergy path. The CED is often used as a screening impact indicator and can also be applied to distinguish between renewable and non-renewable energy demands (Fritsche et al. 1999; Huijbregts et al. 2006; Schmitz and Paulini 1999). The smaller the CED value, the better the assessment result.

(4.2) Minerals (demand for phosphate)

After discussions were held with ecology experts, the conclusion was that the analysis of the scarcity of mineral resources within bioenergy systems should concentrate on the consumption of phosphate as a fertiliser for energy crop cultivation.

Phosphate's scarcity is already a primary problem, especially in agriculture (Cordell et al. 2009). Therefore, the less phosphate is used for energy crop cultivation, the better this is for phosphate resource preservation.

(4.3) Land area consumption

Land area – especially in an unsealed and not built-up state – is a scarce good. Consequently food production, energy crop cultivation and nature conservation compete for it (DEIAGR 2008; Delzeit et al. 2010). As little space use as possible should therefore be assigned to bioenergy to reduce these land use conflicts.

(4.4) Water consumption

As a resource, water should be conserved – not only its quality, but also the quantity used for energy crop cultivation should be taken into consideration. The less water needed throughout a life-cycle, the more sustainable the bioenergy concept.

(5) Protecting biodiversity

Biodiversity is an important factor for a stable ecosystem (Millennium Ecosystem Assessment 2005); consequently, conserving the variety of life forms should be a relevant aspect in environmental assessments. During the project discussions, nature conservationists identified three parameters regarding a bioenergy path's contribution to biodiversity protection. All the parameters focus on agricultural production of bioenergy crops.

(a) Number of different cultivated crops

The first parameter is the number of different cultivated crops needed as a substrate input for a biogas plant. The larger the number of energy crops cultivated on arable land, the more positive their impact on biodiversity.

(b) Mass of applied pesticides

The mass of applied pesticides reduces the variety of living organisms in an agricultural area. Since a high amount of pesticides represses biodiversity, this parameter should be minimised with a view to sustainable development.

(c) Nitrogen fertiliser type

The nitrogen fertiliser type also has an effect on biodiversity. Providing the soil with nitrogen through cultivated legumes is considered better for biodiversity than digestive manure. On the other hand, digestive manure leads to a more active soil life than mineral fertiliser. This consideration leads to the following assessment points: cultivating legumes = one point; digestive manure = two points; mineral fertiliser: three points. The fewer points allocated, the better the evaluation.

The ecological criteria are presented in Table 12.1 below.

12.4.2 Economic Criteria

The broadest differentiation of economic criteria is into investments and operating costs: Investments are defined as the sum of all incurred expenses until plant operation readiness, while operating costs occur during operation and depend on the capacity utilisation (Geldermann and Rentz 2004). In our case, the following cost components need to be considered: investments (one-time), biomass (annual), wages (annual), transport (annual), interest on borrowed capital (annual), dividends for capital contributions (annual), repairs (annual) and miscellaneous (e.g., accounting, trade tax or bookkeeping; annual). Incoming payments result from electricity and heat sales (annual), sponsor funding (annual), residual value (one-time after 20 years, or after the expected plant lifetime).

To develop meaningful economic criteria for the bioenergy concept assessment, the following stakeholder group perspectives have to be considered:

- operating company
- employees
- heat clients
- farmers
- region.

The regional perspective reveals further aspects. Although there are many definitions of a region, and the region around a possible bioenergy village cannot be clearly defined (see Box 12.2), it affects all these stakeholder groups' interests plus those of their neighbours. It can also be referred to as the administrative department's perspective.

(1) Operating company's perspective

The operating company's corporate boundaries begin with biomass supply and end with electricity and thermal energy sales. The bioenergy plant owner can be a single investor, an investor group, or a village cooperative.

(a) Net present value

Net present value (NPV) is a key financial management indicator. An investment's NPV is the difference between the sum of the discounted cash flows expected from an investment and the initial amount invested. An interest rate is chosen to adjust for time and risk (see Chap. 10). The project with the highest NPV should be selected if the NPV is the only criterion.

(b) Supply contract duration

The longer the running time of the contracts between the operating company and the agriculture and forestry suppliers, the higher the operating company's planning security. The operationalisation was undertaken by means of points: running time 0–3 years = 0 points; 4–10 years = 1 point; 11–15 years = 2 points; >15 years = 3 points.

(2) Employee's perspective

The operating company employees are the plant manager, the operator, account staff and the unskilled workers. Their wage level is not part of this list, because it would not distinguish between the different technical and organizational plant forms in Germany.

(a) Profit sharing

If the employees participate in profit realisation, or if there are other incentive systems, this criterion is assessed positively (one point). If there is no possibility of profit realisation participation, it is assessed negatively (0 points).

(b) Possibility of additional fee

If there is the possibility for more people besides full-time employees to work at the plants on a fee basis, this criterion is assessed positively (yes = 1 point, no = 0 points).

(3) Heat clients' perspective

Heat clients are people whose homes are connected to the public hot water grid and who are associated with the operating company. In many considered bioenergy concept cases, this is a cooperative. The heat clients are interested in paying moderate prices for their heat.

(a) Annual heat supply costs

Different biomass energy paths have different price tags for their clients. In a bioenergy village (for the definition see Box 12.1 above), for example, people in connected households pay less for their energy than people using fossil fuel energy (the base is the mean of the basic charge and the heat price per kWh) (Ruppert et al. 2008). The lower the annual heat prices, the better for the clients.

(b) Minimum deposit

If people have the opportunity to participate in an operating company, for instance a cooperative, they have to pay a deposit. The lower the minimum deposit, the better for the clients, because the threshold for people to take this step is then lower.

(c) Connection and conversion fee (one-time pay-offs)

The lower the fee for connection, to the operating company, the conversion costs and deposit, the better this is for heat clients.

(4) Farmers' perspective

Farmers can have different roles in bioenergy projects: They can simultaneously be raw material suppliers who earn money with this and heat clients with an interest in getting heat at a low price. These are opposing targets.

(a) Influencing the price of biomass

If farmers involved in bioenergy plants can influence the price of biomass during the running contract, this criterion is assessed positively with 3 points on a 3-point scale. By having input in the pricing, these farmers can incorporate the agricultural market trend and avoid suffering financial setbacks that no other farmers encounter.

(b) Operational flexibility

The shorter a contract is, the greater the flexibility for farmers. The contract length is assessed by means of points: duration > 15 years $= 0$ points; 11–15 years $= 1$ point; 4–10 years $= 2$ points; < 4 years $= 3$ points.

(5) Regional perspective: Regional net product

The regional value added is the value at which the regional output is bigger than the input. In this case, it refers to the region within a radius of approximately 50 km around the bioenergy plant's location. Behind this is the assumption that, within this radius, all important technical crews and service contractors can be obtained.

(a) Regional value added (investment, one-time)

Examples of regional value added relate to the investment in engineers, in craftsmanship, or in the construction industry (civil engineering). The higher the sum of the regional value added for investment, the better for the region (see for a definition of a region Box 12.3).

(b) Regional value added (current)

Examples of regional value added concerning the scope of current issues are: maintenance and repair work, notary fees, insurance companies, raw materials from farmers, etc. The higher the regional value added for current issues, the better for the region.

(c) Tax revenue

The council and the administrative district obtain trade tax and income tax from the operating company. The higher these earnings are, the better for the region. The assessment takes place by means of points: 1 point if the council has such income; no point if there is none.

The economic criteria can be viewed in Table 12.2 below.

Box 12.3 What Is a Region?

"Region" is a very broad concept. Different authors and scientists use a range of definitions. Among others, the definition depends on the discipline: Natural scientists often rely on other definitions than social scientists do. For example, a classification can be divided into two aspects:

(a) functional assignments that are grown historically. Examples include:

- job market regions (connected by commuter streams)
- business market regions (the catchment area of single contractors).
- regions for nature protection (spatial links between single ecosystems)

Such classifications, which are dependent on functional coherences, are too imprecise to encompass a region's administration. Therefore, there is a group of:

(b) administrative regions:
NUTS (Nomenclature des Unites Territoriales Statistiques) makes provision for the following regions: states (0), federal states (I), districts (II) and communal districts (III). Regional aggregations (such as the Metropolregion Hannover) are not accounted for (NUTS 2007).

There is also the concept of a "region's identity", which comprehends a subjective "identity for the region", which can differ per individual in the same region (Ipsen 1993).

12.4.3 Social Criteria

Social criteria for assessing different bioenergy paths' sustainability can be divided into four sub-categories: acceptance, participation, psychological consequences and employment. Table 12.3 below contains the list of criteria.

(1) Acceptance

Chapter 28 of Agenda 21 (BMU 1992) addresses the participation of the council and people in order to solve environmental problems. Furthermore, to promote sustainable bioenergy use, the population's acceptance of the technical systems and the context of the production and logistics are extremely important. Bioenergy use acceptance refers to the following four aspects:

(a) Scenery of cultivation concepts (aesthetics)

Different cultivation concepts have different impacts on the landscape aesthetics: The landscape can be very colourful and heterogeneous if farmers practice crop rotation (such as triticale, rapeseed, rye and sugar beet, which are perhaps mixed with poppies and other weeds), whereas the scenery can, for example, be boring in the case of maize monoculture. The more positive people's response to a landscape's aesthetics, the better it is. The criterion is operationalised via a five-point scale. The higher the assessment, the better.

(b) Scenery of the technical plants (aesthetics)

Production plants (biogas plant, wood-fuelled heating plant) can also be assessed as either more or less aesthetically pleasing. The more positive this assessment, the better. The criterion is also operationalised via a five-point scale.

(c) Smell

When comparing biomass alternatives, we start with compliance with odour nuisance limit values. Nevertheless, there is a subjective smell regarding biomass use for energy supply (e.g., the storage of silage next to biogas plants, and the transport of liquid manure) and this can impact acceptance. It is important to note that this does not have to be an objective criterion, it can be the *perceived* or *suspected* smell related to biomass use in someone's imagination (it can also be a prejudice). The less the perceived stench, the better.

(d) Noise (factory and transport)

The block heat and power plant in the factory and transport (via truck or tractor) produce some noise. The larger the technical plant and the more biomass it requires, the greater the possibility that people will perceive the noise as annoying. Noise depends on perception: If the purpose of the noise is considered meaningful, it is assessed as positive and vice versa; consequently, this criterion is also subjective (Guski 2000). The less annoying the noise is perceived, the better. The noise is further categorised into perceived factory noise and perceived transport noise.

(2) Participation

Participation refers to different mechanisms through which people can express their opinions and, ideally, can exert influence on political, economic, management, cultural, family or other social decisions, to which the Agenda 21 action plan refers (BMU 1992, Chapter 28). From an administrative perspective, participation can build public support for activities. It can educate people about an agency's activities. Participation can also facilitate useful information exchange regarding local conditions. Furthermore, participation is often legally mandated. From citizens' perspective, participation enables individuals and groups to influence agency decisions in a representational manner (Girschner and Girschner-Woldt 2007).

In terms of diversity management, the following groups should be considered in the planning of biomass use and in the decision process:

- farmers
- heat clients
- men and women equally
- the communal and regional administration
- villagers and the general public
- nature conservationists
- scientists.

The more opportunities for stakeholders to participate, the better. Such participation can comprise different content aspects associated with different participation intensities. These aspects are described below.

(a) Participation in the planning process

People from the seven stakeholder groups can be involved in the planning and decision process for the use of a biomass concept. However, in different biomass concepts regard people's needs and wishes to a differing degree: The more people are involved, the more their wishes and anxieties can be considered and conflicts avoided. This can impact the local residents' satisfaction, self-efficacy, etc. (Eigner-Thiel 2005). The different biomass alternatives differ in the extent to which people are involved in their planning process. For example, as many stakeholder groups as possible should be involved in the planning process of a bioenergy village, while a large-scale plant offers less possibility for extensive involvement. The more groups involved, the better.

Since seven stakeholder groups were identified, participation in the planning process is assessed according to eight points. One point is allotted for each participating group and zero if nobody is allowed to participate. The more groups involved, the better.

(b) Participation via information

The lowest form of obtaining participation is simply by providing information. This can be done by means informational events or meetings, information brochures and flyers or stalls at festivities. It is important to inform people not only once about the status of a planning process, but regularly. The communication interval can differ between the biomass options (e.g., it is dependent on the operator type). The more stakeholder groups are informed, the better. Again, there are eight points according to which participation in the form of providing information is assessed. One point is assigned for each participating group and zero if nobody is allowed to participate.

(c) Participation in finance decisions

Biomass paths can be financed by a single investor, communal institutions or administrations, or groups of individuals (e.g., a cooperative). If individuals have the opportunity to contribute to the finances, this can have various positive consequences: First, this is an additional investment from the population. Individuals can also influence the usage of their investment (e.g., determining the price of heat and raw materials; this could include price corridors and upper and lower boundaries). In addition, being consulted can increase an individual's sense of self-efficacy. Finally, people who participate in the finances will receive (at least a small) financial gain, for example, in the form of lower heat costs, or participation in the profits. This means that biomass paths offering more stakeholder group participation in the finances can be considered more sustainable than those without this possibility. The higher the number of participating groups, the better. Again, there are eight points according to which the participation in finances is assessed. One point is allocated for each participating group, and zero if nobody is allowed to participate.

(3) Psychological effects

Different biomass options can have different consequences for people's self-perceptions through the different degrees that people are allowed to participate in the planning and decision processes. This means different degrees of sustainable development. For the study of these factors (data), see also Eigner-Thiel (2005) and Chap. 12 in this book.

(a) Feeling of independence from large electricity suppliers

If a bioenergy plant operator is a local – perhaps collectively organised – institution, the feeling of independence from large energy suppliers can be especially high. This is the result of many discussions in village meetings and interviews with people engaged in a bioenergy village (Eigner-Thiel 2005). The feeling of independence is assessed as more sustainable, because there is more self-reliance and less heteronomy concerning price determination, supply security and other important aspects of their lives, but also concerning accident risks. The feeling of independence is operationalised on a scale of 0–4. The higher the value, the better.

(b) Feeling of independence from fossil resources

If local residents know that their biomass-based heat source is renewable, the feeling of independence from non-renewable resources (such as fossil fuels, oil, natural gas, etc.) can grow. The assumption is that the higher the participation rate, the higher the awareness of this autonomy. The feeling of independence is operationalised on a scale of 0–4. The higher the value, the better it is.

(c) Sense of solidarity

This criterion is associated with the extent of the possible participation. Interviews indicated that active engagement in a collective climate protection project enhances solidarity in a group. This has positive effects on people's well-being and health (Eigner-Thiel and Schmuck 2010). The sense of solidarity is operationalised on a scale of 0–4. The higher the value, the better.

(d) Self-efficacy

Self-efficacy is the measure of one's competence to complete tasks and reach specific expected goals. This expectation influences one's thoughts, emotions, behaviour, ambition, effort and persistence (Bandura 1992, 1997). This criterion is also dependent on the extent of the participation, as indicated in interviews (Eigner-Thiel and Schmuck 2010). The feeling of self-efficacy is assessed on a scale of 0–4. The higher the value, the better.

(e) Feelings of pride, fun and meaning

There are also coherences between the degree of participation and the pride and joy at planning and implementation, learning success, satisfaction and a positive feeling of meaning. This relationship is described in interviews (Eigner-Thiel and Schmuck 2010; Wüste and Schmuck 2012). The feeling of pride, fun and meaning is operationalised on a scale of 0–4. The higher the value, the better.

(f) Image of the village or town

The existence of bioenergy technologies in a village or a town can affect its image positively (if it is associated with progress or eco-friendliness) or negatively (if it is associated with smell, noise, or low plant aesthetics). This can positively or negatively influence a whole region. The higher the image of a village or a town with regard to bioenergy plants, the better, because people enjoy living in a well-known locality.

(g) Subjective assessment of accident risk

People often associate accident risks or disaster risks with technical plants. This can differ, depending on the technology type, plant type, plant size, etc. The less the assessed risks on a scale of 0 (no risk at all) to 4 (very high risk), the better.

(4) Employment

(a) Additional workplaces

With the usage of local bioenergy for electricity and heat supply, jobs such as a biogas plant manager and also administrative jobs, may be generated. On the other hand, jobs like that of a bioenergy village's chimney sweep are also replaced, because the number of individual heating systems will decrease. The difference between the number of jobs in a village or a region before and after an energy conversion process is another social criterion. The higher the number of additionally created jobs, the better.

(b) Possibility to work part-time

The provision of part-time jobs can contribute to the family life and professional life of men and women being more compatible (OECD 2002, 2005; SEK 2006; Caspar et al. 2005; BMFSJ 2008). Therefore, if a specific biomass concept can provide more part-time jobs than others, it is allocated more points. The crucial value is the number of potential part-time jobs per 1,000 inhabitants.

The social criteria can be viewed in Table 12.3 below.

12.4.4 Technical Criteria

Technical criteria do not follow directly from the three-pillar model of sustainable development as with the other three groups of criteria. However, the authors found it important for the biomass conversion process to report technical criteria separately. In principle, one could also assign the technical criteria to the other three pillars, but this might result in a loss of information. It is important that, within the technical criteria, the ratio of the criteria for the three different pillars – ecology, economy and social aspects – is balanced, or, if not, the weightings of the criteria originating from these three pillars are balanced.

(1) Plant efficiency

Efficiency is generally defined as the ratio between the yielded output and the yielded effort (input). Furthermore, efficiency is an objective of sustainable development in order to minimise the usage of energy and raw materials.

(a) Thermal efficiency factor

The thermal efficiency factor is the ratio of the delivered thermal output to the input energy. The larger this efficiency factor, the better.

(b) Electrical efficiency factor

This is the ratio of the delivered electrical output to the input energy. The larger this efficiency factor, the better.

(c) Heat use in summer

If the heat from the bioenergy plant is used, for example, to warm water (e.g.. for an existing swimming pool), or to dry i.e. wood, corn, clinkers instead of being released into the air, the concept is considered more sustainable. Points are allotted according to a postive answer (one point) and a negative answer (no point).

(d) Modularity

Multiple kettles or biogas plant parts have advantages, because partial workload operation can take place during maintenance, which is positive with regard to emissions and efficiency. The criterion is assigned via points for classes: The higher the number of points, the better.

(2) Biomass transport

Biomass is transported from the fields to the plant and the digested residue is deployed in the fields. Consequently, one needs tractor-drawn trailers or trucks; these lead to noise, energy consumption, emissions and accident risks. The following criteria are relevant here:

(a) Frequency

The less transportation (number of vehicles) needed, the better, as less noise and emissions are produced, fewer accidents occur and the less energy is needed.

(b) Point in time

The more transportation is done during the day (1 point) instead of at night (2 points), the better, due to less noise at bedtime.

(3) Administrative effort (licence duration)

Different approvals are necessary for different kinds and sizes of bioenergy plants; these have associated costs. For example, in Germany, a biogas plant – depending on its size – must comply with a building law and an emissions law (BImSchG). Different approvals have different time-frames. The more complex an energy system, the longer its approval takes. Therefore, the fewer days required for approval, the better.

The technical criteria are shown in Table 12.4 below.

12.5 Bioenergy Alternatives

The research effort's overall aim is to compare the sustainability of different bioenergy concepts regarding a specific geographical location. The socio-geographical framework conditions are defined by the local characteristics of the plant cultivation site (climate, soil type, elevation and air temperature) and the village (population, age distribution and community activities). These characteristics primarily affect energy crop selection, impact the nature and landscape, the yields and the social and economic aspects. The characteristics of such a village influence, inter alia, the demand for energy for electricity and heating. For example, if there is a public swimming pool, or industrial heat customers, the village needs more energy than a "normal" village.

Within these preconditions, two spatial dimensions are considered (Fig. 12.3):

(a) alternative regional bioenergy concepts (regional dimension)
(b) different bioenergy village types as local bioenergy concepts (local dimension).

All the technical bioenergy concepts described are well established and have been available on the market for a long time. The lack of suitable data on new and innovative technologies, i.e. biomass gasification systems has led to their omission from this comparison. We define our base area as farmland; the conversion technologies therefore mostly utilise agricultural products. The woodchip heating plant is only fuelled by local forest products.

For the realisation of bioenergy concepts in a village, the following alternatives are possible:

- the agricultural energy crop cultivation system: conventional farming vs. organic farming vs. crops from contaminated soils[1]

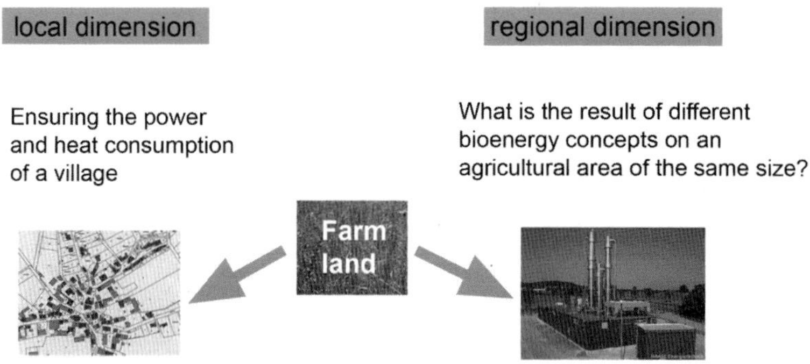

Fig. 12.3 Different spatial dimensions for the comparison of different bioenergy concepts

[1] Cultivation of energy crops on contaminated sites might reduce the competition for agricultural land with food crops (see also Chap. 14).

Table 12.5 Possible combinations for bioenergy villages (local scale)

Cultivation system	conventional	organic	conventional at contaminated site
Biomass fuel	wood	straw	wood from contaminated site
Operation company	people of village	outside investor	farmer group
Participation	yes	no	

- biomass fuel for the woodchip heating plant: wood, straw or wood from contaminated soils
- the operating company: one outside investor, or a corporation with collective investors of whom the majority come from the village, or an investor group comprised of farmers
- the possibility for people to participate in the planning process: yes or no.

The combination of these four aspects (with their particular specifications: 3x3x3x2) leads to 54 theoretically possible alternatives. In an actual bioenergy project, the combinations that are feasible under actual circumstances need to be determined. See Table 12.5 for the various combinations.

For the local dimension, the following presumptions must be met for a comparison of the different approaches to bioenergy villages: The approaches are based on the bioenergy village Jühnde concept. Therefore, the energy conversion techniques will be a biogas plant, a combined heat and power station, a heating plant fuelled by woodchips, a hot water grid, and a boiler fuelled by oil or biodegradable diesel (as a contingency reserve). Furthermore, it is assumed that 70 % of households have a pipe connection to the local hot water grid (see Box 12.1; Ruppert et al. 2008).

On a regional scale, the bioenergy concepts are oriented towards a general bioenergy supply. As a shared reference value for the comparison, the required land area is chosen for a bioenergy village's energy crop cultivation. For example, a bioenergy village needs 300 ha of agricultural land to supply its inhabitants with electricity and heating. These 300 ha will be the land area to be used when comparing the regional biomass concepts. For a large-scale plant, these 300 ha are just a percentage of the whole area that is needed. For a small-scale plant, 300 ha might be more than the plant actually needs. The bioenergy alternatives based on the biogas techniques vary in scale and the type of biogas used. The defined alternatives are listed in Table 12.6.

The challenge is to define appropriate and meaningful alternatives. Data collection and compilation are laborious tasks.

The data relating to the criteria need to be collected and documented with regard to the different alternatives. In the current project, the data will be obtained from the other sub-projects, from databases such as GEMIS and from a literature review.

Table 12.6 Alternative regional bioenergy concepts

Concept	B1: bioenergy village	Large biogas plant	Single biogas plant
Biomass input	Energy crop + wood	Energy crop	Energy crop
Conversion technology	Anaerobic digestion, combustion	Anaerobic digestion, feeding into the gas grid	Anaerobic digestion
Products	Biogas → power, heat	Biogas → power, heat	Biogas → power, heat
Power	716 kW	2.5 MW	225 kW
Arable land area or land use for energy crop cultivation (ha)	~300	~900	~60
Electricity production per year (MWh/a)	4,500	50,000	1,900

Since the focus is on bioenergy villages and their sustainability implications, the investigated bioenergy concepts are deliberately not compared with other renewable energy forms (e.g., wind energy or photovoltaic energy), because the scope would then be too broad and it would be impossible to provide the users with an in-depth differentiated comparison. Nonetheless, Oberschmidt et al. (2010) offer an exemplary comparison of different forms of renewable energy.

12.6 Weighting Process

Once the criteria hierarchy has been established and data on the alternatives have been compiled in the decision table, the weighting process can take place. Weighting factors express the relevance or importance of each attribute. The weighting or valuation of different criteria is a subjective element in the assessment of techniques. It addresses the relative importance of the different criteria of a given decision problem for the decision-maker, or the stakeholder group. The weighting factors thus constitute the preferential information between the criteria (Belton and Stewart 2002). There are several weighting techniques (e.g., direct ratio, SWING (v. Winterfeldt and Edwards 1986), SMART (Edwards 1977; Winterfeldt and Edwards 1986), SMARTER (Barron and Barret 1996; Edwards and Barron 1994), eigen vector method (Saaty 1980), etc.). The discussion of weighting issues leads to the following fundamental questions, especially regarding the valuation of the different criteria of sustainable development:

- Should there be a weighting at all?
- If so, which weighting method should be used?
- Which weights should the different criteria be given?

Scientific research can support decision-makers' quest to better understand the interdependencies in the weighting of environmental criteria. However, this

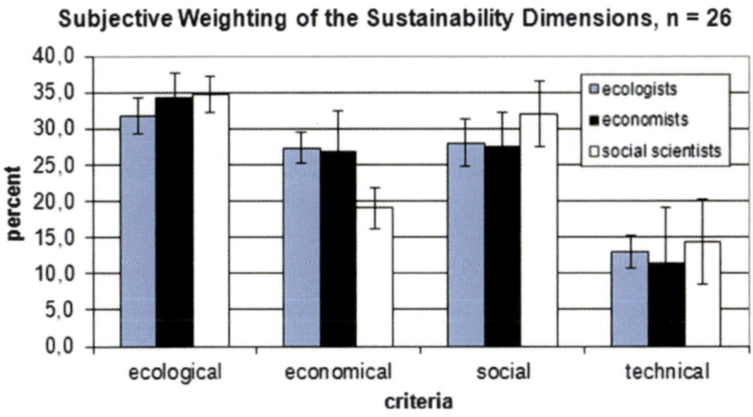

Fig. 12.4 Results of the weighting process of the first criteria level to assess the sustainability of bioenergy paths (n: number of people who weighted the importance)

discussion is very controversial, and it should be noted that some authors favour a more technical approach, while others stress the importance of detailed stakeholder involvement due to context sensitivity and the significant influence this has on the overall results.

In our case study, the SWING method (see Winterfeldt and Edwards 1986) was used for the weighting process. The method comprises everybody in a group awarding a number between 0 and 100 to each criterion. The most important criterion is weighted with a high number such as 100, and the others with smaller numbers. The numbers are converted into percentages so that the sum is 100 %, and the percentages of the (in our case between four and seven) experts are averaged (the mean value is calculated).

In this study, the weighting process occurred in three experts groups:

- social experts
- economic experts
- ecological experts.

Each of the expert groups first weighted the relative importance of four sustainability dimensions: the ecology, economy, social aspects and technical aspects. The weighting process occurred in a moderated group. Firstly, each person weighted the criteria individually, thereafter the appraisals were presented to the others and discussed. After this, everyone had an opportunity to change their judgement. The results of the first preliminary weighting process (the decision table is not yet complete) are shown in Fig. 12.4 as a box-and-whisker diagram. The bars show the average of the expressed weighting factors of the four most important criteria in the criterion hierarchy, while the ends of the whiskers represent the minimum and maximum weighting factors allocated by a group of experts.

Figure 12.4 shows that the experts agreed that the ecological criteria should have the highest priority, followed by the social and economic criteria. The technical criteria were seen as the least important. The primacy of the ecological criteria confirms the underlying assumption that the described approach is a strong sustainability concept (see the next section). It should be noted that no engineers or deciders (farmers or designers) participated in this weighting process, only the three expert groups.

Experts may want to change their assigned weighting factors when they see the complete decision table and the applied MCDA algorithm results. Belton and Stewart (2002) state, for example, that the MCDA is an iterative process, and sensitivity analyses will provide further insights into the decision problem, possibly leading to an adjustment of the stated preferences.

12.7 Data Sets for Criteria Specification

12.7.1 Types of Data to Compare Bioenergy Concepts

Complex decision problems call for the involvement of various groups of experts with different scientific or professional backgrounds. Table 12.7 shows the scientific disciplines that contribute to the research project (as described above in the different criteria's sections) and the data type they usually deliver: Natural scientists and engineers mostly produce quantitative or quantifiable data, while social scientists also deliver qualitative results.

Table 12.7 Overview of the participating disciplines in this bioenergy project and the quality of the data used in the MCDA

Disciplines	Data quality	Quantitative or quantifiable data	Qualitative results
Geography	Spatial data (GIS), temporal data (climate data)	X	
Chemistry	Numerical chemical analysis (concentration)	X	
Soil sciences	Numerical chemical analysis (concentration)	X	
Environmental sciences	Spatial data (habitat, biodiversity)	X	X
Psychology	Interviews, questionnaires (motivation, acceptance)	X	X
Economy	Numerical analysis (cash value)	X	
Crop cultivation	Numerical analysis (crop yield, amount of fertiliser)	X	
Agricultural economics	Numerical analysis (contract design), questionnaires (acceptance)	X	X

12.7.2 Data Format

The format of the ecological data is mostly quantitative. Quantitative data are derived from empirical studies as well as from databases, as described in the following sections:

Knowledge about the local availability of biomass for energy is an important aspect in the context of local and regional bioenergy concepts. Therefore, geo-referenced input data on radiation, precipitation, evapotranspiration, temperature and soil properties have to be compared with the cultivation-specific requirements of the different crops on the site to allow the biomass production yield to be modelled (Bauböck 2009; see also Chap. 7 in this book).

The evaluation criteria also consider agricultural activities' impact on the protection of species and biotopes, the landscape, erosion prevention, and other environmental impacts (van Haaren and Bathke 2007; Wiehe et al. 2009). We will also obtain data from other research studies of this type (see Chap. 8 in this book).

An example of a specific bioenergy concept is the use of biomass from contaminated sites for energy. Contaminated sites that may be polluted with hazardous substances (e.g., heavy metals) due to mining activities or flooding by contaminated water offer an interesting option for energy conversion (Deicke et al. 2006; see also Chap. 14 of this book).

The format of the economic and social data is quantitative as well as qualitative. The data are mostly derived from empirical studies, as described in the following section:

Besides economic reasons, farmers' willingness to cultivate energy crops depends on many social or psychological factors such as environmental awareness, risk attitudes, knowledge and involvement (Ruppert et al. 2008; Granoszewski et al. 2009; see also Chap. 9 in this book). Thus, the decision support tool should consider the drivers and barriers revealed through interviews, questionnaires and the subsequent statistical analyses of, for example, the social criteria.

The format of the technical data is mostly quantitative and the data are derived from databases and literature reviews.

12.7.3 Data Consolidation

The consolidation of data from the diverse scientific fields should consider several aspects: The data have different reference values (site-related yields, plant-specific operating costs per year, the share of the population, etc.) and are stored in different formats (shapefile, spreadsheets, text file, etc.). Furthermore, the data quality can vary, as the data from a chemical analysis may have small ranges, while data on the operating level may have a much higher deviation margin. As mentioned above, the comparability of the data should be guaranteed, therefore the units and the reference systems should be comparable; this is the greatest challenge. Furthermore, certain data to be gathered within the collective research project refer to special local conditions (soil, climate, etc.), while data on other criteria have be taken from general databases (e.g., GEMIS). Here, comparability must be proved very thoroughly (see Schmehl et al. 2010).

The development of a consistent life-cycle inventory database faces similar challenges. Hischier and Gilgen (2005) emphasise the relevance of standardised, comprehensive and actual life-cycle inventory databases. In the ECOINVENT project, a clearly defined and comprehensive data exchange format is used, which includes meta-information, modelling, validation and administrative information. Furthermore, there are already approaches to implement geographic information in life-cycle databases, and vice versa. On the one hand, the inventory data can be site-specifically assigned in the geographic information system. On the other hand, the geographic data can be used to identify the correct characterisation and weighting factors for the life-cycle assessment (Wei and Carlson 2002).

12.8 Visualisation

Modern information systems and decision support systems not only store and process data and information, but also display them in a user-friendly manner (Geldermann 2010). Currently, the visual representation of data tables, for example, with bar charts, pie charts, or trend lines, is widely used. In interdisciplinary research topics, the derived research results need to be presented to many lay persons in the various scientific disciplines. For instance, social scientists and natural scientists have to communicate their results to each other. Interested public, such as the village community, or the local administration, also seek advice on building a bioenergy village. Thus, it is essential to present the analysed bioenergy concept's expected advantages and disadvantages for a specific village or region in an easily understandable way.

An open scientific question is the visualisation of specific aspects of the problem to show that some aspects are characterised by far more assessment criteria than others. In decision theory, this is called bias, which is generated by highly asymmetrical criteria hierarchies (Hämäläinen and Alaja 2008).

Profiles will be generated to assess the different bioenergy concepts in order to depict the impacts that sustainable development's three pillars have on direct comparison. Methods from operations research and main component analysis, allow the graphical illustration of a high dimensional solution space (Bertsch et al. 2007; Bertsch et al. 2006; Treitz et al. 2008; Geldermann et al. 2009). Figure 12.5 displays four ways to visualise the results of an MCDA algorithm with regard to the same decision problem (with illustrative data). It should be noted that similar graphical representations are being developed for various MCDA algorithm types, such as multi-attribute utility theory (MAUT), or outranking.

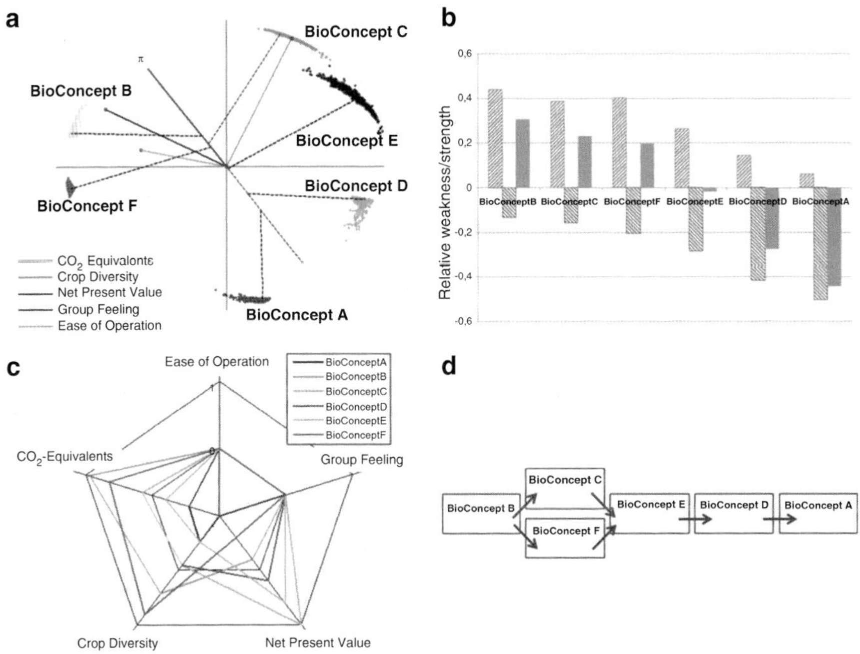

Fig. 12.5 Example of a possible graphical presentation of a PROMETHEE analysis of six biomass concepts (A to F) with regard to five criteria: (**a**) principal components analysis under consideration of uncertainties by means of a Monte Carlo simulation; (**b**) a histogram of the outranking flows; (**c**) a spider diagram; (**d**) a partial pre-order

Geldermann and Schöbel (2011) show that different approaches often have the same mathematical foundation.

Further research is necessary to answer the question on which presentations types most people understand best: Cognitive aspects lead to people perceiving graphically visualised evaluation results in different ways. Exposure to graphical representations is not self-evident or elemental; therefore, their comprehension has to be learned (Cox and Brna 1995; Petre and Green 1993; Weidenmann 1994; Ainsworth 1999). The use of graphics for the visualisation of non-spatial, abstract information – for example, economic data – has only been in common practice in the West since the eighteenth century (Tversky 2000; Roth and Bowen 1999). Consequently, graphical representations can easily be misunderstood, especially by inexperienced or lay persons, and are therefore likely to be interpreted superficially (Weidenmann 1994; Cheng et al. 2001). It is therefore important to edit multi-criteria decision support results graphically, thus allowing perception-psychological knowledge to be considered. This is open to further research by specific psychological studies.

Box 12.4 Representation of Information in the Human Memory

Representation is the illustration of an issue in the mind (Palmer 1978). According to Larkin and Simon (1987), a concept can be differentiated into propositional representations (like language, logical statements, and linearly arranged information) and graphical representations (the use of spatial relations and the availability of information at a glance). On the other hand, individuals' different cognitive styles are also relevant. There are verbalisers and visualisers, i.e. people with different preferences for different types of illustration and, therefore, with a different understanding of them (Cox et al. 1994). Schmuck et al. (1998) examined the intelligibility of various symbols for specific product groups and company groups' assessment according to sustainable development aspects. They showed that various symbols can have very different effects on perception speed and clarity.

12.9 Conclusions

In this chapter, we described the progress of a possible process to choose a special biomass option that is as sustainable as possible. We used an interdisciplinary approach to illustrate the process. The aim of the developed approach is to aggregate different bioenergy concepts' sustainability strengths and weaknesses within a ranking order.

The Multi-Criteria decision analysis (MCDA) helps to structure a decision problem (Department for Communities and Local Government 2009; Geldermann and Rentz 2001; Wilkens and Schmuck 2012). This method seeks to gain insight into the decision problem and to learn more about the investigated alternatives. It increases the transparency of the assessment of the various criteria and alternatives; and therefore reduces the complexity of the decision process. Further, the goal is to distinguish between the subjective and objective preferences specified during the decision process. The alternatives have to be comparable. The itemisation of the indicators (especially the sustainability indicators; see Sect. 12.3) is crucial. In this regard, the development of social criteria within the regional biomass paths is specifically a new research aspect.

We showed that decision support system development is always site-specific for villages and regions. The requirements of local and regional deciders should therefore be considered to support the best choice of a suitable and sustainable concept.

As described, an information system for the assessment of different bioenergy concepts with regard to sustainable development has to manage data on ecological, economic, social and technical aspects. On the synthesis side, there are data on geo-referenced environmental information, acceptance surveys, bioenergy plants' technical characteristics, the documentation of interviews and questionnaires, chemical analysis results, etc. The weighting process is also a differentiated step that should be undertaken by experts who are deeply involved in sustainable development. The challenge is therefore not only the vast amount of data in very non-homogeneous formats, but especially the mastering of the logical coherence of the data from different sources and scenarios. In addition, a vast base of experience with and knowledge of sustainable development is essential, which will inevitably lead to an interdisciplinary discussion.

After the interdisciplinary effort to establish a criteria system and to calculate the criteria values, or the relative strengths and weaknesses of individual biomass alternatives, it is crucial that the results should be understood by as many people as possible, because people need to accept one of the alternatives. Again, this participatory aspect is part of a sustainable biomass concept. Wilkens and Schmuck (2012) describe this process as follows: The MCDA process offers a platform for the exchange of arguments and different perspectives, provides data that can answer residents' questions, and combines scientific data with the actors' perspectives, thus making well-balanced decision-making possible. This requires the professional preparation of the MCDA process in the form of well-understood visual presentations. Only then will the theoretic scientific effort lead to the successful application – also by lay people – and support of a more sustainable life on earth.

To date, the spatial and temporal scaling problematic is unsolved. Specifically, in the field of interdisciplinary research, in which economists, natural scientists and social scientists work and collect data on different scales, further research is needed to extrapolate and model the data from one scale to another in the different research fields and to combine them properly. Ensuring the comparability of different data sources is another great challenge.

The results of the process depicted here can contribute to sustainable development. The scientific findings we describe can help preserve biodiversity, reduce global warming, rekindle village life, strengthen the democratic will, consolidate regions' economic potential and strengthen rural development.

References

Ahl, C., Eigner-Thiel, S., Girschner, W., Karpenstein-Machan, M., Roland, F., Ruppert, H., Ruwisch, V., Sauer, B., Scheffer, K., & Schmuck, P. (2007). *Bioenergiedörfer – Dörfer mit Zukunft,* Information brochure. Göttingen: Interdisciplinary Centre for Sustainable Development.

Ainsworth, S. (1999). *Designing effective multi-representational learning environments* (Report no. 58). Nottingham: University of Nottingham, ESRC Centre for research in Development, Instruction and Teaching.

Bandura, A. (1992). Exercise of personal agency through the self-efficacy mechanism. In R. Schwarzer (Ed.), *Self-efficacy: Thought control of action* (pp. 3–38). Washington, DC: Hemisphere.

Bandura, A. (1997). *Self-efficacy: The exercise of control.* New York: Freeman.

Barron, F. H., & Barret, B. E. (1996). Decision quality used ranked attribute weights. *Management Science, 42*, 1515–1523.

Bauböck, R. (2009). Bioenergie im Landkreis Göttingen – GIS-gestützte Biomassenpotenzialabschätzung anhand ausgewählter Energiepflanzen, Triticale und Mais (Band 18). In

M. Kappas (Ed.), *Erdsicht – Einblicke in geographische und geoinformationstechnische Arbeitsweisen*. Stuttgart: Ibidim-Verlag.

Belton, V., & Stewart, T. J. (2002). *Multiple criteria decision analysis: An integrated approach*. Dordrecht: Kluwer Academic Publishers.

Bertsch, V., Geldermann, J., Rentz, O., & Raskob, W. (2006). Multi-criteria decision support and stakeholder involvement in emergency management. *International Journal of Emergency Management, 3*(2/3), 114–130.

Bertsch, V., Treitz, M., Geldermann, J., & Rentz, O. (2007). Sensitivity analyses in multi-attribute decision support for off-site nuclear emergency and recovery management. *International Journal of Energy Sector Management, 1*(4), 342–365.

BioSt-NachV. (2009). Verordnung über Anforderungen an eine nachhaltige Herstellung von flüsssiger Biomasse zur Stromerzeugung. *Bundesgesetzblatt Jahrgang 2009, Teil 1*, Nr. 46. Bonn.

BMFSJ. (2008). *Familienfreundlichkeit als Erfolgsfaktor für die Rekrutierung und Bindung von Fachkräften. Ergebnisse einer repräsentativen Umfrage unter Arbeitgebern und Beschäftigten*. Retrieved August 23, 2010, from: http://www.bmfsfj.de/bmfsfj/generator/BMFSFJ/Service/Publikationen/publikationsliste,did=112440.html

BMU (Federal Ministry for the Environment, Nature conservation and Nuclear Safety). (1992). *Umweltpolitik*. Agenda 21. Konferenz der Vereinten Nationen für Umwelt und Entwicklung im Juni 1992 in Rio de Janeiro. Berlin, Germany.

BMU (Federal Ministry for the Environment, Nature conservation and Nuclear Safety). (2009). *German cabinet has approved national biomass action plan*. Press release no. 122/09. Berlin, Germany.

Bouwman, A. F., van Vuuren, D. P., Derwent, R. G., & Posch, M. (2002). A global analysis of acidification and eutrophication of terrestrial ecosystems. *Water, Air, and Soil Pollution, 141*, 349–382.

Breitschuh, G., Breitschuh, T., & Eckert, H. (2008). Einfluss einer erhöhten Biomasseproduktion auf die Nachhaltigkeitsparameter landwirtschaftlicher Betriebe. In Kuratorium für Technik und Bauwesen in der Landwirtschaft e.V. (KTBL) (Ed.), *Ökologische und ökonomische Bewertung nachwachsender Energieträger* (pp. 104–116). Darmstadt: KTBL.

Buchholz, T., Rametsteiner, E., Volk, T. A., & Luzardis, V. A. (2009). Multi-criteria analysis for bioenergy systems assessments. *Energy Policy, 37*(2), 484–495.

Caspar, S., Kirchmann, A., Seibold, B., & Stieler, S. (2005). *Kinder, Konflikt, Karriereknick. Notwendigkeiten und Ansatzpunkte für eine bessere Vereinbarkeit von Familie und Beruf*. Tübingen: IAW Forschungsberichte, 65. IMU-Infodienst (2).

Cheng, P. C.-H., Lowe, R. K., & Scaife, M. (2001). Cognitive science approaches to understanding diagrammatic representations. *Artificial Intelligence Review, 15*, 79–94.

Cordell, D., Drangert, J.-O., & White, S. (2009). The story of phosphorus: Global food security and food for thought. *Global Environmental Change, 19*, 292–305.

Cox, R., & Brna, P. (1995). Supporting the use of external representations in problem solving: The need for flexible learning environments. *Journal of Artificial Intelligence in Education, 6*(2/3), 239–302.

Cox, R., Stenning, K., & Oberlander, J. (1994). *Graphical effects in learning logic: reasoning, representation and individual differences*. In Proceedings of the 16th annual conference of the Cognitive Science Society. Hillsdale: Lawrence Erlbaum Associates.

Daly, H. (1999). *Wirtschaft jenseits von Wachstum: Die Volkswirtschaftslehre nachhaltiger Entwicklung*. Salzburg: Pustet.

DEIAGRA (Dipartimento di Economica ed Ingeneria Agrarie). (2008). *The competition between food crops and non food crops for energy*. European Parliament, Brussels.

Deicke, M., Ruppert, H., & Schneider, J. (2006). Mining and smelting in the Harz mountains (Germany) – A neverending environmental story. *Schriftenreihe der Deutschen Gesellschaft für Geowissenschaften, 45*, 237–256.

Delzeit, R., Gömann, H., Holm-Müller, K., Kreins, P., Kretschmer, B., Münch, J., & Peterson, S. (2010). *Analysing bioenergy and landuse competition in a coupled modelling system: The role of bioenergy in renewable energy policy in Germany*. Kiel: Kiel Institute for the World Economy.

Department for Communities and Local Government. (2009). *Multi-criteria analysis: A manual*. London: Department for Communities and Local Government.

Dörner, D. (2003). *Die Logik des Misslingens – Strategisches Denken in komplexen Situationen*. Hamburg: Rowohlt.

Edenhofer, O., Pichs-Madruga, R., Sokona, Y., Seyboth, K., Arvizu, D., Bruckner, T., Christensen, J., Devernay, J. M., Faaij, A., Fischedick, M., Goldstein, B., Hansen, G., Huckerby, J., Jäger-Waldau, A., Kadner, S., Kammen, D., Krey, V., Kumar, A., Lewis, A., Lucon, O., Matschoss, P., Maurice, L., Mitchell, C., Moomaw, W., Moreira, J., Nadai, A., Nilsson, L. J., Nyboer, J., Rahman, A., Sathaye, J., Sawin, J., Schaeffer, R., & Schei, T. (2011). Summary for policy makers. In S. Schlömer, R. Sims, A. Verbruggen, C. von Stechow, K. Urama, R. Wiser, F. Yamba, T. Zwickel, O. Edenhofer, R. Pichs-Madruga, Y. Sokona, K. Seyboth, P. Matschoss, S. Kadner, T. Zwickel, P. Eickemeier, G. Hansen, S. Schlömer, & C. V. Stechow (Eds.), *IPCC special report on renewable energy sources and climate change mitigation*. Cambridge: Cambridge University Press.

Edwards, W. (1977). How to use multiattribute utility measurement for social decision making. *IEEE Transactions on Systems, Man, and Cybernetics, SMC-7*, 326–340.

Edwards, W., & Barron, F. H. (1994). SMARTS and SMARTER: Improved simple methods for multiattribute utility measurement. *Organizational Behaviour and Human Decision Processes, 60*, 306–325.

Eigner-Thiel, S. (2005). *Kollektives Engagement für die Nutzung erneuerbarer Energieträger – Motive, Mobilisierung und Auswirkungen am Beispiel des Aktionsforschungsprojektes "Das Bioenergiedorf"*. Hamburg: Kovac.

Eigner-Thiel, S. (2010). Psychologie. In Projektgruppe Bioenergiedörfer (Hrsg.), *Das Bioenergiedorf – Voraussetzungen und Folgen einer eigenständigen Wärme- und Stromversorgung durch Biomasse für Landwirtschaft, Ökologie und Lebenskultur im ländlichen Raum*. Göttingen: Institut für Bioenergiedörfer Göttingen e.V., Schriftenreihe "Fortschritt neu denken", Heft 1, 106–149.

Eigner-Thiel, S., & Geldermann, J. (2009). Entscheidungsunterstützung bei der Planung eines Bioenergiedorfes. In J. Geldermann & L. Lauven (Eds.), *Einsatz von OR-Verfahren zur Entscheidungsunterstützung*. Aachen: Shaker.

Eigner-Thiel, S., & Schmuck, P. (2010). Gemeinschaftliches Engagement für das Bioenergiedorf Jühnde: Ergebnisse einer Längsschnittstudie zu psychologischen Auswirkungen auf die Dorfbevölkerung. *Umweltpsychologie, 14*(2), 27, 98–120.

European Parliament. (2009). *Directive 2009 / 28 EC of the European Parliament and of the Council on the promotion of the use of energy from renewable sources and amending and subsequently repealing Directives 2001/77/EC and 2003/30/EC*. Official Journal of the European Union.

Ferge, T., Maghun, J., Hafner, K., Mühlberger, F., Davidovic, M., Warnecke, R., & Zimmermann, R. (2005). On-line analysis of gas-phase composition in the combustion chamber and particle emission characteristics during combustion of wood and waste in a small batch reactor. *Environmental Science and Technology, 39*, 1393–1402.

Figueira, J., Greco, S., & Ehrgott, M. (2005). *Multiple criteria decision analysis – State of the art. Surveys*. New York: Springer.

Fleury, A. (2005). *Eine Nachhaltigkeitsstrategie für den Energieversorgungssektor – dargestellt am Beispiel der Stromversorgung in Frankreich*. Karlsruhe: University of Karlsruhe (TH), French-German Institute for Environmental Research.

Fritsche, U. R., Jenseit, W., & Hochfeld, C. (1999). *Methodikfragen bei der Berechnung des Kumulierten Energieaufwands (KEA)*. Darmstadt: Öko-Institut e.V.

Fritsche, U. R., Hennenberg, K., Hermann, A., Hünecke, K., Schulze, F., Wiegmann, K., et al. (2009). *Sustainable bioenergy: Current status and outlook*. Darmstadt/Heidelberg: Öko-Institut e.V./Institute for Energy and Environmental Research.

Gallego Schmid, A. (2009). *Calculation of LCA characterization factors for terrestrial eutrophication at regional scale*. Presentation at the IX Conference "toward the global life cycle economy" 28th September–2nd October 2009. Boston: American Center for Life Cycle Assessment.

Gamba, L. (2008). *Erste Modellentwicklung zur nachhaltigen Nutzung von Biomasse*. Doctoral dissertation, Technical University Berlin, Germany.

Geldermann, J. (2010). Explanation systems. In D. Rios Insua & S. French (Eds.), *e-Democracy: A group decision and negotiation perspective* (pp. 241–259). Berlin: Springer.

Geldermann, J., & Rentz, O. (2001). Integrated technique assessment with imprecise information as a support for the identification of best available techniques (BAT). *OR Spectrum, 23*, 137–157.

Geldermann, J., & Rentz, O. (2004). The reference installation approach for the techno-economic assessment of emission abatement options and the determination of BAT according to the IPPC-directive. *Journal of Cleaner Production, 12*, 389–402.

Geldermann, J., Bertsch, V., Treitz, M., French, S., Papamichail, K. N., & Hämäläinen, R. P. (2009). Multi-criteria decision support and evaluation of strategies for nuclear remediation management. *OMEGA – The International Journal of Management Science, 37*(1), 238–251.

Geldermann, J., & Schöbel, A. (2011). On the similarities of some MCDA methods. *Journal of Multi-Criteria Decision Analysis, 18*(3–4), 219–230.

Geldermann, J., Jahn, C., Spengler, T., & Rentz, O. (1999). Proposal for an integrated approach for the assessment of cross-media aspects relevant for the determination of "best available techniques" BAT in the European Union. *International Journal of Life Cycle Assessment, 4*(2), 94–106.

Girschner, W., & Girschner-Woldt, I. (2007). Diaphane Planung als Modell nachhaltigkeitsorientierter Planungspraxis. In E. J. Krauß, M. Müller, & R. Münchmeier (Eds.), *Soziale Arbeit zwischen Ökonomisierung und Selbstbestimmung*. Kassel: University Press.

Granoszewski., K., Reise, C., Spiller, A., & Mußhoff, O. (2009). *Entscheidungsverhalten landwirtschaftlicher Betriebsleiter bei Bioenergie-Investitionen* (Discussion Paper No. 0911). Göttingen: University of Göttingen, Department of Agricultural Economics and Rural Development.

Gromke, U., & Detzel, A. (2006). *Ecological comparison of office papers in view of fibrous raw material*. Heidelberg: Institute for Energy and Environmental Research.

Guski, R. (2000). *Wahrnehmung – eine Einführung in die Psychologie der menschlichen Informationsaufnahme*. Stuttgart: Kohlhammer.

Hämäläinen, R. P., & Alaja, S. (2008). The threat of weighting biases in environmental decision analysis. *Ecological Economics, 68*(1/2), 556–569.

Heiland, S., Tischer, M., Döring, T., Pahl, T., & Jessel, B. (2003). *Indikatoren zur Zielkonkretisierung und Erfolgskontrolle im Rahmen der Lokalen Agenda 21*. Berlin: Umweltbundesamt.

Heijungs, R., Guinée, J. B., Huppes, G., Lankreijer, R. M., Udo de Haes, H. A., & Wegener Sleeswijk, A. (1992). *Environmental life cycle assessment of products – Backgrounds*. Leiden: Centrum voor Milieukunde.

Helsinki Conference. (1993). *Second ministerial conference on the protection of forests in Europe*. 16–17 June 1993, Helsinki/Finland.

Hewitt, C. N., & Jackson, A. V. (Eds.). (2003). *Handbook of atmospheric science*. Malden: Blackwell.

Hischier, R., & Gilgen, P. W. (2005). Life cycle assessment databases as part of sustainable development strategies: The example of ecoinvent. In L. M. Hilty, E. K. Seifert, & R. Treibert (Eds.), *Information systems for sustainable development* (pp. 15–24). Hershey: Idea Group Publishing.

Hoffmann, D. (2007). *Regionale Wertschöpfung durch optimierte Nutzung endogener Biomassepotenziale*. Doctoral dissertation, Saarland University, Germany.

Hofstetter, P. (1998). *Perspectives in life cycle impact assessment; a structured approach to combine models of the technosphere, ecosphere and valuesphere*. Dordrecht: Kluwer Academic Publishers.

Hughes, L. (2003). Climate change and Australia: Trends, projections and impacts. *Austral Ecology, 28*(4), 423–443.

Huijbregts, M. A. J., Rombouts, L. J. A., Hellweg, S., Frischknecht, R., Hendriks, J., van de Meent, D., Ragas, A. M. J., Reijnders, L., & Struijs, J. (2006). Is cumulative fossil energy demand a useful indicator for the environmental performance of products? *Environmental Science and Technology, 40*(3), 641–648.

Hwang, C. L., & Yoon, K. (1981). *Multiple attribute decision making, methods and applications – A state-of-the-art survey*. Heidelberg: Springer.

IEA (International Energy Agency). (2004). *Biofuels for transport: An international perspective*. Paris: IEA.

IPCC. (2007). Fourth assessment report: Climate change 2007. In S. Solomon, D. Qin, M. Manning, Z. Chen, M. Marquis, K. B. Averyt, M. Tignor, & H. L. Miller (Eds.), *The physical science basis*. New York: Cambridge University Press.

Ipsen, D. (1993). Regionale Identität. *Überlegungen zum politischen Charakter einer psychosozialen Raumkategorie. Raumforschung und Raumordnung, 51*, 9–18.

Jessel, B. (2008). *Biodiversity and bioenergy*. Presentation at the 9th meeting of the Conference of the Parties (COP 9), 19.-30.05. 2008. Bonn, Germany.

Kempener, R., Beck, J., & Petrie, J. (2009). Design and analysis of bioenergy networks. *Journal of Industrial Ecology, 13*, 284–305.

Kimming, M., Sundberg, C., Nordberg, A., Baky, A., Bernesson, S., Norén, O., & Hansson, P. A. (2011). Biomass from agriculture in small-scale combined heat and power plants – A comparative life cycle assessment. *Biomass and Bioenergy, 35*(4), 1572–1581.

Kort, J., Collins, M., & Ditsch, D. (1998). A review of soil erosion potential associated with biomass crops. *Biomass and Bioenergy, 14*(4), 351–359.

Larkin, J. H., & Simon, H. A. (1987). Why a diagram is (sometimes) worth ten thousand words. *Cognitive Science, 11*, 65–99.

Leitl, T. (2007). *Biomasse, Bestandteil der RWE-Strategie im Bereich erneuerbare Energien*. Presentation at the 2nd bioEnergy conference Carinthia, 06.11.2007. Velden, Austria.

Lenz, V. A. (2010). *Feinstaubminderung im Betrieb von Scheitholzkaminöfen unter Berücksichtigung der toxikologischen Relevanz*. DBFZ Report Nr. 3. Leipzig.

Lobb, D. A., Gary Kachanoski, R., & Miller, M. H. (1999). Tillage translocation and tillage erosion in the complex upland landscapes of southwestern Ontario, Canada. *Soil and Tillage Research, 51*(3–4), 189–209.

Malczewski, J. (1999). *GIS and multicriteria analysis*. New York: Wiley.

Miller, G. A. (1956). The magic number seven plus or minus two: Some limits on our capacity for communicating information. *Psychological Review, 63*, 81–97.

Millennium Ecosystem Assessment. (2005). *Ecosystems and human well-being. Current state and trends*. Washington, DC: Island Press.

Monier V., & Labouze E. (2001). *Critical review of existing studies and life cycle analysis on the regeneration and incineration of waste oils*. In European Commission, DG Environment.

Munda, G. (1995). *Multicriteria evaluation in a fuzzy environment: Theory and applications in ecological economics*. Heidelberg: Physika-Verlag.

Mustajoki, J., Hämäläinen, R. P., & Martunen, M. (2003). Participatory multicriteria decision analysis with Web-HIPRE: A case of lake regulation policy. *Environmental Modelling & Software. Vol., 19*(6), 537–547.

Nixon, S. C., Lack, T. J., Hunt, D. T. E., Lallana, C., & Boschet, A. F. (2000). *Sustainable use of Europe's water?* Copenhagen: European Environment Agency.

NUTS (2007). Verordnung (EG) Nr. 105/2007 der Kommission vom 1. Februar 2007 zur Änderung der Anhänge der Verordnung (EG) Nr. 1059/2003 des Europäischen Parlaments und des Rates über die Schaffung einer gemeinsamen Klassifikation der Gebietseinheiten für die Statistik.

Oberschmidt, J., Ludwig, J., & Geldermann, J. (2009). Ein modifizierter PROMETHEE-Ansatz zur Lebenszyklus-orientierten Bewertung der Strom- und Wärmeversorgung. In J. Geldermann & L. Lauven (Eds.), *Einsatz von OR-Verfahren zur Entscheidungsunterstützung*. Aachen: Shaker.

Oberschmidt, J., Geldermann, J., Ludwig, J., & Schmehl, M. (2010). Modified PROMETHEE approach to assessing energy technologies. Special issue: Uses of frontier efficiency methodologies and multi-criteria decision making for performance measurement in the energy sector. *International Journal of Energy Sector Management, 4*(2), 183–212.

OECD. (2002). *Family-friendly policy can generate a range of benefits to society*. Retrieved August 23, 2010, from http://www.oecd.org/document/10/0,3343,en_21571361_44315115_1837194_1_1_1_1,00.html

OECD. (2005). *Trends and determinants of fertility rates. The role of policies, social, employment and migration* (Working Paper No 27). Paris: OECD.

Ott, K. (2003). The case for strong sustainability. In K. Ott & P. Thapa (Eds.), *Greifswald's environmental ethics*. Greifswald: Steinbecker Verlag Ulrich Rose.

Page, B., & Rautenstrauch, C. (2001). Environmental informatics-methods, tools and applications in environmental information processing. In C. Rautenstrauch & S. Patig (Eds.), *Environmental information systems in industry and public administration* (pp. 2–11). Hershey: Idea Group Publishing.

Palmer, S. E. (1978). Fundamental aspects of cognitive representation. In E. Rosch & B. B. LLoyd (Eds.), *Cognition and categorization* (pp. 259–303). Hillsdale: Erlbaum.

Petre, M., & Green, T. R. G. (1993). Learning to read graphics: Some evidence that 'seeing' an informational display is an acquired skill. *Journal of Visual Languages and Computing, 4*, 55–70.

Pimentel, D. (2006). Soil erosion: A food and environmental threat. *Environment, Development and Sustainability, 8*, 119–137.

Projektgruppe Bioenergiedörfer. (2010). *Das Bioenergiedorf – Voraussetzungen und Folgen einer eigenständigen Wärme- und Stromversorgung durch Biomasse für Landwirtschaft, Ökologie und Lebenskultur im ländlichen Raum*. Göttingen: Institut für Bioenergiedörfer Göttingen e.V., Schriftenreihe "Fortschritt neu denken", Heft 1.

Rautenstrauch, C. (1999). *Betriebliche Umweltinformationssysteme, Grundlagen, Konzepte und Systeme*. Berlin/Heidelberg: Springer.

Reul, F. (2002). *Entwicklung einer Nachhaltigkeitsstrategie für den Stadtverkehr – das Beispiel Berlin*. Doctoral dissertation, Humboldt-University Berlin, Germany. Retrieved July 13, 2009, from http://edoc.hu-berlin.de/dissertationen/reul-frithjof-2002-12-17

Rodrígues, L., Reuter, H. I., & Hengl, T. (2008). A framework to estimate the distribution of heavy metals in European soils. In G. Tóth, L. Montanarella, & E. Rusco (Eds.), *Threats to soil quality in Europe* (pp. 79–86). Ispra: Joint Research Centre, European Commission.

Roedl, A. (2010). Production and energetic utilization of wood from short rotation coppice – A life cycle assessment. *International Journal of Life Cycle Assessment, 15*(6), 567–578.

Rösch, C., Skarka, J., Raab, K., & Stelzer, V. (2009). Energy production from grassland – Assessing the sustainability of different process chains under German conditions. *Biomass and Bioenergy, 33*(4), 689–700.

Roth, W.-M., & Bowen, G. M. (1999). Of cannibals, missionaries, and converts: Graphing competencies from grade 8 to professional science inside (classrooms) and outside (field/laboratory). *Science, Technology, and Human Values, 24*(2), 179–212.

Ruppert, H., Eigner-Thiel, S., Girschner, W., Karpenstein-Machan, M., Roland, F., Ruwisch, V., Sauer, B., & Schmuck, P. (2008). *"Wege zum Bioenergiedorf" – Leitfaden für eine eigenständige Wärme- und Stromversorgung auf Basis von Biomasse im ländlichen Raum*. Gülzow: Fachagentur Nachwachsende Rohstoffe.

Rusco, E., Montanarella, L., & Bosco, C. (2008). Soil erosion: A main threats to the soil in Europe. In G. Tóth, L. Montanarella, & E. Rusco (Eds.), *Threats to soil quality in Europe European Commission* (pp. 37–45). Ispra: Joint Research Centre.

Saaty, T. L. (1980). *The analytic hierarchy process*. New York: McGraw Hill.

Schmehl, M., Eigner-Thiel, S., Hesse, M., Ibendorf, J., & Geldermann, J. (2010). Development of an information system for the assessment of different bioenergy concepts regarding sustainable development. In M. G. Teuteberg (Ed.), *Corporate environmental management information systems: Advancements and trends* (pp. 318–336). Hershey: IGI Global.

Schmehl, M., Hesse, M., & Geldermann, J. (2012). *Ökobilanzielle Bewertung von Biogasanlagen unter Berücksichtigung der niedersächsischen Verhältnisse*. Research Paper der Georg-August-Universität Göttingen, Wirtschaftswissenschaftliche Fakultät, Schwerpunkt Unternehmensführung, Professur für Produktion und Logistik, Nr.11, Göttingen.

Schmitz, S. (1995). *Ökobilanz für Getränkeverpackungen*. Berlin: Umweltbundesamt.

Schmitz, S., & Paulini, I. (1999). *Valuation as an element of life cycle assessment – German Federal Environmental Agency method for impact indicator standardization, impact category grouping (ranking), and interpretation in accordance with ISO 14042 and 14043*. In Federal Environmental Agency, Berlin.

Schmuck, P., Eigner, S., Kaufhold, A., & Krapoth, S. (1998). *Psychologische Aspekte von Unternehmensbewertung*. Schriftenreihe des Koordinations- und Studienzentrums Frieden und Umwelt, Heft 13. Göttingen, Germany.

SCOPE (Scientific Committee on Problems of the Environment). (1995). *Environmental indicators: systematic approach to measuring and reporting on the environment in the context of sustainable development*. Discussion paper presented at the workshop "International Consultation of Sustainable Development Indicators", Ghent. Bruxelles, Belgium: Bureau du Plan.

SEK. (2006). *Mitteilung der Kommission an den Rat, das Europäische Parlament, den Europäischen Wirtschafts- und Sozialausschuss und den Ausschuss der Regionen – Ein Fahrplan für die Gleichstellung von Frauen und Männern 2006–2010*. Retrieved August 23, 2010, from http://eur-lex.europa.eu/LexUriServ/LexUriServ.do?uri=COM:2006:0092:FIN:DE:HTML

SETAC. (1996). Towards a methodology for life cycle impact assessment. In *Society of environmental toxicology and chemistry*, Brussels.

Thuiller, W., Lavergne, S., Roquet, C., Boulangeat, I., Lafourcade, B., & Araujo, M. (2011). Consequences of climate change on the tree of life in Europe. *Nature, 470*(7335), 531–534.

Treitz, M., Bertsch, V., Schollenberger, H., Geldermann, J., & Rentz, O. (2008). *Multi-criteria decision support for process design*. International Transactions in Operational Research (accepted), Vol. Special, Issue of the IFORS conference in Hawaii 2005, 10–15 July Honolulu, Hawaii.

Tversky, B. (2000). Some ways maps and diagrams communicate. In C. Freksa, W. Brauer, C. Habel, & K. F. Wender (Eds.), *Spatial cognition II*. Berlin: Springer.

Udo de Haes, H. A. (1996). *Towards a methodology for life cycle impact assessment*. Brussels: SETAC-Europe.

UNCED. (1992). *Earth Summit, United Nations conference on environment and development*, Rio de Janeiro, 3–14 June 1992

United Nations. (1987). General Assembly Resolution, 42/187.

Van Haaren, C., & Bathke, M. (2007). Integrated Planning and remuneration of agri-environmental goods and services for the management of sustainable landscapes – Results of a case study in the Fuhrberg region of Germany. *Journal of Environmental Management, 89*, 209–221.

Van Loo, S., & Koppejan, J. (Eds.). (2008). *Handbook of biomass combustion and co-firing*. London: Earthscan.

VDI (1997). *Cumulative energy demand – Terms, definitions, methods of calculation*. In Verein Deutscher Ingenieure (VDI) (Ed.). Düsseldorf: Beuth Verlag.

Vester, F. (2003). *Die Kunst, vernetzt zu denken Ideen und Werkzeug für einen Umgang mit Komplexität Der neue Bericht an den Club of Rome*. München: dtv.
Vis, M. W., Vos, J.,& van den Berg, D. (2008). *Sustainability criteria & certification systems for biomass production*. Final report no. 1386 for the European Commission. Brussels
von Winterfeldt, D., & Edwards, W. (1986). *Decision analysis and behavioral research*. Cambridge: Cambridge University Press.
WBGU. (2008). *Wissenschaftlicher Beirat der Bundesregierung Globale Umweltveränderungen. Zukunftsfähige Bioenergie und nachhaltige Landnutzung*. Berlin: Springer.
Weber-Blaschke, G., Frieß, H., Peichl, L., & Faulstich, M. (2002). Aktuelle Entwicklungen bei Umweltindikatorensystemen. *Zeitschrift für Umweltwissenschaften und Schadstoffforschung, 14*(3), 187–193.
Wei, B., & Carlson, R. (2002). *Practical Pre-Study of GIS application with LCA database SPINE*. Presentation at the fifth international conference on Ecobalance, November 6–8, 2002. Tsukuba, Japan.
Weidenmann, B. (Ed.). (1994). *Wissenserwerb mit Bildern*. Bern: Huber.
Werheit, M. (1996). *Indikatoren für eine nachhaltige Entwicklung*. Information letter No. 29 of the Independent Institute for Environmental Concerns (UfU), Berlin, Germany.
Wiehe, J., Ruschkowski, E., Rode, M., Kanning, H., & Haaren, C. v. (2009). Auswirkungen des Energiepflanzenanbaus auf die Landschaft am Beispiel des Maisanbaus für die Biogasproduktion in Niedersachsen. *Naturschutz und Landschaftsplanung, 41*(4), 107–113.
Wilkens, I. (nee Braune), & Schmuck. P (2012). Transdisciplinary evaluation of energy scenarios for a German village using multi-criteria decision analysis. *Sustainability, 4*(4), 604–629.
World Health Organization. (2006). *WHO Air quality guidelines for particulate matter, ozone, nitrogen dioxide and sulfur dioxide*. Geneva: World Health Organization.
Wuppertal-Institut für Klima, Umwelt, Energie. (1997). *Zukunftsfähiges Deutschland – ein Beitrag zu einer global nachhaltigen Entwicklung*. Basel: Birkhäuser.
Wüste, A., & Schmuck, P. (2012). Bioenergy villages and regions in Germany: An interview study with initiators of communal bioenergy projects on the success factors for restructuring the energy supply of the community. *Sustainability, 4*(2), 244–256. doi:10.3390/su4020244. Retrieved December 3, 2012, from http://www.mdpi.com/2071-1050/4/2/244/
Yue, C. Y., & Li, H. (1998). Multi-attribute evaluation for environment, resource and sustainable development. *Monitoring and Assessment, 12*, 28–30.

Part VI
Combustion of Biomass for Heat and Power

Chapter 13
Emissions of Organic and Inorganic Pollutants During the Combustion of Wood, Straw and Biogas

Torben Seidel, Jürgen Orasche, Hans Ruppert, Jürgen Schnelle-Kreis, and Hans Hartmann

Abstract In Europe, wood combustion in stoves and boilers is widely applied for residential heating. In Germany, approximately 15 million of 40 million households own small-scale furnaces, which deliver 7 % of Germany's heat consumption. Using state-of-the-art small-scale combustion systems, we investigated how the air quality changes due to the emissions of harmful elements and organic pollutants during the combustion of wood and straw.

Heavy metals: Beside the fuel, we analysed all the originating ashes – grate ash, heat exchanger ash, and fly ash – to reconstruct element fluxes. As the input/output balance calculations show, some elements – such as cadmium, zinc, tin, thallium, lead, bismuth and antimony – may also be retained within the cooler zones of the furnace, in the chimney, or in the refractory lining material where samples could not

T. Seidel • H. Ruppert (✉)
Department of Sedimentology and Environmental Geology, Interdisciplinary Centre for Sustainable Development, University of Göttingen, Goldschmidtstr. 3, D-37077 Göttingen, Germany
e-mail: hrupper@gwdg.de

J. Orasche
Department of Sedimentology and Environmental Geology, Interdisciplinary Centre for Sustainable Development, University of Göttingen, Goldschmidtstr. 3, D-37077 Göttingen, Germany

Joint Mass Spectrometry Centre, Cooperation Group "Comprehensive Molecular Analytics", Helmholtz Zentrum München, Munich, Germany

Joint Mass Spectrometry Centre, Institute of Chemistry, Division of Analytical and Technical Chemistry, University of Rostock, Rostock, Germany

J. Schnelle-Kreis
Joint Mass Spectrometry Centre, Cooperation Group "Comprehensive Molecular Analytics", Helmholtz Zentrum München, Munich, Germany

H. Hartmann
Technology- and Support Centre of Renewable Raw Materials (TFZ), Schulgassse 18, Straubing D-94315, Germany

H. Ruppert et al. (eds.), *Sustainable Bioenergy Production - An Integrated Approach*,
DOI 10.1007/978-94-007-6642-6_13,

be taken. Only the elements contained in the filter ash are emitted; at most, these element portions represent 30 % of the amount contained in the fuel.

Organic pollutants: The concentration of organic compounds strongly depends on the fuel type, the furnace and the combustion conditions. The emission of, for instance, polycyclic aromatic hydrocarbons (PAHs; especially Benzo(a)pyrene) can only barely be detected in wood pellet boilers, is more in wood chip furnaces and is more than a factor 100 higher in wood-log-fuelled fireplaces, indicating inappropriate conditions for complete oxidation. This situation is critical, considering that there are now six million wood log fireplaces in Germany.

The pollutants are bound in fine (<1 μm) particles or gascous compounds and may enter the lungs' alveoli and contaminate the body. Clearly, effective emission reduction measures are necessary.

Keywords Combustion • Wood • Straw burning • Ash composition • Concentrations and balances of inorganic and organic pollutants • Heavy metals • PAH • Emission • Air pollution • Health

Abbreviations

CHE	condensing heat exchanger
CHP plant	combined heat and power plant
EC	elemental carbon
ESP	electrostatic precipitator
ICP-OES	inductively coupled plasma optical emission spectrometer
ICP-MS	inductively coupled plasma mass spectrometer
MJ	Megajoule = 10^6 Joule
OC	organic carbon
PAH	polycyclic aromatic hydrocarbon
PJ	Petajoule = 10^{15} Joule
PM	particulate matter
$PM_{2.5}$	particulate matter with a diameter of smaller than 2.5 μm
SOA	secondary organic aerosol
TEQ	toxic equivalent
TSP	total suspended particulate matter (the total of all particles in the air)
TTC	threshold of toxicological concern
WSOC	water-soluble organic compounds

13.1 Introduction

Socially accepted as well as ecologically and economically feasible concepts for our energy supply should be sought to counteract the greenhouse gas effect, the shortage of non-renewable fossil energy sources and dependence on energy suppliers. Wood residues from forests, wood from short rotation tree plantations and straw are interesting alternative sources to replace fossil fuels. In Germany, of which approximately 30 % is covered in forests, 54.7 million solid cubic meters of wood, corresponding to 43 % of the wood production, were used for energy purposes in 2008 (FNR 2012), mostly for heat production. The energy potential of wood from forestry is estimated to be 360 PJ in 2050 (FNR 2012). The use of wood fuel offers many advantages if forests are not cut down or overexploited:

- Wood from sustainable forestry provides a well-balanced CO_2 cycle (CO_2 neutrality). Energy wood use can promote forest care and allows for an optimum use of wood that is fairly low in quality and value.
- In contrast to other renewable energy sources, such as wind and solar power, these natural fuels are available when needed.
- The "grey energy" – i.e. the energy necessary to utilise the raw material – is low. In relation to its energy value, the energy consumed in wood pellet production is below 6 %, while the energy consumption required to produce natural gas, liquid petroleum gas and heating oil is between 10 % and 14.5 % (FNR 2012).
- Wood processing, transportation, storage and combustion are low risk activities. Accidents and environmental damage (e.g., leakage of tank trucks) have been reduced to a minimum.
- Applying renewable energy sources saves fossil resources for future use.
- Because wood is available locally and regionally, the added value remains in the region, which helps stabilise the local economy and jobs.

The disadvantages of wood combustion are the emission of gaseous, liquid and solid substances that may impact our health negatively. Given the increasing use of wood for heating, this aspect should not be neglected, since it may lead to some parts of the population having problems accepting it. Fortunately, due to technical advancement – partly mandated by the amendment of the Ordinance on Small and Medium Firing Installations (Deutscher Bundestag 2009) –, modern firing facilities' emission of harmful gases and particulate matter has decreased considerably.

In this chapter, we focus on organic and inorganic pollutants' concentration in the dust and their fluxes from the chimney into the environment. Ideally, complete combustion should be achieved; the emitted particulate matter should then mainly consist of soluble salts – such as chlorides, sulphates and nitrates but also of carbonates and oxides – of the elements potassium, calcium and sodium – as well as low concentrations of harmful elements, which will depend on the fuel composition. If the combustion is not perfect, the flue gas also consists of organic matter that may contain a wide range of harmful organic compounds.

In this chapter, we specifically concentrate on heat production through biomass (wood and straw) and show how the quality of different fuels and heating systems' burning conditions affect the composition of emitted exhaust gases and ashes. All the burning experiments were performed under typical conditions to ensure results that mirror reality. By combining the concentrations of elements in the fuel, the formed ashes and the emitted aerosols (fly ash) with the amount of burnt fuel, the amount of the various ash fractions and the amount of emitted aerosols, we were able to calculate the energy-normalised mass fluxes of every element in different combustion facilities (large-scale and small-scale furnaces). In addition to the elements, we analysed many organic compounds in the fly ashes resulting from incomplete combustion and calculated the emission factors for all the compounds. This allows an evaluation of the emissions of both the inorganic and organic compounds released from different sources.

We additionally analysed the exhaust air produced by a combined heat and power (CHP) plant relying on biogas as a low-emission reference.

13.2 Particulate Matter Emission

The amount of emitted particulate matter, its grain size distribution and its composition determine how health is affected by harmful substances released into the air during the combustion. Respirable particulate matter smaller than a few micrometers is considered a strong potential hazard for health. It can therefore be assumed that the acceptance of biomass fuels depends on the amount of particulate matter released into the air. The combustion of straw or wood fuels with a high bark content specifically leads to high amounts of ash residues. Wood pellets are mostly produced from by-product of wood processing which usually do not contain any bark, what leads to low ash residues and low ash emissions.

Particulate matter (PM) comprises particles that do not sediment immediately and remain in the atmosphere for longer. According to Brunner et al. (2006), a distinction is made between coarse-mode particles (from approximately 40 to 1 μm) and fine-mode particles (<1 μm).

Since airborne particles do not have a uniform shape and density, a dust particle's size is characterised by its aerodynamic diameter. This diameter describes a spherical particle with a diameter of 1 g/cm^3 that drops in air at the same velocity as the monitored particle. Definitions:

TSP:	Total suspended particulate matter (the total of all particles in the air)
PM_{10}:	Particles with an aerodynamic diameter < 10 μm
$PM_{2.5}$:	Particles with an aerodynamic diameter < 2.5 μm
$PM_{1.0}$:	Particles with an aerodynamic diameter < 1 μm
Ultrafine particle:	Particles with an aerodynamic diameter < 0.1 μm.

According to Klippel and Nussbaumer (2007), aerosols formed during combustion and identified in flue gas can be classified as follows:

- Soot, which is generated as a synthesis product during the incomplete combustion of organic matter under oxygen depletion
- Inorganic, salt-like particulate matter, which is easily visible during complete combustion

- Particulate and condensable organic compounds from incomplete combustion. They are formed, for instance, in poorly operating wood stoves as decomposition or synthesis products and may contain strong enrichments of polycyclic aromatic hydrocarbons (PAH).

Depending on their size, aerosols have different deposition maxima on the way to or in lungs. The group PM_{10} is called the respirable group. Particles larger than 10 μm deposit in the extrathoracic respiratory tract (mouth, nose, throat), conversely, particles between 2.5 and 10 μm usually deposit in the respiratory tract (mouth, nose, throat, bronchia). Particles smaller than 2.5 μm deposit in the respiratory tract and the alveoli. A particle's penetration depth depends on different factors, such as the particle size, form and hygroscopy. Physiological and anatomical factors, such as the anatomy of the respiratory tract, pathological changes, mouth/nasal respiration, inhalation volume and the breathing space, also play a central role (US-EPA 2004). To date, the limiting values for particle emissions from wood combustion are provided as mass concentrations, but do not consider any concentrations of their toxic and nontoxic substances.

One of the best-known studies on the health relevance of fine dust is the Harvard Six Cities Study (Dockery et al. 1993). More than 8,000 people were surveyed for 15 years and their illnesses, discomforts and symptoms recorded. The influence that TSP, ozone, SO_2, SO_4 and $PM_{2.5}$ had on their health was examined. Although the smoking, overweight and school education were included as factors, the town with the highest particulate matter load associated with sulphate still had a significant 26 % extra deaths. Based on this study, Laden et al. (2006) investigated six towns for more than 8 years; here, the air's particulate matter load was lower. While this study verified the results of the previous study, it also showed a decrease in mortality with a decreasing aerial load. An additional load of 10 μg/m^3 of $PM_{2.5}$ increases the mortality rate by a factor of 1.16, the lung cancer death risk by a factor of 1.27, and the heart circulatory disease death risk by a factor of 1.28. Mortality was also associated with sulphate. No association was verified between mortality and TSP, CO and NO_2.

A study by Pope and Dockery (2006) shows the influence of a short-term exposure to $PM_{2.5}$. Such exposure leads to the blockage of the coronary arteries and causes heart circulatory illnesses. In this study, 12,865 patients were examined for circulation problems in the heart arteries. The risk of such an illness increases by 4.5 % if the concentration of $PM_{2.5}$ rises by 10 μg/m^3.

A study on children in Switzerland links TSP to upper airway symptoms and coughing (Braun-Fahrländer et al. 1992; Braun-Fahrländer 2001). Dockery et al. (1996) ascertained an increase in the prevalence of bronchitis and a significant association between particulate matter and lung function parameters. Ackermann-Liebrich et al. (1997) as well as Zemp et al. (1999) associate TSP, PM_{10} and NO_2 with chronic cough, chronic sputum, breathlessness and reduced lung function. There is a higher risk of diabetics and cardiovascular complications (Zanobetti et al. 2000).

The health risks of chronic exposure to particulate matter are manifold and seem to be associated with shorter life expectancy. That is why the EU guideline 1999/30/EG aimed at a daily mean PM_{10} concentration of 50 μg/m^3 and an annual mean of

40 μg/m³ in the first stage (Council of the European Union 1999). The second stage (from 2010) retains the daily mean, but reduces the annual mean to 20 μg/m³. The daily mean of 50 μg/m³ may not be exceeded on more than 35 days per year. This guideline was converted to a national right in the Ordinance for the Implementation of the Federal Immission Control Act (Deutscher Bundestag 2010). The Federal Environment Agency documents address the exceeding of particulate matter limits in the Germany (UBA 2009).

13.3 Pollutant Formation Mechanisms in Biomass Combustion

Combustion products are formed during combustion of woody and non-woody biomass. These can be categorised as follows:

- Products from the complete combustion of organic compounds, such as carbon dioxide (CO_2) and water (H_2O)
- Products from the incomplete combustion of organic compounds, such as carbon monoxide (CO), hydrocarbons (C_nH_m, tar), polycyclic aromatic hydrocarbons (PAK), soot (C contaminated by organic groups), etc.
- Non-ignitable emissions of dust and ash containing inorganic heavy metals such as Cu, Pb, Zn, Cd as well as nitrogen, sulphur, chlorine, potassium and calcium compounds (NO, NO_2, HCN, NH_3, N_2O, SO_2, HCl, KCl, $CaSO_4$, etc.).

The only residues after complete combustion are CO_2, water and ash. Nussbaumer (1989) presents a scheme with important products emitted during the combustion of wood (see Fig. 13.1).

The CO_2 released during biomass combustion is considered climate-neutral, because only the amount of CO_2 taken up by photosynthesis from the atmosphere during the growth of the plant is released. The water originates from fuel moisture content and from the oxidation of the hydrogen contained in biomass.

The picture changes, however, during incomplete combustion. Unburnt or partially burnt components are then delivered into the air; but owing to their toxicity, these should be minimised. Optimal combustion conditions are therefore a priority. However, this optimum can be reached in different ways, depending on the fuels and the fireplaces' characteristics. According to Kaltschmitt et al. (2009), the properties of the combustion chamber and furnace operation should be optimally harmonised with the burning conditions of the biogenic fuels and should fulfil the following requirements:

- *The division of the added combustion air into primary air and secondary air.* Primary air is required for fuel and carbon gasification, while secondary air supports the burning of gases. Primary air influences the heat input, while secondary air influences the burnout of the combustion gases.
- *An excess of oxidising agents.* If the excess air supply (lambda value) is lower than the optimum value, this results in oxygen depleted zones; if the excess air

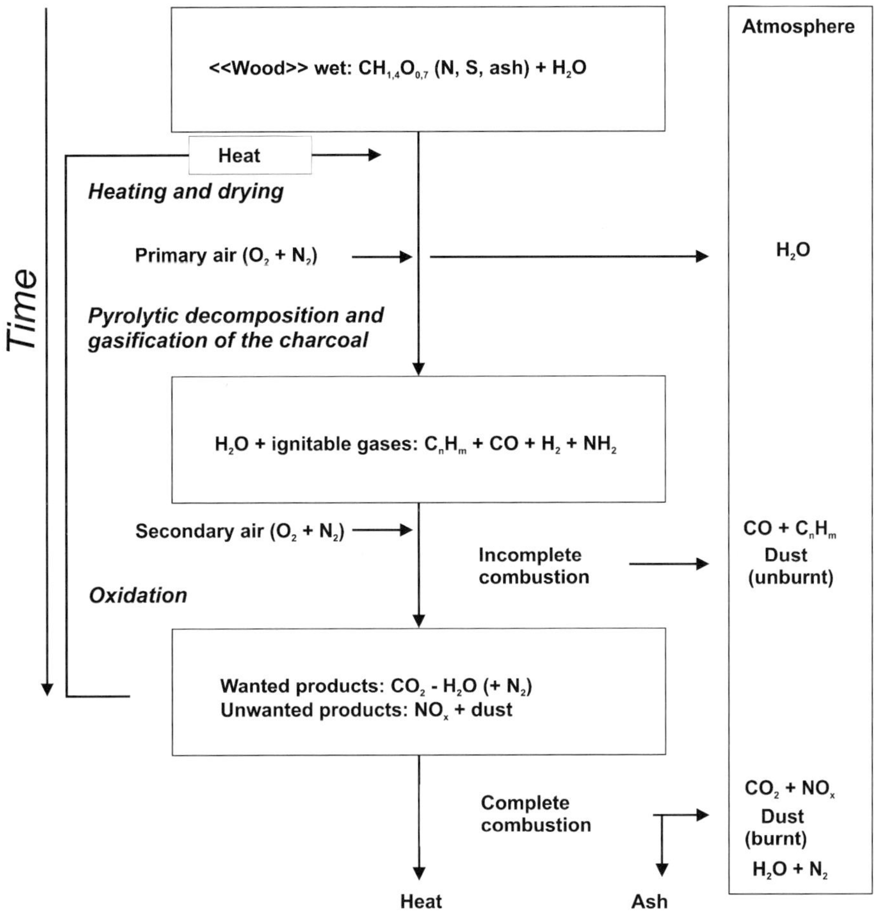

Fig. 13.1 Process chart of products formed during wood combustion (following Nussbaumer 1989)

supply is too high, this results in an undercooling of the flame, accompanied by incomplete combustion.

- *A homogeneous mixture of secondary air and ignitable gases.*
- *A high temperature and the ignitable gases should remain in the hot zone for a sufficient time* (at least 850 °C for 0.5 s). In short, a combustion plant must be optimised according to the 3T criteria (time, temperature, and turbulence) as well as the available oxygen. The constructional conditions can have an adverse influence on these factors and can thus result in elevated C contents in the ash as well as in the enhanced emissions.

The nitrogen oxide NO and NO_2 – which are summarised as NO_x – originate during the biomass combustion. The source of the nitrogen is the nitrogen in the combustion air (79 Vol.-%) and the nitrogen contained in the biomass. NO_x can develop in different ways. The most significant formation occurs at a high temperature (1,300–1,400 °C) and a high oxygen content by means of the oxidation of the nitrogen (Zeldovich 1946).

In addition to NO_x gases, emissions of sulphur, chlorine and potassium compounds occur. These emissions are also low due to the relatively low content of these components in biomass. Sulphur contained in the fuel can be released from the ash as $CaSO_4$ or K_2SO_4, or can be released into the air with the exhaust gases as SO_2 and SO_3, or as H_2S in special cases.

After the combustion, chlorine can be found in the ash, predominantly in the form of salts (KCl, NaCl). Very small amounts of HCl can also be emitted. The Technical Instructions on Air Quality Control (TA-Luft; BMU 2002) is seeking an HCl emissions limit of 30 mg/m^3 (at 11 % O_2 content), which would be exceeded with grass combustion, which typically has emissions of 20–120 mg/m^3; grass combustion should therefore be omitted. Potassium, which is especially present in the form of KCl and K_2SO_4 in exhaust gas, plays an important role by enhancing slagging and the corrosion of boilers.

The aerosol products originating from combustion are not only solid and gaseous, but also liquid. These aerosols can be formed during complete and incomplete combustion. One can distinguish between primary and secondary particles. Primary particles are released directly from processes, while secondary particles originate from nucleation, condensation, etc. Unburned particles can specifically be released in large amounts if combustion does not follow the 3T rule and produces particles larger than 1 μm with a high carbon content. These organic particles are formed during pyrolysis and may still contain oxidisable carbon compounds. One can distinguish between C-containing solid or liquid decomposition products and C-containing condensed synthesis products. The organic compounds originating from decomposition during the pyrolysis of fuel are considered C-containing decomposition products.

Owing to the increasing supply of energy from biomass, especially wood, the importance of the emission of particles and nitrous gases has increased steadily (van Loo and Koppejan 2002). Studies by Hinds (1998) and Friedlander (2000) provide a detailed summary of aerosol formation. In comparison to other heating systems, wood also emits high amounts of particles at almost complete combustion (Nussbaumer and Hasler 1999; Kessler et al. 2000). Particle-shaped salts are emitted from the fuel. The following particle-forming path is suggested as the origin of these salts: The vaporisation of the salts in the glow bed, carried away in the flue gas stream and then recondensation (Livbjerg 2001). The chemical compounds formed did not originally exist in the wood. The originating particles' diameters are around 0.1 μm (Hasler and Nussbaumer 1998; Oser et al. 2003). Volatile salts can also be carried along with the flue gas (Livbjerg 2001). These salts consist predominantly of potassium and calcium as sulphate, chloride, carbonate, oxide and hydroxide compounds (Oser et al. 2003).

The principle of particles forming during combustion is shown in Fig. 13.2. Five particle-forming paths can be distinguished. The most important path during appropriate combustion is the solid-gaseous-particle path, in which the vaporisation of

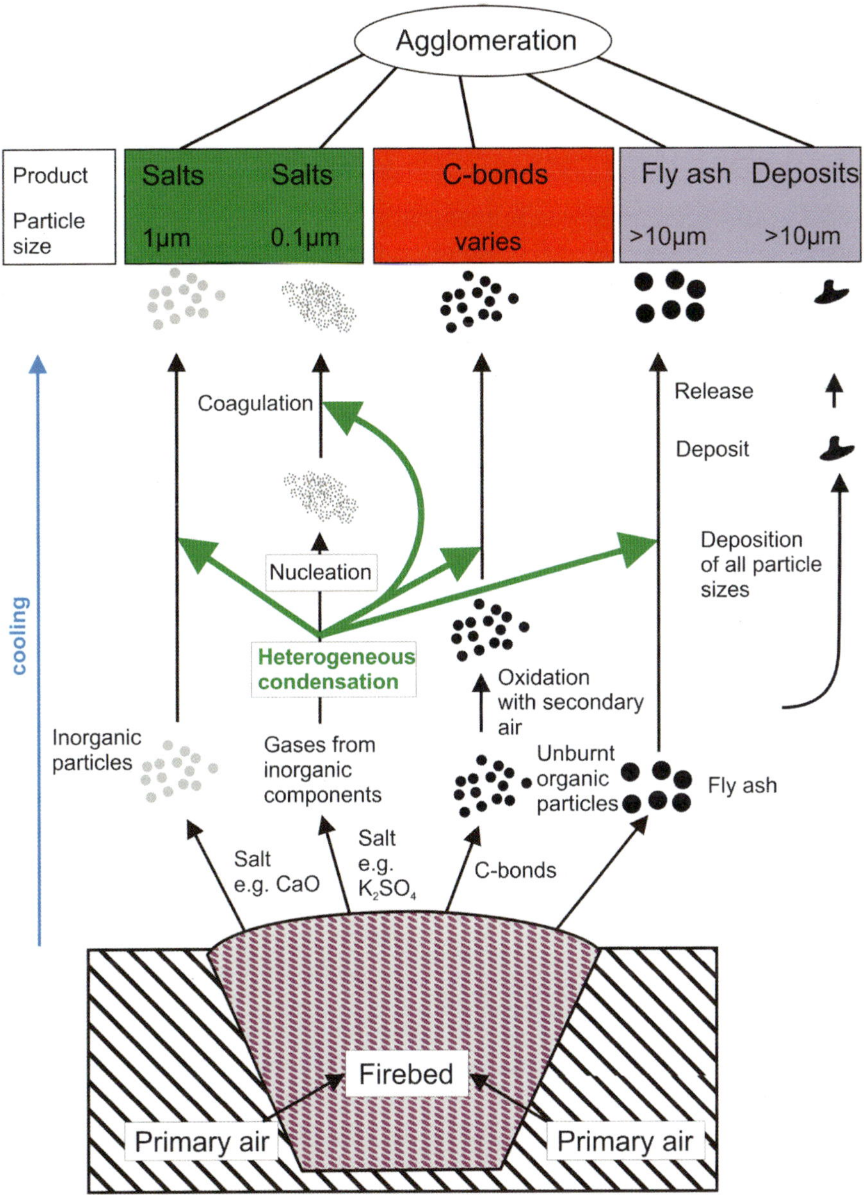

Fig. 13.2 Formation of particles and the typical particle size range (following Oser et al. 2003)

the inorganic wood components plays a central role. If complete combustion is achieved, aerosols are formed by the gas-to-particle conversion of ash-forming vapours. The vaporisation of these components depends on the added air and is determined by the following effects (Oser et al. 2004):

- The oxidation of some chemical elements and organic compounds into more volatile types.
- Locally higher temperatures which promote the vaporisation process, especially at a high oxygen concentration.
- The amount of primary air, e.g. a low primary air supply leads to a low air flow speed in the glow bed and a reduction of the ash particles being carried away.

Oxygen depletion in the glow bed can cause a lower particle concentration in the exhaust fumes (Oser et al. 2004). However, the achieved minimum temperature should be high enough to convert the organic components.

13.4 Emission of Particulate Matter Through Biomass Burning

Over the past decade, particulate matter emissions – especially owing to wood combustion – have increased steadily. In 2003, 24,000 tonnes of dust from wood combustion were released into the atmosphere in Germany alone (Struschka et al. 2003). Although wood is a renewable energy source, these emissions were much higher than emissions from heating oil or gas, which caused approximately 2,000 tonnes of PM in Germany in 2003. The main contributors were usually single-room residential heaters, often used as an additional heating source. Beside their lower efficiency and poorer heat utilisation (e.g. bad combustion chamber design, single-room installation, heat loss due to hot exhaust release without heat exchange), wood stoves are responsible for the major PM mass due to residential heating. Compared to 2003, the German Federal Environmental Agency (UBA 2007) noticed a trend towards reduced mean emission values (Table 13.1; Struschka et al. 2003, 2008). The related dust emissions are an average at 120 mg/MJ for small-scale furnaces and at 1.7 mg/MJ for heating oil boilers (UBA 2007).

The Bioenergy Combustion Task of the International Energy Agency (IEA) calculated the characteristic emission values of different countries (Nussbaumer 2008). In Germany, the specified values of the overall PM emissions of open fireplaces were on average 160 mg/MJ. PM emissions of wood stoves were on average 94 mg/MJ. The averages of the inspected European countries were 250 mg/MJ for fireplaces and 47–83 mg/MJ for closed inset appliances. Norwegian and Finnish studies found 910 and 860 mg/MJ as partial load operation after dilution and cooling of the exhaust stream in a dilution tunnel (Nussbaumer et al. 2008). In the dilution tunnel approach, particle sampling can be carried out more realistically. The dilution of exhaust gases with cool and particle-free air decelerates particle interactions, while cooling to a temperature below 51 °C leads to gas-phase condensation, mainly of organic compounds. Condensation results in particle formation or occurs on existing particles (condensation nuclei). Finnish studies have shown that condensable organic matter may increase the mass of the primary particulate matter up to fivefold. Especially log furnaces are responsible for particle growth due to the release of much organic vapour during combustion (Nussbaumer 2008).

Table 13.1 Emission factors of particles and PAHs for different furnace types and a comparison of their 2000 and 2005 stock in Germany (Struschka et al. 2003, 2008)

Furnace type	Stock in 2000	Stock in 2005	Mean particle emission (kg TJ^{-1})	Mean PAHs emission (kg TJ^{-1})
Tiled stoves	4,300,000	3,890,000	125	0.99
Fireplaces	2,870,000	3,140,000	146	0.55
Log stoves	4,830,000	5,240,000	74–106	0.55–0.99
Cooking stoves	1,750,000	1,350,000	75	0.16
Others	889,000	757,000		
Total	14,639,000	14,377,000		

13.5 Chemical Composition of Emissions During Biomass Burning

The chemical composition of flue gas during biogas burning can be categorised into three groups: inorganic content, elemental carbon (EC) and particulate organic matter (organic carbon OC). Organic matter contains mainly soot and organic residues from incomplete combustion. The organic carbon content depends on the formation of pyrolysis breakdown products, including condensable ones. Combustion in simple log stoves is particularly affected by areas with low temperatures or/and a lack of oxygen, while optimised combustion chambers (e.g. pellet boilers) favour complete combustion. Pyrolysis breakdown products are indicators of incomplete oxidation. In the following the main wood combustion products are described.

The organic matter of wood combustion consists of unspecific and source-specific compounds. Source-specific emissions are related to wood constituents. Cellulose, hemicellulose and lignin are the wood's main components and are responsible for its mechanical properties. Additional wood constituents can also contribute to a fingerprint of biomass combustion products. Constituents are released through evaporation and distillation during the flaming phase. The molecules become chemically degraded or transformed by oxidation, hydrolyses and dehydration. Volatile products are evaporated. Highly flammable gaseous matter, such as methanol, ignites completely; in the inflaming phase breakdown products from cellulose, hemicellulose, lignin, or resin will be released with partial conversion or without any further decomposition. The water content of the fuel especially affects the flaming phase's combustion conditions: The higher the water content, the more energy is needed in the flaming phase to dry the wood, which usually results in lower temperatures and incomplete combustion. In contrast, water contents less than about 10 % could lead to a formation of pyrolysis products caused by oxygen deficiency during spontaneous ignition.

The anhydrous sugars levoglucosan, galactosan and mannosan are omnipresent constituents of ambient aerosols. They are relatively stable in the atmosphere and directly affect particles' hydrophilic properties. Levoglucosan is a breakdown product of cellulose, while mannosan and galactosan are breakdown products of

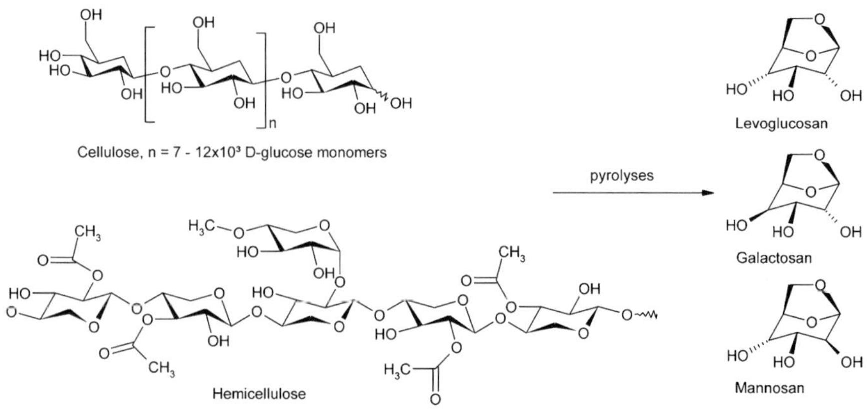

Fig. 13.3 Pyrolytic formation of anhydrous sugars

Table 13.2 Concentrations of potassium and levoglucosan originating during wood combustion (units are given in the last column)

Reference	Wood type	Potassium	Levoglucosan	Unit
Schauer et al. (2001)	Pine	0.28	14	% of PM
	Oak	0.65	14	% of PM
Fine et al. (2004a)	Hardwood	0.4–1.9[*1]	8–18[*2]	[*1] % of PM
	Softwood	0.2 – 0.4[*1]	1–27[*2]	[*2] % of OC
Fine et al. (2004b)	Hardwood	1–2.7[*1]	11–22[*2]	[*1] % of PM
	Softwood	0.5–1.1[*1]	25–41[*2]	[*2] % of OC
Iinuma et al. (2007)	Pine	0.1	22	% of PM
Schmidl et al. (2008a)	Hardwood	0.21–0.41	4.1–13.3	% of PM_{10}
	Softwood	0.07–0.19	10.7–15.1[*1]	[*1] % of PM
Schmidl et al. (2008b), Jankowski et al. (2009)	70 % spruce 20 % beech 10 % briquettes	0.174	9.3	% of PM

hemicellulose (Simoneit 2002) (Fig. 13.3). Levoglucosan is the main constituent of organic carbon. As shown in Table 13.2, depending on combustion conditions, concentration values of more than 20 % of levoglucosan are observed.

Besides cellulose, lignin is the main constituent of wood. The construction of lignin monomers is very specific in different wood types; lignin breakdown products are also very specific depending on the type of wood. Lignin composition depends on the different structures of hardwood from broad leaves (angiosperms), softwood from conifers (gymnosperms) and grass/straw (gramineae). The three originating monomers are p-coumaryl alcohol (grass/straw), coniferyl alcohol (conifers) and sinapyl alcohol (hardwood). A variety of breakdown products is released owing to lignin's unstable composition and depending on the burning conditions. The emitted phenols are coumaryl derivatives (gramineae), guaiacyl derivatives (gymnosperms)

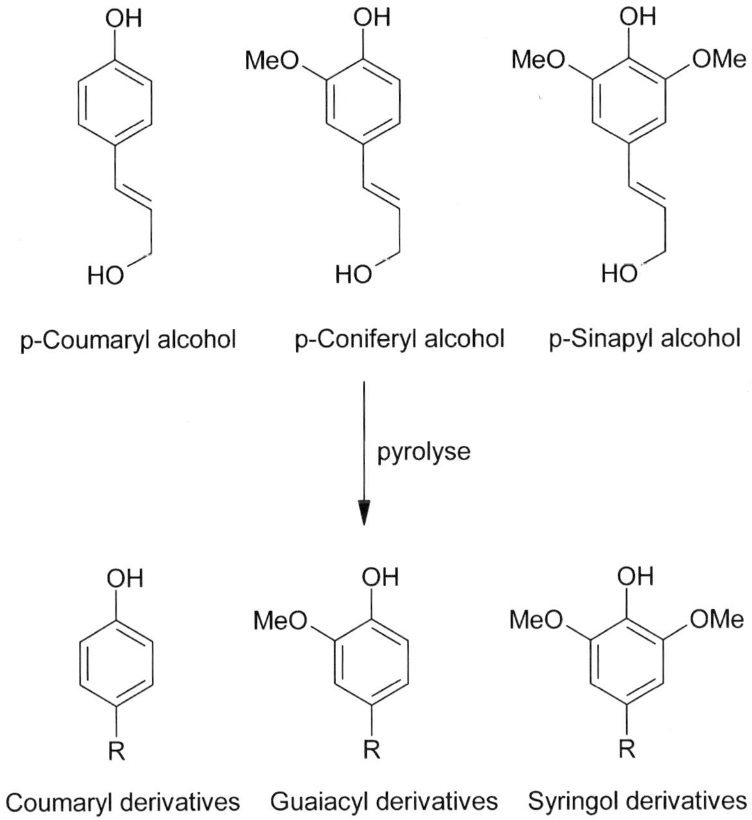

Fig. 13.4 Lignin monomers and their related pyrolysis breakdown products with R = alcohols, aldehydes and carboxylic acids with saturated or unsaturated aliphatic chains

and syringyl derivatives (angiosperms). All can contain a multitude of functional groups preferred in para-orientation. These methoxyphenols are also responsible for wood smoke's characteristic smell; the best-known methoxyphenol is the guaiacyl derivative vanillin. Lignin breakdown can also result in the formation of dimers from p-coumaryl, coniferyl and sinapyl derivatives – the lignans (Fig. 13.4).

The resin acids provided by colophony from conifers are even more source-specific. These diterpenoids can be released unaltered or in dehydrated, oxidised alteration products, including retene (Fig. 13.5). Retene itself is often classified in the literature as a polycyclic aromatic hydrocarbon (PAH).

In contrast to retene, other PAHs are not specific to biomass combustion. The PAH compound group is generally related to the incomplete combustion of organic matter. Owing to their high toxicological potential, PAHs are described in Sect. 13.6. Besides the PAHs, there are also PAH derivatives such as alkylated PAHs, oxygenated (o-PAHs) and – as a minority – nitrated PAHs. All the described PAHs and PAH derivatives are direct decomposition products from lignin or reaction products of parent PAHs (Fig. 13.6).

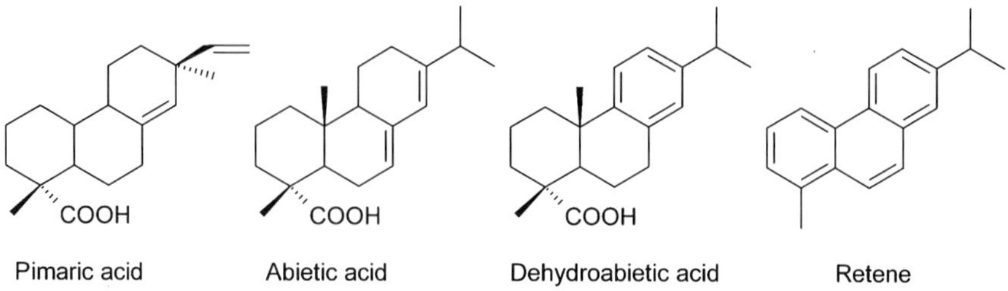

Fig. 13.5 A small selection of diterpenoids

Both the PM load and its organic portion vary strongly during combustion. High concentrations of organic compounds can be especially found during and shortly after ignition. The same effect can be observed immediately after the reloading of logs on an existing blaze (Gaegauf et al. 2008). Furthermore, the amount and composition of the emitted particles depend on specific fuel parameters, such as the appearance and size (logs, wood chips, briquettes and pellets), the type of wood (softwood or hardwood) and the water and bark content, in addition to the furnaces' construction parameters such as their air stream navigation (primary and secondary air) and the temperature (Hartmann et al. 2008). Consequently, it is hardly surprising that the particle compositions analysed in different studies show highly varying concentrations of the components potassium and levoglucosan. Potassium concentrations was found from <0.1 to 2.7 %, while up to 22 % of the levoglucosan content was related to the emitted particle mass (Table 13.2). To date, there has been no international standard sampling procedure for collecting particles for subsequent inorganic or organic analysis. In this study, we applied a simultaneous sampling on large filter areas (a filter with 150 mm diameter) with suitable filter material for both the organic and inorganic analysis (quartz fibre or PTFE membrane; PTFE = polytetrafluoroethylene) under ambient air conditions (cooling and dilution in a dilution tunnel to ensure sampling temperatures below 50 ° C).

13.6 Organic Pollutants' Toxicity

The toxicity of PM-bounded organic pollutants is mainly related to one chemical compound class: the polycyclic hydrocarbons (PAHs) and their modified, analogue-like alkylated, oxidised, or nitrogenised forms of parent PAHs, are responsible for most health effects associated with PM. The PAHs are the result of the incomplete combustion of organic materials. The reaction mechanisms can therefore be the direct formation of PAHs through the breakdown of wood compounds, or the indirect formation of PAHs through the rebuilding of molecules from small primary radicals (e.g., acetylene fragments). In addition to the described biomass emissions, PAHs are the products of the incomplete combustion of nearly all anthropogenic

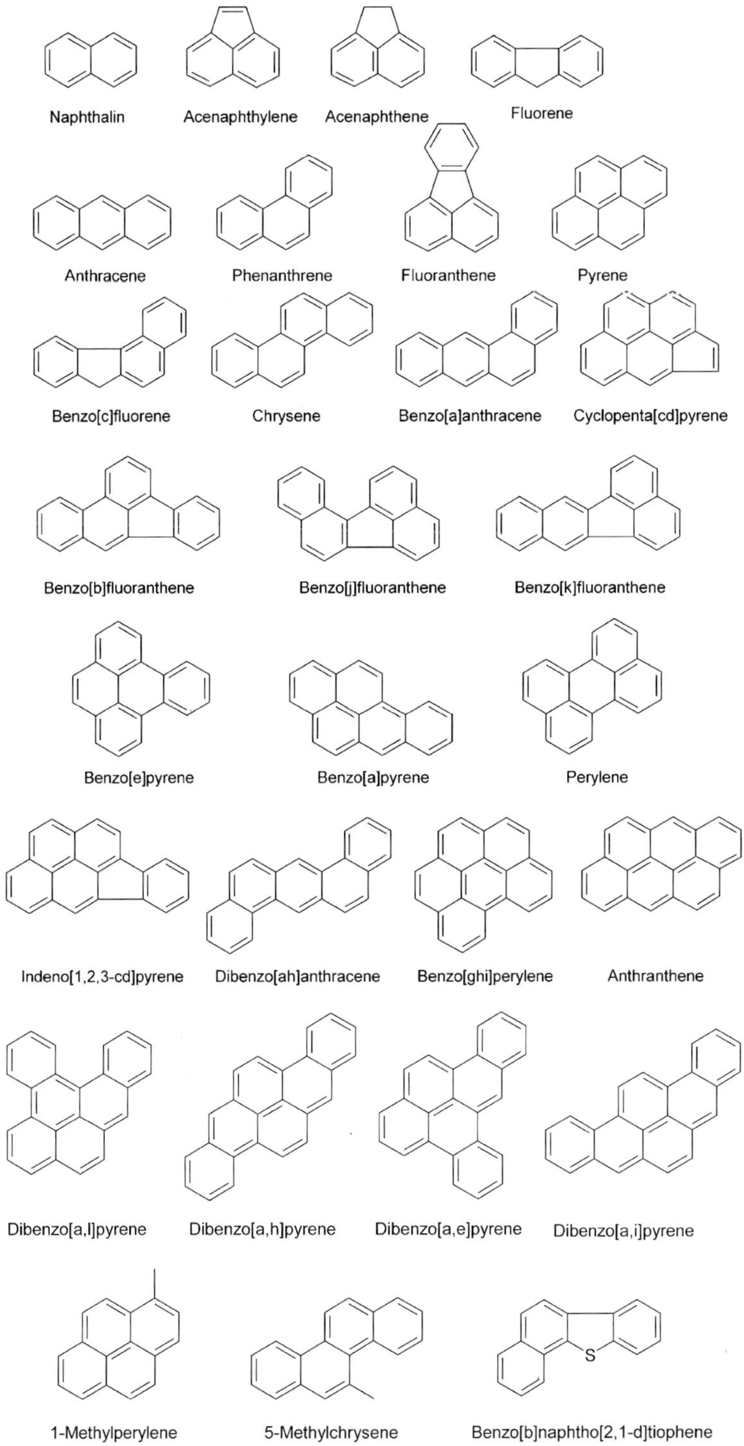

Fig. 13.6 Lists of PAHs with potential health risks

combustion sources. Traffic, the metal industry, the petrochemical industry and energy plants are – besides residential heating – the main sources of particle-bounded PAHs in our atmosphere.

PAHs are human carcinogens without a set impact threshold due to their genotoxic properties. Nevertheless, environmental and health protection agencies utilise the Threshold of Toxicological Concern (TTC) to impose legislative and statutory limits. The TTC is based on the socially accepted lifetime tumour risk of one case per million. Some PAHs are mutagenic and carcinogenic, while some are even believed to be teratogenic. Benz[a]pyrene (BaP) serves as a PAH toxicity marker. BaP is a strong carcinogenic and mutagenic compound and is also considered teratogenic (Straif et al. 2005). The tumour risk potential is based on the DNA adduct formation of PAH metabolites, which are caused by the binding of PAHs with the aromatic hydrocarbon (AH) receptors. The formation of PAH-rebuilding enzymes is thereby enforced and the human body can excrete the resulting phenols and epoxies. In contrast to persistent polychlorinated biphenyls and dioxins, PAHs are not enriched in body fat and breast milk. Nevertheless, reactive PAHs induce epoxy metabolites, which are partially transferred to dihydrodiol epoxides before excretion. These epoxy derivatives can be bounded on DNA bases by ring-opening reactions, which may lead to cancer formation. The strength of the health effects caused by the diols from epoxides is influenced by the PAH molecule's bay region (Jerina et al. 1976; Yagi and Jerina 1982; Weis et al. 1998). The bay region is indicated by three aromatic carbon bonds that enclose a "bay".

Table 13.3 lists potential health risks of PAHs sorted into three lists with different approaches. The first column shows the 16 PAHs indicated as potential environmental toxics by the US Environmental Protection Agency (US EPA 1984). Discussions were held on new PAH lists containing critical PAHs in food from the 1990s onward. The result was the European Food Safety Authority's list, which involves some non-volatile aromatic six-rings PAHs (EFSA 2008). These PAHs contain two of the characteristic bay regions and their impact on health is considered ten times higher than BaPs with a one bay region (DFG 2008). Additionally, the list of Germany's largest research funding organisation – Deutsche Forschungsgemeinschaft (DFG) – contains volatile PAHs that are also relevant in workplace situations (DFG 2008). Furthermore, the DFG supported the development of a tool with which to assess health risks resulting from PAHs. This tool is based on the proposal that PAHs' health risks can be estimated by summarising the concentrations of every single PAH compound after multiplying it with a factor corresponding to its health risk potential – the toxic equivalence factors (TEF). The sum corresponds to the toxic equivalent (TEQ). It has been established that this type of approach also applies to dioxins as a pollutant group. In the case of a PAH, BaP's TEF is set to 1. Relative to BaP, the other PAHs have a higher or lower health risk potential. This approach allows for a detailed description of a mixture of PAHs on health. Although this approach is a good approximation, an underestimation of their impact on health is possible. Therefore, from a present-day perspective, a chemical and physical characterisation is not sufficient to evaluate wood combustion emissions' possible health effects.

Table 13.3 PAH lists indicated by their application area; occupational health toxicity equivalence factors (TEF) arranged by the DFG (2008)

Compound	CAS. No.	US-EPA	EFSA	DFG	DFG-TEF
Year announced		1984	2008	2008	
Application area		Environment	Food exposure	Workplace exposure	
Acenaphthene	83-32-9	✓			
Acenaphthylene	208-96-8	✓			
Anthanthrene	191-26-4			✓	0.1
Anthracene	120-12-7	✓			
Benzo[a]anthracene	56-55-3	✓	✓	✓	0.1
Benzo[b]fluoranthene	205-99-2	✓	✓	✓	0.1
Benzo[j]fluoranthene	205-82-3		✓	✓	0.1
Benzo[k]fluoranthene	207-08-9	✓	✓	✓	0.1
Benzo[c]fluorene	205-12-9		✓		
Benzo[b]naphtho[2,1-d] thiophene	239-35-0			✓	0.01
Benzo[ghi]perylene	191-24-2	✓	✓		
Benzo[a]pyrene	50-32-8	✓	✓	✓	1
Chrysene	218-01-9	✓	✓	✓	0.01
Cyclopenta[cd]pyrene	27208-37-3		✓	✓	0.1
Dibenzo[a,h]anthracene	53-70-3	✓	✓	✓	1
Dibenzo[a,l]pyrene	191-30-0		✓	✓	10
Dibenzo[a,e]pyrene	192-65-4		✓	✓	1
Dibenzo[a,h]pyrene	189-64-0		✓	✓	10
Dibenzo[a,i]pyrene	189-55-9		✓	✓	10
Fluoranthene	206-44-0	✓			
Fluorene	86-73-7	✓			
Indeno[1,2,3-cd]pyrene	193-39-5	✓	✓	✓	0.1
Naphthalin	91-20-3	✓		✓	0.001
Phenanthrene	85-01-8	✓		✓	0.001
Pyrene	129-00-0	✓		✓	0.001
1-Methylpyrene	2381-21-7			✓	0.1
5-Methylchrysene	3697-24-3		✓		

Some methylated PAHs or Nitro-PAHs are more carcinogenic and bioavailable than the related PAHs that often serve as precursors. Their carcinogenic and mutagenic activity varies with the methyl position in the aromatic system. Methylated PAHs or Nitro-PAHs play a minor role in direct emissions from wood combustion (e.g., diesel engines are a significant source of Nitro-PAH), whereas oxidised PAHs (o-PAHs) generally occur in direct emissions from wood burning sources. Our study shows o-PAHs have concentration ranges similar to that of a PAH. Especially log stoves produce PAHs and o-PAHs in high concentration levels. Redox-active PAH quinones are believed to be responsible for oxidative stress (Squadrito et al. 2001; Xia et al. 2004), which is commonly due to ambient aerosols' potential to lead to inflammation in lungs after inhalation exposure.

13.7 Biogas

Emissions from biogas plants vary strongly for different technical reasons. The major constituents of exhaust gases are methane, sulphur dioxide (SO_2), nitrogen oxides (NO_x) and formaldehyde (HCHO). The release of the pollutants SO_2, NO_x, CO and HCHO is related to the type of engine employed and the fermentation gas preparation. Emissions of methane are also possible through leaks in the digester and storage facility, which depend on the sealing measures. The emitted PM is lower than 1 mg/MJ, which is comparable to the values of residential heating boilers using domestic gas. The emissions of 30 biogas-driven engines were measured on behalf of the German Federal Environmental Agency (Degel and Jörß 2009). The averaged values were as follows: 89 mg/MJ NO_2, 58 mg/MJ SO_2, 133 mg/MJ CO and 10 mg/MJ HCHO. Since these results are related to the fermentation gas's heating value, real emissions related to the utilised energy are higher. The utilised energy depends on engine efficiency, but mostly on the full utilisation of the combination of heat and power. In plants producing only electricity and not using the heat, the emission values per MJ are much higher. Common electrical efficiency values are approximately 35 %, whereas the efficiencies of combined heat and power plants can be as high as 80 %.

13.8 Particulate Matter in the Atmosphere and Its Climatic Effects

The exhaust air from wood combustion contains gaseous components when leaving the chimneys. Physical processes, such as condensation and adsorption during existing particles' liquid or solid phases, can contribute to the PM mass. Atmospheric reactions lead to less volatile organic compounds, both from gaseous and particle-bound precursors (aerosol ageing). Less volatile compounds condense on existing particles, resulting in a secondary organic aerosol (SOA) (Kanakidou et al. 2005; Hallquist et al. 2009). Atmospheric processes triggered by OH radicals, NO_2 and O_3 are responsible for the formation of water-soluble organic compounds (WSOC). An increase in particles' hydrophilic properties is responsible for increasing their water contents. These hydrophilic particles are the cloud condensation nuclei (CCN), which have a direct climatic effect.

Other climatic effects caused by the PM in the atmosphere are physical effects influenced by the general particle composition. Particles rich in inorganic matter have pronounced light-scattering effects, thus reflecting a part of infrared light and reducing global warming. On the other hand, particles rich in carbonaceous matter can adsorb infrared light and contribute to global warming.

13.9 Determining Organic and Inorganic Constituents of PM

Systematic combustion experiments were carried out by burning different fuels in various state-of-the-art furnaces. Some furnaces were run with secondary arrangements, such as an electrostatic precipitator (ESP) or a condensing heat exchanger (CHE).

Experiments with the following equipment were performed in this project:

- A pellet boiler. Nominal load: 25 kW. Fuel: pellets from spruce wood. Variation: with or without a condensing heat exchanger.
- A pellet stove: Nominal load: 13 kW. Fuel: pellets from spruce wood.
- A wood log stove. Nominal load: 8 kW. Fuel: spruce wood logs, beech wood logs. Variation: with or without an electrostatic precipitator.
- A wood chip boiler. Nominal load: 30 kW. Fuel: wood chip from spruce. Variation: with or without an electrostatic precipitator.
- A wood chip boiler. Nominal load: 30 kW. Fuel: pellets from chaffed straw. Variation: without/with a condensing heat exchanger.
- A wood chip boiler. Nominal load: 30 kW. Fuel: pellets from Miscanthus.
- A wood log boiler. Nominal load: 30 kW. Fuel: spruce wood logs, beech wood logs.

The combustion experiments were carried out in Straubing, Bavaria at the TFZ combustion test stand using a flue gas tract with a dilution tunnel (Fig. 13.7). The emitted fly ashes were collected using an innovative plane filter device with a diameter of 150 mm (Fig. 13.8). Subsequently, 1/12 of the filter was punched out using a titanium tool especially constructed to cut the filter. The large portion was used for the element analysis and the small portion to determine organic compounds (Fig. 13.9). Usually, a heated sampling tract and filter casing with a 45 mm diameter filter is applied for out-stack measurements (Fig. 13.10). We conversely used the unheated plain filter casing of a 150 mm filter and the clean material of a new filter holder (PTFE) and of the filter itself (quartz fibre) to ensure enough sample material for the inorganic and organic analyses. This kind of sampling guarantees good detection limits with low background concentrations of the elements. It further enables longer sampling times, e.g. over the full batch charging of a log wood stove. Compared to a conventional 45 mm filter, the 150 mm filter's area is 11 times larger.

For the inorganic analysis, all the collected ashes, such as the grate ash, the electrostatic precipitator (ESP) ash and the internal heat exchanger ash, were digested in ultraclean closed PTFE vessels (Picotrace, Göttingen, Germany) with a mixture of ultrapure concentrated $HF\text{-}HClO_4\text{-}HNO_3$. The fly ashes were only digested with $HClO_4$ and HNO_3 to prevent the quartz filters from dissolving. All the main and trace elements were quantified at the Geosciences Centre at the University of Göttingen, using an ICP-OES (inductively coupled plasma optical emission spectrometer; Optima 3300 DC Perkin Elmer) and an ICP-MS (inductively coupled plasma mass spectrometer; Perkin Elmer DRC-II). Blind samples assured that handling during digestion and measurement was clean and that no notable contamination occurred. The excellent agreement between our reference samples'

Fig. 13.7 Dilution tunnel at TFZ in Straubing (Bavaria, Germany)

values and the internationally published data ensures the reliability and comparability of the obtained data. Based on input and output data, it was thus possible to calculate the element fluxes normalised to the energy units and to compute the recovery rates.

In the first phase of our study, only the directly emitted PM was determined (primary particles). Even below 50 °C, semi-volatile compounds show lower findings when the PM collection is only undertaken from the filters. In the second phase, applicable methods were developed to ascertain the volatile and semi-volatile organic compounds considered as possible precursors of SOA formation. As described above, SOA can be a considerable fraction of the total PM originating from wood combustion. While exact determinations of the total PM mass are not yet possible, approximations were done by means of modelling (Nopmongcol et al. 2007).

Fig. 13.8 Innovative 150 mm plane filter holder consisting of PTFE (Developed by TFZ)

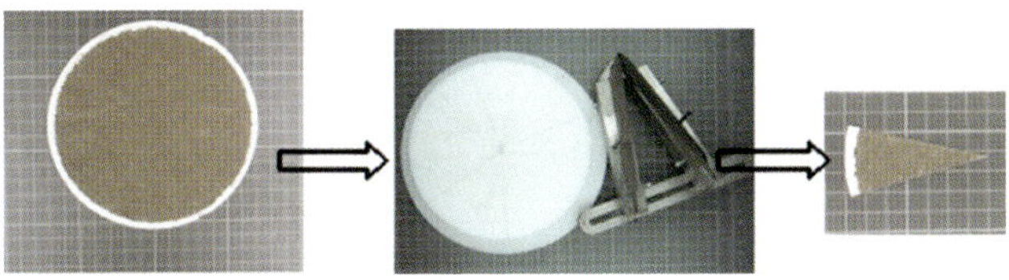

Fig. 13.9 Pathway from the loaded filter (*left*) to the punched out segment (30° angle; *right*) for organic analysis by means of a punch basis made of PTFE and a punch tool made of titanium (*centre*)

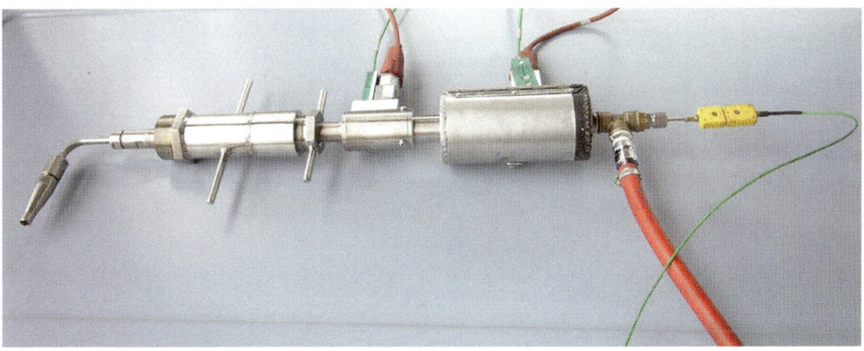

Fig. 13.10 Heated sampling tract and filter casing for outstack measurements

13.9.1 Inorganic Element Concentrations

Biomass consists of carbon (C), oxygen (O) and hydrogen (H), which represent approximately 95–97 % of a plant's dry matter. The macro-elements sulphur (S), phosphorous (P), magnesium (Mg), nitrogen (N), potassium (K) and calcium (Ca) comprise 5 % or less of a plant's dry matter. The elements boron (B), chlorine (Cl), copper (Cu), iron (Fe), manganese (Mn), molybdenum (Mo), nickel (Ni), zinc (Zn), silicon (Si) and sodium (Na) are micronutrients.

The element concentrations in the fuels used are presented in Table 13.4. Thc clements are arranged according to their properties: elements with high plant availability, elements mainly bound on adhering dust or soil material (pedogenic) and critical heavy metals. There are transitions between the groups: Mo and Cu are heavy metals, have a certain plant availability and some parts of them are integrated into minerals. The macro-elements like K, Ca, Mg, Cl, P and S in the fuel determine the inorganic ash amount formed during the combustion. The composition of the ash residues differs strongly at different locations within a furnace. It can also be assumed that every fuel type shows unique element fluxes and deposition behaviours, which can be identified by analysing the different ash fractions. Non-wood-like biomass fuels such as straw and Micanthus have higher concentrations of Cu, Pb, and Sb than wood has. The individual associations with the different ash fractions depend on each element's particle formation or sorption behaviour, on the element tendency to vaporise and on combustion conditions such as turbulences in the combustion chamber and the disturbance of the fire bed by adding new fuel.

Table 13.4 Element concentrations (mg/kg dry matter) of the utilized fuels

	Fuel					
Element	Beech logs	Spruce logs	Spruce pellets	Spruce chips	Straw pellets	Miscanthus pellets
Plant available elements						
P	90	75	57	65	264	428
S	114	65	56	81	731	491
K	1,057	558	382	902	4,924	6,139
Ca	1,393	1,388	780	1,359	2,942	910
Mg	191	137	132	204	888	729
Mn	141	263	120	116	18	9
Li	0.08	0.02	0.059	0.06	2.61	0.08
Na	3	14	10	12	185	34
Rb	4.98	1.37	1.85	3.46	1.95	1.32
Cs	0.001	0.006	0.023	0.021	0.031	0.004
Sr	11.1	3.93	3.53	8.65	15.4	10.3
Ba	29.7	18.0	9.25	20.4	21.6	9.91

(continued)

Table 13.4 (continued)

	Fuel					
Element	Beech logs	Spruce logs	Spruce pellets	Spruce chips	Straw pellets	Miscanthus pellets
Pedogenic elements						
Fe	14	6	68	25	287	77
Al	12.9	11.1	46.5	44.8	297	41.8
Be	0.006	0.0024	0.0026	0.0038	0.013	0.0015
Sc	<0.01	<0.01	<0.004	0.005	0.045	<0.003
Ti	1.1	0.9	3.5	3.1	24.7	6.3
V	0.80	0.86	0.08	<0.1	0.22	<0.08
Y	0.007	0.006	0.022	0.017	0.117	0.014
Zr	0.05	1.81	0.14	0.1	1.06	0.13
La	0.012	0.025	0.042	0.033	0.201	0.028
Ce	0.0238	0.0224	0.0758	0.0495	0.3461	0.0509
Pr	0.0025	0.0023	0.0084	0.0056	0.0386	0.005
Nd	0.0094	0.0082	0.032	0.0218	0.149	0.0194
Sm	0.0018	0.0017	0.0062	0.0043	0.0293	0.0037
Eu	0.0012	0.0008	0.0011	0.0009	<0.00002	0.0006
Gd	0.003	0.0029	0.0062	0.0056	0.0344	0.0043
Tb	0.0002	0.0002	0.0006	0.0005	0.0036	0.0004
Dy	0.0012	0.0011	0.0044	0.0033	0.024	0.0027
Ho	0.0002	0.0001	0.001	0.0007	0.0047	0.0005
Er	0.0007	0.0005	0.0025	0.0018	0.0145	0.0016
Tm	0.0001	0.0001	0.0003	0.0002	0.0019	0.0002
Yb	0.0048	0.0015	0.0029	0.0046	0.0182	0.0034
Lu	0.0003	0.0001	0.0003	0.0003	0.0021	0.0003
Th	0.003	0.001	0.012	0.005	0.055	0.008
Heavy metals						
Cr	2.2	2.1	1.4	1.4	12.1	2.7
Co	0.07	0.39	0.16	0.53	0.14	0.06
Ni	1.08	0.7	0.6	0.9	9.3	0.63
Cu	1.23	1.03	1.43	1.49	1.81	5.37
Zn	3.58	10.2	9.06	15.2	6.02	17.4
Mo	0.198	0.05	0.05	0.07	0.366	0.39
Cd	0.05	0.12	0.13	0.24	0.04	0.08
Sn	0.007	0.011	0.014	0.021	0.059	0.244
Sb	0.002	0.004	0.011	0.005	0.033	0.058
Tl	0.001	0.021	0.018	0.025	0.005	0.001
Pb	0.097	0.216	0.251	0.374	0.564	0.512
Bi	0.00046	0.00075	0.00075	0.00070	0.00309	0.0114
U	0.0007	0.0005	0.0033	0.0021	0.0172	0.0023

Unfortunately, elements that are harmful to our health and the environment are strongly enriched in fly ashes. An ESP's influence on the element concentration in filter fly ash is insignificant. The concentrations in a variety of ashes are shown in Table 13.5.

Table 13.5 Element concentrations (mg/kg dry matter) in different ash fractions in a 30 kW wood chip boiler. Fuel: spruce chips

Element	Grate ash	ESP ash	Fly ash without ESP	Fly ash with ESP
Plant available elements				
P	11,847	6,429	1,125	1,000
S	3,534	68,192	41,770	30,269
K	95,553	395,752	423,453	335,808
Ca	330,400	18,986	3,885	3,336
Mg	38,065	3,761	785	688
Mn	20,012	2,080	927	854
Li	11.1	15.5	12.7	14.3
Na	2,136	3,650	3,494	2,731
Rb	183	935	777	850
Cs	1.25	10.44	10.6	10.3
Sr	1,027	109.4	26.4	35.1
Ba	2,291	315	330	857
Pedogenic elements				
Fe	9,713	25,359	929	1,730
Al	12,667	1,891	458	381
Be	0.607	0.119	0.111	0.126
Sc	1.75	0.22	<0.129	<0.229
Ti	832	115	459	3,947
V	6.1	3.5	1.37	2.31
Y	4.59	0.58	0.12	<0.17
Zr	27.3	3.5	2.21	7.08
La	7.86	1.1	0.32	0.5
Ce	12.49	1.61	0.46	0.70
Pr	1.41	0.17	0.05	0.075
Nd	5.48	0.68	0.18	0.29
Sm	1.07	0.13	<0.08	<0.15
Eu	0.123	0.006	<0.005	<0.008
Gd	1.19	0.112	<0.2	<0.4
Tb	0.122	0.016	<2	<4
Dy	0.835	0.102	0.0298	0.0282
Ho	0.149	0.018	<0.1	<0.2
Er	0.486	0.058	0.0160	0.0294
Tm	0.0497	0.0063	<0.002	<0.004
Yb	0.581	0.053	0.0137	0.0262
Lu	0.065	0.007	<0.08	<0.2
Th	2.01	0.23	0.048	0.074
Heavy metals				
Cr	47	51	50	288
Co	18.6	4.2	4.6	10.5
Ni	80	30	26.6	135
Cu	132	214	160	244
Zn	145	10,324	8,096	9,568
Mo	0.9	5.9	7.1	14.2
Cd	0.25	49.66	58.6	62.0

(continued)

Table 13.5 (continued)

Element	Grate ash	ESP ash	Fly ash without ESP	Fly ash with ESP
Sn	0.88	5.27	6.11	6.25
Sb	0.35	3.38	2.68	3.01
Tl	0.031	19.118	15.6	16.9
Pb	1.59	249	213	232
Bi	0.02	1.25	1.85	1.32
U	0.621	0.096	0.050	0.123

Ashes from wood burning can be used as fertilisers, but must comply with the German Fertiliser Regulations (DüMV 2008) that only categorise ashes into "grate ash" and "ashes from the final filtering unit in the flue gas path". Generally, ash from the final filtering unit and from condensing heat exchangers may not be used as fertilisers. Ashes used as fertilisers must conform to the limiting values for critical elements such as As (20 mg/kg), Pb (100 mg/kg), Cd (1 mg/kg), Cr (300 mg/kg), Ni (40 mg/kg), Hg (0.5 mg/kg) and Tl (0.5 mg/kg). In certain cases, the limiting values may be exceeded by up to 50 % and the ash still used as fertiliser (DüMV 2008). The required methods of analysis are found in the Fertiliser Sampling and Analysis Ordinance (DüngMProbV 2006). Our analytical data show that grate ash exceeds the limiting values for Ni, and the ash from the ESP the limits for Cd, Pb and Tl (Table 13.5). Accordingly, both types of ash may not be used as fertiliser. The elements Hg and As are not determined.

13.9.2 *Inorganic Elements Recovery Rates*

Recovery rates [%] can be calculated according to the output of an element (= the amount in all collected residues, i.e. ashes and flue gas particles) divided by its input (= the amount in the fuel) multiplied by 100. The recovery rates are important for assessing the reliability of the fluxes of elements through the system. Deviations from 100 % require an explanation.

Recovery rates greater than 100 % indicate either a contamination of the grate ash by an exterior source, for instance during fuel preparation, or can be ascribed to sampling errors due to inhomogeneous fuels. More importantly, recovery rates less than 100 % indicate losses. Figure 13.11 shows the recovery rates [%] of some elements from a wood chip boiler with a nominal load of 30 kW. The furnace was operated without and with an ESP and was fuelled by spruce chips. Unfortunately, the elements with low recovery rates – such as Cd, Pb, Tl, Sn and Zn – are harmful to our health and the environment. These elements are presumably lost in unknown sinks (e.g. deposition in the cold zones of the furnace, reaction with the refractory lining materials, etc.), which are not accessible due to the furnace's construction. Impinger flask experiments behind the filters show that these critical elements are not released into the atmosphere. Only very small portions of the elements sampled

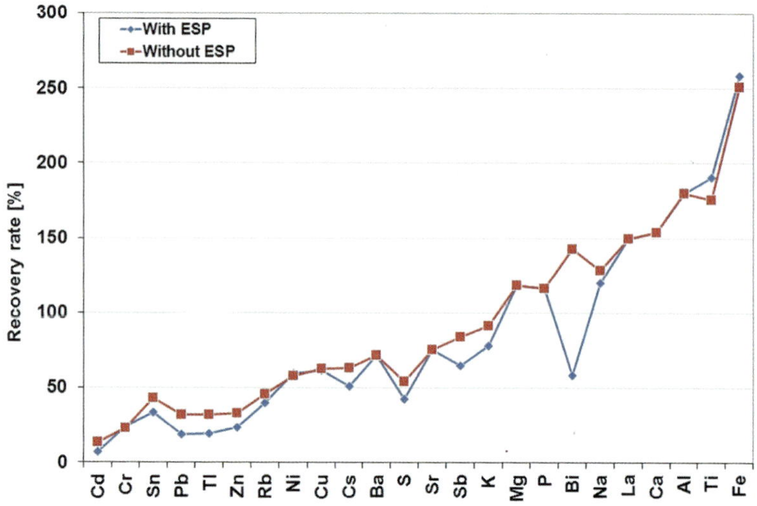

Fig. 13.11 Recovery rates [%] (sum of the amounts of an element in ashes divided by the amount of the element in the burnt fuel multiplied by 100); 30 kW wood' chip boiler with and without an ESP, fuelled by spruce chips

as filter fly ash leave the chimney (Table 13.6). In relation to the amount of the elements contained in the fuel, less than 1 % to approximately 30 % of the harmful elements contained in wood chips are released into the atmospheric environment during combustion (first number after the element: % without an ESP; second number: % with an ESP): Cd (11.8; 4.3), Pb (28.9; 10.6), Zn (25.7; 11.4), Tl (30.6; 13.2), Ni (1.2; 2.9) and Bi (141; 27.1). The values for Bi are not certain as the measured concentrations are close to the detection limit.

13.9.3 Inorganic Element Fluxes

Flux calculations normalised to the used energy units (e.g. in mg of an element per MJ; Table 13.6) are very useful to compare the fuel, the furnace properties and the secondary devices' influences on the element mobilisation during biomass burning and the element fixation in different ashes. Fluxes can also be used to calculate the amount of an element leaving a chimney. Normalised fluxes should also be the primary basis on which to compare the emissions of different fuels, such as biomass and conventional fossil fuels. For example, the input fluxes of elements added by the fuel and the output fluxes with fly ash are presented in Table 13.6.

The fluxes should be related to usable energy output, but not to the primary energy contained in the fuel, because after applying efficiency measures such as a CHE, more realistic energy normalised emission can be achieved for single elements. The application of an ESP can considerably reduce the emissions of

Table 13.6 Amount of elements in spruce chips and corresponding fly ash obtained with and without an ESP normalized to the usable heat output (mg/MJ)

				% of an element in the fly ash relative to its amount in the fuel[a]	
Element	Spruce chips	Fly ash without ESP	Fly ash with ESP	Without ESP	With ESP
Plant available elements					
P	3.52	0.035	0.0094	1.00	0.27
S	4.41	1.35	0.37	30.84	8.39
K	48.8	12.5	4.045	25.73	8.29
Ca	73.49	0.114	0.0266	0.16	0.04
Mg	11.0	0.0231	0.00559	0.21	0.05
Mn	6.31	0.0270	0.0085	0.43	0.14
Li	0.00342	0.000375	0.000160	10.94	4.68
Na	0.649	0.102	0.0331	15.83	5.11
Rb	0.187	0.0213	0.00909	11.42	4.86
Cs	0.00114	0.000304	0.000118	26.66	10.36
Sr	0.468	0.00070	0.000265	0.15	0.06
Ba	1.10	0.00738	0.00510	0.67	0.46
Pedogenic elements					
Fe	1.38	0.0287	0.0164	2.08	1.19
Al	2.42	0.0138	0.00298	0.57	0.12
Be	0.000208	0.0000034	0.0000017	1.63	0.81
Sc	0.000278	a	a	a	a
Ti	0.168	0.0138	0.0377	8.26	22.45
V	a	0.0000378	0.0000295	a	a
Y	0.000930	0.0000032	a	0.35	0.00
Zr	0.00515	0.0000538	0.0000335	1.04	0.65
La	0.00181	0.0000087	0.0000029	0.48	0.16
Ce	0.00267	0.0000129	0.0000045	0.48	0.17
Pr	0.000306	0.0000014	0.0000005	0.45	0.15
Nd	0.00118	0.0000052	0.0000018	0.44	0.15
Sm	0.000234	a	a	a	a
Eu	0.0000520	a	a	a	a
Gd	0.000303	a	a	a	a
Tb	0.0000286	a	a	a	a
Dy	0.000182	0.0000008	0.0000004	0.46	0.23
Ho	0.0000352	a	a	a	a
Er	0.0001006	0.0000004	0.0000002	0.43	0.17
Tm	0.0000134	a	a	a	a
Yb	0.000251	0.0000004	0.0000001	0.15	0.06
Lu	0.0000215	a	a	a	a
Th	0.000306	0.0000014	0.0000004	0.45	0.15
Heavy metals					
Cr	0.0795	0.00181	0.00284	2.28	3.57
Co	0.0288	0.000128	0.000128	0.44	0.45
Ni	0.0491	0.000567	0.00141	1.15	2.88
Cu	0.0809	0.00431	0.00249	5.33	3.08

(continued)

Table 13.6 (continued)

Element	Spruce chips	Fly ash without ESP	Fly ash with ESP	% of an element in the fly ash relative to its amount in the fuel[a]	
				Without ESP	With ESP
Zn	0.823	0.211	0.094	25.70	11.43
Mo	0.00379	0.000201	0.0000974	5.30	2.57
Cd	0.0130	0.00152	0.000559	11.76	4.30
Sn	0.00117	0.000194	0.0000682	16.64	5.82
Sb	0.000249	0.0000807	0.0000230	32.39	9.22
Tl	0.00133	0.000409	0.000176	30.62	13.22
Pb	0.0202	0.00584	0.00214	28.86	10.61
Bi	0.0000412	0.0000581	0.0000112	141.04	27.10
U	0.000116	0.0000014	0.0000007	1.23	0.58

Last columns: % of an element leaving the chimney as fly ash relative to the element input by wood. Wood chip boiler

[a]Concentrations below the detection limit. Consequently, the fluxes and % of the element in the fly ash cannot be calculated

elements in fly ashes compared to a situation without an ESP (Table 13.6). For instance, the Cd flux can be reduced significantly from 0.00159 mg/MJ without an ESP to 0.00058 mg/MJ with an ESP, and the Zn flux in the fly ash can drop from 0.218 to 0.086 mg/MJ.

13.9.4 Results of Health-Relevant Flue Gas Components

Table 13.7 shows the results of the major health-relevant components, comprising the TEQ of the analysed PAHs, PM, organic gaseous carbon, and carbon monoxide. A selection is shown of the results of the experiments with residential wood stoves and boilers. Orasche et al. (2012, 2013) provide more detailed data. Generally, the boilers all showed fairly good results. The TEQ of nominal load emissions from the pellet boiler was on average 0.118 μg/MJ and from the wood chip boiler 0.160 μg/MJ. The log wood boiler emitted particles with a somewhat higher TEQ of 0.297 μg/MJ. The combustion of wood logs in comparison to wood chips and pellets was more critical – the TEQ increased to 27 μg/MJ. The data on the utilisation of an appropriate wood chip boiler to burn both straw pellets and miscanthus pellets are ambivalent. Straw combustion showed satisfactory results, with emissions in the same order of magnitude as observed in the log wood stove exhaust air. However, no satisfying combustion parameters were found during the experiments with miscanthus pellets. Although the grate's feed rate was increased, grate slagging could not be avoided. The concentrations of PAHs, the amount of soot and PM started at a high level and increased with on-going slagging. The miscanthus experiments' results are therefore not representative and are not

Table 13.7 Comparison of different combustion systems. Different wood types, partly with or without an electrostatic precipitator Orasche et al. (2012, 2013)

Furnace	Unit	Pellet boiler		Pellet stove with heat exchanger		Chip/Pellet boiler		Chip/Pellet boiler with electrostatic precipitator		Chip/Pellet boiler with electrostatic precipitator		Log boiler		Log stove		Log stove with electrostatic precipitator		Log stove		Log stove with electrostatic precipitator	
Conditions of operation		Nominal load with cold starting ignition	Nominal load	Nominal load with cold starting ignition	Nominal load	Nominal load with cold starting ignition	Nominal load	Nominal load with cold starting ignition	Nominal load	Nominal load with cold starting ignition	Nominal load	Nominal load with cold starting ignition	Nominal load	Nominal load with cold starting ignition	Nominal load	Nominal load with cold starting ignition	Nominal load	Nominal load with cold starting ignition	Nominal load	Nominal load with cold starting ignition	Nominal load
Fuel type		Spruce pellets				Spruce chips				Straw pellets		Spruce logs						Beech logs			
TEQ	(μg/MJ)	**0.746**	**0.118**	**2.38**	**0.16**	**2.97**	**0.346**	**2.09**	**0.115**	**43.6**	**8.14**	**27**	**0.297**	**27.6**	**20.1**	**9.52**	**7.15**	**14.4**	**18.3**	**5.95**	**7.23**
PM	(mg/MJ)	**13**	**11**	**92**	**22**	**73**	**30**	**23**	**11**	**121**	**37**	**64**	**17**	**105**	**67**	**56**	**25**	**119**	**93**	**24**	**40**
Organic gaseous carbon	(mg/MJ)	**9**	**2**	**90**	**10**	**34**	**1**	**36**	**1**	**222**	**25**	**120**	**2**	**425**	**155**	**773**	**337**	**1,190**	**162**	**148**	**436**
Carbon monoxide	(mg/MJ)	**112**	**17**	**1,390**	**413**	**1,030**	**75**	**1,030**	**92**	**1,670**	**284**	**888**	**27**	**2,140**	**1,940**	**3,810**	**2,430**	**3,530**	**1,850**	**2,350**	**1,800**

The amounts of selected PAHs released with fly ash is shown normalized to the usable amount of energy (mg/MJ). At the end of the table, the resulting TEQ is calculated (excluding naphthalene due to its high gas-phase partitioning) and compared to the PM mass demonstrated here. The emitted PM mass of straw pellet combustion was 37 mg/MJ with an electrostatic precipitator, while the contamination of the emitted dust with toxic-relevant substances was TEQ = 8.14 μg/MJ. The TEQ of the PM emitted by the log stoves fired with spruce was about 200 times higher and beech 150 times higher than that of the pellet boiler.

The organic analysis led to two major conclusions:

1. Log wood stoves emit high amounts of toxic substances into the atmosphere. Owing to its high prevalence in Europe, this type of furnace is responsible for a considerable mass of harmful particulate matter. The applied prototypes of electrostatic precipitators do not yet have a sufficient deposition rate. The PM and the TEQ were, however, reduced by approximately 60 %. Nevertheless, the emitted TEQ values of log stoves were still far beyond the TEQ level that the pellet boilers emitted. Improvements of combustion design (primary measures) or precipitator development (secondary measures) should therefore be speeded up. Electrostatic precipitation may not be the appropriate technique for all furnace types due to the large differences in the composition of the emitted dust. Currently, electrostatic precipitation techniques are most appropriate for dust with major components of inorganic material. Soot and organic material have physical properties that are poorly suitable for electrostatic precipitation.
2. The best boiler results were obtained when they were run at a nominal load. Frequent restarts in the combustion chambers should be avoided because ignition events produce pollution rates that are twice as high, or higher, than during the flaming phase. Every ignition is a highly incomplete combustion phase, even if wood combustion boilers have technically optimised furnaces (see also the comparison of the nominal load and ignition phases in Table 13.7). It should therefore be the goal to improve emissions by applying well-dimensioned heat reservoirs in order to reduce the number of boiler starts. For boilers, too, the option of using electrostatic precipitators is given (it reduces the PM mass by about 70 %), but they need to be further improved.

13.9.5 Chemical Characterisation of Flue Gas Over the Course of a Heating Event

Owing to the great number of log wood stoves in use and their importance to the ambient air pollution, one experiment is described in detail; it demonstrates the characterisation of a heating event typical for a single family house evening operation during winter time. The heating event was separated into four periods. Each period took approximately 45 min of sampling time. The starting point of flue gas sampling was the loading of the wood stove furnace with approximately three beech logs. The number of logs depended on their weight; each load was about 1.5 kg. Figure 13.12 shows the sequence and the results of this experiment.

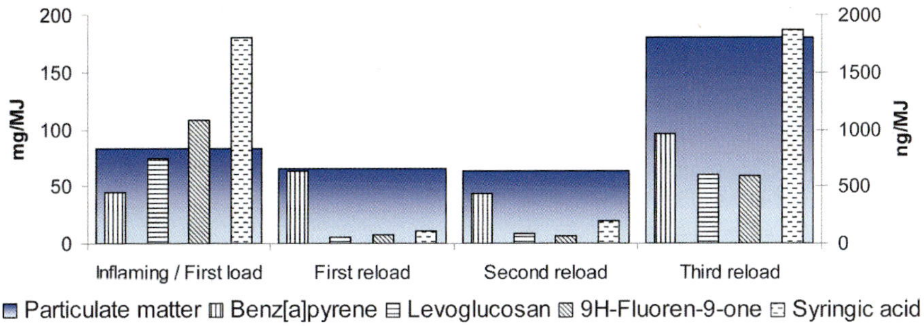

Fig. 13.12 PM emissions and PM-component emissions from a chimney stove fired with beech logs. Typical operation time: 3 h (45 min per batch). *Left* y-axis: particulate matter (mg/MJ); *right* y-axis: benz[a]pyrene (μg/MJ), levoglucosan (1,000 μg/MJ), 9H-Fluoren-9-one (10 μg/MJ) and syringic acid (10 ng/MJ)

The first sampling was done during the inflaming phase (ignition batch). Ignition was performed by igniton blocks made of wood wool coated with paraffin wax. Two further batches were measured, each starting immediately after recharging the wood logs. The logs for the ignition batch were placed on the cold grate, whereas each further reload was placed on the hot grate, directly onto the embers. The inflaming of the reloaded logs took place immediately due to the high temperatures and a sufficient combustion air. In this experiment, the third reload was done on somewhat colder embers and there was less available air in the combustion chamber. Under these conditions, the inflaming of fresh fuel needs approximately 10 min. During smouldering, white and black smoke evolved before the white smoke ignited. The black smoke was reduced while the white smoke constituents were burned. The smouldering phase occurs when users apply the "last log of the day" and close the air entrance to retain heat for as long as possible during the night.

However, besides the inflaming process, the last log of the day showed the highest emissions of organics (Fig. 13.12). Emissions of PM and organics declined significantly when the reload occurs at high temperatures with sufficient air for the fast drying of wood and the distillation of flammable gases. After a short peak with higher particle emissions immediately after the reload, the combustion process was eased. The emissions of nearly all compounds were at a relatively low level. Especially levoglucosan, syringic acid (specific hardwood tracer originated from lignin degradation) and the oxygenated PAH 9H-Fluoren-9-one were strongly influenced by the inflaming and smouldering phases during the combustion of the third reload. Benz[a]pyrene was not affected to the same extent.

This is interesting, because it means that the potential health effects related to PM or EC/OC content may be underestimated. The cause and effect principle calls for health effects to be studied by exposing living material in-vitro. Therefore, in the second part of the study, an approach will be undertaken to combine the results of the chemical characterisation and the examination of health effects through in-vitro tests by using human tissue equivalents of respiratory epithelia.

13.10 Conclusions

The exploitation of the energetic potentials of wood and straw as residual biomasses complements the energy extracted from crops. Both regenerative energy sources should be incorporated into the planning of energy concepts for an area or region. The population should be convinced that the use of state-of-the-art furnaces for wood and straw burning is not hazardous to our health and the environment, provided that modern furnaces are used, the operation by the user is done carefully and the quality criteria (e.g. water content) for the fuel are fulfilled. The results of our research will help objectify the discussion of "real" particulate matter and its potential risks and possibly influences by the Federal Emission Control Act for biomass burning facilities. Modern facilities are also believed to be much easier to adjust to optimal combustion conditions than older furnaces. We hope that communicating the study results will increase the acceptance of and ease the way to a consensus-orientated extension of bioenergy utilisation.

References

Ackermann-Liebrich, U., Leuenberger, P., Schwartz, J., Schindler, C., Monn, C., & Bolognini, G. (1997). Lung function and long term exposure to air pollutants in Switzerland. *American Journal of Respiratory and Critical Care Medicine, 155*, 122–129.

BMU (Federal Ministry for the Environment, Nature Conservation and Nuclear Safety 2002). *First general administrative regulation pertaining the Federal Immission Control Act.* Retrieved December 5, 2012, from http://www.bmu.de/files/pdfs/allgemein/application/pdf/taluft_engl.pdf

Braun-Fahrländer, C. (2001). *Health effects of aerosols: What is the epidemiologic evidence? Aerosols from biomass combustion* (pp. 11–18). International Energy Agency IEA und Bundesamt für Energie, ISBN 3-908705-00-2, 2001.

Braun-Fahrländer, C., Ackermann-Liebrich, U., Schwartz, J., Gnehm, H. P., Rutishauser, M., & Wanner, H. U. (1992). Air pollution and respiratory symptoms in preschool children. *American Review of Respiratory Disease, 145*(1), 42–47.

Brunner, T., Bärnthaler, G., & Obernberger, I. (2006). *Feinstaubemissionen aus modernen Pellet-Kleinfeuerungsanlagen – Charakterisierung, Bildung und Möglichkeiten zur Emissionsminderung.* Submitted to: Österreichische Ingenieur- und Architektenzeitschrift ÖIAZ, Heft 10–12, Wien, Österreich. Retrieved November 28, 2012, from http://www.bios-bioenergy.at/uploads/media/Paper-Brunner-FeinstaubemissionenKleinfeuerungsanlagen-OEIAZ-2006.pdf

Council of the European Union. (1999). Council Directive 1999/30/EC of 22 April 1999 relating to limit values for sulphur dioxide, nitrogen dioxide and oxides of nitrogen, particulate matter and lead in ambient air. *Official Journal of the European Communities of 29.06.1999*; L 163/41 – L163/60. Retrieved December 7, 2012, from http://eurlex.europa.eu/LexUriServ/LexUriServ.do?uri=OJ:L:1999:163:0041:0060:DE:PDF

Degel, M., & Jörß, W. (2009). Aufbereitung von Daten der Emissionserklärungen gemäß 11. BImSchV aus dem Jahre 2004 für die Verwendung bei der UNFCCC- und UNECE-Berichterstattung. *Umweltbundesamt Texte 45/2009.* 150 pp. Retrieved December 5, 2012, from http://www.umweltdaten.de/publikationen/fpdf-l/3887.pdf

Deutscher Bundestag. (2009). Erste Verordnung zur Durchführung des Bundes-Immissionsschutzgesetzes (Verordnung über kleine und mittlere Feuerungsanlagen – 1.

BImSchV). *Deutscher Bundestag Drucksache 11/74*, 25.11.2009. Retrieved August 29, 2012, from http://dipbt.bundestag.de/dip21/btd/17/000/1700074.pdf

Deutscher Bundestag. (2010). Neununddreißigste Verordnung zur Durchführung des Bundes-Immissionsschutzgesetzes (Verordnung über Luftqualitätsstandards und Emissionshöchstmengen – 39. BImSchV). *Deutscher Bundestag Drucksache 17/508*, 25.01.2010. Retrieved December 7, 2012, from http://dip21.bundestag.de/dip21/btd/17/005/1700508.pdf

DFG (Deutsche Foschungsgemeinschaft 2008). Greim, H. (Ed.), *Gesundheitsschädliche Arbeitsstoffe – Toxikologisch-arbeitsmedizinische Begründungen von MAK-Werten und Einstufungen*. 45. Lieferung, 534 pp.

Dockery, D., Pope, C. A., Xu, X., Spengler, J., Ware, J., Fay, M., Ferris, B., & Speizer, F. (1993). An association between air pollution and mortality in six U.S. cities. *The New England Journal of Medicine, 329*, 1753–1759.

Dockery, D. W., Cunningham, J., Damokosh, A. I., Neas, L. M., Spengler, J. D., Koutrakis, P., et al. (1996). Health effects of acid aerosols on north American children: Respiratory symptoms. *Environmental Health Perspectives, 104*(5), 500–505.

DüMV (Fertilisers Regulation 2008). Verordnung über das Inverkehrbringen von Düngemitteln, Bodenhilfsstoffen, Kultursubstraten und Pflanzenhilfsmitteln (Düngemittelverordnung – DüMV) vom 16. Dezember 2008. *BGBl. I Nr. 60, p. 2524*. Retrieved December 10, 2012, from www.gesetze-im-internet.de/bundesrecht/d_mv_2008/gesamt.pdf

DüngMProbV (Fertilizer-Sampling and Analysis Ordinance 2006). Verordnung über Probenahmeverfahren und Analysemethoden für die amtliche Düngemittelüberwachung (Düngemittel–Probenahme– und Analyseverordnung); 19 Dec 1977 in der Fassung der Bekanntmachung vom 27. Juli 2006. *BGBl. I p. 1822*. Retrieved December 10, 2012, from http://www.gesetze-im-internet.de/bundesrecht/d_ngmprobv/gesamt.pdf

EFSA (European Food Safety Authority 2008). Polycyclic aromatic hydrocarbons in food, scientific opinion of the panel on contaminants in the food chain. *The EFSA Journal 724*, 1–114 (Question N° EFSA-Q-2007-136), Adopted on 9 June 2008. Retrieved December 3, 2012, from www.efsa.europa.eu/en/efsajournal/doc/724.pdf

Fine, P. M., Cass, G. R., & Simoneit, B. R. T. (2004a). Chemical characterization of fine particle emissions from the fireplace combustion of wood types grown in the Midwestern and Western United States. *Environmental Engineering Science, 21*, 387–409.

Fine, P. M., Cass, G. R., & Simoneit, B. R. T. (2004b). Chemical characterization of fine particle emissions from the wood stove combustion of prevalent United States tree species. *Environmental Engineering Science, 21*, 705–721.

FNR (Fachagentur Nachwachsende Rohstoffe, 2012). *Bioenergy in Germany: Facts and figures*. Retrieved November 28, 2012, from http://mediathek.fnr.de/media/downloadable/files/samples/b/a/basisdaten_engl_web_neu.pdf

Friedlander, S. (2000). *Smoke, dust and haze: Fundamentals of aerosol dynamics* (2nd ed., 432 pp). New York: Wiley. ISBN-10: 0195129997.

Gaegauf, C., Sattler, M., Meyer, N. K., & Heringa, M. F. (2008, September 11/12). Partikelemissionen in den instationären Verbrennungsphasen von Holzfeuerungen. 15. Schweizerisches Status-Seminar "Energie- und Umweltforschung im Bauwesen" (pp. 481–486). Switzerland: ETH-Zürich. Retrieved March 24, 2013, from http://www.brenet.ch/pdfstat_2008/10_ht_wp_gaeg.pdf

Hallquist, M., Wenger, J. C., Baltensperger, U., Rudich, Y., Simpson, D., Claeys, M., Dommen, J., Donahue, N. M., George, C., Goldstein, A. H., Hamilton, J. F., Herrmann, H., Hoffmann, T., Iinuma, Y., Jang, M., Jenkin, M. E., Jimenez, J. L., Kiendler-Scharr, A., Maenhaut, W., McFiggans, G., Mentel, T. F., Monod, A., Prévôt, A. S. H., Seinfeld, J. H., Surratt, J. D., Szmigielski, R., & Wildt, J. (2009). The formation, properties and impact of secondary organic aerosol: Current and emerging issues. *Atmospheric Chemistry and Physics, 9*(14), 5155–5236.

Hartmann, H., Ellner-Schubert, F., Turowski, P., Lenz, V., & Gerth, J. (2008). *Quantification and characterisation of particle emissions from residential wood stoves and boilers*. Presented at the 16th European biomass conference, Juni 2008, Valencia, Spanien.

Hasler, P., & Nussbaumer, T. (1998). *Particle size distribution of the fly ash from biomass combustion. Biomass for energy and industry*. Presented at the 10th European conference and technology exhibition, June 8–11, 1998, Würzburg, Germany, pp 1330–1333.

Hinds, W. (1998). *Aerosol technology. Properties, behavior, and measurements of airborne particles* (2nd ed., 504 pp). New York: Wiley, ISBN-10: 0471194107.

Iinuma, Y., Brüggemann, E., Gnauk, T., Müller, K., Andreae, M. O., Helas, G., Parmar, R., & Herrmann, H. (2007). Source characterization of biomass burning particles: The combustion of selected European conifers, African hardwood, savanna grass, and German and Indonesian peat. *Journal of Geophysical Resarch, 112*(D08209), 1–16.

Jankowski, N., Sageder, M., Bauer, H., Marr, I., Kasper-Giebl, A., Limbeck, A., Caseiro, A., Handler, M., Klatzer, B., Kationova, P., Pouresmaeil, P., Schmidl, C., Ramirez-Santa Cruz, C., Dattler, A., Andrade-Sanchez, L., Puxbaum, H., & Aquella-Team (2009). *Kurzfassung des Endberichts für das Projekt: "AQUELLA Linz – Oberösterreich" Aerosolquellenanalyse für Linz – Oberösterreich*. Technisch Universität Wien. Retrieved November 28, 2012, from http://www.land-oberoesterreich.gv.at/cps/rde/xbcr/SID-CA6A670F-B8B71E0A/ooe/Kurzfassung_Endbericht_Aquella_OOE.pdf

Jerina, D. M., Selander, H., Yagi, H., Wells, M. C., Davey, J. F., & Mahadevan, V. (1976). Dihydrodiols from anthracene and phenanthrene. *Journal of the American Chemical Society, 98*(19), 5988–5996.

Kaltschmitt, M., Hartmann, H., & Hofbauer, H. (2009). *Energie aus Biomasse – Grundlagen, Techniken, Verfahren.*, 2. Auflage (1032 pp). Heidelberg: Springer, ISBN-10: 3540850945.

Kanakidou, M., Seinfeld, J. H., Pandis, S. N., Barnes, I., Dentener, F. J., Facchini, M. C., van Dingenen, R., Ervens, B., Nenes, A., Nielsen, C. J., Swietlicki, E., Putaud, J. P., Balkanski, Y., Fuzzi, S., Horth, J., Moortgat, G. K., Winterhalter, R., Myhre, C. E. L., Tsigaridis, K., Vignati, E., Stephanou, E. G., & Wilson, J. (2005). Organic aerosol and global climate modelling: A review. *Atmospheric Chemistry and Physics, 5*, 1053–1123.

Kessler, F., Knechtle, N., & Frischknecht, R. (2000). Heizenergie aus Heizöl, Erdgas oder Holz. *BUWAL-Bericht, Schriftenreihe Umwelt Nr. 315*, 161 pp., Bern.

Klippel, N., & Nussbaumer, T. (2007). *Wirkung von Verbrennungspartikeln – Vergleich der Umweltrelevanz von Holzfeuerungen und Dieselmotoren.* Bundesamt für Energie und Bundesamt für Umwelt, Schlussbericht (65 pp), ISBN 3-908705-16-9, Zürich. Retrieved November 29, 2012, from http://www.bfe.admin.ch/php/modules/enet/streamfile.php?file=000000009137.pdf&name=000000270014.pdf

Laden, F., Schwartz, J., Speizer, F. E., & Dockery, D. W. (2006). Reduction in fine particulate air pollution and mortality: Extended follow-up of the Harvard Six cities study. *American Journal of Respiratory and Critical Care Medicine, 173*(6), 667–672.

Livbjerg, H. (2001). *Aerosol formation from straw combustion – Danish experiences. Aerosols from biomass combustion.* International seminar, International Energy Agency (IEA) and Swiss Federal Office of Energy, Zürich, 29–30, ISBN 3-908705-00-2

Nopmongcol, U., Khamwichit, W., Fraser, M. P., & Allen, D. T. (2007). Estimates of heterogeneous formation of secondary organic aerosol during a wood smoke episode in Houston, Texas. *Atmospheric Environment, 41*, 3057–3070.

Nussbaumer, T. (1989). *Schadstoffbildung bei der Verbrennung von Holz.* Dissertation. Laboratorium für Energiesysteme, ETH Zürich., 193 pp., Schriftenreihe "Forschungsberichte", Nr. 6. Zürich: Juris Druck + Verlag.

Nussbaumer, T. (2008). *Feinstaub-Emissionsfaktoren von Holzheizungen: Übersicht aus Ländern der Internationalen Energie Agentur – Bioenergy combustion task.* Presented at the 10. Holzenergie-Symposium, 12 Sept. 2008, ETH Zürich.

Nussbaumer, T., & Hasler, P. (1999). Bildung und Eigenschaften von Aerosolen aus Holzfeuerungen. *Holz als Roh- und Werkstoff, 57*, 13–22.

Nussbaumer, T., Czasch, C., Klippel, N., Johansson, L., & Tullin, C. (2008). *Particulate emissions from biomass combustion in IEA Countries – Survey on measurements and emission factors.* International Energy Agency (IEA) Bioenergy Task 32 and Swiss Federal Office of Energy

(SFOE); 40 pp, ISBN 3-908705-18-5, Zürich. Retrieved December 5, 2012, from http://www.ieabcc.nl/publications/Nussbaumer_et_al_IEA_Report_PM10_Jan_2008.pdf

Orasche, J., Seidel, T., Hartmann, H., Schnelle-Kreis, J., Chow, J., Ruppert, H., & Zimmermann, R. (2012). Comparison of emissions from wood combustion – Emission factors and characteristics from different small scale residential heating appliances considering particulate matter and PAH related toxicological potential of particle bound organic species. *Energy & Fuels, 26*(11), 6695–6704.

Orasche, J., Schnelle-Kreis, J., Schön, C., Hartmann, H., Ruppert, H., & Arteaga-Salas, J. M. (2013). Comparison of emissions from wood combustion – 2 The impact of combustion conditions on emission factors and characteristics of particle bound organic species and PAH related toxicological potential. *Energy & Fuels, 27*(3), 1482–1491.

Oser, M., Nussbaumer, Th., Müller, P., Mohr, M., & Figi, R. (2003). *Grundlagen der Aerosolbildung in Holzfeuerungen: Beeinflussung der Partikelemissionen durch Primärmassnahmen und Konzept für eine partikelarme automatische Holzfeuerung (Low-Particle-Feuerung)*. Bundesamt für Energie, Bundesamt für Umwelt, Wald und Landschaft, ISBN 3-908705-02-9, Bern. Retrieved December 5, 2012, from www.verenum.ch/Publikationen/Oser_Aerosol_2003_V2.pdf

Oser, M., Nussbaumer, Th., Müller, P., Mohr, M., & Figi, R. (2004). Mechanisms of particle formation in biomass combustion. In E. T. A. Florence & W. I. P. Munich (Eds.), *Proceedings of the second world biomass conference, 1246–1249*, 10–14 May 2004; ISBN 88-89407-04-2, Rome, Italy.

Pope, C. A., III, & Dockery, D. W. (2006). Health effects of fine particulate air pollution: Lines that connect. *Journal of the Air & Waste Management Association (1995), 56*, 709–742.

Schauer, J. J., Kleeman, M. J., Cass, G. R., & Simoneit, B. R. T. (2001). Measurement of emissions from air pollution sources. 3. C1-C29 Organic compounds from fireplace combustion of wood. *Environmental Science and Technology, 35*, 1716–1728.

Schmidl, C., Marr, I. L., Caseiro, A., Kotianová, P., Berner, A., Bazuer, H., Kasper-Giebl, A., & Puxbaum, H. (2008a). Chemical characterisation of fine particle emissions from wood stove combustion of common woods growing in mid European Alpine regions. *Atmospheric Environment, 42*, 126–141.

Schmidl, C., Bauer, H., Dattler, A., Hitzenberger, R., Weissenboeck, G., Marr, I. L., & Puxbaum, H. (2008b). Chemical characterisation of particle emissions from burning leaves. *Atmospheric Environment, 42*, 9070–9079.

Simoneit, B. R. T. (2002). Biomass burning – A review of organic tracers for smoke from incomplete combustion. *Applied Geochemistry, 17*, 129–162.

Squadrito, G., Cueto, R., Dellinger, B., & Pryor, W. A. (2001). Quinoid redox cycling as a mechanism for sustained free radical generation by inhaled airborne particulate matter. *Free Radical Biology and Medicine, 31*(9), 1132–1138.

Straif, K., Baan, R., Grosse, Y., Secretan, B., El Ghissassi, F., & Cogliano, V. (2005). Carcinogenicity of polycyclic aromatic hydrocarbons. *The Lancet Oncology, 6*(12), 931–932.

Struschka, M., Zuberbühler, U., Dreiseidler, A., Dreizler, D., Baumbach, G., Hartmann, H., Schmid, V., & Link, H. (2003). *Ermittlung und Evaluierung der Feinstaubemissionen aus Kleinfeuerungsanlagen im Bereich der Haushalte und Kleinverbraucher sowie Ableitung von geeigneten Maßnahmen zur Emissionsminderung*. UFOPLAN Forschungsbericht 299 44 140, UBA-FB 000477, Universität Stuttgart/Umweltbundesamt.

Struschka, M., Kilgus, D., Springmann, M., & Baumbach, G. (2008). *Effiziente Bereitstellung aktueller Emissionsdaten für die Luftreinhaltung*. UFOPLAN Forschungsbericht 205 42 322, UBA-FB 001217, Universität Stuttgart/Umweltbundesamt (UBA 2008).

UBA (Umweltbundesamt 2007). *Hintergrundpapier: Die Nebenwirkungen der Behaglichkeit: Feinstaub aus Kamin und Holzofen*, Umweltbundesamt. Dessau. Retrieved November 28, 2012, from http://www.umweltdaten.de/publikationen/fpdf-l/3556.pdf

UBA (Umweltbundesamt 2009). *Trends in air quality in Germany*. 18 pp. Retrieved December 6, 2012, from http://www.umweltdaten.de/publikationen/fpdf-l/3943.pdf

US-EPA (US Environmental Protection Agency 1984). Polynuclear aromatic hydrocarbons. Federal register. *Rules and Regulations*, 49(209), Method 610.

US-EPA (US Environmental Protection Agency 2004). *Air quality criteria for particulate matter* (Final report, Oct 2004). U.S. Environmental Protection Agency, Washington, DC, EPA 600/ P-99/002aF-bF. Retrieved December 10, 2012, from http://cfpub.epa.gov/ncea/cfm/ recordisplay.cfm?deid=87903#Download

Van Loo, S., & Koppejan, J. (2002). *Handbook of biomass combustion and co-firing*. Twente: International Energy Agency, Twente University Press.

Weis, L. M., Rummel, A. M., Masten, S. J., Trosko, J. E., & Upham, B. L. (1998). Bay or baylike regions of polycyclic aromatic hydrocarbons were potent inhibitors of gap junctional intercellular communication. *Environmental Health Perspectives, 106*, 17–22.

Xia, T., Korge, P., Weiss, J. N., Li, N., Venkatesen, M. I., Sioutas, C., & Nel, A. (2004). Quinones and aromatic chemical compounds in particulate matter induce mitochondrial dysfunction: Implications for ultrafine particle toxicity. *Environmental Health Perspectives, 112*, 1347–1358.

Yagi, H., & Jerina, D. M. (1982). Absolute configuration of the trans-9,10-dihydrodiol metabolite of the carcinogen benzo[a]pyrene. *Journal of the American Chemical Society, 104*(14), 4026–4027.

Zanobetti, A., Wand, M. P., Schwartz, J., & Ryan, L. M. (2000). Generalized additive distributed lag models: Quantifying mortality displacement. *Biostatistics, 1*, 279–292.

Zeldovich, Y. A. (1946). The oxidation of nitrogen in combustion and explosions. *Acta Physicochimica USSR, 22*, 577.

Zemp, E., Elsasser, S., Schindler, C., Künzli, N., Perruchoud, A. P., Leuenberger, P., & Team, S. A. P. A. L. D. I. A. (1999). Long-term ambient air pollution and respiratory symptoms in adults. *American Journal of Respiratory and Critical Care Medicine, 159*, 1257–1266.

Part VII
Bioenergy from Polluted Soils

Chapter 14
Bioenergy Production as an Option for Polluted Soils – A Non-phytoremediation Approach

Benedikt Sauer and Hans Ruppert

Abstract Contaminated arable land should not be used for the production of food or forage crops but for bioenergy production. As it takes thousands of years for plants to extract enough harmful metals from contaminated soils to reach acceptable low levels (phytoremediation), it is more feasible to leave the toxic elements in the soil. The metal transfer from various kinds of polluted soils to a variety of energy crops was investigated to identify crops with a low metal uptake. The advantages of using slightly contaminated crops for biogas production are that the heavy metals in the biogas plant will not impair the fermentation process. Furthermore, the residues of the biogas production can be returned to the fields where the crops were harvested. All important nutrients are recycled back into the fields (except nitrogen) without exceeding the maximum permissible values for heavy metal of farm fertilisers. Possible energy crops that show a low uptake of toxic elements are: the maize cultivars Padrino and Amadeo, the rye cultivar Vitallo and the barley cultivar Christelle. In contrast, amaranth (spec.) sunflower, the energy beet Kyros, the grass hybrid *Miscanthus giganteus* and sunchoke should not be cultivated on contaminated soils for bioenergy production due to their high cadmium uptake.

Keywords Heavy metals • contamination • soil to plant transfer factor • remediation

B. Sauer (✉) • H. Ruppert
Department of Sedimentology and Environmental Geology and Interdisciplinary Centre for Sustainable Development, University of Göttingen, Goldschmidtstr. 1-3, 37077 Göttingen, Germany
e-mail: bsauer@gwdg.de; hrupper@gwdg.de

H. Ruppert et al. (eds.), *Sustainable Bioenergy Production - An Integrated Approach*, DOI 10.1007/978-94-007-6642-6_14,

14.1 Introduction

Water, air, sunshine and fertile soils are the basic needs for food cultivation. Soil formation usually takes thousands of years and the area of soil suitable for food production is limited and decreasing. Arable soils comprise only about 12 % of the Earth's land area (FAO 2012). On a global scale, many threats have been identified to soils, such as sealing, erosion, compaction, a decline in biodiversity and organic matter content, salinization and contamination. Young et al. (2003) estimate that, globally, approximately 11.6 t of soil are anually lost per hectare (ha) due to erosion. The decline in fertile soil is occurring simultaneously with the growth of the world's population and an increasing need for food. Consequently, we should not only protect soil, but also use previously deteriorated soils beneficially.

Bioenergy production competes for arable land. In Germany, about 2.5 million ha (about 20 %) of arable land was used to produce energy crop in 2012 (FNR 2012). On the other hand, up to 10.4 % of the arable land in Germany is potentially contaminated (the estimate is based on the German soil map data; Knappe et al. 2008). These areas mainly comprise the flood plains of rivers, but also industrialised and city areas that had high emissions in the past, peri-urban fields with sewage irrigation, etc. Heavily contaminated soil is a danger to human and animal health; it should not be used for food or fodder production but for the production of energy crops.

More than a thousand years of mining and smelting of sulfidic polymetallic ores in the Harz Mountains in Germany have contaminated many soils in this area with heavy metals. Heavy rain falls and melting snow washed heavy-metal-contaminated soil and smelting waste material into rivers. During floods, these materials are deposited on alluvial areas (Deicke et al. 2006). Later in the Industrial Age since the nineteen century , almost all of the big German rivers' floodplains have been contaminated by the uncontrolled discharge of sewage water. At the end of the 1960s, Müller and Förstner already revealed that the clay fraction of the large western Germany rivers' sediments was extremely contaminated with heavy metals (Förstner and Müller 1974). While this situation has clearly improved, many floodplains are still contaminated due to the earlier depositions of contaminated sediments.

Remediation is normally not possible as these areas are too vast. Moreover, a lack of suitable methods, the amount of contaminated soil, ecologic reasons and the extremely high costs involved make technical remediation unrealistic (Salt 1998; Wenzel et al. 1999; Suthersan 1997; Iskandar 2001). The phytoremediation of heavy metal contaminated soils takes thousands of years, as the example of cadmium, one of the most mobile heavy metals, shows (see Box 14.1). An alternative utilisation of contaminated areas is therefore required, such as growing crops for technical products or renewable energy.

The systematic cultivation of different crops with a good potential for energy production on contaminated soil has shown that heavy metals are transferred from the soil to these plants. Many pot experiments were performed in Germany during the 1970s (e.g. Sauerbeck 1989). However, the conditions in these experiments differed significantly from the real situation, because the soils were spiked with mostly highly soluble heavy metal salts. It is highly questionable whether adding heavy metals in a soluble form to soils has the same phytoavailability as the heavy metals in contaminated soils. In addition, factors that influence plants' metal uptake, such as the pH-value, were not properly investigated in these experiments.

In contrast to scientists proclaiming the phytoremediation of heavy-metal-contaminated soils (Salt 1998; Raskin and Ensley 2000), our approach does not search for accumulator or hyperaccumulator crops, but for high yield energy crops with a minimum uptake of heavy metals. If this approach is followed, the dangerous elements remain in the contaminated soil and will not enter the biogas fermenter. The risk of declining biogas formation in the fermenter due to high loads of toxic heavy metals can then be avoided. After these crops have been used for biogas production, the residues in the fermentation plant should be used to fertilise the areas from which the energy crops were taken, thereby completing the cycle of all the elements, including the nutrients extracted from the soil.

In our study, the number of commonly analysed elements, such as chromium (Cr), nickel (Ni), copper (Cu), zinc (Zn), molybdenum (Mo), cadmium (Cd) and lead (Pb), was extended to include the elements arsenic (As), antimony (Sb), bismuth (Bi), thallium (Tl), uranium (U), thorium (Th), cobalt (Co), tin (Sn), vanadium (V) and beryllium (Be). The latter elements may also be part of the contamination. Additionally, we examined how fertilising soil with biogas digestate influences the soil-plant transfer of elements.

14.2 Methods

14.2.1 Open Land Pot Trial

Different crops were cultivated in 130 big pots – containing ten soils from various types of contaminated sites in Lower Saxony – in an open park-like area close to the Geoscience Centre at Göttingen University. Crops were also cultivated in an open plot with a highly contaminated soil near Harlingerode, which is close to the Harz Mountains (Lower Saxony).

The soil samples for the pots represented the top horizons of arable soils and contained material to a depth of about 30 cm. Before filling the pots, the soil material from each location was homogenised by means of a new cement mixer. Six pots were filled with soil material from each of the locations and positioned in the park the Geosciences Centre. In the first year, the soils were already conventionally fertilised. In the second and third years, biogas residues from the bioenergy plant in Jühnde near Göttingen were used as fertiliser. During very dry summer periods, the pots were irrigated with tap water. The following crops were cultivated and analysed in a 2-year period (the crops' cultivars are indicated in parentheses):

- Summer 2009: clover (Heusers), durum wheat (Duramar), field bean (Espresso), sunflower (Salut), sorghum (Maja), maize (Amadeo)
- Winter 2009/2010: winter triticale (Tulus) winter rye (Vitallo), winter wheat (Mulan), winter rape (Elektra), hairy vetch (Welta), winter barley (Christelle)
- Summer 2010: maize (Amadeo), ryegrass (Gisel), phacelia (Amerigo), maize (Padrino), amaranth (*Amaranthus spec.*)
- Winter 2010/2011: winter wheat (Isengrain), winter triticale (Tulus), winter barley (Malwinta), winter barley (Souleky), winter rye (Protector), winter rye (Minello)
- Summer 2011: white mustard (Semper), maize (Revolver), maize (Ronaldinio), maize (Sulexa), maize (Amadeo), sunflower (Metharoc)

For five locations, 40 additional and bigger pots with a capacity of about 100 l were filled with soil. In these pots, sunchoke (*Helianthus tuberosus*), igniscum (knotweed; *Fallopia sachalinensis*), cup plant (*Silphium perfoliatum*), *Miscanthus giganteus,* summer oat species, and short rotation wood trees (poplar Max 1, willow Tordis, robinia and alder) were cultivated. The short rotation trees will be harvested and analysed in the winter of 2012/2013; the results are thus not yet available.

14.2.2 Analytical Methods

Soil samples were taken from each pot and the six samples of each individual location were mixed. The samples were air dried and sieved <2 mm. Their pH values were measured according to the DIN ISO 10390 with 0.01 mol/l $CaCl_2$ (10 g soil + 25 ml $CaCl_2$ solution). The samples' grain size was analysed using Beckman Coulter's laser diffraction particle size analyzer LS 13320. The soil samples were dried at 105 °C for the geochemical analysis and ground in an agate ball mill until a grain size of <63 μm was achieved. In total, 200 mg of the ground samples were fully digested using a mixture of ultrapure, concentrated HNO_3, $HClO_4$ and HF in closed, ultra clean PTFE vessels (Picotrace, Göttingen, Germany) under pressure and then filled up to 100 ml with ultrapure water.

In the milk-ripe stage (the stage when plants are harvested for biogas production), the plants were cut approximately 10 cm above the soil surface with a sharp knife. At this stage, plants consist of about 30 % dry matter (at 105 °C) and are suitable for making silage. Next, the samples were finely ground in an agate ball mill. In total, 700 mg of the sample powders were digested, using the same mixture of concentrated acids as above, and then filled up to 50 ml with pure water.

All main and trace element measurements were conducted at the Geosciences Centre at the University of Göttingen, using ICP-OES (inductively coupled plasma optical emission spectrometer; Optima 3300 DC Perkin Elmer) and ICP-MS (inductively coupled plasma mass spectrometer; Perkin Elmer DRC-II). Blind samples assured that the handling during the digestion and measurement was

Table 14.1 Percentage of grain sizes within the soils

		Fraction [μm]					
Location	Soil type	<2	2–<63	63–<200	200–<630	630–2,000	Sum
Trögen	Stagnosol-cambisol	6.7	75.9	11.0	6.2	0.2	100
Schwülper	Arenosol	1.1	12.0	25.8	54.8	6.3	100
Ohrum	Gleyic fluvic cambisol	5.9	61.1	22.2	9.3	1.5	100
Harlingerode	Luvisol	5.7	50.9	13.8	23.3	6.3	100

clean and that no notable contamination occurred. The excellent agreement between our reference samples' values and the internationally published data assures that the measurements are correct and internationally comparable.

14.2.3 Results

The Schwülper site has sandy soil contaminated with sewage sludge from the city of Braunschweig over several decades. This soil is mainly enriched with Bi, Cd, Ni, Sn and Zn. The low pH value and the sandy character of the soil led to very high Ni and Zn contents in some of the plants.

The silt-rich soil near Ohrum comes from the floodplain soil of the Oker river in the northern foreland of the Harz Mountains. Over several centuries, metal industries in the Harz mountains contaminated – and continue to contaminate – the floodplains of the draining rivers. The soil from Ohrum is contaminated with As, Cd, Cu, Pb, Sb, Tl and Zn.

The sandy-silty soils of the Harlingerode site are close (700 m) to a former zinc and lead smelter installed in 1935 for armament purposes. Production stopped in 2000. The soils from Harlingerode are contaminated with As, Cd, Cu, Pb, Sb, Tl and Zn.

The silt-rich soil from Trögen serves as an unpolluted reference site.

Table 14.1 shows the soil type and the grain sizes of the soils. Table 14.2 lists the pH values and contents of some elements in the soils, which ascend according to their Cd content. The Trögen soil can be regarded as uncontaminated.

14.2.4 Physiological Element Content: Correction for Soil Material Adhering to Plant Samples

Small amounts of local soils and of dust material adhere to outdoor plants. The concentration of most elements is much higher in soils than in plants. Therefore, a correction is required for the element concentration in the adhering material in order to calculate plants' physiological element content. Some elements, such as titanium (Ti) and aluminium (Al), are not, or only slightly, mobile in soils with pH values between 5.5 and 7.4. Aluminium and titanium are highly enriched in soils and are barely transferred to plants at these pH values. Consequentially, it is

Table 14.2 pH values and element concentrations in the examined soils; data in mg/kg dry matter

Location	pH	Al	As	Bi	Cd	Co	Cr	Cu	Mo	Ni	Pb	Sb	Sn	Ti	Tl	Zn
Trögen	5.2	43,421	9	0.1	0.3	7	77	9	0.5	16	29	0.7	2	4,898	0.5	52
Schwülper	4.4	14,586	4	4.2	3.6	1.3	16	72	1.4	13	88	3	20	954	0.3	121
Ohrum	6.8	54,905	50	2.3	13	28	95	458	1.5	54	2,232	22	9	4,477	2.1	2,760
Harlingerode	6.9	35,879	39	1.7	20	11	49	126	3.0	23	815	23	9	3,288	2.2	820

Table 14.3 Example of the calculation of the physiological cobalt and nickel content in plants; correction for adhering soil material according to the titanium content; location Ohrum; data in mg/kg dry matter

Location Ohrum	measured values			physiological contents (Ti-corrected values)		
	Ti	Co	Ni	Ti	Co	Ni
Soil Ohrum	**4536**	**27.5**	**52.6**	0	0	0
Amaranth	16	0.43	0.8	0	0.33	0.6
Barley Christelle	0.8	0.04	0.2	0	0.03	0.2
Clover Heusers	44	0.19	1.0	0	<0.02	0.5
Maize Amadeo	16	0.07	0.4	0	<0.02	0.2
Rye Vitallo	1.2	<0.02	0.2	0	<0.02	0.1
Cup plant	11	0.15	1.2	0	0.08	1.1
Oat Aragon	3	0.03	2.2	0	0.01	2.2
Oat Zorro	75	0.45	3.4	0	<0.02	2.5
Sunflower Salut	7	0.09	0.7	0	0.05	0.6

possible to use the titanium concentrations in plants and in soil samples to calculate the amount of adhering soil material. We can also correct the measured concentration of an element (El) in the plant with regard to the adhering element concentration in order to calculate the physiological element concentration in the plant. Formula:

$$\mathrm{El}_{\text{physiological content plant}} = \mathrm{El}_{\text{total content measured in plant}} - \left(\mathrm{Ti}_{\text{total content measured in plant}}/\mathrm{Ti}_{\text{content in soil}}\right) \times \mathrm{El}_{\text{content in soil}}$$

During heavy rains or storms, soil particles splash against the plants, leaving residues on them. In this contribution, we calculate the physiological element concentration using the residue concentration in the corresponding soil as a reference, because the atmospheric dust composition is usually not known. This correction is only an approximation. This calculation would be rather unrealistic in terms of elements that are very highly concentrated in the soil but have a low concentration in the dust.

Table 14.3 shows examples of the measured total contents of Co and Ni in different plants in the soil from Ohrum and their resulting physiological concentrations after normalization to titanium.

Tables 14.4, 14.5, 14.6, and 14.7 show the physiological element content of the different analysed crops grown on differently polluted soils. They are arranged in ascending concentrations of the critical element cadmium.

Table 14.4 Physiological element concentrations in the crops after correcting for adhering soil material by Ti in mg/kg dry matter

	Physiological element contents of crops											
Location Harlingerode	Ti	As	Bi	Cd	Co	Cr	Mo	Ni	Pb	Sb	Tl	Zn
Soil Harlingerode	3,365	39	1.74	20.1	10.6	48.5	3.0	27.3	815	22.6	2.19	1,226
Rye Vitallo	0	<0.4	0.0005	0.7	**0.22**	<0.5	5.5	0.5	0.8	0.015	0.0069	48
Cup plant	0	0.7	<0.0005	0.7	0.06	0.8	0.7	**0.7**	<0.3	0.047	0.0747	24
Rye Protector	0	0.4	<0.0005	1.2	0.01	<0.5	3.8	0.1	<0.3	0.012	0.0054	52
Clover Heusers	0	0.4	0.0025	1.2	**0.06**	0.8	**24.2**	0.6	0.2	**0.076**	0.0182	64
Maize Padrino	0	<0.4	0.0040	1.3	0.01	0.6	1.6	0.2	0.3	0.027	0.0152	89
Sunchoke Rozzo tuber	0	1.4	0.0014	1.5	0.02	<0.5	0.3	**0.8**	**2.0**	**0.093**	0.0955	25
Maize Sulexa	0	<0.4	0.0027	1.6	0.01	<0.5	3.5	0.1	<0.3	<0.001	0.0025	83
Barley Christelle	0	0.4	<0.0005	1.7	0.02	0.5	4.9	0.1	0.6	0.006	0.0145	80
Wheat Mulan	0	0.5	<0.0005	2.1	**0.20**	<0.5	7.6	0.5	0.2	0.007	0.0173	40
Maize Ronaldinio	0	<0.4	0.0031	2.2	0.01	<0.5	2.9	0.1	<0.3	<0.001	0.0011	83
Maize Revolver	0	<0.4	0.0010	2.9	0.03	<0.5	1.0	0.2	<0.3	0.008	0.0088	61
Triticale Tulus	0	0.5	<0.0005	3.2	0.04	<0.5	5.3	0.1	<0.3	0.022	0.0137	90
Rye Minello	0	**4.2**	<0.0005	3.4	0.01	0.5	3.9	0.2	<0.3	0.003	0.0028	63
Ryegrass Gisel	0	0.5	0.0016	4.0	0.03	0.6	**12.5**	0.3	0.3	0.054	0.0954	55
Hairy vetch Welta	0	0.6	0.0014	4.1	**0.10**	**1.5**	**21.1**	**0.8**	**3.2**	**0.073**	0.0970	**192**
Maize Amadeo	0	<0.4	**0.0076**	4.2	0.02	0.7	1.3	0.2	<0.3	0.026	0.0081	92
Rape Elektra	0	0.6	0.0013	5.1	0.05	<0.5	**19.9**	0.2	0.2	0.028	**22.1**	53
Sorghum Maja	0	0.4	**0.0085**	5.4	0.03	**1.4**	2.5	**0.8**	0.8	0.045	0.032	78
Igniscum	0	0.5	**0.0042**	6.4	0.05	0.6	2.6	0.3	<0.3	**0.071**	0.163	71
Sunchoke Rozzo above ground	0	**1.4**	0.0031	7.9	0.03	1.0	0.7	0.3	<0.3	0.040	0.148	90
White Mustard Semper	0	**1.6**	0.0015	8.1	0.04	0.6	**25.8**	0.7	1.0	**0.113**	**0.771**	**139**
Miscanthus gig.	0	<0.4	<0.0005	**8.1**	0.01	<0.5	1.4	0.2	<0.3	<0.001	0.0010	**199**
Sunflower Salut	0	0.5	**0.0076**	**9.0**	0.03	1.2	1.5	0.8	1.0	0.048	**0.315**	110
Beet Kyros body	0	0.4	0.0036	**13.3**	0.04	0.6	0.4	0.3	1.5	0.012	**0.606**	**168**
Sunflower Metharoc	0	0.6	0.0025	**20.0**	0.06	0.8	1.0	0.4	**2.1**	0.020	0.296	**199**
Amaranth spec.	0	0.7	0.0033	**30.3**	**0.27**	**1.5**	**9.5**	0.5	<0.3	0.066	**0.580**	119
Beet Kyros head + leaves	0	**2.0**	0.0003	**80.3**	**0.07**	**2.0**	7.6	0.5	<0.3	0.008	**0.573**	**589**

Pot trial with soil from Harlingerode (pH = 6.9). The top 20 % (quantile 0.8) of the values are in bold

Table 14.5 Physiological element concentrations in crops after Ti correction in mg/kg dry matter

	Physiological element contents of crops											
Location Ohrum	Ti	As	Bi	Cd	Co	Cr	Mo	Ni	Pb	Sb	Tl	Zn
Ohrum soil	4,536	50.1	2.28	12.8	27.5	95	1.5	52.6	2,232	21.8	2.08	4,142
Cup plant	0	0.6	0.0042	0.1	0.1	0.8	0.4	**1.2**	<0.3	**0.046**	0.0063	19
Rye Vitallo	0	<0.4	<0.0005	0.2	<0.02	<0.5	3.3	0.1	<0.3	0.007	<0.0001	101
Rye Protector	0	**0.8**	<0.0005	0.3	<0.02	<0.5	1.6	0.2	<0.3	<0.001	<0.0001	62
Maize Padrino	0	<0.4	0.0041	0.5	<0.02	0.6	1.1	0.2	<0.3	0.039	0.0084	121
Sunchoke tuber	0	<0.4	<0.0005	0.5	<0.02	<0.5	0.1	**1.6**	**1.1**	0.023	0.0166	34
Barley Christelle	0	<0.4	<0.0005	0.6	0.03	0.9	3.2	0.2	0.4	0.005	0.0010	137
Maize Sulexa	0	<0.4	0.0025	0.6	<0.02	<0.5	1.8	0.2	<0.3	0.006	<0.0001	160
Maize Ronaldinio	0	<0.4	0.0022	0.8	<0.02	<0.5	2.3	0.2	<0.3	0.010	<0.0001	85
Maize Amadeo	0	<0.4	0.0015	0.8	<0.02	0.7	1.1	0.2	<0.3	0.006	<0.0001	95
Wheat Mulan	0	<0.4	<0.0005	0.9	0.02	<0.5	**4.2**	0.3	<0.3	0.006	0.0020	81
Triticale Tulus	0	<0.4	0.0003	0.9	0.02	<0.5	3.1	0.2	<0.3	0.007	0.0004	179
Rye Minello	0	<0.4	<0.0005	0.9	<0.02	<0.5	2.4	0.4	<0.3	0.009	<0.0001	147
Clover Heusers	0	<0.4	<0.0005	1.2	<0.02	<0.5	**8.0**	0.5	<0.3	<0.001	<0.0001	95
Ryegrass Gisel	0	0.5	0.0016	1.3	0.05	0.8	**4.7**	**1.0**	<0.3	0.036	0.0125	84
Rape Elektra	0	0.7	0.0008	1.5	<0.02	0.5	**8.0**	0.7	<0.3	0.003	**6.1473**	106
Hairy vetch Welta	0	<0.4	0.0013	1.6	**0.11**	**1.1**	**8.7**	**1.1**	0.8	0.016	0.0236	**232**
White Mustard Semper	0	**0.9**	**0.0091**	2.0	0.07	<0.5	**6.2**	0.9	<0.3	**0.214**	**0.0807**	192
Shorgum Maja	0	<0.4	0.0040	2.6	<0.02	0.8	0.9	0.3	<0.3	0.014	0.0043	137
Beet Kyros body	0	<0.4	0.0021	3.2	0.04	0.6	0.1	0.3	0.6	0.013	**0.1499**	187
Maize Revolver	0	<0.4	0.0010	3.4	<0.02	<0.5	1.3	0.2	<0.3	<0.001	0.0029	142
Sunchoke above plant	0	0.5	0.0024	3.4	0.03	1.0	0.4	**5.9**	<0.3	**0.039**	0.0183	172
Sunflower Salut	0	0.4	**0.0042**	**3.5**	0.05	1.0	0.7	0.6	1.0	0.025	0.0295	**209**
Miscanthus gig.	0	<0.4	0.0006	**3.9**	<0.02	<0.5	0.4	0.2	<0.3	0.030	<0.0001	**283**
Igniscum	0	0.6	**0.0057**	**4.2**	**0.08**	0.7	1.7	0.8	**1.7**	**0.069**	0.0157	**210**
Sunflower Metharoc	0	0.5	<0.0005	**8.7**	0.05	0.6	0.6	0.6	<0.3	0.021	0.0482	**326**
Amaranth spec.	0	0.7	**0.0044**	**9.1**	**0.33**	**1.3**	1.6	0.6	<0.3	**0.078**	0.0641	204
Beet Kyros head + leaves	0	**2.8**	0.0022	**15.3**	0.04	**2.3**	0.9	**1.3**	<0.3	0.026	**0.1451**	**977**

Pot trial with soil from Ohrum (pH = 6.8). The top 20 % (quantile 0.8) of the values are in bold

Table 14.6 Physiological element concentrations in crops after Ti correction in mg/kg dry matter

Location Schwülper	Physiological element contents of crops											
	Ti	As	Bi	Cd	Co	Cr	Mo	Ni	Pb	Sb	Tl	Zn
Schwülper soil	967	4.0	4.23	3.6	1.26	15.9	1.4	13.3	88	2.6	0.31	165
Maize Padrino	0	<0.4	<0.0005	0.5	0.02	0.5	0.4	0.8	<0.3	0.022	0.0214	206
Rye Vitallo	0	<0.4	<0.0005	0.5	0.36	0.6	1.3	2.6	<0.3	0.009	0.0088	129
Maize Ronaldinio	0	<0.4	<0.0005	0.5	0.03	<0.5	0.8	1.2	<0.3	0.017	0.0082	123
Maize Sulexa	0	<0.4	<0.0005	0.5	0.02	0.5	0.5	0.5	<0.3	0.025	0.0134	192
Maize Amadeo	0	<0.4	<0.0005	0.6	0.02	0.6	0.4	1.2	<0.3	0.019	0.0218	120
Sunchoke Rozzo tuber	0	<0.4	**0.0070**	0.7	0.03	0.7	0.2	15.1	<0.3	0.023	0.0359	51
Barley Christelle	0	<0.4	<0.0005	0.8	0.02	0.6	1.2	1.5	<0.3	0.010	0.0060	179
Rye Protector	0	<0.4	<0.0005	0.9	0.07	<0.5	1.8	1.5	<0.3	0.019	0.0097	166
Rye Minello	0	<0.4	<0.0005	0.9	0.08	<0.5	3.3	3.0	<0.3	0.011	0.0024	163
Ryegrass Gisel	0	<0.4	<0.0005	1.2	0.03	0.8	3.6	9.8	<0.3	0.039	0.0229	99
Triticale Tulus	0	<0.4	<0.0005	1.4	0.02	<0.5	1.3	2.9	<0.3	0.010	0.0093	188
Maize Revolver	0	<0.4	<0.0005	1.4	0.03	<0.5	0.4	0.9	<0.3	0.022	0.0109	193
Shorgum Maja	0	<0.4	<0.0005	1.6	0.03	1.0	0.5	2.0	<0.3	0.024	0.0081	142
Hairy vetch Welta	0	<0.4	<0.0005	2.0	0.17	0.6	**13.0**	12.7	<0.3	0.043	0.0831	**535**
Cup plant	0	0.7	<0.0005	2.3	**0.29**	**1.2**	0.5	**25.8**	<0.3	**0.139**	0.0674	207
Rape Elektra	0	<0.4	<0.0005	2.4	0.11	0.6	**5.3**	**16.8**	<0.3	0.013	**1.6653**	317
White Mustard Semper	0	0.4	<0.0005	2.6	0.08	0.5	**4.2**	9.0	**3.1**	**0.168**	**0.1816**	399
Clover Heusers	0	0.4	<0.0005	2.8	**0.36**	1.1	**35.2**	15.3	<0.3	**0.089**	0.0086	**578**
Miscanthus gig.	0	<0.4	<0.0005	4.0	0.05	0.2	0.5	0.6	<0.3	0.034	0.0015	316
Sunflower Salut	0	<0.4	<0.0005	4.5	0.06	**1.3**	0.6	**16.1**	<0.3	0.049	0.0630	521
Sunflower Metharoc	0	<0.4	<0.0005	4.6	0.07	0.9	1.0	**17.7**	<0.3	0.034	0.0447	**627**
Beet Kyros body	0	0.4	<0.0005	**5.9**	0.05	0.6	0.1	5.7	**0.3**	0.010	**0.1023**	341
Sunchoke Rozzo above ground	0	**1.3**	<0.0005	**5.9**	0.10	**1.2**	0.5	6.4	<0.3	0.068	0.0792	486
Phacelia Amerigo	0	**0.9**	<0.0005	**7.0**	**0.22**	1.1	**11.9**	**15.4**	<0.3	0.061	**0.4448**	**592**
Igniscum	0	0.5	<0.0005	**8.1**	0.27	0.9	2.5	7.8	<0.3	0.081	0.0523	381
Amaranth spec.	0	0.4	<0.0005	**13.7**	**0.38**	**1.2**	**4.0**	**16.7**	<0.3	**0.068**	**0.1895**	**749**
Beet Kyros head + leaves	0	**0.9**	<0.0005	**20.2**	0.13	**2.0**	2.4	10.7	<0.3	**0.082**	**0.1921**	**812**

Pot trial with soil from Schwülper (pH = 4.4). The top 20 % (quantile 0.8) of the values are in bold

Table 14.7 Physiological element concentrations in crops after Ti correction in mg/kg dry matter

	Physiological element contents of crops											
Location Trögen	Ti	As	Bi	Cd	Co	Cr	Mo	Ni	Pb	Sb	Tl	Zn
Trögen soil	**4,959**	9.1	0.14	0.32	6.61	76.9	0.5	29.6	29	0.74	0.46	61
Rye Protector	0	<0.4	0.0013	0.03	<0.02	<0.5	0.2	0.2	<0.3	0.027	<0.0001	17
Rye Minello	0	<0.4	0.0010	0.04	<0.02	<0.5	0.4	0.4	<0.3	0.025	<0.0001	20
Rye Vitallo	0	<0.4	0.0008	0.04	<0.02	<0.5	0.3	0.2	<0.3	0.012	0.0006	18
Barley Christelle	0	<0.4	0.0011	0.04	0.09	<0.5	0.4	0.2	<0.3	0.010	0.0008	24
Triticale Tulus	0	<0.4	0.0009	0.04	0.02	<0.5	0.3	0.4	<0.3	0.014	0.0006	25
Hairy vetch Welta	0	0.4	0.0019	0.04	**0.37**	0.5	**2.0**	0.4	<0.3	0.020	0.0110	20
Cup Plant	0	0.5	0.0064	0.06	0.08	0.9	0.1	**1.3**	0.4	**0.094**	0.0095	19
Ryegrass Gisel	0	0.4	0.0031	0.06	**0.10**	0.5	**0.8**	0.3	<0.3	0.052	0.0075	15
Maize Sulexa	0	<0.4	0.0069	0.07	0.03	0.5	0.3	0.2	0.5	0.054	0.0028	33
Clover Heusers	0	<0.4	0.0036	0.07	0.06	0.8	0.2	**1.2**	**0.7**	0.046	0.0033	33
Maize Revolver	0	<0.4	0.0049	0.07	0.03	<0.5	0.3	0.2	0.4	0.051	0.0032	38
Sunchoke Rozzo tuber	0	<0.4	0.0002	0.08	0.02	0.7	0.0	**1.9**	<0.3	0.012	0.0061	10
Maize Ronaldinio	0	<0.4	0.0072	0.08	0.03	0.5	0.3	0.2	0.4	0.045	0.0026	37
Maize Padrino	0	<0.4	0.0064	0.08	0.03	<0.5	0.2	0.2	<0.3	0.048	0.0169	48
Maize Amadeo	0	<0.4	**0.0091**	0.09	0.03	1.0	0.2	0.5	0.5	0.062	0.0024	49
Shorgum Maja	0	<0.4	0.0073	0.25	0.03	0.7	0.4	0.4	0.3	0.034	0.0032	**51**
Rape Elektra	0	<0.4	0.0030	0.33	0.05	0.6	**1.1**	0.6	<0.3	0.019	**1.7054**	42
Sunflower Metharoc	0	0.4	0.0027	0.42	0.06	0.6	0.1	0.6	0.3	0.052	0.0089	**53**
White Mustard Semper	0	<0.4	0.0028	0.48	0.05	0.6	**1.0**	0.4	**0.6**	0.039	0.0146	**53**
Sunchoke Rozzo above ground	0	0.5	0.0068	0.48	0.05	**1.1**	0.2	0.7	0.7	**0.082**	0.0103	38
Amaranth spec.	0	**0.7**	0.0073	**0.51**	**0.44**	**1.6**	0.5	**1.7**	0.5	**0.087**	**0.0191**	64
Igniscum	0	0.6	**0.0096**	**0.52**	**0.10**	0.6	**0.5**	0.6	0.6	**0.098**	0.0103	40
Sunflower Salut	0	<0.4	**0.0100**	**0.62**	0.07	**1.5**	0.1	**1.3**	0.5	0.047	0.0149	51
Miscanthus gig.	0	<0.4	0.0012	**0.86**	<0.02	<0.5	<0.1	0.1	<0.3	0.046	0.0016	41
Beet Kyros body	0	<0.4	0.0025	**0.88**	0.06	0.6	0.1	**2.6**	0.4	0.018	**0.0164**	**91**
Beet Kyros head + leaves	0	**1.2**	**0.0118**	**3.50**	**0.18**	**2.0**	**0.5**	**4.4**	**1.9**	**0.148**	**0.0259**	**325**

Pot trial with soil from Trögen (pH = 5.1). The top 20 % (quantile 0.8) of the values are in bold

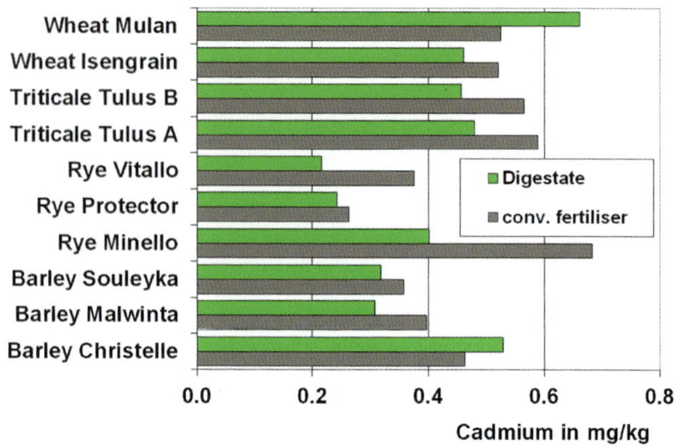

Fig. 14.1 Comparison between the physiological cadmium content in energy crops after conventional and biogas digestate fertilisation. Location: open test field near Harlingerode. Cd in soil: 15 mg/kg; pH: 7.0

In all the concentrations, the critical elements in the path soil to crop, to man or animal and to a biogas plant are: cadmium, molybdenum, nickel, thallium and zinc, with cadmium the most critical. The very high thallium transfer to rape in all soils is especially alarming. This observation needs further scientific examination.

Since the tables are given in ascending concentrations of cadmium, the crops in the top rows are the most suitable for bioenergy production. These are: cup plant, rye (Vitallo), maize (Padrino/Amadeo) and barley (Christelle).

In pot trials with soils from Harlingerode, it became evident that even different breeds of maize differ in their uptake of cadmium up to a factor two. Owing to their high uptake of cadmium, energy beet (Kyros), amaranth (spec.), sunflower (both Salut and Metharoc), phacelia (Amerigo) and Igniscum should not be grown on cadmium-contaminated locations for bioenergy production.

Pellets made from *Miscanthus gigantheus* are being considered for heat production. However, the high uptake of cadmium by this plant, even from uncontaminated soils, is a critical issue, as potentially elevated cadmium emissions may occur when the plant is burnt. It is thus inadvisable to use it for heat production.

On a strongly contaminated open field plot near Harlingerode, we tested the impact of biogas digestate as a fertiliser on the element uptake. Winter crops were planted side by side on small divided plots. Normal amounts of farm fertiliser were applied on one half of the plot, while 5 l of uncontaminated biogas digestate was applied per square meter on the other half. The cadmium results are visualised in Fig. 14.1. With the exception of winter wheat (Mulan) and barley (Christelle), the cadmium uptake decreased by an average of 22 % after applying biogas digestate fertilisation compared to the percentage after conventional fertilisation.

The distribution of cadmium differs significantly within the different crop parts such as straw and grain as well as in the whole crop. The winter crops results are presented in Fig. 14.2. The grains contain only about one-third of the cadmium

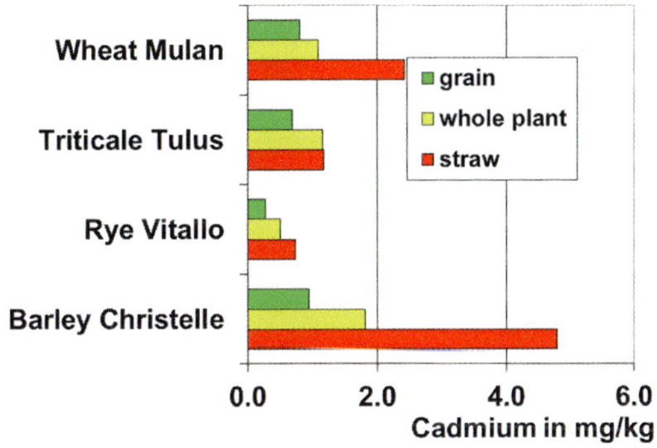

Fig. 14.2 A comparison between different winter crops' cadmium content in their grain and straw as well as in the whole plant. Location: open test field near Harlingerode. Cd content in soil: 15 mg/kg; pH: 7.0

concentration in straw. The value for the whole crop was logically somewhere in the middle. There has been talk of using the grains for animal fodder and the straw for heat production. However, our results suggest that, if no additional reduction measures are installed in the heating system, elevated cadmium emissions can be expected when burning straw.

14.3 Interpretation

14.3.1 Soil to Crop Transfer Factors

In order to estimate how much of an element a plant will absorb from a certain soil, we first need to calculate the soil to plant transfer factor. This is defined as the ratio of an element's concentration in a plant to its concentration in the soil based on dry matter. The transfer factors and bioavailability are influenced by the plant species and many soil parameters. These parameters include the soil's physical and chemical properties, such as its texture, organic matter, Fe/Mn oxides, pH-value and the presence of complexing agents that affect the bioavailability of metals (Sauerbeck and Lübben 1991; Li et al. 2004; Luo et al. 2010).

In Fig. 14.3, the cadmium transfer factors of the five examined breeds of maize are shown for the four locations. In contaminated soils, the Padrino cultivar always shows the lowest transfer factor. If the soil's cadmium content is low, like that at Trögen, a relatively high transfer factor follows.

Figure 14.4 shows the soil to plant transfer factors of cadmium in the examined three rye breeds. Again, one breed (Vitallo) has the lowest transfer factors in all the contaminated locations.

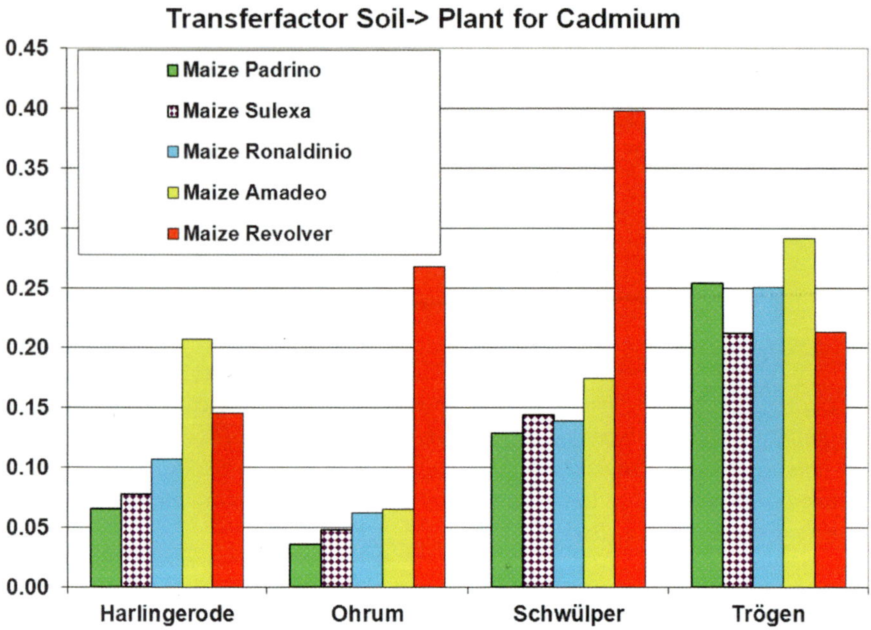

Fig. 14.3 Soil to plant transfer factors of cadmium in different maize breeds

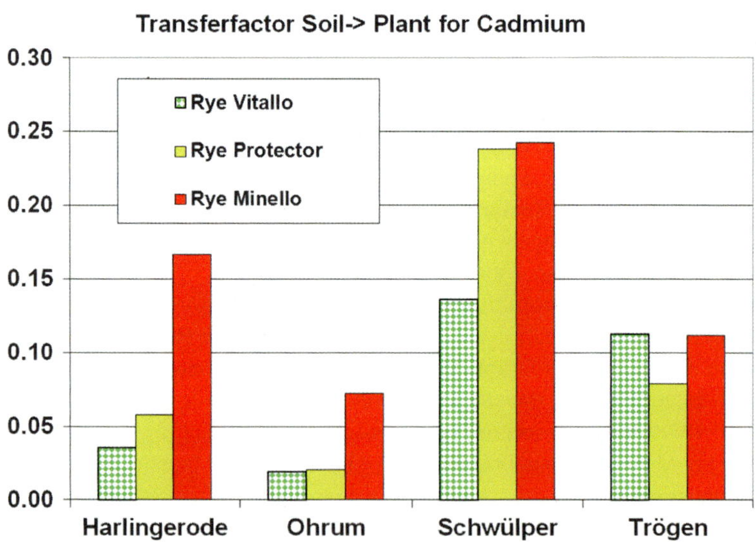

Fig. 14.4 Soil to plant transfer factors of cadmium in different rye breeds

14.3.2 Element Enrichment During Fermentation

During biogas production, contaminants in the fermentation residue (often called digestate or biogas manure) are enriched by approximately a factor three, which is a rather big problem for biogas plants fuelled by silage from contaminated soils (data based on dry matter; Sauer 2009). Likewise, in biogas plants, the volume drops by 20–30 % if wet silage is used with about 30 % dry matter (Amon et al. 2006; Breitschuh et al. 2006). Only carbon, hydrogen, oxygen and small amounts of nitrogen and sulphur escape during the transformation of the organic material into the biogas. Owing to these major elements being lost to the biogas, the remaining elements in the liquid digestate undergo a concentration enrichment. The application of digestate is very positive with regard to desired nutrients like potassium, phosphorus, nitrogen and trace elements, as it substitutes conventional mineral fertiliser. However, the contaminants in the digestate pose a great danger.

Digestates with a higher toxic element content should only be returned to the contaminated sites from which the energy crops were originally harvested. Distributing recycled residues to unpolluted sites may cause contamination over time, even if the biogas residue is only slightly contaminated. Since both the heavy metals and the important nutrients are fully recycled, the application of additional fertilisers (except for some nitrogen) is no longer necessary. If biogas plants' residues are used as fertiliser, they have to fulfil the legal fertiliser regulation. A fertiliser not complying with the regulation – especially with regard to critical elements – should not be applied. For example, the maximum permissible value for cadmium in fertilisers is 1.5 mg/kg dry matter in Germany. The mean dry matter concentration in the input feed for the biodigester is not allowed to exceed 0.5 mg/kg cadmium, because a factor three enrichment during the transformation of crop material into the biogas residue would lead to a violation of the limiting value. This narrows down the choice of energy crops that could be produced on contaminated soils. Tables 14.4, 14.5, 14.6, and 14.7 provide information on crops (whole plants) that contain less than 0.5 mg/kg cadmium.

14.3.3 Phytoremediation

As the simple calculation in Box 14.1 shows, it is not feasible to use plants to extract heavy metals (by means of phytoremediation) from contaminated sites, because it will take centuries or millennia to bring the elements down to acceptable levels. This applies to even the most mobile heavy metals, such as cadmium. In addition to the calculation in Box 14.1, we estimate how effectively amaranth, a moderately high-yield plant with large cadmium extraction rates, would remediate the strongly cadmium-contaminated Harlingerode soil (see Table 14.5).

Box 14.1 Is the Phytoremediation of Heavy Metal-Contaminated Soils Possible?

Phytoremediation uses plants to extract harmful substances from contaminated soil in order to clean a site. The transfer of elements from soil to plants depends on many factors, such as the:

- pH value of the soil,
- kind of crop,
- yield of the crop,
- nutrient status of the soil,
- amount of reactive solids in the soil, such as the organic material, iron and aluminium oxide-hydroxides and clay minerals,
- climate (temperature, water supply, etc.),
- form in which the element is bound (sorbed at surfaces, adhering to precipitation products and part of the mineral framework), and
- concentration of complexing organic and inorganic ligands that affect elements' mobility.

Many plant extraction experiments are carried out worldwide to decontaminate soils of harmful substances. While some organic pollutants can be extracted, it is difficult to extract inorganic pollutants, such as heavy metals, as we demonstrate here. On the basis of studies and compilations by Kloke (1984), Kloke et al. (1984), Sauerbeck (1989), Kabata-Pendias (2010) and of own investigations, we can provide a rough estimation of the soil-plant transfer factors of elements in agricultural soils (pH values between 5.3 and 7.5):

- factor <0.01–0.1 (nearly immobile): lead, uranium, mercury, cobalt, chromium (III),
- factor <0.1–1 (intermediate): mercury, arsenic, antimony, manganese, nickel, copper,
- factor <1–10 (mobile): selenium, molybdenum, zinc, thallium, cadmium.

To approximate how much of an element plants can remove from the soil and how efficient phytoremediation is, we use the most mobile and one of the most toxic heavy metals – cadmium. The aim of our example calculation is to clean an arable soil containing high concentrations of cadmium by means of a high yield crop plant.

Assumptions:
Cadmium (*Cd*) *concentration* in dry arable soil: 10 g/t (mg/kg) (= 1 t of soil contains 10 g Cd)

Transfer factor TF of 1 for Cd from soil to plants.

Mass of topsoil: 3,000 t/ha. This is calculated by multiplying the thickness of the topsoil (0.30 m = thickness of the plough horizon) by the area (1 ha = 10,000 m^2) and the dry density of the horizon (1 t/m^3).

Crop yield: 10 t of biomass annually per ha

(continued)

Box 14.1 (continued)

Calculation:
Amount of Cd in 1 ha of topsoil before remediation:

$$3,000 \text{ t of soil/ha} * 10 \text{ g Cd/t of soil} = 30,000 \text{ g Cd/ha}$$

Amount of Cd extracted in the first year through the harvest of 10 t biomass/ha and TF = 1:

$$10 \text{ g Cd/t biomass} * 10 \text{ t biomass/(ha} * \text{year)} = 100 \text{ g Cd/(ha} * \text{year)}$$

Simple but unrealistic Model 1: *linear extraction of cadmium* (the same amount of Cd is extracted every year): [30,000 g Cd/ha] / [100 g Cd/(ha*year)] = 300 years (time required to remove cadmium completely from the soil).

More realistic Model 2: *exponential extraction of Cd* (the amount of extracted Cd decreases every year, as the Cd concentration in the soil decreases every year; see Fig. 14.5).

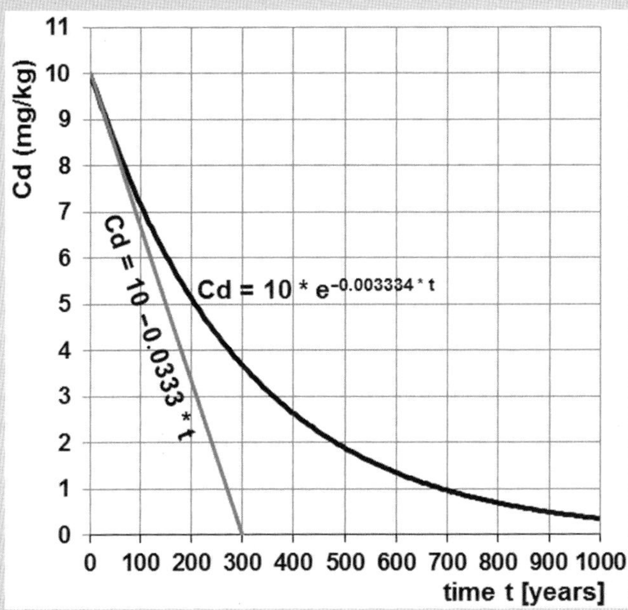

Fig. 14.5 Linear and exponential depletion of cadmium in a contaminated soil by means of phytoremediation

(continued)

Box 14.1 (continued)

This gradual decrease in pollutants through extraction means that it will take a very long time to achieve a suitably low concentration. Moreover, it is impossible to achieve a noxious element concentration of zero. The decrease is exponential according to the following equation (similar to the law of radioactive decay):

$$C = C_o * e^{-\lambda t};\ \ln\left(C_{\text{after t years}}/C_0\right) = -\lambda * t;$$
$$\text{half of the content left}: t_{1/2} = \ln\ 2/\lambda$$

Calculation of λ for Cd:

$$\ln\ (C_{\text{after 1 year}}/C_0) = \ln\ (29{,}900/30{,}000) = -0.00334 = -\lambda * 1$$

$$\lambda = 0.00334;\ t_{1/2} = 0.693/0.00334 = 208\ \text{years}.$$

This means that after 208 years, 5 g Cd/t will be left in the soil; after 416 years, 2.5 g/t, etc.

If, over the years, the soil-plant transfer factor of cadmium decreases (which is very likely due to the decreasing portion of phytoavailable cadmium), it would take even longer to decrease its concentrations.

Conclusion:
The sufficient phytoremedition of contaminated soils by means of high yield crops cannot be achieved within only a few generations. It will take centuries to millennia to even remove contaminations by mobile heavy metals such as cadmium. Phytoremediation will take significantly longer if less mobile elements, such as lead, are involved.

The soil near Harlingerode contains 20 mg/kg of cadmium. The topsoil has a density of 1.3 g/cm^3. Thus, the 30 cm deep topsoil has a weight of about 3,900 t/ha. The amount of cadmium in the topsoil is 78 kg/ha. The amaranth we harvested contained 30 mg cadmium per kg. With a harvest of about 10 t of amaranth per ha, we extracted 0.3 kg of cadmium in the first year. In our static model with a constant amount of extraction, we would need 260 years to achieve a cadmium concentration of zero. According to our dynamic model with decreasing amounts of extracted cadmium (Box 14.1), a much longer time of exclusive amaranth cultivation would be necessary to remediate the soil. Other, less mobile elements such as lead and arsenic, need a much longer time before soil is remediated.

Another argument for cultivating crops with low transfer factors is that contaminated crops would interact in an unknown manner with microorganisms, thus influencing the biogas production in the fermentation plant. This influence has not yet been investigated.

14.4 Conclusions

We propose using contaminated areas for bioenergy production by harvesting high yield crops with low pollutant extraction rates for the following reasons:

1. Heavy-metal-contaminated areas cannot be cleaned in a timely manner – not even by crops with a high extraction efficiency.
2. If contaminated land is used for the production of energy crops, the content of critical elements in fermentation plants' residues should be lower than the critical values determined by fertiliser legislation. Consequently, only crops with low transfer factors should be used.
3. Energy crops richer in critical elements may lead to undesired reactions and problems (disturbances in the biogas production) in the fermenter.

Energy crop production is an intelligent way of using contaminated land. However, more research is needed to explore this finding further. Future studies could look into, for example, the element uptake of popular new breeds, the element uptake of fast growing trees in short rotation forestry and the role of biogas residue input in soils to mobilise or fix elements.

References

Amon, T., Machmüller, A., Kryvoruchko, V., Milovanovic, D., Hrbek, R., Eder, M. W., & Stürmer, B. (2006). *Endbericht Forschungsprojekt: Optimierung der Methanausbeute aus Zuckerrüben, Silomais, Körnermais, Sonnenblumen, Ackerfutter, Getreide, Wirtschaftsdünger und Rohglyzerin unter den Standortbedingungen der Steiermark*. Retrieved August 29, 2012, from http://www.noest.or.at/intern/dokumente/098_ZB_Optimierung_Biogaserzeugung.pdf

Breitschuh, G., Reinhold, G., & Breitschuh, T. (2006). *Ökologische Konsequenzen (einschließlich CC) des Anbaus und der Verwendung nachwachsender Biogasrohstoffe*. Thüringer Landesanstalt für Landwirtschaft (TLL) Jena-Zwätzen, 10 p. Retrieved August 29, 2012, from http://www.tll.de/ainfo/pdf/bio20206.pdf

Deicke, M., Ruppert, H., & Schneider, J. (2006). Mining and smelting in the Harz mountains (Germany) – A never-ending environmental story. *Schriftenreihe der Deutschen Gesellschaft für Geowissenschaften, 45*, 237–256.

FAO (Food and Agriculture Organization of the United Nations). (2012). *FAO statistical Yearbook 2012*. Retrieved July 30, 2012, from http://www.fao.org/economic/ess/ess-publications/ess-yearbook/yearbook2012/en/

FNR (Fachagentur Nachwachsende Rohstoffe). (2012). *Basisdaten Bioenergie Deutschland*. 51 pp. Retrieved August 29, 2012, from http://mediathek.fnr.de/media/downloadable/files/samples/f/n/fnr_basisdaten_2012_web.pdf

Förstner, U., & Müller, G. (1974). *Schwermetalle in Flüssen und Seen als Ausdruck der Umweltverschmutzung* (225 pp.). Berlin: Springer.

Iskandar, I. K (Ed., 2001). *Environmental restoration of metals-contaminated soils* (304 pp.). Boca Raton: Lewis Publishers.

Kabata-Pendias, A. (2010). *Trace elements in soils and plants* (4th Rev. ed., 520 pp.). Boca Raton/London/New York: CRC Press/Taylor & Francis Group.

Kloke, A. (1984). *Problematik von Orientierungs-, Richt- und Grenzwerten für Schwermetalle in biologischen Substanzen*. In Schutz des Umweltmediums Boden. Evangelische Akademie Loccum, 10.-12.02.1984, 61–119.

Kloke, A., Sauerbeck, D. R., & Vetter, H. (1984). The contamination of plants and soils with heavy metals and the transport of metals in terrestrial food chains. In J. O. Nriagu (Ed.), *Changing metal cycles and human health* (pp. 113–142). Berlin, Heidelberg, New York, Tokyo: Springer.

Knappe, F., Möhler, S., Ostermayer, K., Lazar, S., & Kaufmann, C. (2008). Vergleichende Auswertung von Stoffeinträgen in Böden über verschiedene Eintragspfade. *Umweltbundesamt Texte 36/08*, 410 pp., Umweltbundesamt, Dessau (Germany); ISSN 1862–4804; Retrieved August 29, 2012, from http://www.umweltdaten.de/publikationen/fpdf-l/3644.pdf

Li, Z., Mclaren, R. G., & Meiherell, A. K. (2004). The availability of native and applied soil cobalt to ryegrass in relation to soil cobalt and manganese status and other soil properties. *New Zealand Journal of Agricultural Research, 47*, 33–43.

Luo, D., Zheng, H., Chen, Y., Wang, G., & Fenghua, D. (2010). Transfer characteristics of cobalt from soil to crops in the suburban areas of Fujian province, southeast China. *Journal of Environmental Management, 91*, 2248–2253.

Raskin, I,. & Ensley, B. D. (2000). *Phytoremediation of toxic metals – Using plants to clean up the environment* (304 pp.). New York: Wiley.

Salt, C. (1998). Schwermetalle in einem Rieselfeld-Ökosystem. Landschaftsentwicklung und Umweltforschung. *Schriftenreihe des Fachbereichs Landschaftsentwicklung der TU Berlin, 53*, 214 pp.

Sauer, B. (2009). Elementgehalte und Stoffströme bei der Strom- und Wärmegewinnung im Bioenergiedorf Jühnde. *Schriftenreihe Fortschritt neu denken, Heft 2* (124 pp.). Göttingen, Institut für Bioenergiedörfer Göttingen e. V. (Germany). ISSN 2190–7722.

Sauerbeck, D. (1989). Der Transfer von Schwermetallen in die Pflanze. – In Beurteilung von Schwermetallkontaminationen in Böden. In D. Behrens & J. Wiesner (Eds.), *Dechema-Fachgespräche Umweltschutz* (pp. 281–316). Dechema Gesellschaft für Chemische Technik und Biotechnologie e. V, Frankfurt.

Sauerbeck, D., & Lübben, S. (1991). Auswirkungen von Siedlungsabfällen auf Böden, Bodenorganismen und Pflanzen. *Berichte aus der Ökologischen Forschung, Band 6*. Jülich: Forschungszentrum Jülich GmbH, ISSN 0939–4931.

Suthersan, S. S. (1997). *Remediation engineering* (362 pp.). Boca Raton: Lewis Publishers.

Wenzel, W. W., Lombi, E., & Adriano, D. C. (1999). Biogeochemical processes in the rhizosphere: Role in phytoremediation of metal-polluted soils. In M. N. V. Prasad & J. Hagemeyer (Eds.), *Heavy metal stress in plants* (pp. 273–304). Heidelberg: Springer.

Young, D., Kanae, S., Oki, T., Koike, T., & Musiake, K. (2003). Global potential soil erosion with reference to land use and climate changes. *Hydrological Processes, 17*, 2913–2928.

Index

H. Ruppert et al. (eds.), *Sustainable Bioenergy Production - An Integrated Approach*, DOI 10.1007/978-94-007-6642-6, © Springer Science+Business Media Dordrecht 2013

Printed by Printforce, the Netherlands